PHOTOGRAPHIC GUIDE TO

LONGHORNED BEETLES OF BOLIVIA

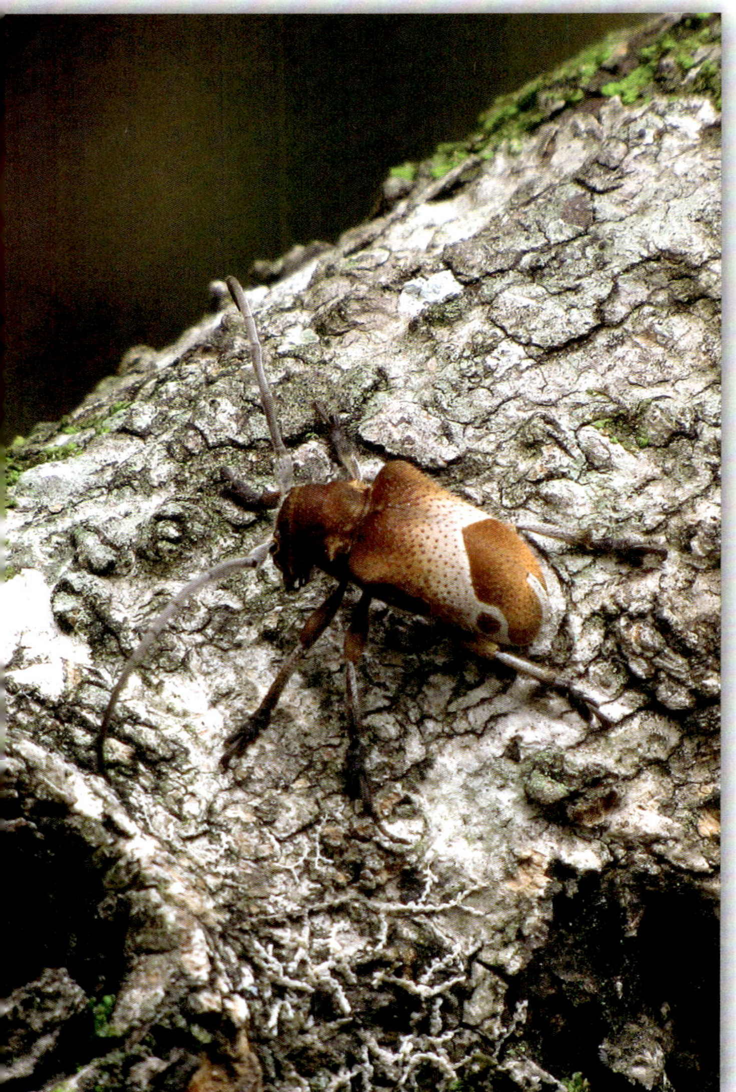

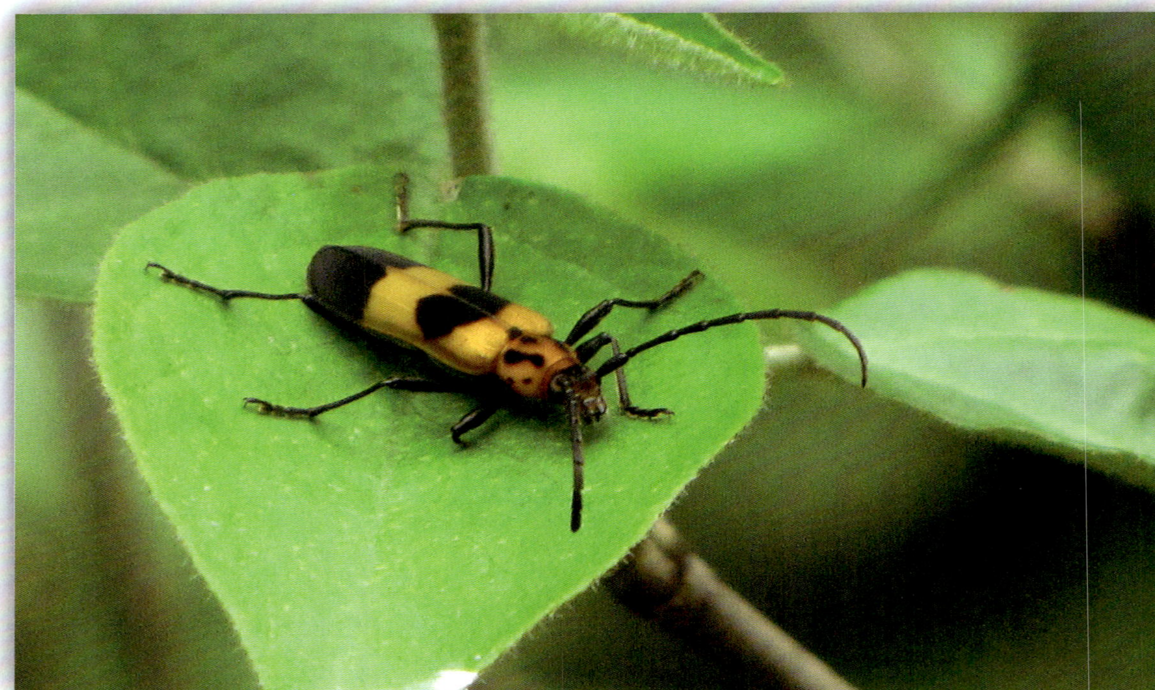

FRONTISPIECE. Left: *Compsosoma oculata* Lingafelter & Wappes. This 10–14 mm long species is most similar to *C. monnei* Martins & Galileo in size, shape, color, and pubescence. It differs in having the elytron with the angled postmedian white fascia extending posterolaterally into a circular eyespot. The species is endemic to Bolivia and known only from Refugio Los Volcanes in Santa Cruz Department. Above: *Erythrochiton jucundum* (Gounelle). This is a medium-sized (10–14 mm), orange and black, diurnal species with distinctive aposematic coloration. It can be locally abundant on *Croton* blossoms in the humid Chaco of Argentina, Uruguay, and Bolivia.

PHOTOGRAPHIC GUIDE TO

LONGHORNED BEETLES OF BOLIVIA

GUÍA FOTOGRÁFICA DE ESCARABAJOS LONGICORNIOS DE BOLIVIA

Steven Wayne Lingafelter // James Earl Wappes // Julieta Ledezma Arias

A Smithsonian Contribution to Knowledge

Smithsonian Institution
Scholarly Press

Washington, D.C.
2017

ABSTRACT. Nearly 1,900 species of longhorned woodboring beetles (Disteniidae, Vesperidae, and Cerambycidae) are known from Bolivia (with more than 200 additional and yet to be determined, or described as new, species). We have featured 500 of them, representing the breadth of morphological evolution, for this work. For each species, we present dorsal habitus photographs, detailed Bolivia distribution maps, and morphological, biological, taxonomic, and general distributional information. The subfamilies included in this work include Anoplodermatinae, Disteniinae, Parandrinae, Prioninae, Lepturinae, Cerambycinae, and Lamiinae.

Published by
SMITHSONIAN INSTITUTION SCHOLARLY PRESS
P.O. Box 37012, MRC 957, Washington, D.C. 20013-7012
https://scholarlypress.si.edu

Cover images (left to right): *Apoaerenica martinsi*, *Blabicentrus alberti*, *Charmallaspis pulcherimma*, *Ischionodonta iridipennis*, *Pathocerus wagneri*, and *Calocomus desmarestii*

Back cover and spine image: *Phaedinus lanio*

Original cover design by Taina Litwak, Systematic Entomology Laboratory, Agriculture Research Service, National Museum of Natural History, MRC-168, Washington, DC 20013-7012, USA

Library of Congress Cataloging-in-Publication Data:
Names: Lingafelter, Steven W., author. | Wappes, James E. (James Earl), author. | Ledezma A., Ma. Julieta, author. | Smithsonian Institution Scholarly Press, publisher.
Title: Photographic guide to longhorned beetles of Bolivia = Guía fotográfica de escarabajos longicornios de Bolivia / Steven Wayne Lingafelter, James Earl Wappes, Julieta Ledezma Arias.
Other titles: Smithsonian contribution to knowledge.
Description: Washington, D.C. : Smithsonian Institution Scholarly Press, [2017] | Series: A Smithsonian contribution to knowledge | Includes bibliographical references and index.
Identifiers: LCCN 2016053311 (print) | LCCN 2016058101 (ebook) | ISBN 9781944466053 (hardcover) | ISBN 9781944466060 (ebook)
Subjects: LCSH: Longhorned beetles—Bolivia—Pictorial works. | Cerambycidae—Bolivia—Identification. | Disteniidae—Bolivia—Identification. | Vesperidae—Bolivia—Identification.
Classification: LCC QL596.C4 L557 2017 (print) | LCC QL596.C4 (ebook) | DDC 595.76/48—dc23 | SUDOC SI 1.60:P 56
LC record available at https://lccn.loc.gov/2016053311

ISBN-13: 978-1-944466-05-3 (print)
ISBN-13: 978-1-944466-06-0 (ebook)

Printed in Canada

♾ The paper used in this publication meets the minimum requirements of the American National Standard for Permanence of Paper for Printed Library Materials Z39.48–1992.

DEDICATION

DEDICATION TO TWO FRIENDS

To Fray Andrés Langer, a strong man, determined and forceful in his convictions, whose impressive efforts to understand biodiversity and conservation were undertaken in combination with his religious convictions, often against a resistant population who did not understand and appreciate nature. Through his positive, daily insistence on conservation and the importance of each organism—whether insect, vertebrate, or plant—he preached the value of our biodiversity. He let no organism go to waste that lost its life, either naturally or by impact with a car or windshield, by studying it and extracting all information possible. Each sample collected by Brother Andrés is deposited for further study in the collections of the Noel Kempff Museum.

To Paolo Bettella, whose everlasting enthusiasm toward the development of scientific collections and biological research impressed all who knew him, and who, through his passion, developed the entomological, botanical, and vertebrate collections at the Noel Kempff Museum and made them available for short- and long-term study. The first samples of the Noel Kempff Museum were collected by Paolo from trips to very remote locations and under extreme danger from disease and accidents. He left us with a great legacy of courage, spirit, and optimism to continue the development of the scientific collections and understand the role and importance of each specimen present in the scientific collections.

DEDICACIÓN

ESTE LIBRO VA DEDICADO A DOS AMIGOS

A Fray Andres Langer un hombre fuerte, decidido y fuerte en sus convicciones cuyo impresionante espíritu de lucha por la biodiversidad y la conservación de la naturaleza a pesar de su competencia religiosa, hizo recapacitar y entender a toda una población hostil sobre las bondades de la naturaleza cuando no se podía. A través de su actitud positiva y cotidiana en la insistencia a la conservación e importancia de cada uno de los organismos tratese de insecto, vertebrado o planta; recordando en cada una de sus predicas sobre lo valioso que representa nuestra biodiversidad además de la importancia de recuperar cada uno de los organismos muertos naturalmente, atropellados en las carreteras o estrellados en los parabrisas con el único propósito el de estudiar y extraer la información de cada uno de las muestras colectadas y depositadas en el Museo Noel Kempff por el hermano Andrés.

A Paolo Bettella que nos impresiono a cada instante con su entusiasmo y carisma al desarrollo de las colecciones científicas y la investigación Biológica transmitiendo informacion y concentrando cada una de las muestras entomológicas, botánicas y de vertebrados presentes en las colecciones del Museo NKM, haciendo posible y disponible la materia prima para estudios a corto, mediano y largo plazo. Las primeras muestras de las colecciones del MNKM son parte de las colectas de nuestro amigo obtenida en viajes a lugares muy lejanos muchos de los cuales en situaciones complejas por la inaccesibilidad a ciertos lugares exponiendo su vida a los peligros de la naturaleza como enfermedades y accidentes, mismos que trascendieron para dejarnos este gran legado de coraje, espíritu y optimismo para continuar con el desarrollo de las colecciones científicas para entender la función y la importancia de cada uno de los organismos presentes en las colecciones científicas.

CONTENTS

PREFACE

With loss of habitats throughout the world occurring at a staggering rate, it is critical to document what is being lost. This book strives to do that by focusing on the longhorned woodboring beetles in Bolivia. In the Amazon region of South America, a forest area equal in size to the state of Delaware is destroyed each year due to ranching, illegal logging, or agricultural expansion (World Wildlife Fund, 2016). This wholesale clearing of large tracts of land kills everything or forces species to move quickly to other areas, disrupting the balance of the ecosystem. We hope this book will help people discover and appreciate some of the amazing diversity of life that exists in undeveloped and/or remote natural areas. Much of this diversity is still unknown, or was until a few years ago, as revealed here. We believe it is important that even more effort be made to limit the destruction of these habitats and explore sustainable ways to use our rich natural resources so that humans and the millions of other species in these habitats can survive and coexist.

Longhorned woodboring beetles include four families totaling nearly 40,000 species (Tavakilian and Chevillotte, 2016). They get their English common name from the fact that most species have very long antennae (usually at least as long as the body) and bore in woody plant tissue as larvae. In Bolivia they are referred to as *chaqueadores* since they are often seen “slashing” wood with their mandibles in cut and burned areas, or *chaqueos*. Of the four families, Vesperidae (including Anoplodermatinae), Disteniidae (often treated as a subfamily), and Cerambycidae (including the subfamilies Parandrinae, Prioninae, Lepturinae, Cerambycinae, and Lamiinae) are known from Bolivia (Monné, 2016a, 2016b, 2016c). Although most species in these groups are woodborers whose larvae develop in dead wood, the grubs of some species develop in living tissue of woody or herbaceous plants, and some others develop in the soil after feeding in the root crown of various woody or nonwoody plants. Bolivia, with its large land area, rugged and mountainous topography, heterogeneous landscapes, and tropical location, harbors a very high number of plant species (Killeen et al., 1993; Ibisch and Merida, 2004). Therefore, these beetles have a wide range of potential host plants and are abundantly represented in tropical and subtropical localities where plant diversity is highest (Svacha and Lawrence, 2014).

Historically, the first significant scientific expedition to Bolivia, with an entomology component, was the Mulford Expedition in 1921–1922, chronicled in MacCreagh (2001). This expedition had a team of biologists that included William Mann, an entomologist, who collected some of the first longhorned beetles known from Bolivia. But it was not until the 1990s that expeditions with a focus on surveying for longhorned beetle species (notably that by Edmund F. Giesbert) started in earnest. This was followed in the early twenty-first century by the Bolivian Cerambycidae Project (BCP) developed by the Museo de Historia Natural Noel Kempff Mercado in Santa Cruz de la Sierra, and cooperators in the United States, including the American Coleoptera Museum, the Florida State Collection of Arthropods, and the Smithsonian Institution. Since then, the BCP has conducted more than 30 expeditions to various areas in Bolivia to survey for species. In addition to the collecting of thousands of specimens, the process included the exportation of collected material so that it could be properly curated, labeled with important collection data, and disseminated to appropriate specialists throughout the New World for identification and scientific description. This process has been time consuming but has resulted in an extremely high degree of accuracy regarding species identification and allowed for the description of species new to science by qualified individuals.

These recent survey efforts have resulted in a tremendous increase of our knowledge of the Bolivian longhorned woodborer fauna (Aramayo and Ledezma Arias, 2004; Perger and Guerra, 2013; Wappes et al., 2006, 2011, 2013). This is evidenced by the fact that less than 500 species of Cerambycidae were known from Bolivia 20 years ago. Currently, nearly 1,900 are known (Wappes et al., 2013), with almost half of this number added since 2005 (Wappes et al., 2006, 2011, 2013).

These investigations have discovered more than 230 species new to science, species that exhibit a remarkable range of morphologies and colors. They range in size from less than 2 to over 120 mm in length (one-eighth of an inch to over 5 inches).

Many of the newly discovered and described species are known only from Bolivia and some from only one locality. These are endemics and collectively represent slightly more than 22% of the total species known from Bolivia (Wappes et al., 2011, 2013). This percentage is among the highest in the New World and is indicative of the uniqueness of the fauna. As more surveying is done at areas between 1,500 and 3,000 m it is expected that the level of endemism will likely increase as endemism and elevation are known to correlate in other animals (see Herzog et al., 2013, fig. 1). It also supports the need for and importance of further survey work in expanded areas of Bolivia.

In this book, we feature 500 of the most notable species. We have chosen the species based on taxonomic diversity (representing the greatest number of subfamilies, tribes, and genera of longhorned beetles), morphology (representing the spectrum of shape, size, and color), geography (representing all parts of Bolivia, including the Altiplano, Amazonian lowlands, Chaco, Chiquitano, Yungas, and transitional areas), and their abundance (representing common, widespread species and rare species, sometimes known from only a few specimens). For each species, we include a dorsal photograph, a map of known distribution in Bolivia, a brief discussion of morphological characteristics and habits (where known), along with comments on distribution outside Bolivia, if applicable.

ACKNOWLEDGMENTS

The success of our survey work in Bolivia and the production of this photographic guide would not have been possible without the help and collaboration of many people. We appreciate each and every one of them and thank them for their individual contribution. They include Taina Litwak (USDA, Smithsonian Institution) for the book cover design, layout, photography, and digital enhancement of images in Adobe Photoshop; Gary Ouellette (USDA) for specimen photography, image enhancement, and page layout in Adobe Illustrator; Alicia Hodson (formerly USDA, Smithsonian Institution) for specimen photographs and enhancing their images; Elisabeth Roberts (USDA, Smithsonian Institution) and Lucrecia Rodriguez (USDA, Smithsonian Institution) for image enhancement in Adobe Photoshop; and Andrés Garzón-Moreno (USDA) for preparing maps for the subfamily Lamiinae and for Spanish translation. Additionally, we appreciate all the participants who helped in the survey and contributed specimens they collected to the project. They include Toni Bonaso (Potrerillo del Guenda Reserve), Norman Woodley (formerly USDA, Smithsonian Institution), Alexander Konstantinov (USDA, Smithsonian Institution), Fred Skillman, Eugenio Nearns (Purdue University), Charyn Micheli (Smithsonian Institution), Andrés Garzón-Moreno (USDA, Smithsonian Institution), Richard Leschen (Landcare Research, New Zealand Arthropod Collection), Robin Clarke (Flora and Fauna Hotel), Tiziano Bettella (Potrerillo del Guenda Reserve), Jose Luis Aramayo (Noel Kempff Museum), Thomas Henry (USDA, Smithsonian Institution), Donald Windsor (Smithsonian Tropical Research Institute), Robert Perger (Colección Boliviana de Fauna), Ana Caroli Hamel-Leigue, Roy Morris, Mike Thomas (formerly Florida State Collection of Arthropods), Donald Thomas (USDA), Paul Skelley (Florida State Collection of Arthropods), Byrd Dozier, William King, Kenneth Kuckartz, Andrew Cline (California Department of Food and Agriculture), and George Ball (University of Alberta). We also sincerely thank Tiziano Bettella and Toni Bonaso, Albert Schweining (Refugio Los Volcanes), and Robin Clarke for availing their properties for our investigations and providing logistical support while there. We thank Vincent Vos for his logistical support and collecting suggestions in Riberalta. For their collaboration and identifications, special thanks to Miguel Monné and Marcela Monné (Museu Nacional, Universidade Federal do Rio de Janeiro), Ubirajara Martins and Antonio Santos-Silva (Museu de Zoologia, Universidade de São Paulo), Maria Helena M. Galileo (Museu de Ciências Naturais, Fundação Zoobotânica do Rio Grande do Sul), Dilma Solange Napp (Departamento de Zoología, Universidade Federal do Paraná), and Angel Arial Cespedes (Departamento de Entomología, Universidad de Sucre).

MATERIALS AND METHODS

HABITATS AND COLLECTING LOCALITIES

Most of the surveying took place in the expansive and diverse Santa Cruz Department. Primary collecting localities are listed here with their coordinates and elevations and depicted on the map below (Figure 1) by the associated numbers (note that “a” and “b” are used for the same number when the specific localities are so close to one another that they would not show up on the map as distinct). There is a relatively large cluster of localities in west-central Santa Cruz Department around Amboró National Park. This region has been the most productive, providing over 90% of the Bolivian species of longhorned beetles:

1. Santa Cruz Botanical Garden, Andrés Ibáñez Province, 17°47'S, 63°04'W (375 m)

2. Potrerillo del Guenda Reserve, Andrés Ibáñez Province, 17°40'S, 63°27'W (370 m; Figure 2c)

3. Flora and Fauna Hotel, Ichilo Province, 4–6 km south-southeast of Buena Vista, 17°30'S, 63°38'W (400 m; Figure 2d)

4. Concepción vicinity, Ñuflo de Chávez Province, 16°08'S, 62°01'W (380 m)

5a. Cuevas vicinity, Florida Province, 18°12'S, 63°41'W (1,310 m; Figure 2a)

5b. Refugio Los Volcanes Reserve, Florida Province, 18°06'S, 63°36'W (1,000–1,200 m; Figure 2b)

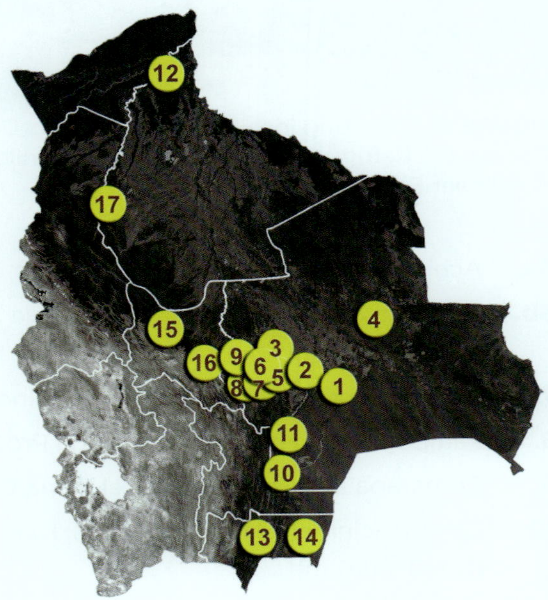

FIGURE 1. Map of Bolivia showing primary survey localities, numbered as described in text.

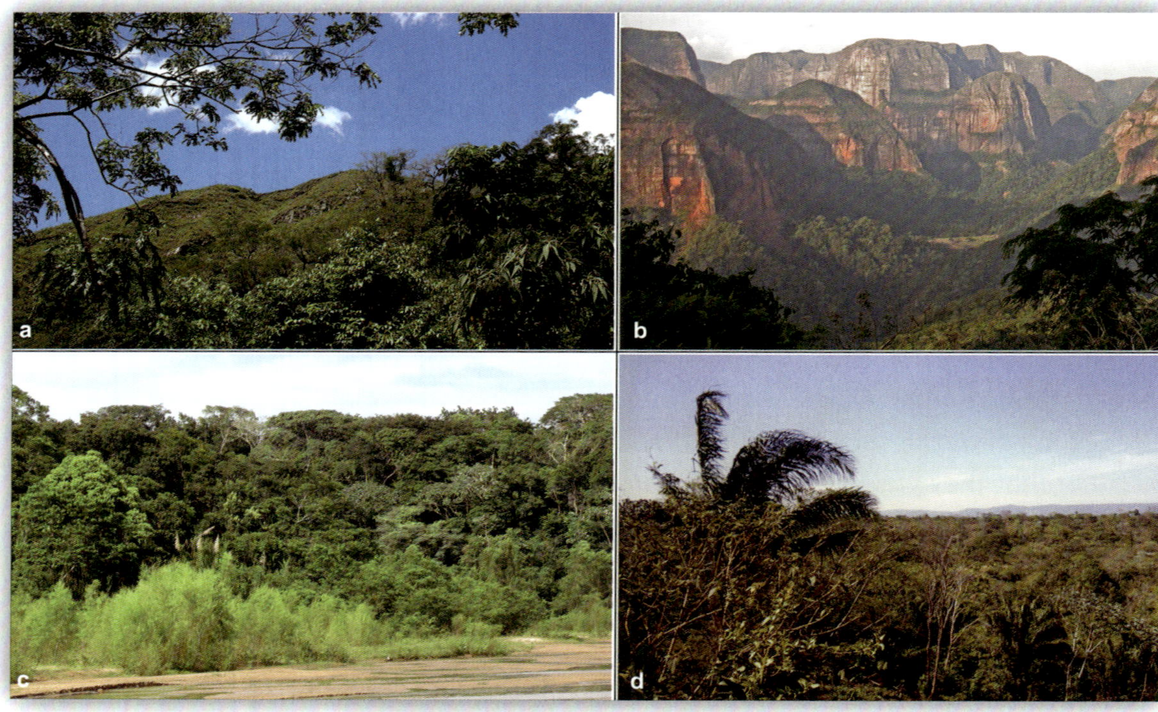

FIGURE 2. Collecting localities in Santa Cruz Department, Bolivia: (a) road to Bella Vista near Cuevas; (b) Refugio Los Volcanes; (c) Potrerillo del Guenda Reserve; and (d) Flora and Fauna Hotel near Buena Vista.

6a. Achira Sierra Resort, Florida Province, 18°09′S, 63°49′W (1,350 m)

6b. Vicoquin area, road to Amboró above Achira, Florida Province, 18°07′S, 63°47′W (1,700–2,000 m; Figure 3c)

7. Samaipata and vicinity, Florida Province, 18°09′S, 63°52′W (1,620 m; Figure 3b)

8. Pampagrande and vicinity, Florida Province, 18°06′S, 64°15′W (1,700 m; Figure 3a)

9. Comarapa and vicinity, Manuel María Caballero Province, 17°58′S, 64°29′W (1,725 m)

10. Camiri vicinity (22 km northeast), road to Eyti, Cordillera Province, 19°53′S, 63°30′W (1,050–1,140 m)

11. Camiri vicinity (83 km north), road to Itaí, Cordillera Province, 19°21′S, 63°26′W (885 m; Figure 3d)

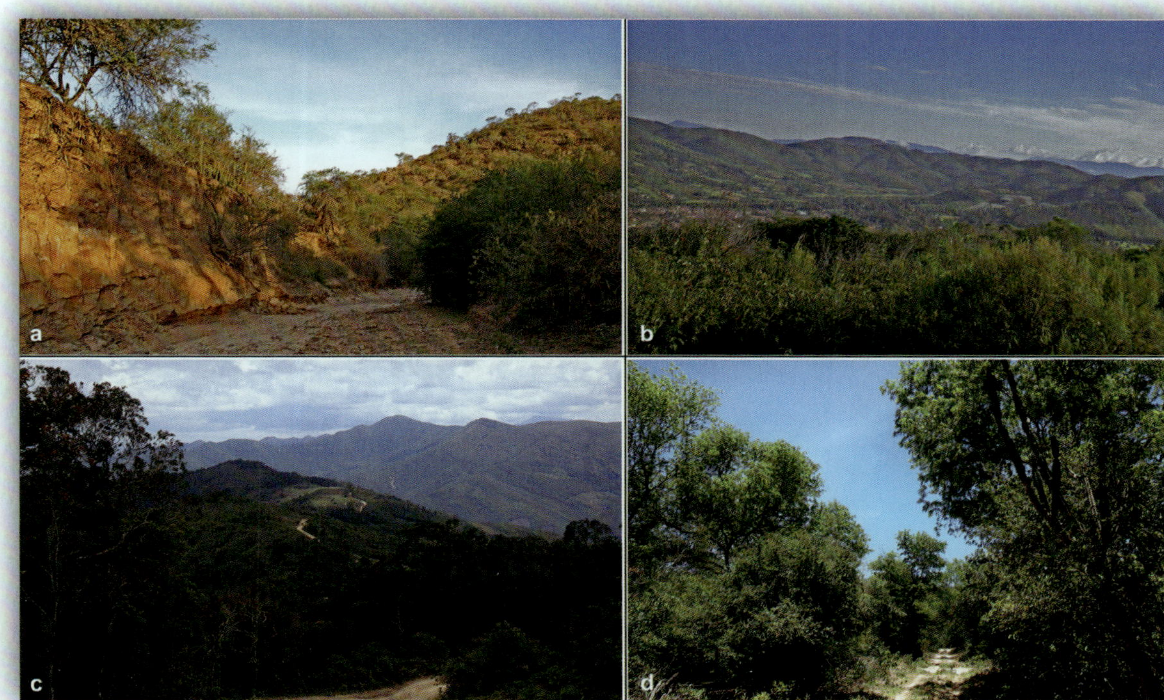

FIGURE 3. Collecting localities in Santa Cruz Department, Bolivia: (a) Pampagrande; (b) Samaipata; (c) above Achira (Vicoquin Region); and (d) north of Camiri.

Additional localities have been sampled in other departments including Tarija, Beni, Pando, Cochabamba, and La Paz, the most productive of which are listed here and depicted on Figure 1. These localities have shown the greatest percentage of new country records for Bolivia of species previously known from countries adjacent to Bolivia (Argentina, Brazil, Paraguay, and Peru):

12a. Riberalta vicinity, Beni Department, Vaca Díez Province, 11°00′S, 66°04′W (130 m)

12b. Las Piedras and Agua Dulce vicinity, Pando Department, 11°03′S, 66°11′W (129 m)

13. Villamontes vicinity, Tarija Department, Gran Chaco Province, 21°19′S, 63°15′W (360 m)

14. Ibibobo, Tarija Department, Gran Chaco Province, 21°32′S, 62°09′W (325 m)

15. Villa Tunari vicinity (Hotel de Selva El Puente), Cochabamba Department, Chaparé Province, 16°59′S, 65°25′W (340 m)

16. El Sacta, Cochabamba Department, Carrasco Province, 17°30'S, 65°00'W (220 m)

17. Rurrenabaque, Beni Department, José Ballivián Province, 14°26'S, 67°32'W (220 m)

Lack of logistics and difficulty of access dictated that most survey work be done at localities below 2,000 m. This combined with the need to work during the rainy season (the most important biological trigger for insect emergence) when primitive roads can become impassable makes it easy to understand the challenge of surveying mountainous terrain. Nevertheless, in the future such areas are likely to receive more emphasis from Bolivian Cerambycidae Project surveyors.

COLLECTING METHODS

Longhorned woodboring beetles have been collected using many different techniques. Species that are large or brightly colored and diurnal are easily seen moving on tree trunks and branches, fallen trees, or leaves and flowers during the day, when they can be hand collected or netted. However, most species are nocturnal and stay hidden during the day. The collecting technique that is most successful for many of these is lighting via mercury vapor and ultraviolet lights suspended over or in front of white sheets (or walls). Once the seasonal rains begin, it is common on warm, humid, dark nights to attract thousands of insects to these lights and not unusual for them to include 50 to 100 species of longhorned woodborers.

In the daytime, beating trail-side or forest-edge tree foliage and branches and other vegetation is one of the best methods of collecting. By beating vegetation over a sheet, the specimens are dislodged and easily seen against the white cloth, where a fast hand or aspirator can collect desired specimens. This is a very productive means to collect many of the cryptic or smaller species, including nocturnal species that are not attracted to lights. Other productive techniques include Malaise, flight-intercept, and fruit traps. A Malaise trap is a strong tent-like apparatus that intercepts flying insects and funnels them into collecting bottles. These traps can be mounted on the ground or hung high from trees. A flight-intercept trap is a fine-mesh fabric stretched taught across a potential "fly zone" through a forest or trail; collecting trays below are filled with some type of preservative. Flying insects, not seeing the obstruction, upon encountering it drop into the collecting trays. Fruit traps filled with ripe bananas are successful for capturing species that are attracted to the smells of fermentation. The traps are constructed to allow attracted insects to enter the trap but not leave. Each of these trapping methods is likely to collect species not normally encountered by other means.

Last, when woody shrubs and some forest trees are in peak bloom they will attract a diverse fauna of insects. Among them are many species of Cerambycidae. As an example of the productivity of blooms, in 2013 we encountered a small tree along a forested road that was in peak bloom. It was buzzing with insects. By bagging the blossoms with a long-handled net we captured some 20 species of cerambycids in a few minutes. An hour later we repeated the process and were almost as successful. Two days later we revisited the tree and found it devoid of insects as the blooms were then wilting. These techniques, and others, are fully described and illustrated in Evans (2014).

SPECIMEN DEPOSITION

All the holotypes and approximately half of all recently collected material from Bolivia are deposited in the Museo de Historia Natural Noel Kempff Mercado in Santa Cruz de la Sierra, Bolivia. Other specimens, including paratypes, are deposited in various museums in the United States and Brazil, including the Smithsonian Institution's National Museum of Natural History (Washington, D.C., USA), the American Coleoptera Museum (San Antonio, Texas, USA), the Florida State Collection of Arthropods (Gainesville, Florida, USA), the Museu Nacional, Universidade Federal do Rio de Janeiro (Rio de Janeiro, Brazil), and the Museu de Zoologia, Universidade de São Paulo (São Paulo, Brazil), as well as in author Steven W. Lingafelter's personal collection (Hereford, Arizona, USA).

PREPARATION OF PHOTOGRAPHS AND FIGURES

For imaging, specimens were first cleaned using alcohol and a fine camel-hair brush. Habitus images were obtained using a Visionary Digital imaging system. The system consists of a 64 mm fixed macro lens affixed to a Canon EOS 40D digital SLR camera. A Dynalite M2000ER power pack and Microptics ML1000 light box provide illumination. Image capture software is a proprietary application of Visionary Digital; images were saved as TIF files, and the RAW conversion occurred in Adobe Photoshop Lightroom 1.4. Image stacks were montaged with Helicon Focus 4.2.1. Further image processing and enhancement, including all maps, was done in Adobe Photoshop CS4 and CS6.

GUIDE TO SPECIES ACCOUNTS

The 500 species accounts that follow are grouped by subfamily and tribe (each noted in the heading) and are alphabetically arranged within those groups. Authorities are given for all species. Parentheses around the author and year indicate that the genus name is different from the original species description; no parentheses indicate that the species is in the same genus as originally described. Each record includes a photo of a pinned specimen (digitally enhanced and cosmetically improved to remove the pin and foreign objects). This photo is not necessarily to scale for individual specimens or relative to other photos in the book; rather, it is printed at the largest size that will show fine detail and still accommodate the text of the species account.

Each account includes a distribution map for Bolivia that shows each locality from which specimens were collected. The text of the species account includes, in approximate order: the size range; distinctive characteristics such as color, pubescence, and other structural features; comments on known biology, such as hosts, seasonality and behavior; indications of how specimens are collected, such as with lights or by sweeping vegetation; and an overview of the distribution of the species with specific reference to known localities in Bolivia.

Measurements (in mm) in the species accounts represent the length of the insect from the front of the head (or mandibles if they extend past the head) to the end of the elytra (or any protruding structures if they have them) but exclude the antennae. The diagram in Figure 4 and the glossary that follows label and define parts of the cerambycid anatomy that appear in the species accounts.

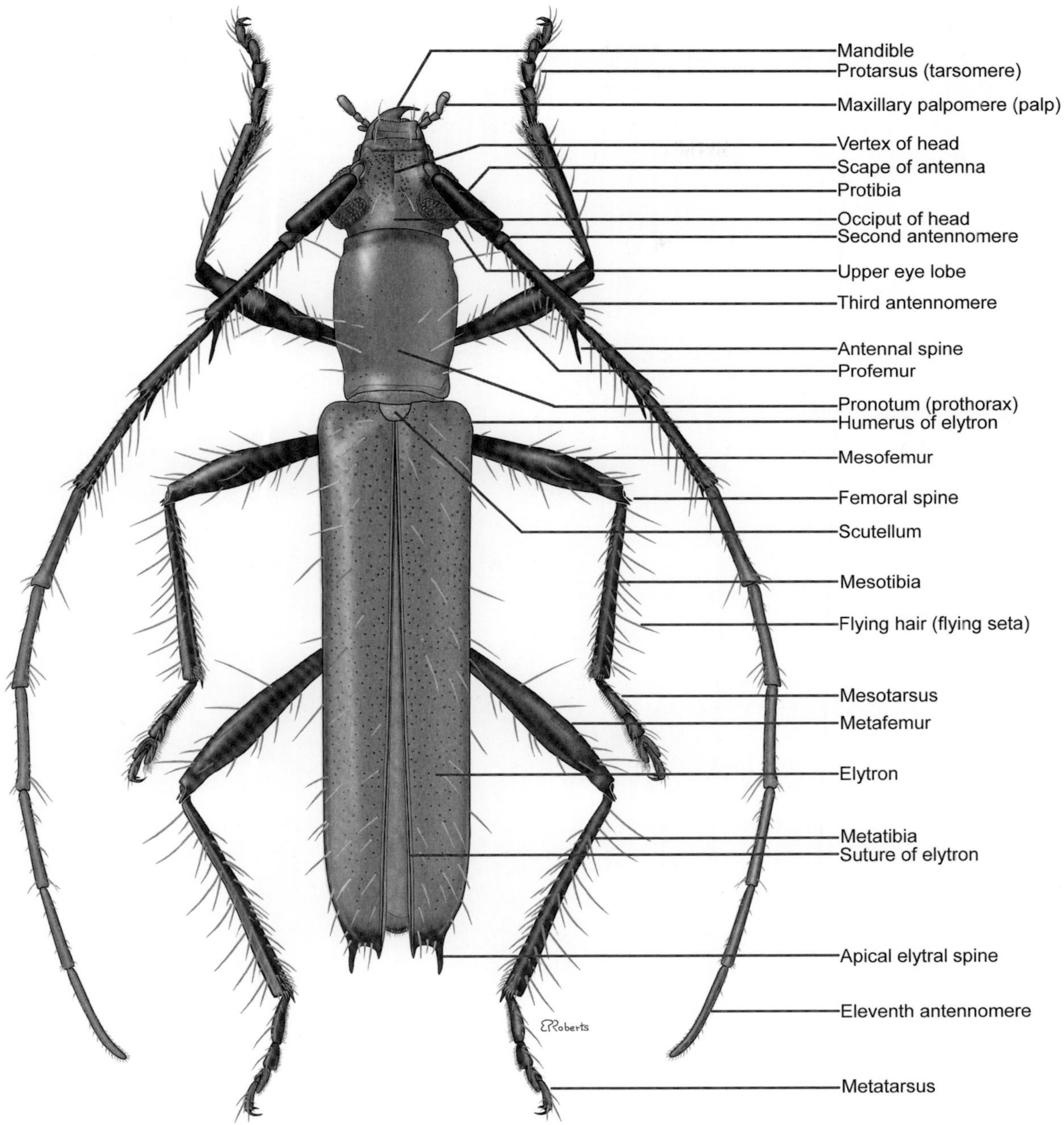

FIGURE 4. Anatomy of a typical longhorned beetle. *Stizocera delicata* Lingafelter, from Flora and Fauna Hotel, Santa Cruz, Bolivia.

GLOSSARY

COMMONLY USED TERMS

The following glossary defines the most common descriptive terms used in the text of the species accounts. Listed terms and their definitions follow standard accepted use and practice.

ABDOMEN posterior division of the insect body.

ALUTACEOUS covered with a network of minute cracks; typically relating to the surface of the elytra or pronotum.

ALVEOLATE covered with closely placed shallow cells; typically occurring on the surface of the pronotum.

ANNULATIONS, ANNULATE ringed with a different color; typically occurring on antennal segments or legs.

ANTENNA, ANTENNAE a multisegmented sensory appendage arising from the head, most often with 11 segments.

ANTENNOMERE the segments of the antenna, which, in Cerambycidae, usually consists of a scape, pedicel, and nine flagellomeres.

APICAL, APICES away from the base, directed toward the apex or end of a structure.

BASAL that part of a structure at the beginning or point of attachment, nearest the body.

BULBOUS swollen, like a bulb, evenly rounded.

CALLOSITY, CALLUS, CALLI a flattened elevation or thick and swollen structure; often used in relation to pronotal modification.

CARINA, CARINATE an elevated ridge or keel, with several linear ridges; often referring to modifications of the antennae.

CHEVRON an inverted V-shaped marking; often used to refer to elytral maculations.

CLAVATE clublike, thickened gradually, as in the femora or scape.

COALESCING merged or coming together at least in part.

COMPRESSED flattened laterally.

CONCAVE hollowed out, sunken, opposite of convex.

CONGENER a species belonging to the same genus as another.

CONTIGUOUS side by side, so close to each other as to touch; often used to describe elytral or pronotal punctation.

CONVEX swollen or rounded outwardly.

COXA, COXAE basal segment of a leg where it is attached to the body.

CREST a prominent longitudinal carina on the dorsal surface of the body.

DENUDED without covering, lacking setae, pubescence, or other vestiture.

DISK the central surface area of the elytra or pronotum, sometimes slightly elevated from the surrounding area.

DISTAL, **DISTALLY** away from the base, toward the free end, further from the body.

DIURNAL active in the daytime.

DORSUM, DORSAL the upper surface, topside, or back.

EBURNEOUS ivory white, generally shining, color; often applied to markings on the elytral surface.

ELYTRON, ELYTRA the two chitinous wing covers over the abdomen protecting the membranous flight wings.

EMARGINATE a notched or obtusely rounded edge or margin; often referring to the shape of the elytral apices.

ENDEMIC native, known only from a particular area or region.

EXPLANATE having an outspread or flattened margin as applied to antennal segments or other structures.

FASCIA, FASCIAE a transverse band or broad line; typically referring to an elytral macula.

FEMUR, FEMORA principal segment of the insect leg distal to the coxa.

FLABELLATE fan shaped, with long, thin processes lying flat on each other, as often applied to the shape of antenna or antennal segments.

FLYING HAIR very long, slender, typically upright setae.

FRONS the upper, anterior portion of the head.

GLABROUS smooth, without hairs or punctures.

GLOBOSE, GLOBULAR spherical or evenly rounded.

HUMERUS, HUMERI basal exterior angle of the elytra.

IMPUNCTATE not marked with punctures.

INFLATED blown up or tumid.

INTEGUMENT chitinous surface or outer body covering in insects.

INTERNAL inside edge or angle.

LATERAL, LATERALLY related to or attached to the side of a structure.

MACULA, MACULAE a relatively large, colored mark typically occurring on the pronotum or elytra.

MANDIBLES tooth-like mouthpart used to chew, bite, or cut.

MARGIN, MARGINED narrow part of a surface within the edge.

MEDIAN at the middle.

MESO prefix used to denote the middle or middle structure (e.g., mesocoxa, mesosternum, mesothorax).

META prefix used to denote posterior part or structure (e.g., metacoxa, metasternum, metathorax).

NOCTURNAL active at nighttime.

NOMINATE SPECIES the original named form versus subspecies.

PECTINATE comb-like, as typically applied to the shape of an antenna or antennal segments.

PICEOUS brown and black or black with reddish tinge.

PILOSE, PILOSITY soft down or fine short hairs.

PRO prefix to denote first or before.

PRONOTUM, PRONOTAL dorsal surface of the prothorax directly behind the head.

PROTHORAX first thoracic segment, which bears the anterior legs ventrally and pronotum dorsally.

PUBESCENCE, PUBESCENT short, typically fine, closely set hair.

PUNCTATION, PUNCTURE, PUNCTATE impressed points in the integument, of various shapes and sizes.

RECUMBENT reclining or lying down; typically applied to setae.

RETICULATE, RETICULATED a network of lines, mesh-like.

RUGOSE, RUGOSITY wrinkled, can be narrow or broad, shallow or deep; typically referring to integument surface, most often in reference to elytra or pronotum.

SAPPING feeding on plant juices flowing from wounds or naturally from joints.

SCAPE first or basal segment of a beetle antenna.

SCUTELLUM structure (usually triangular in shape) located between the elytra at their base, next to the pronotum.

SERRATE notched, like the teeth of a saw; often applied to the shape of antenna or antennal segments.

SETA, SETAE, SETOSE slender, stiff hair or bristle-like vestiture on the surface of insects.

SETIGEROUS bearing setae or bristle-like hairs.

BOLIVIAN LONGHORNED BEETLES

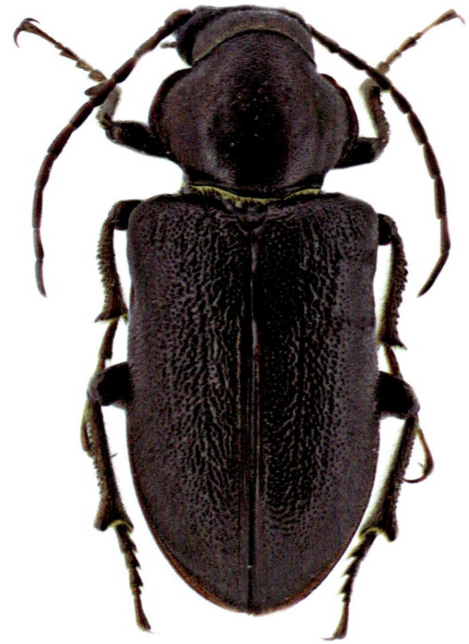

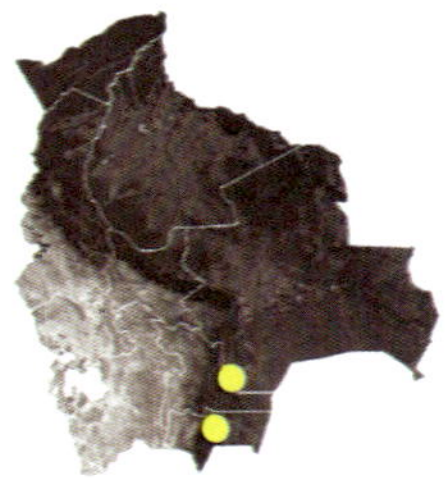

Anoploderma breueri Lameere, 2012

Anoplodermatinae: Anoplodermatini

The peculiar subfamily Anoplodermatinae includes a disparate assemblage of species, including this one. Members of this subfamily have a fully exposed fourth tarsomere, unlike most Cerambycidae. Only three species of anoplodermatines are known from Bolivia. This species (20–26 mm) is recognized by its somewhat globose appearance with a deflexed, rounded pronotum and head. Both sexes have sharp, sickle-shaped mandibles, with those of the male significantly larger. It is known from Brazil, northern Argentina, Paraguay, and Bolivia. In Bolivia it has been collected in southern Santa Cruz Department near Camiri and in Tarija Department near Villamontes.

Pathocerus wagneri Waterhouse, 1901

Anoplodermatinae: Mysteriini

This is another unusual member of the diverse but small subfamily Anoplodermatinae. This species (24–30 mm) is immediately recognized by the pale brown color, pectinate antennae, small, narrow head and pronotum, and long tarsi. Like other members of Anoplodermatinae, the usually hidden fourth tarsomere is fully exposed. It is the only member of the tribe Mysteriini recorded from Bolivia. It is also distributed in Brazil, Paraguay, and Argentina. In Bolivia it is known from near Villamontes in Tarija Department and the Camiri region in Santa Cruz Department.

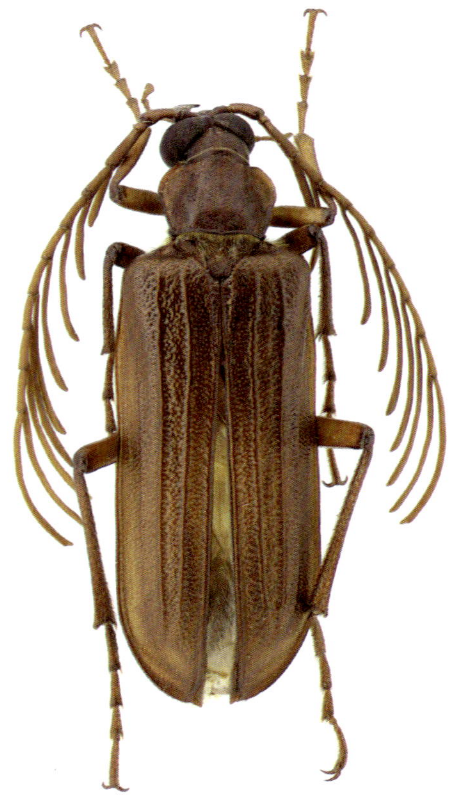

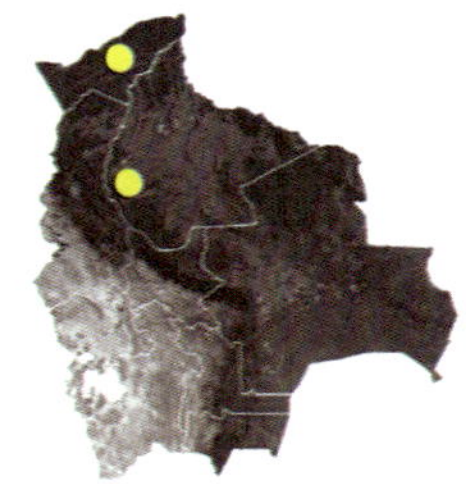

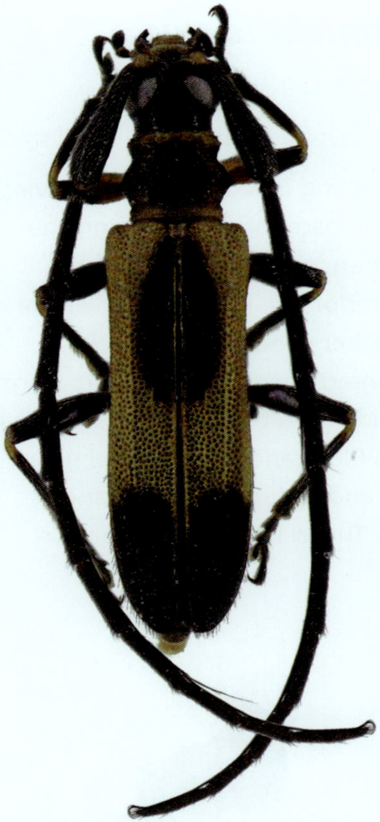

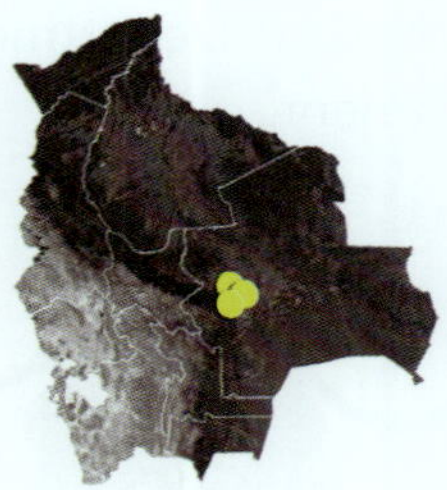

Cupecuara soledari (Martins & Galileo, 2001)

Disteniinae: Disteniini

Like the subfamily Lepturinae, the Disteniinae is poorly represented for Bolivia, with only 13 species known. This is a small (11–14 mm) species, formerly placed in the genus *Cometes*, with a distinctive yellow and black elytral pattern. The antennae have very long setae along the inner margin of the last antennomeres. Collecting localities in Bolivia include Refugio Los Volcanes, adjacent to Amboró National Park, Potrerillo del Guenda Reserve, and the grounds of the Flora and Fauna Hotel near Buena Vista in Santa Cruz Department.

Distenia chaparensis Tippmann, 1953

Disteniinae: Disteniini

This small (12–16 mm) species is very distinct with its bluish or purple elytra, dark head and pronotum, and bright yellow legs with apically black femora and basally black tibiae. It is known from Brazil and Bolivia. In Bolivia it has been collected from lower-elevation localities in Santa Cruz Department, including the grounds of the Flora and Fauna Hotel near Buena Vista and Potrerillo del Guenda Reserve.

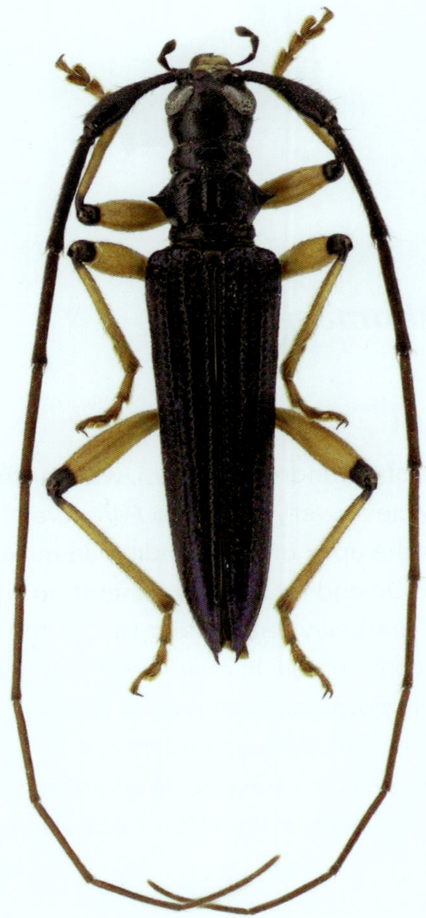

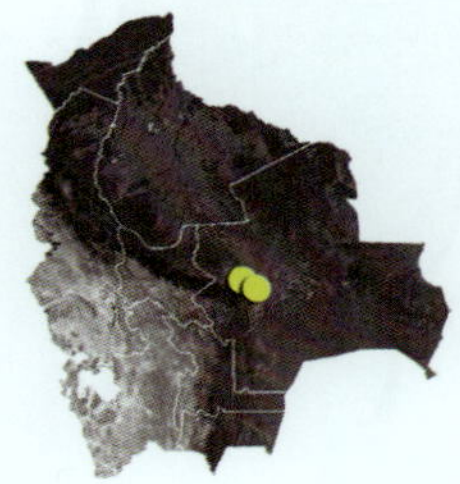

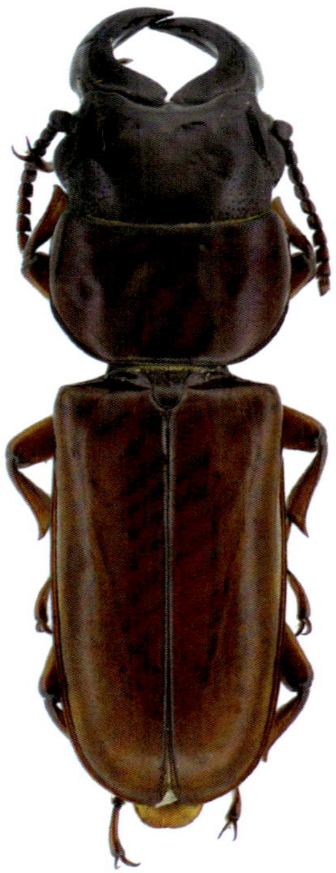

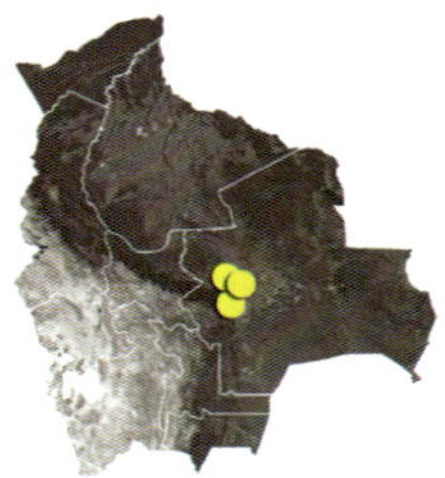

Parandra (Parandra) thomasi (Santos-Silva, 2002)

Parandrinae: Parandrini

This is one of the largest species of Parandrinae known, with large males exceeding 35 mm. The species is very similar to *Parandra glabra* but can be recognized by the apex of the mandible in males (pictured here) being relatively wide and having the punctation area not distinctly bordered and gradually terminating before the apex (*P. glabra* has the apex of the mandible in males relatively narrow and with a distinctly bordered punctation area abruptly terminating before mandible apex). This Bolivian endemic is known from the grounds of the Flora and Fauna Hotel near Buena Vista, Potrerillo del Guenda Reserve, and Refugio Los Volcanes in Santa Cruz Department.

Parandra (Tavandra) longicollis Thomson, 1861

Parandrinae: Parandrini

This species averages around 22 mm and is recognized by having the eye margin abruptly constricted posteriorly and having the ventral carina of the gena distinct, narrow, and elevated (as opposed to a gradually arising posterior eye margin and ventral carina of gena absent or vague in other species of *Parandra*, including *P. thomasi*). It is a widespread species that occurs from Mexico through Central America and south to Argentina. In Bolivia it is known from the grounds of the Flora and Fauna Hotel near Buena Vista, Potrerillo del Guenda, and Refugio Los Volcanes in Santa Cruz Department and Villa Tunari in Cochabamba Department.

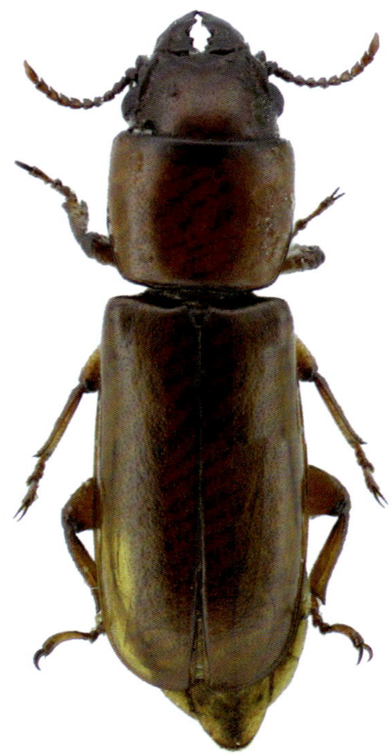

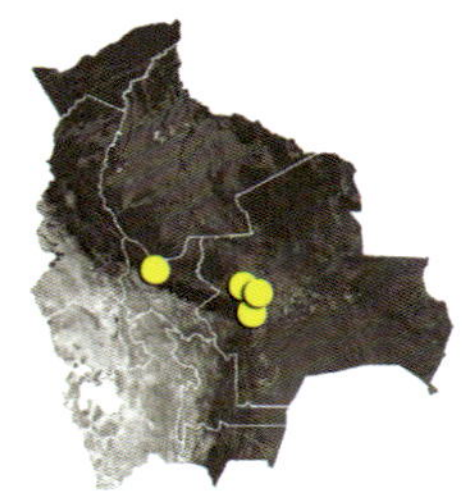

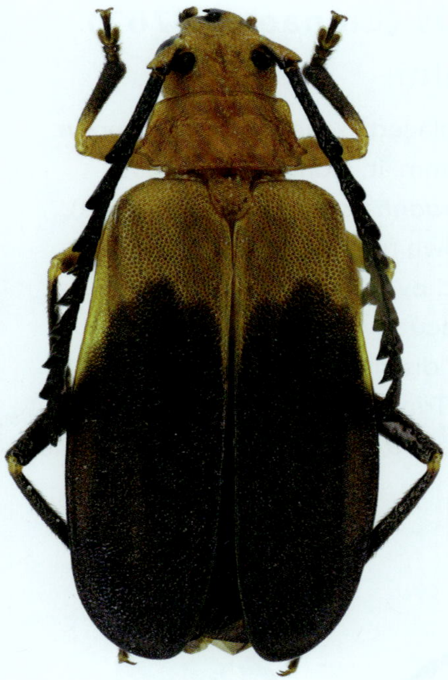

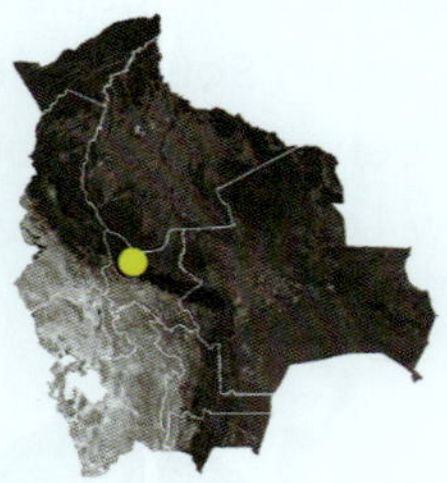

Myzomorphus amabilis (Tippmann, 1960)

Prioninae: Anacolini

This small (8–15 mm) prionine is a highly sexually dimorphic species and one of only three species of the tribe Anacolini recorded from Bolivia. The female (shown here) has fully developed elytra and short, weakly appendiculate antennae, while the male has the elytra short, exposing most of the abdomen and hind wings, and moderately appendiculate antennae nearly the length of the body. It is known only from Brazil and Bolivia. In Bolivia it has been found near Villa Tunari in Cochabamba Province.

Calocomus desmarestii (Guérin-Méneville, 1831)

Prioninae: Calocomini

This unusual prionine is somewhat variable in color. Some specimens are completely black, while others have variably sized yellow or orange maculae on the elytra, sometimes arranged into one or two separate spots before the middle and on the outer edge to the apex. Some specimens have a partially reddish pronotum. The antennae are strongly appendiculate in males, less so in females. Specimens range in size from 32 to 45 mm. This species is widely distributed in Bolivia, Paraguay, and Argentina. In Bolivia it is known from the dry Chaco region of southern Santa Cruz Department near Paraguay.

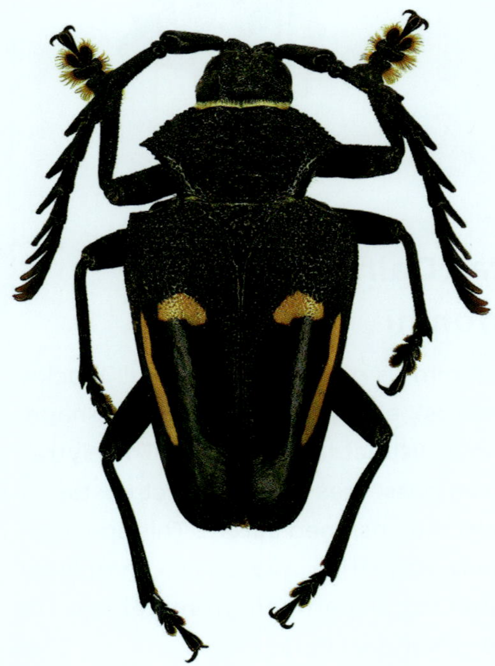

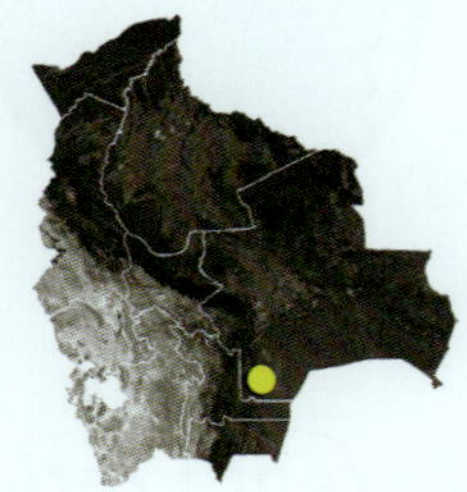

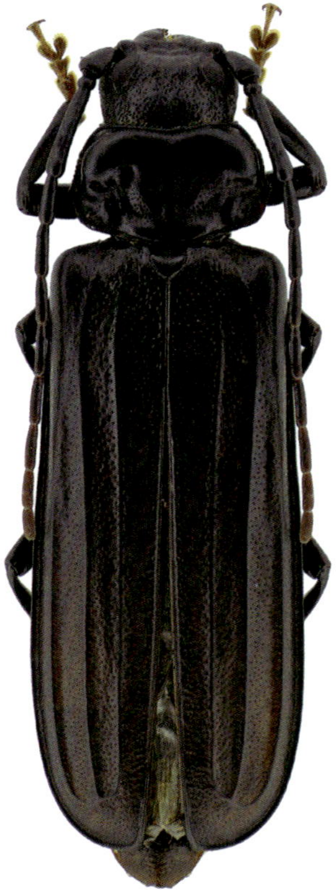

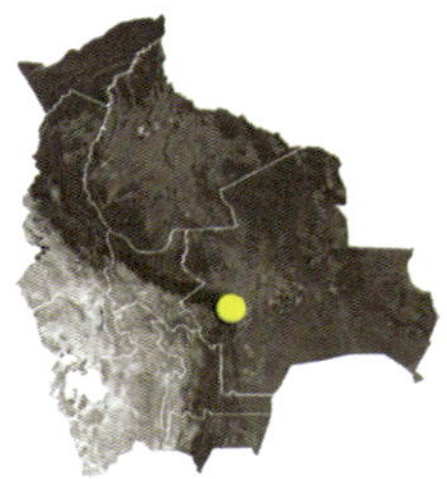

Chorenta reticulata (Dalman, 1817)

Prioninae: Callipogonini

This unusual prionine (38–44 mm) is recognized by its all-black (except yellowish protarsomeres), elongate, parallel-sided shape. The pronotum is short, without distinct lateral spines. The elytra are punctate, and each elytron possesses three distinct costae. The elytral apex is rounded to the unspined suture. This species is known from Brazil, Paraguay, Argentina, and Bolivia. In Bolivia it has been collected in Santa Cruz Department from Refugio Los Volcanes, adjacent to Amboró National Park.

Enoplocerus armillatus (Linnaeus, 1767)

Prioninae: Callipogonini

Until recently, this species was placed in the genus *Callipogon*. With some specimens over 130 mm, it, along with *Titanus giganteus* (Linnaeus) and *Macrodontia cervicornis* (Linnaeus), is one of the largest beetles known from Bolivia. Aside from its robust size, it is characterized by the chestnut brown elytra, laterally spined pronotum, and acutely spined inner apex of the scape. The species is widely distributed from Panama to Amazonian South America and Trinidad and Tobago. In Bolivia it has been recorded from Beni, Pando, and Santa Cruz Departments below 1,000 m in elevation.

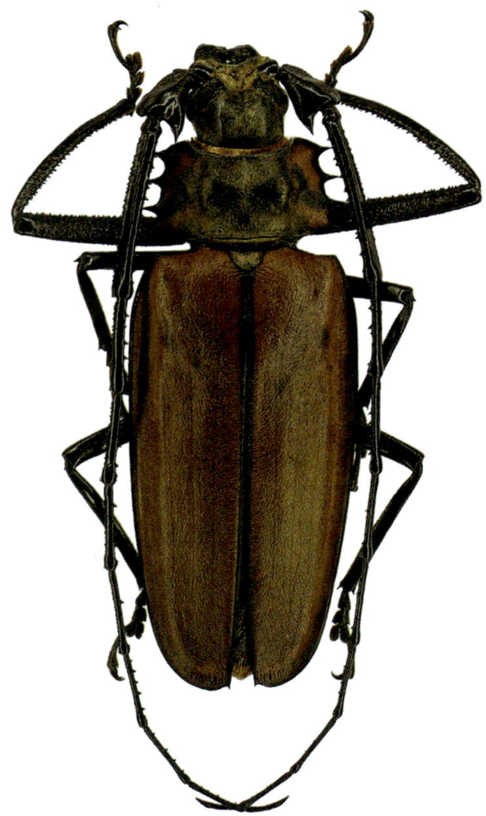

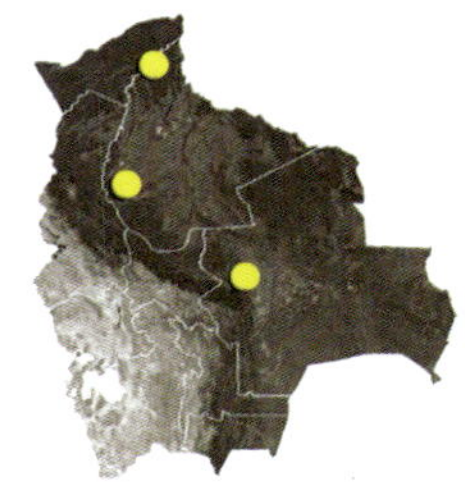

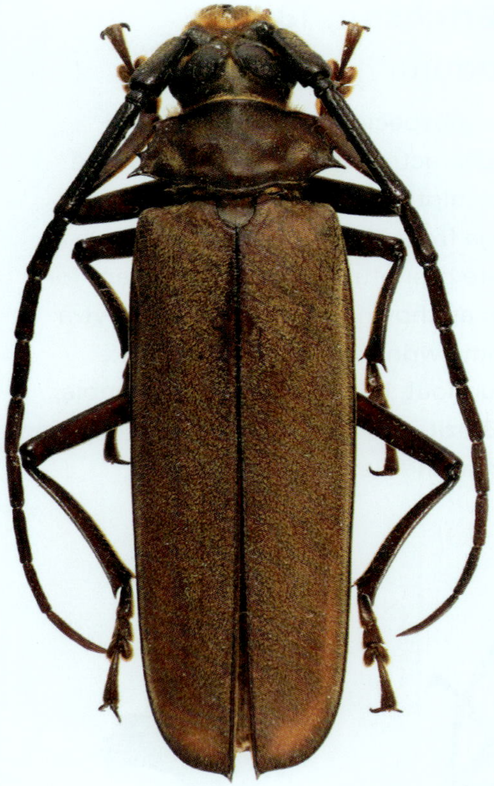

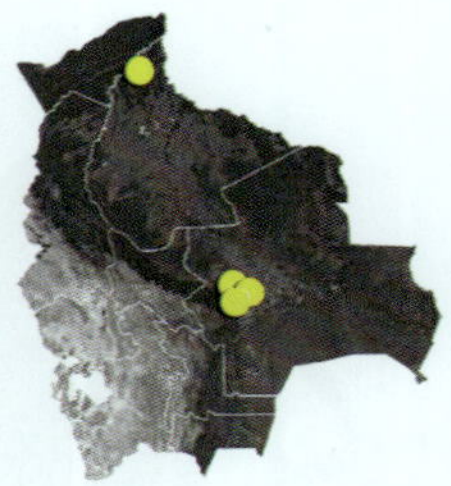

Orthomegas cinnamomeus (Linnaeus, 1758)

Prioninae: Callipogonini

Until recently, this species was placed in the genus *Callipogon*. It is highly variable in size, with large specimens exceeding 70 mm. The dorsum of the body is mostly covered in rusty, reddish-brown pubescence. The pronotum is laterally spined, along with the elytral sutural apex. This is among the most widespread New World species of Prioninae known, and it occurs from Mexico to Bolivia and on several Caribbean Islands, including Grenada, Trinidad, and Tobago. In Bolivia it is known from lowland sites in Beni and Santa Cruz Departments.

Macrodontia cervicornis (Linnaeus, 1758)

Prioninae: Macrodontiini

This is the largest *Macrodontia* species and one of the largest of all beetles found in Bolivia, with some males exceeding 170 mm. Much of this length is due to the intimidating mandibles that may extend forward for almost half the length of the body. The body in *Macrodontia* is compressed, not thick as in other species of prionines. This species is found on maripa palm trees, which is one of its larval hosts. The interconnected network of pale lines and dark brown maculae on the elytra, combined with the acute spines at the apex and base of the pronotum, are very distinctive for this species. It is a widespread species found throughout Amazonian South America. In Bolivia it is known from Pando, Beni, and Santa Cruz Departments.

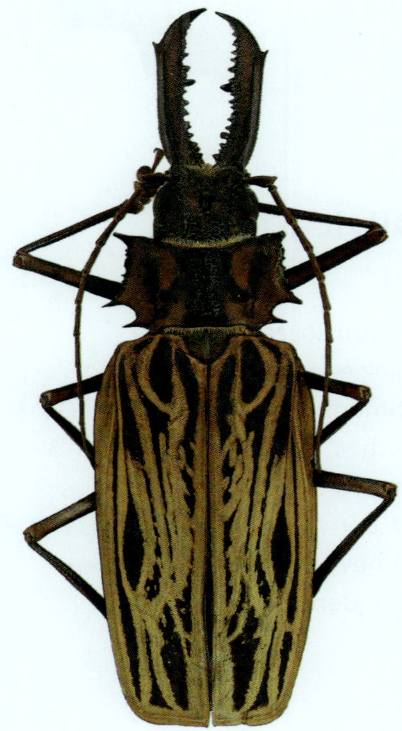

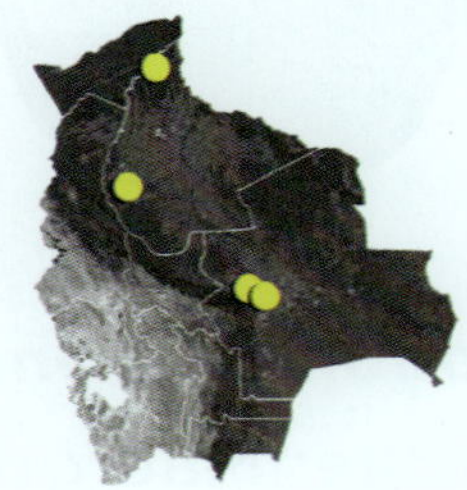

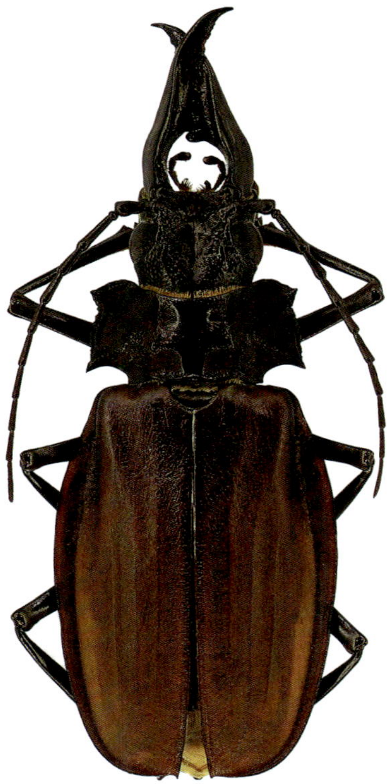

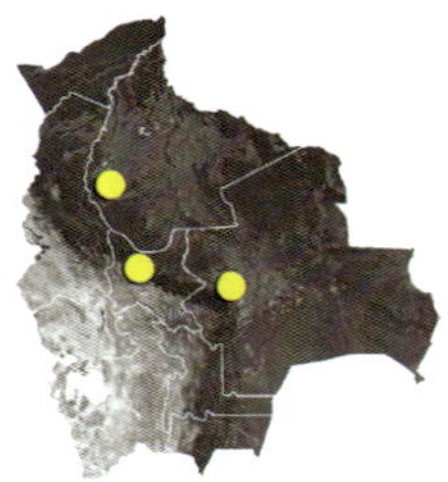

Macrodontia crenata (Olivier, 1795)
Prioninae: Macrodontiini

This species is smaller than its large cousin, *Macrodontia cervicornis*, but also has ominous, sickle-shaped mandibles, particularly in the males. Most specimens range from 60 to 70 mm, but females are shorter due to their smaller mandibles. The integument is uniformly chestnut to dark brown and lacks the interconnected maculae as in *M. cervicornis*. This species is widespread in South America but never common. In Bolivia it has been collected in Cochabamba, Beni, and Santa Cruz Departments.

Mallodon baiulus Erichson, 1847
Prioninae: Macrotomini

This species was formerly a subspecies of the widespread *Mallodon dasystomus*. It is characterized by the shape of the mandibles and the surface sculpturing of the pronotum and elytra. Most specimens range from 35 to 60 mm. Males have a very distinctly micropunctate pronotum throughout except for calli arranged in the pattern as shown in the photo. The elytra are impunctate but have some wrinkling and puncture-like depressions scattered throughout. It is known from Colombia, Peru, Ecuador, Venezuela, Brazil, and Bolivia. In Bolivia it has been collected in Santa Cruz, Tarija, and Beni Departments.

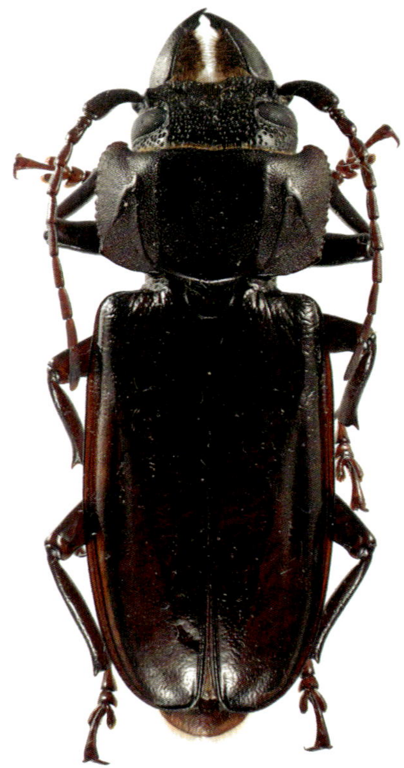

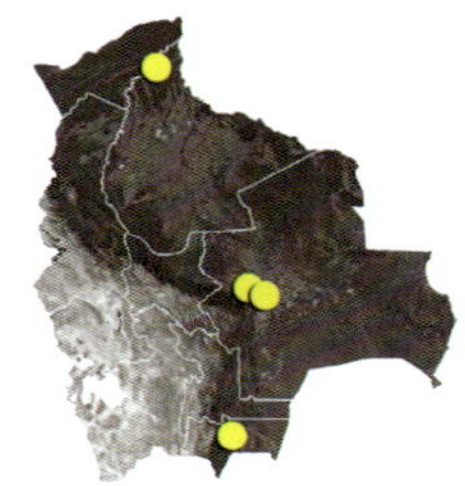

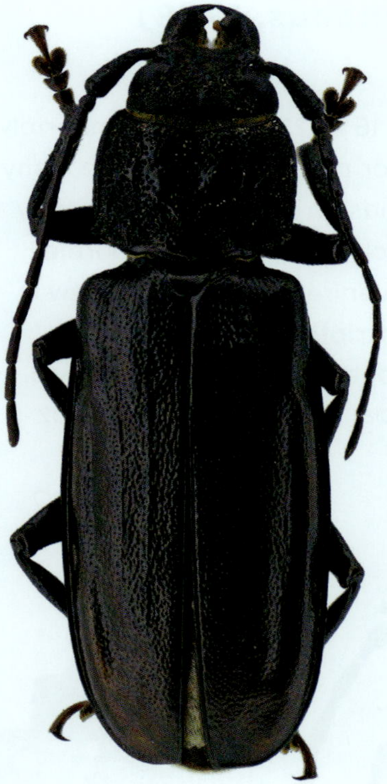

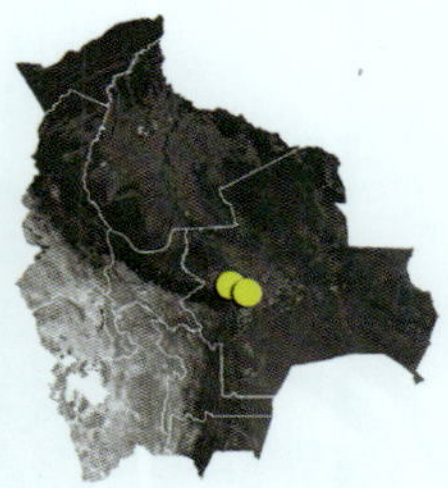

Physopleurus exiguus

Santos-Silva & Martins, 2003

Prioninae: Macrotomini

The all-black coloration, uniformly punctate elytra, pronotum, and head, and pronotal lateral margins lacking spines are all characteristic of this species. Most specimens are 35–45 mm in length. It is very similar to *Physopleurus amazonicus* but can be distinguished by its protibiae having spines (these are lacking in *P. amazonicus*). This species is recorded from Brazil and Bolivia. In Bolivia it is known from the grounds of the Flora and Fauna Hotel near Buena Vista and Potrerillo del Guenda in Santa Cruz Department.

Strongylaspis bolivianus

Monné & Santos-Silva, 2003

Prioninae: Macrotomini

This prionine is robust, although not particularly large (most specimens range in length from 30 to 40 mm) as are many others in the subfamily. It is recognized by the posteriorly widened pronotum with posterolateral spines on each side, the heavily granulate, light brown elytra, and the sculptured and partially yellow-pubescent pronotum. The legs, antennae, head, and pronotum are dark brown and contrast strongly with the elytra. It is known only from Bolivia and has been collected in La Paz Department near Coroico and Santa Cruz Department from the Bella Vista road near Cuevas and Refugio Los Volcanes.

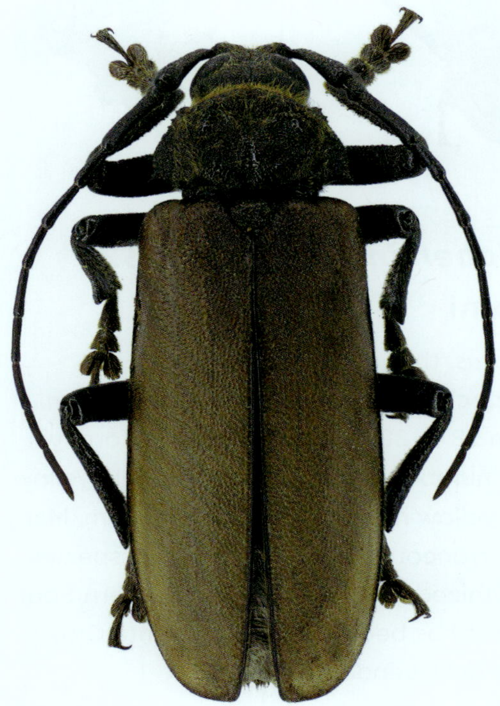

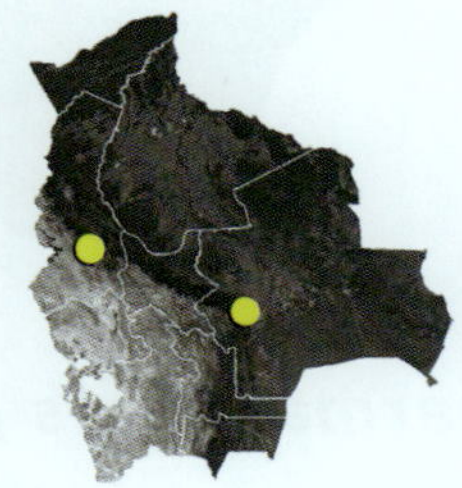

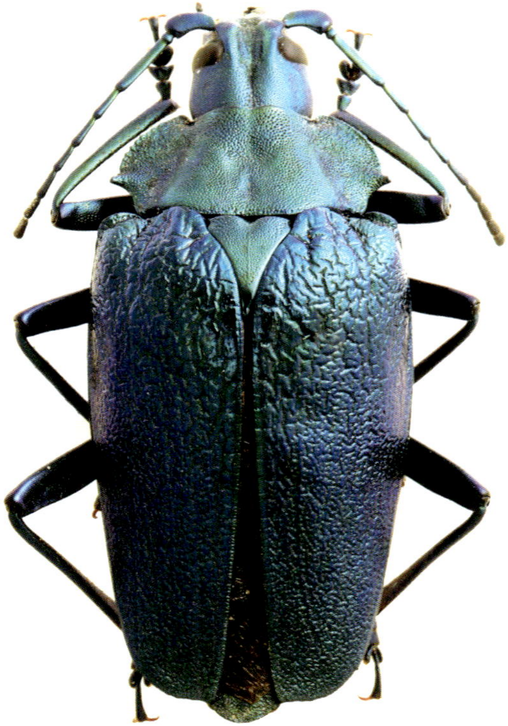

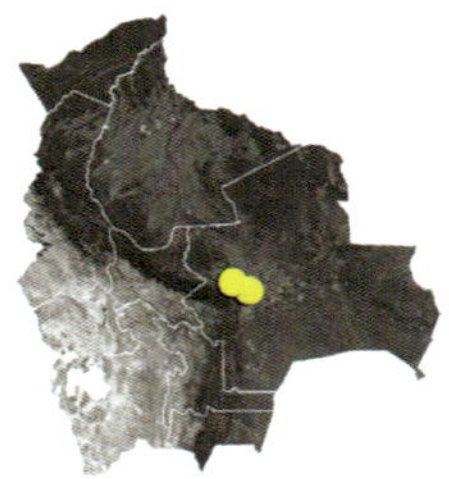

Charmallaspis pulcherrima (Perty, 1832)

Prioninae: Mallaspini

Along with *Hileolaspis auratus*, this is one of the most morphologically variable species of longhorned beetles known. Specimens may range from 28 to 38 mm and can be uniformly green, blue, bronze, or any number of variants with middle yellow elytral maculae and red or yellow maculae on the pronotum. Many of these color forms were erroneously described as new species and have now been synonymized. It is known from northern South America into Bolivia, where it has been collected in Santa Cruz Department from Potrerillo del Guenda Reserve.

Hileolaspis auratus (Linnaeus, 1758)

Prioninae: Mallaspini

This highly variable species is from 16 to 22 mm and is uniformly green, red, yellow, orange, bronze, or purple and can have many variations of yellow maculae or bands on the elytra. Some may have the legs partially black, all black, or all orange or purple. Because of all the variants, it has been redescribed as 11 new species since Linnaeus' original description! It is known from throughout northern and central South America. From Bolivia, it has been collected in Cochabamba, La Paz, and Santa Cruz Departments.

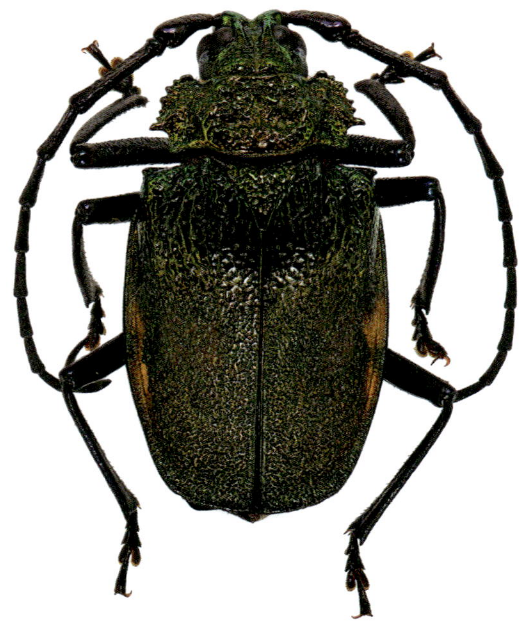

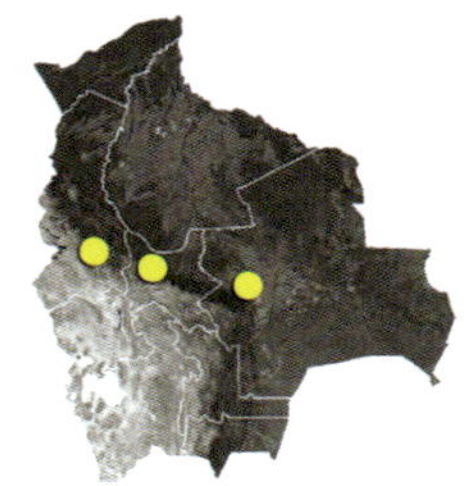

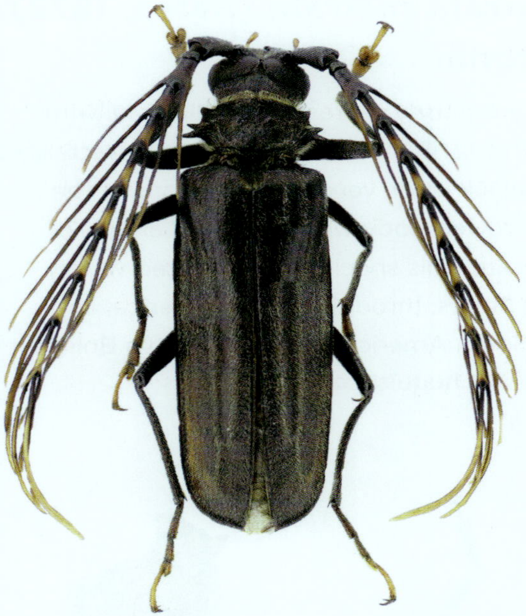

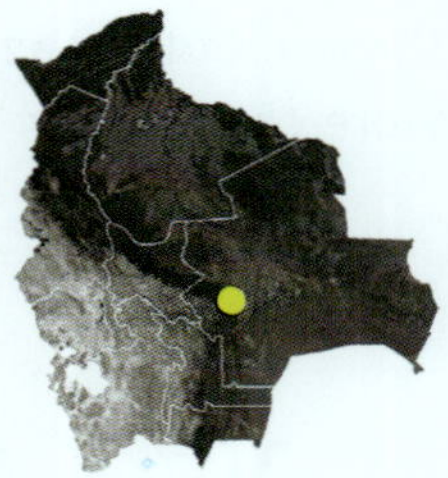

Sarifer flavirameus Kirsch, 1870

Prioninae: Meroscelisini

This is the only member of the tribe Meroscelisini known from South America. Ranging in size from 30 to 45 mm, it is immediately recognized by the beautiful, pectinate antennae of the males. The pronotum is small and laterally spined. The elytra are weakly costate and partially covered in yellowish setae. It is recorded from Colombia, Brazil, Peru, Ecuador, Argentina, and Bolivia. In Bolivia it has been collected only from Refugio Los Volcanes in Santa Cruz Department.

Psalidognathus friendii friendii Gray, 1831

Prioninae: Prionini

With its large size (50–70 mm), menacing mandibles, very long legs, and metallic, heavily sculptured integument, this is one of the most striking prionines known. This species is diurnal, unlike most members of the tribe Prionini, and prefers high, wet, mountainous terrain and hence has rarely been collected in Bolivia, with the only record from north of Villa Tunari in Cochabamba Department. It is also known from Colombia, Venezuela, and Ecuador.

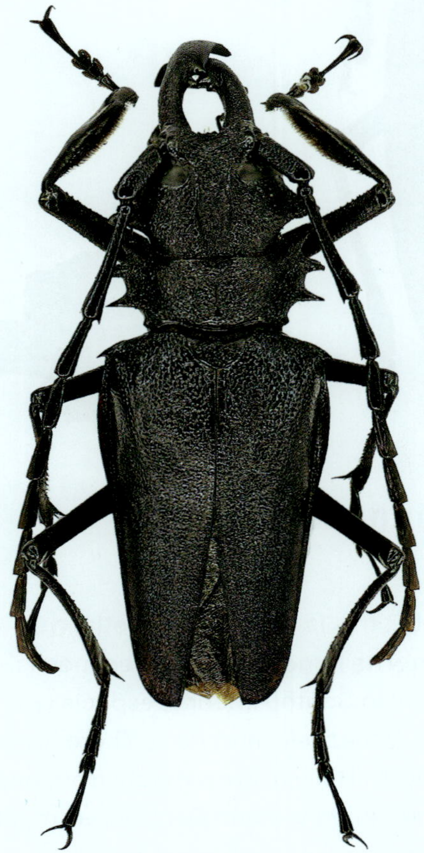

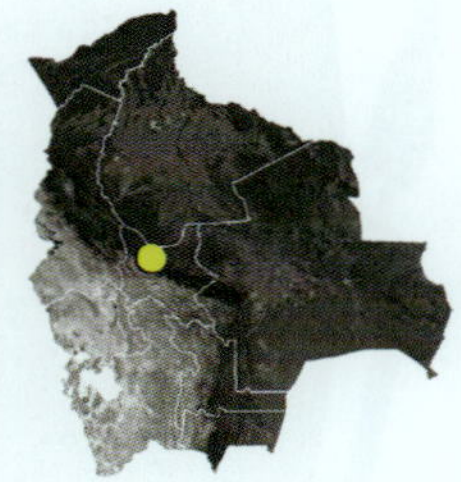

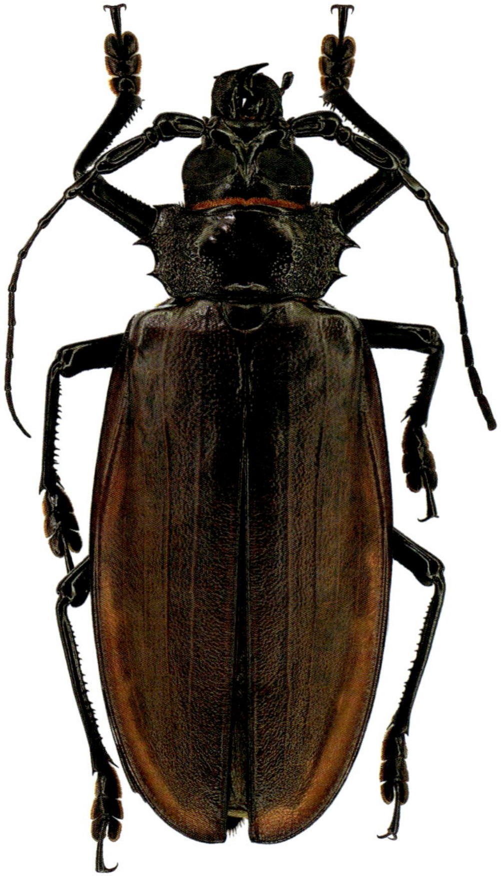

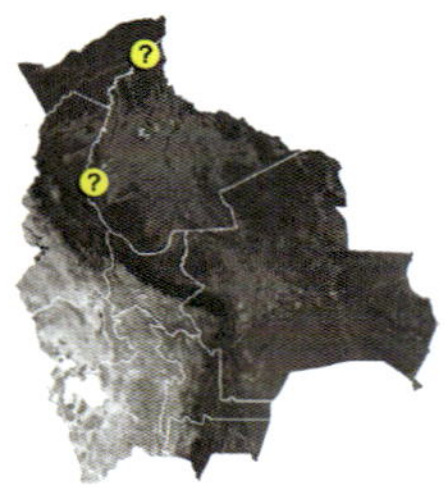

Titanus giganteus (Linnaeus, 1771)

Prioninae: Prionini

Commonly called the Titan beetle, this is the largest beetle in the world based on body length. The largest known specimen is 167 mm It is known from Amazonian South America, especially in French Guiana, Brazil, Colombia, Venezuela, and Peru. There are a few old museum records from north Bolivia, presumably from near the Beni and Madre de Dios Rivers in Pando and Beni Departments. However, no specimens have been recently collected in Bolivia.

Cyphonotida rostrata rostrata (Bates, 1872)

Lepturinae: Lepturini

The subfamily Lepturinae is poorly represented in Bolivia, with only 12 species known. This 12–15 mm long species is characterized by its reddish-brown to black color, very elongate frons of the head, and posteriorly narrowed body with the apical angles of the elytra obliquely truncate. This species is distributed widely from Texas in the United States, through Central America, and throughout Amazonian South America. Specimens from Bolivia were collected in Tarija and Chuquisaca Departments.

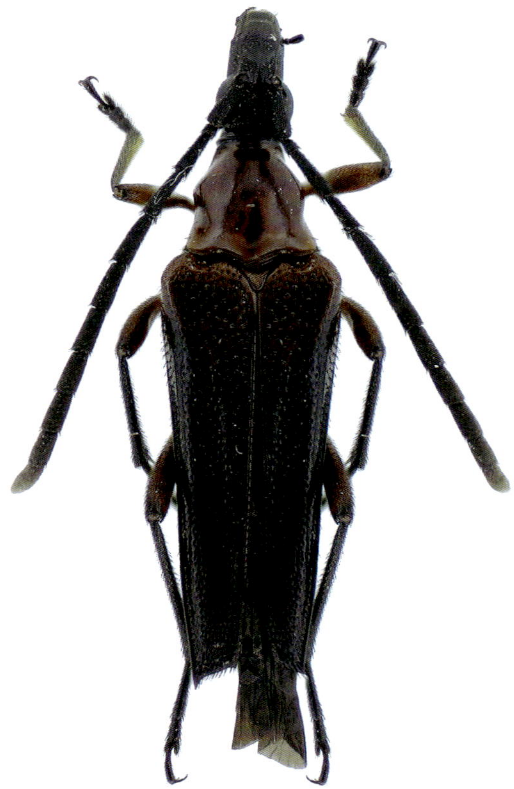

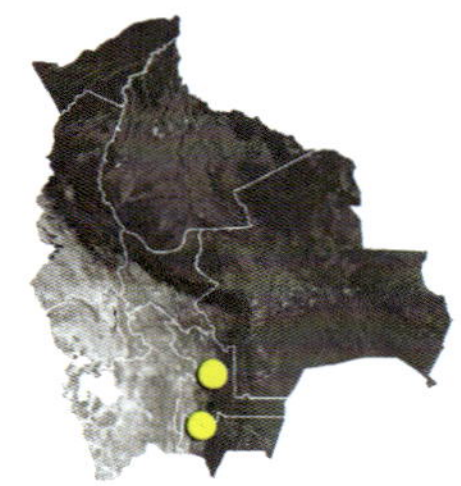

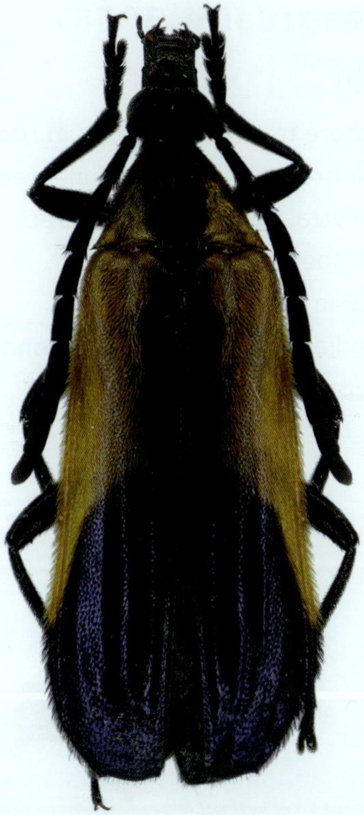

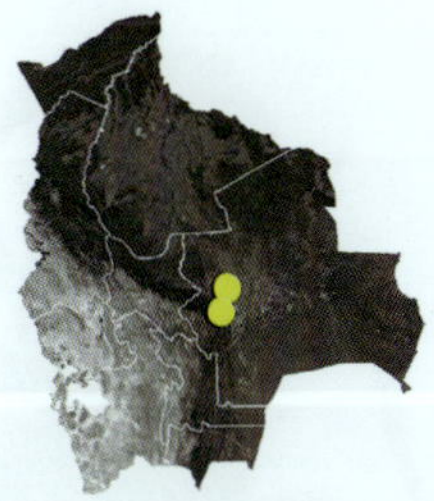

Megachoriolaus clarkei Monné & Monné, 2008

Lepturinae: Lepturini

This attractive species (13–15 mm long) is distinct with its very constricted head posteriorly, elongate frons, and pattern of orange or red markings on the pronotum and elytra. The black apex and central region of the pronotum is characteristic of this species. It is known from Rondônia, Brazil, and Bolivia. In Bolivia it has been collected at 400 m in Santa Cruz Department at the Flora and Fauna Hotel grounds near Buena Vista as well as at over 2,000 m in the Vicoquin area above Achira.

Strangalia albicollis (Pascoe, 1860)

Lepturinae: Lepturini

This lepturine (12–16 mm long) is distinct with its lateral margins of the pronotum, elytra, and legs yellowish but remaining portions black. Like other lepturines, the abruptly, posteriorly constricted head and anteriorly constricted pronotum is distinctive. It is widespread but uncommonly collected. It is known from Brazil, Peru, French Guiana, and Bolivia. It has been collected only a few times in Bolivia and is known from the Flora and Fauna Hotel grounds near Buena Vista at 400 m and from Samaipata at 1,600 m in Santa Cruz Department.

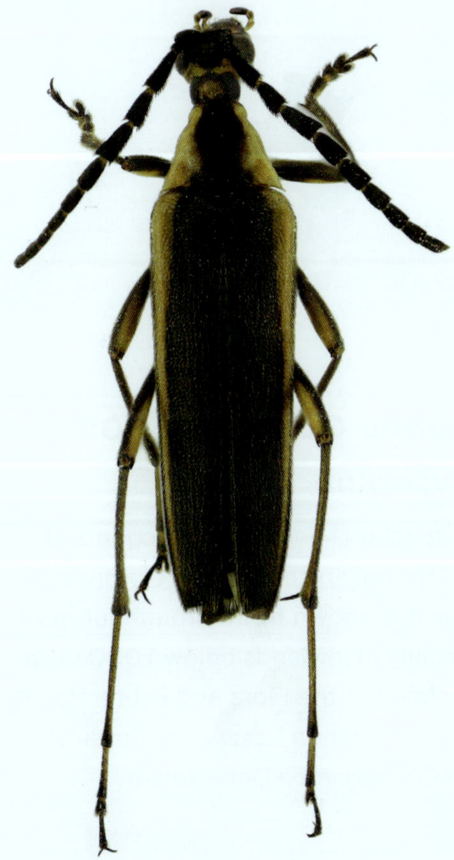

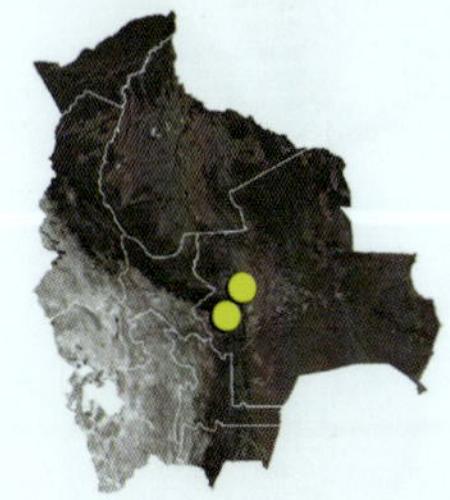

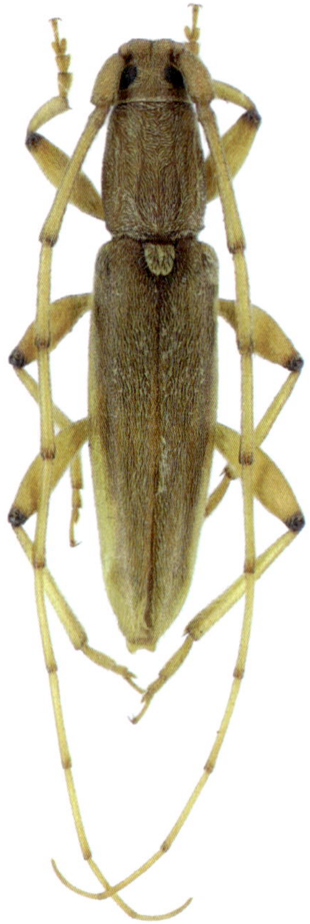

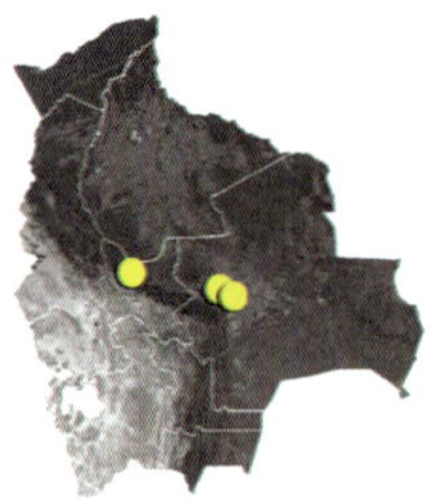

Achryson immaculipenne Gounelle, 1909

Cerambycinae: Achrysonini

This small- to moderate-sized species (14–16 mm) is recognized by its yellow coloration, black femoral apices, elongate pronotum, and large, pubescent scutellum. It is known from throughout most of tropical South America, typically in lowlands below 1,000 m. In Bolivia it is known from the grounds of the Flora and Fauna Hotel near Buena Vista and Potrerillo del Guenda Reserve in Santa Cruz Department and Villa Tunari in Cochabamba Department.

Achryson maculipenne (Lacordaire, 1868)

Cerambycinae: Achrysonini

This species (15–17 mm) is one of the more maculate species in the large genus, *Achryson*. It is mostly dark brown with interconnected yellowish maculae scattered over the elytra. The body is covered in sparse, long, white pubescence. This species is known from Brazil, Paraguay, Argentina, and Bolivia. In Bolivia it has been collected from Refugio Los Volcanes, adjacent to Amboró National Park, and near Camiri in Santa Cruz Department.

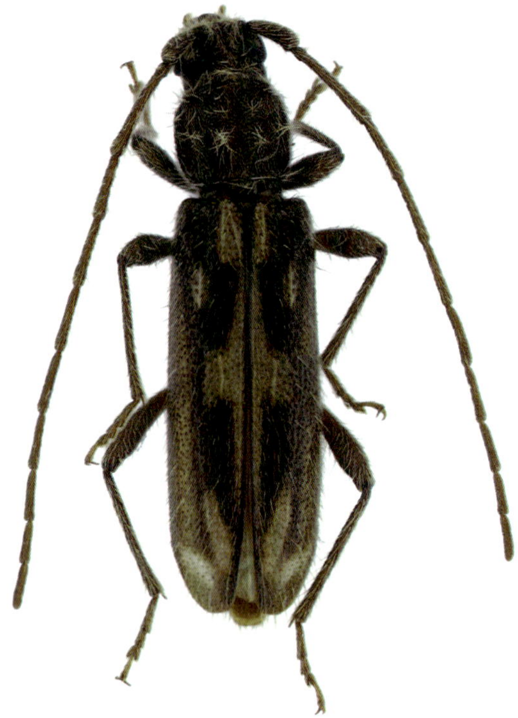

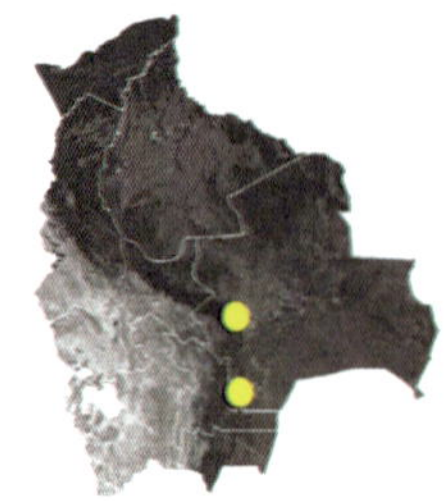

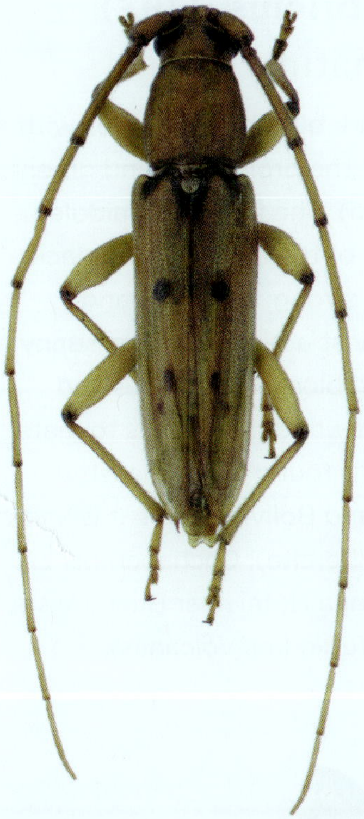

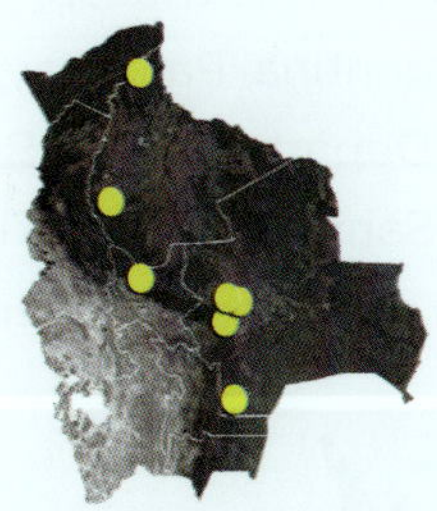

Achryson surinamum (Linnaeus, 1767)

Cerambycinae: Achrysonini

This common species is 11–20 mm long, mostly pale yellow in color, but with several dark brown maculae on the elytra. Specimens are variable in this feature and some have far more markings than others. This is one of the most widespread species of Cerambycidae in the New World. It has been found in every country in North, Central, and South America except Canada. In Bolivia it is known from the grounds of the Flora and Fauna Hotel near Buena Vista, Refugio Los Volcanes, Camiri, and Potrerillo del Guenda in Santa Cruz Department. It has also been found in Riberalta in northern Beni Department, Rurrenabaque in southern Beni Department, and Cristal Mayu in Cochabamba Department.

Chlorida festiva (Linnaeus, 1758)

Cerambycinae: Bothriospilini

This is a moderate to large (23–24 mm), reddish-brown and pale green species with black antennomeres that is commonly attracted to lights. It is one of the most widespread and common cerambycid species in the New World, ranging in distribution from Mexico to Central America, the Caribbean Islands, and most of South America. In Bolivia it has been found at low- to midelevation (200–1,000 m) localities in Cochabamba Department (in Villa Tunari) and Santa Cruz Department from the vicinity of the Flora and Fauna Hotel near Buena Vista, the Potrerillo del Guenda Reserve, Refugio Los Volcanes, adjacent to Amboró National Park, and the Camiri environs. It has also been collected near Riberalta in Beni Department.

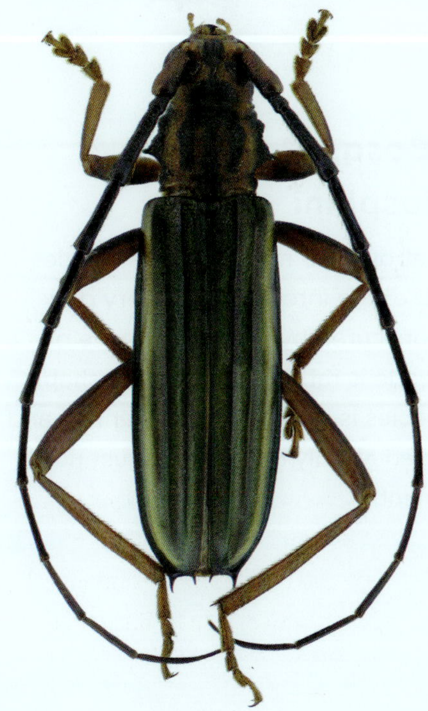

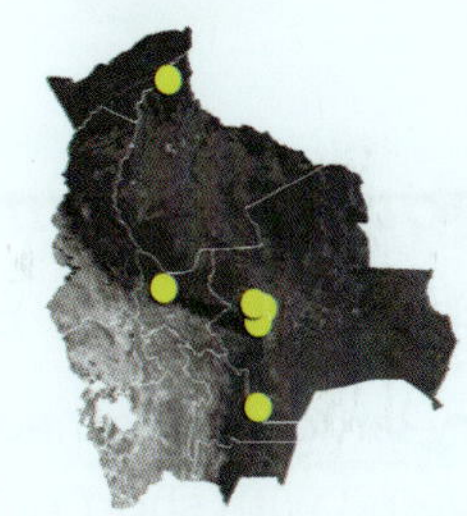

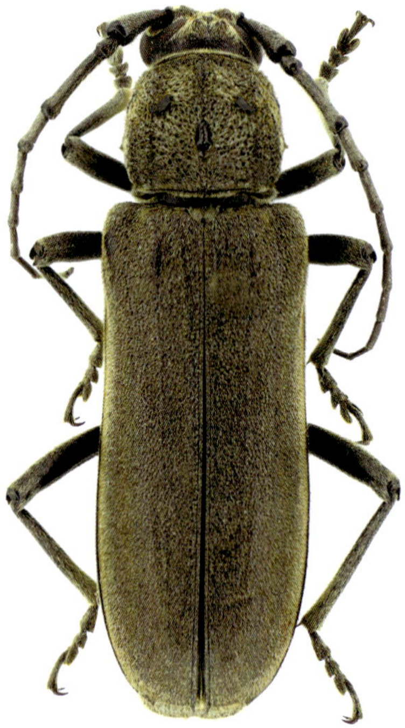

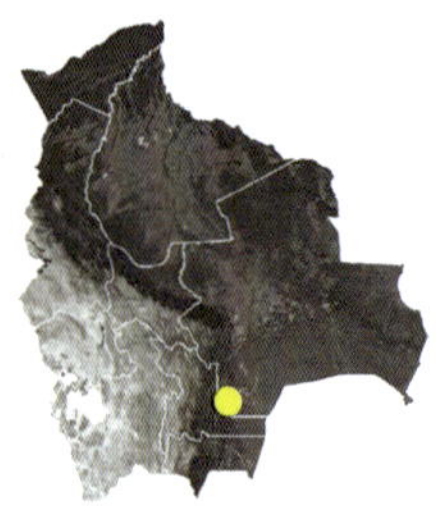

Ranqueles gounellei Bosq, 1947

Cerambycinae: Bothriospilini

A very large longhorned beetle at over 40 mm long, this species is covered with short golden setae. The integument is very dark brown to black, but it is mostly obscured by the pubescence on the pronotum and elytra. The beetle is rather cylindrical in shape, not tapering posteriorly. This species is known from Argentina and Bolivia, where it has been collected at only one locality, just north of Camiri in Santa Cruz Department.

Callichroma sericeum (Fabricius, 1793)

Cerambycinae: Callichromatini

This is a large (25–30 mm), velvety, dark blue-black species with shining metallic reflections of blue (on the pronotum) and green (along the sutural margins of the elytra). The front and middle femora are orange with the remainder of the appendages black. This attractive diurnal cerambycid is a strong, able flyer and easily spooked when disturbed. It is most active on warm, sunny mornings when it can be found visiting blossoms of trees and woody vines or gathering on freshly downed tree trunks to mate and oviposit. This widespread species is found in southeastern Brazil, Argentina, Paraguay, Uruguay, and Bolivia, where it is known from the Departments of Beni (near Riberalta), Cochabamba, La Paz, and Santa Cruz (the Flora and Fauna Hotel near Buena Vista, Potrerillo del Guenda Reserve, and Refugio Los Volcanes).

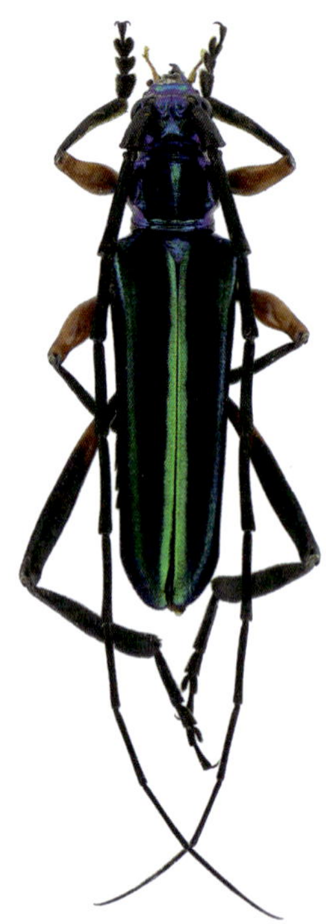

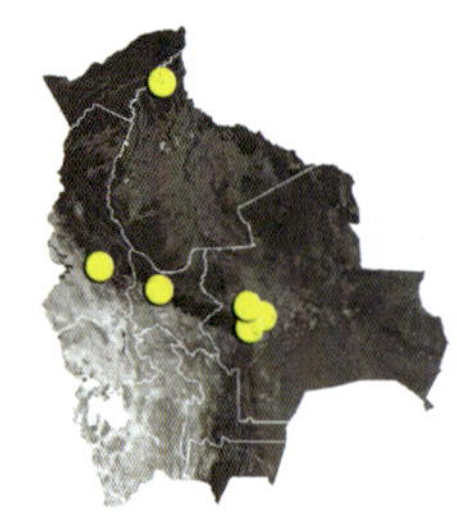

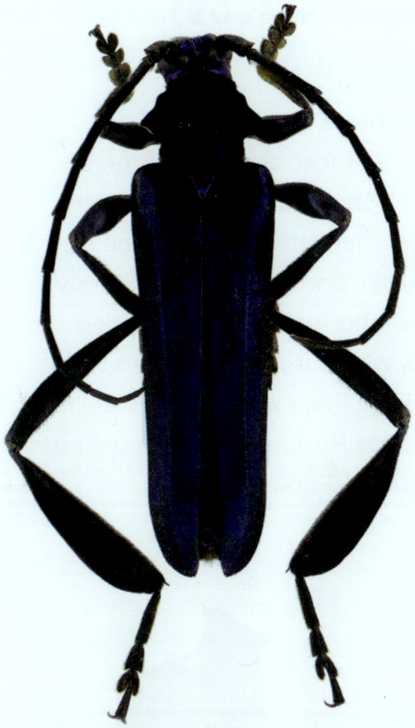

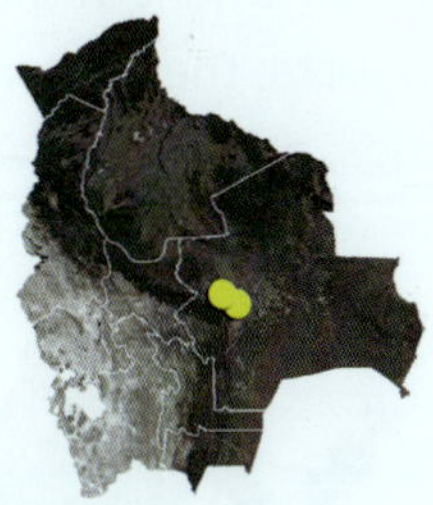

Cnemidochroma buckleyi (Bates, 1879)

Cerambycinae: Callichromatini

A very striking species, this large (24–28 mm), velvety, black and blue callichromatine with greatly expanded, flat hind tibiae is a gem of the lowlands. The pronotum is velvety black as is the rest of the dorsum and appendages except for the minutely punctate, iridescent, shining blue-black elytra. It is one of the most beautiful of the Bolivian Cerambycidae and can be found early in the rainy season, during sunny mornings, on the blossoms of small flowering trees where, unlike the *Callichroma*, it feeds quietly and is easily collected. This species is known from Ecuador, Peru, and Bolivia, where it has been recorded only from Santa Cruz Department in the vicinity of the Flora and Fauna Hotel near Buena Vista and the Potrerillo del Guenda Reserve.

Mionochroma electrinum (Gounelle, 1911)

Cerambycinae: Callichromatini

This is a moderate to large (15–24 mm), brightly colored species of orange, blue, black, and iridescent green. It varies in its appearance depending on the angle from which the observer looks at it and the amount of direct light on its variably textured surface. Typically, the head and pronotum will appear as iridescent blue or centrally green with blue-green to blue-black lateral edges. The elytra are minutely punctate and partially covered in short, dense pubescence arranged in stripes that will reflect green to blue or even black. The femora and scape are shining orange with the tibia darker and the antennomeres black. It is another of the day-flying flower visitors most active on warm, sunny mornings and disappearing during the heat of the day. It is distributed in central and southern South America from Brazil, Argentina, Paraguay, Uruguay, and Bolivia, where it is recorded from southern Santa Cruz Department (near Camiri) and Tarija Department (near Villamontes).

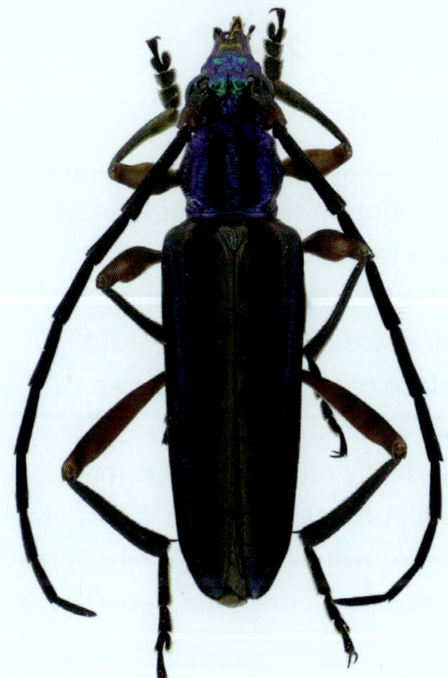

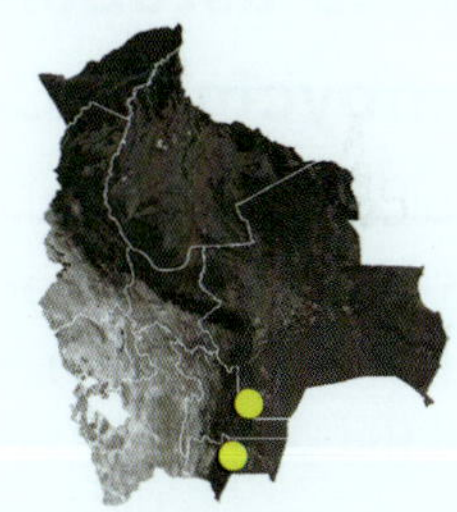

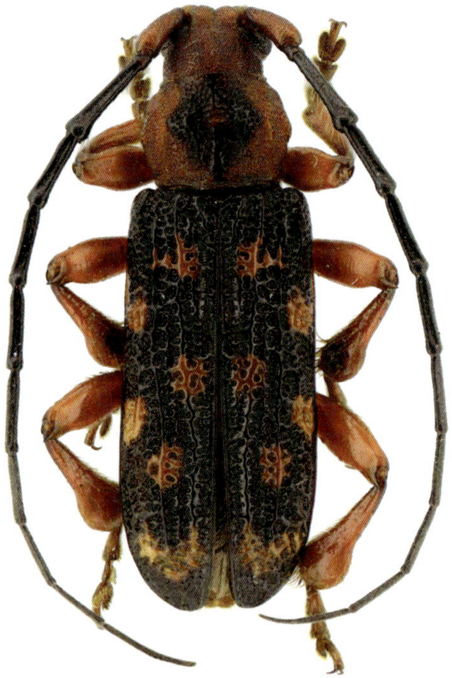

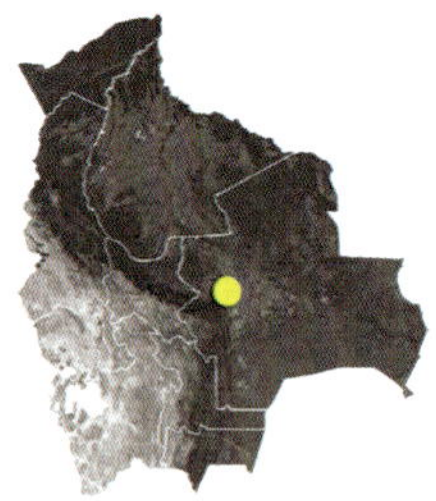

Cevaeria estebani Tavakilian, 2004

Cerambycinae: Cerambycini

An atypical Cerambycini, this is a very bold species of moderate size (13–15 mm). The head, scape, pronotum (except for a middle black diamond macula), and legs are red or orange or reddish. Beyond the scape, the antennae are black. The middle and hind femora are armed internally at their apices with a short, stout spine. The shining black elytral surface has coarse punctures roughly aligned in longitudinal rows and separated by irregular rugose ridges. Each elytron has five to six reddish-orange maculae. It is known only from French Guiana and Bolivia, where it is rarely encountered at lights. In Bolivia it is known from the vicinity of the Flora and Fauna Hotel near Buena Vista in Santa Cruz Department.

Coleoxestia atrata (Gounelle, 1909)

Cerambycinae: Cerambycini

This is a large (20–25 mm), shining black species with a transversely rugose pronotum and alutaceous elytral surface. The antennae are shorter than the body in both sexes, with the distal antennomeres piceous rather than shining black. The elytral apices are doubly spined on each elytron. This species is rarely encountered except at lights in lowland forests. It is recorded from Brazil, Colombia, and Bolivia, where it is known from Santa Cruz Department in the vicinity of the Flora and Fauna Hotel near Buena Vista, the Potrerillo del Guenda Reserve, and Refugio Los Volcanes, adjacent to Amboró National Park.

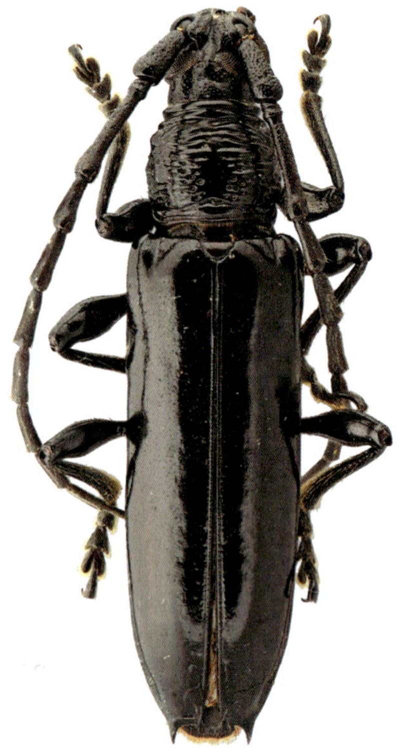

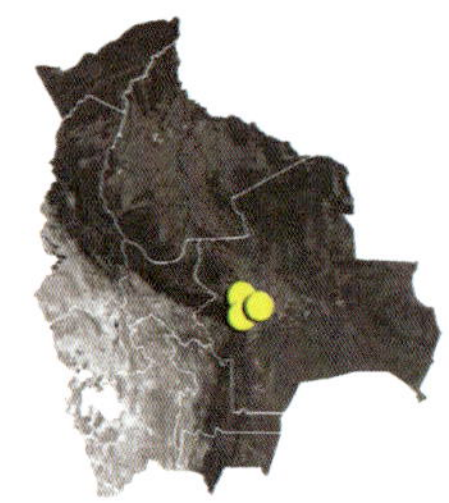

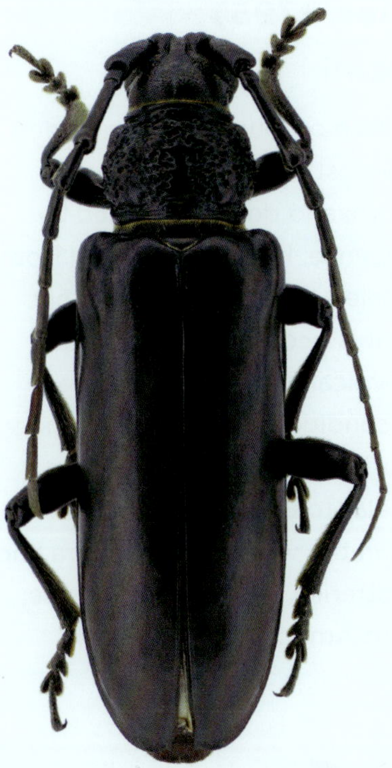

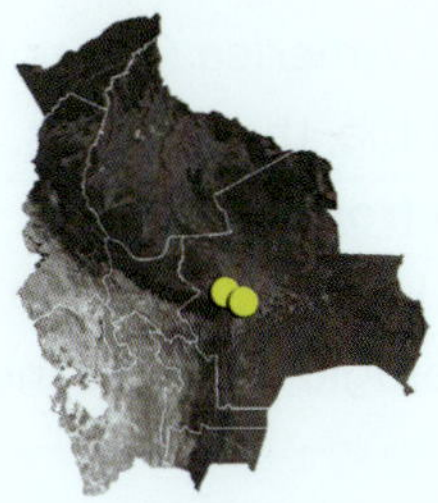

Coleoxestia corvina (Germar, 1824)

Cerambycinae: Cerambycini

This is a large (28–30 mm), robust, slightly shining reddish-brown species with sparse, minute punctation on its alutaceous elytral surface and deep coarse punctures (pock-like) on the pronotum. Unlike *Coleoxestia atrata*, only the internal apex of each elytron is spined, and then only with a small, acute spine. This stout, black species is easily recognizable among the 13 members of *Coleoxestia* found in Bolivia. It is locally abundant at lights in lowland forest locales. It is recorded from central and southeastern Brazil, Argentina, Paraguay, Uruguay, and Bolivia, where it is known from Santa Cruz Department in the vicinity of the Flora and Fauna Hotel near Buena Vista and the Potrerillo del Guenda Reserve.

Coleoxestia julietae Galileo & Martins, 2006

Cerambycinae: Cerambycini

This is a moderately large (18–24 mm), reddish-brown species with light tan antennomeres and orange femora. The elytra are alutaceous with small, scattered punctures. It also has a dark sutural stripe, bordered by a lighter reddish-brown stripe, which is then bordered by a dark linear stripe. The elytral apices are armed with stout spines on both margins and the pronotum is transversely rugose. This species is attracted to lights and can be locally abundant. *Coleoxestia julietae* was only recently described and is a Bolivian endemic known only from Santa Cruz Department in the vicinity of the Flora and Fauna Hotel near Buena Vista and the Potrerillo del Guenda Reserve in lowland forest habitat.

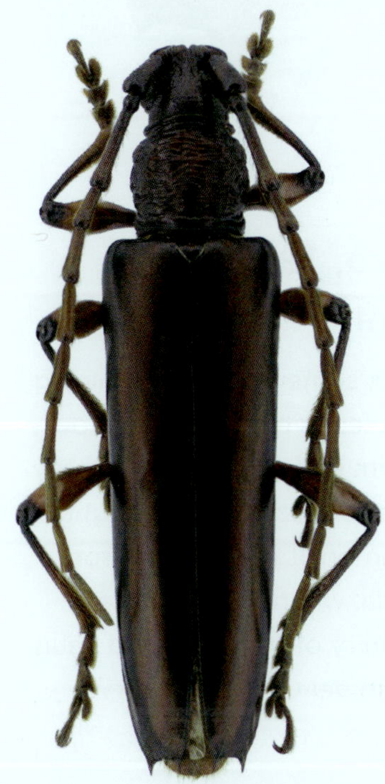

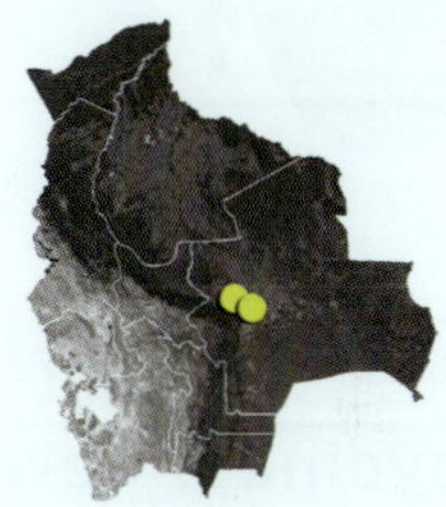

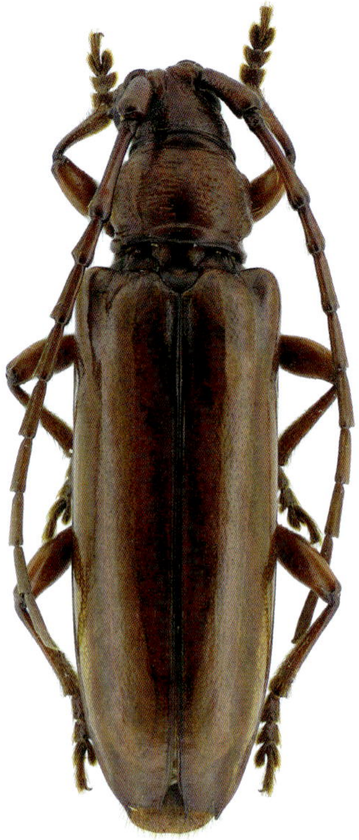

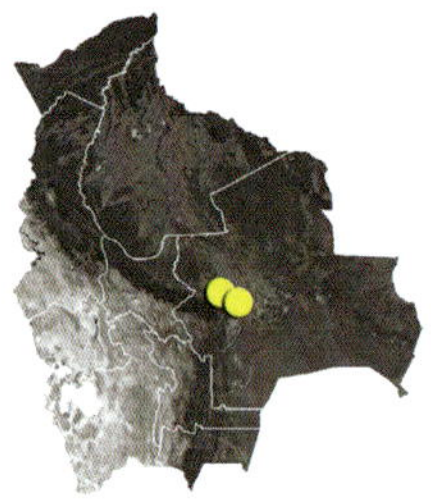

Coleoxestia olivieri Fragoso, 1993

Cerambycinae: Cerambycini

This is a moderate to large (20–30 mm), somewhat stout, shining dark orangish-brown species with alutaceous elytra and shallowly rugose pronotum. Its coloration and form make it easy to identify among the Bolivian *Coleoxestia* species. It is occasional at lights during mid-rainy season but never common. It is recorded from French Guiana, central Brazil, and Bolivia, where it is recorded from Santa Cruz Department in the vicinity of the Flora and Fauna Hotel near Buena Vista and the Potrerillo del Guenda Reserve.

Criodion cinereum (Olivier, 1795)

Cerambycinae: Cerambycini

This easily recognizable genus has three or four species recorded from Bolivia, with this one being the most common. It is very large (40–60 mm), dark reddish brown, and densely clothed by short, recumbent, light brown pubescence. It is distinguished from *Criodion torticolle*, another large species, by its short scape that does not reach the pronotal margin. It is occasionally attracted to lights well into the rainy season. Typically it shows up alone on supersaturated humid nighttime conditions. It is found from Central America (Costa Rica) into northern South America (Colombia, the Guianas, and Venezuela) to Bolivia, where it is known from Santa Cruz Department in the vicinity of the Flora and Fauna Hotel near Buena Vista, the Potrerillo del Guenda Reserve, and Refugio Los Volcanes, adjacent to Amboró National Park.

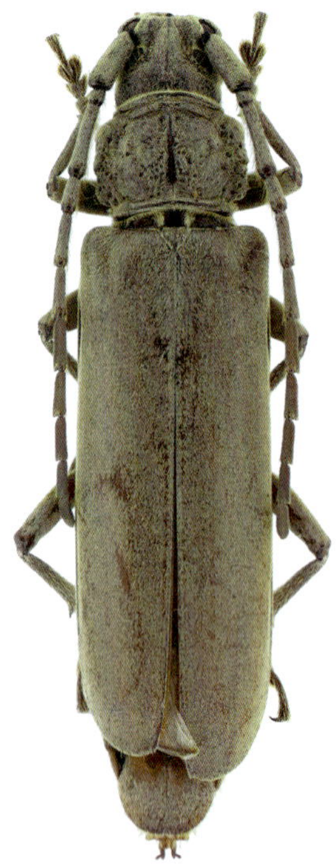

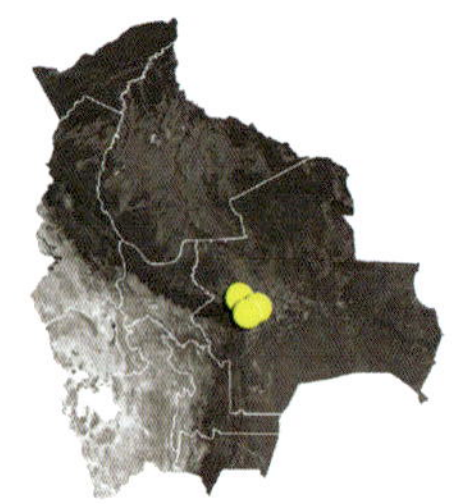

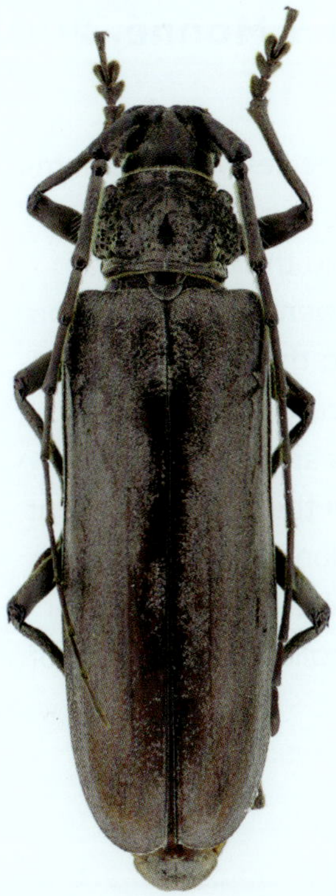

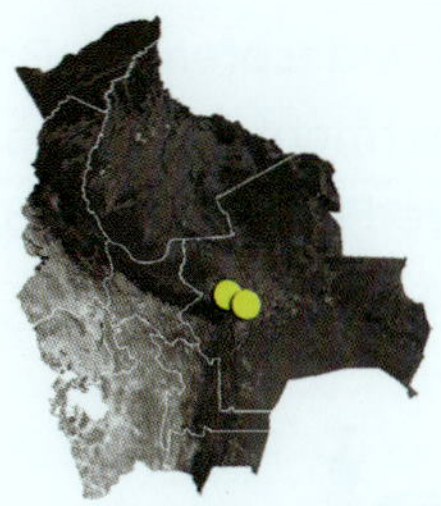

Criodion torticolle Bates, 1870

Cerambycinae: Cerambycini

This is among the largest of the subfamily Cerambycinae occurring in Bolivia, with most specimens exceeding 60 mm. Like its congeners, it has dark reddish-brown integument covered with short, dense, recumbent, light brown pubescence. In addition to its large size this species can be distinguished by its different tuberculate pronotal pattern and longer, stouter scape. It is rare at lights in the rainy season, with females more common than males. Its distribution includes French Guiana, Argentina, Brazil, Paraguay, and Bolivia, where it is known from Santa Cruz Department in the vicinity of the Flora and Fauna Hotel near Buena Vista and the Potrerillo del Guenda Reserve.

Juiaparus batus (Linnaeus, 1758)

Cerambycinae: Cerambycini

This is a large (25–42 mm), dark brown species, covered with fine, short, recumbent light brown pubescence that partially obscures the dorsal surface. It has the basal third and fourth (and sometimes fifth) antennal segments with a long, forward-directed spine at their apices. Large males have very long antennae that can be two and one-half times their body length. They are nocturnal beetles attracted to lights and locally abundant early in the rainy season or found at night on recently cut or partially burned large tree trunks in agricultural *chaqueos* (recently burned and cut forests for agricultural use). It is widely distributed and recorded from Central America to Brazil, Argentina, Paraguay, and Bolivia. In Bolivia it has been collected in Beni Department (near Riberalta), Tarija Department (near Villamontes), and Santa Cruz Department (at nearly every site sampled).

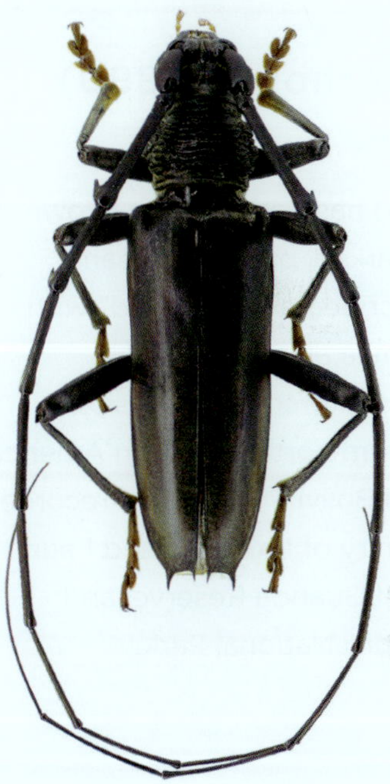

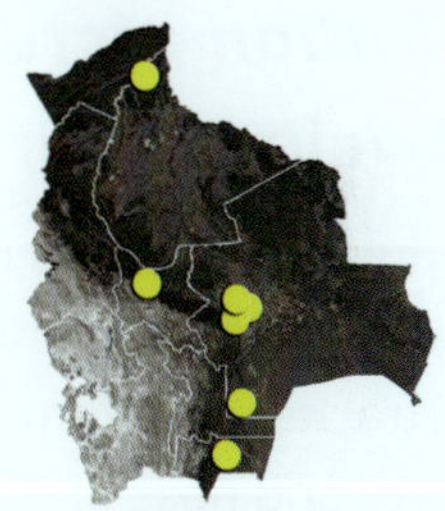

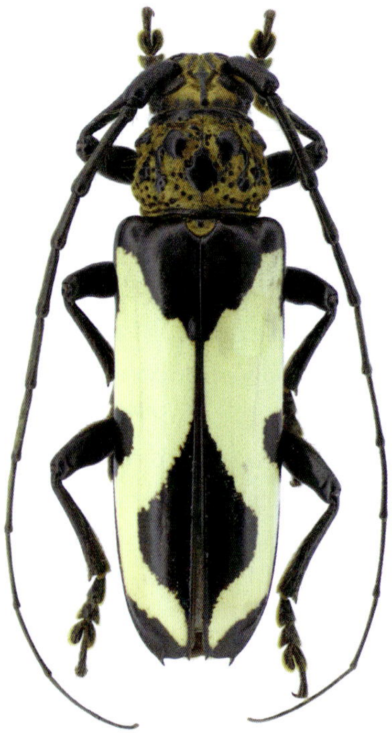

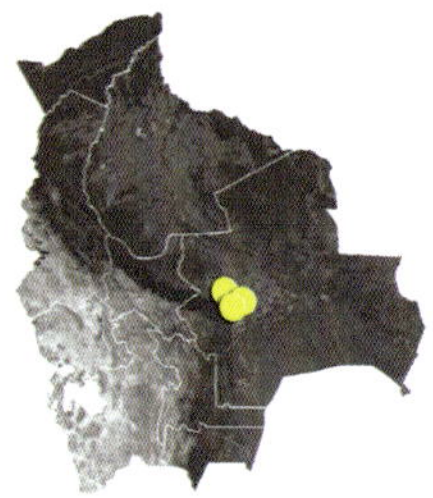

Metacriodion pictum (Waterhouse, 1880)

Cerambycinae: Cerambycini

This beautiful, large species (26–30 mm) has dark reddish-brown integument and shining, distinctively marked, primarily cream-colored elytra. Its pronotum is best described as "cat faced" with glabrous eyes and nose and a rugose smile all surrounded in golden-rust pubescence. It is rarely encountered but occasional and singular at lights. It is known only from northern South America (Ecuador, French Guiana, and Peru) and Bolivia, where it is recorded from Santa Cruz Department in the vicinity of the Flora and Fauna Hotel near Buena Vista, the Potrerillo del Guenda Reserve, and Refugio Los Volcanes, adjacent to Amboró National Park.

Plocaederus confusus Martins & Monné, 2002

Cerambycinae: Cerambycini

This is a medium-sized (14–22 mm), black species with medium-tan elytra and dark reddish-brown antennomeres. The dark pronotum has a transversely rugose disk, and the elytra are minutely punctate with shining gold, recumbent pubescence on the surface. The antennae from the fourth to the tenth segment are externally produced into a flange-like acute spine. This is the most common of the seven *Plocaederus* species recorded from Bolivia and is locally abundant, early in the rainy season, at lights. It is widespread in Brazil and the lowland forests of Santa Cruz Department, Bolivia, and has been recorded from most of the surveyed localities including the lesser collected localities of Samaipata and Camiri.

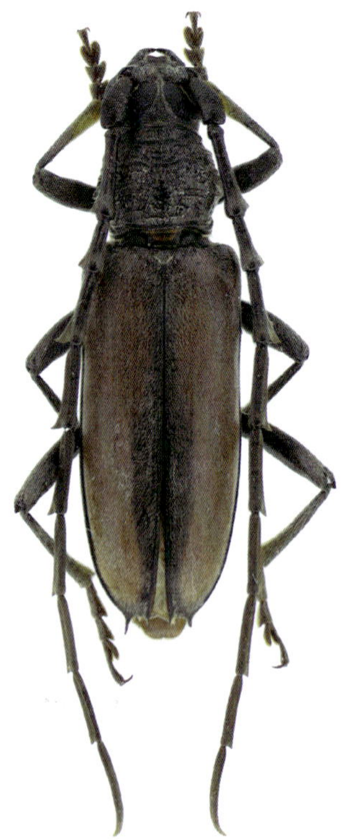

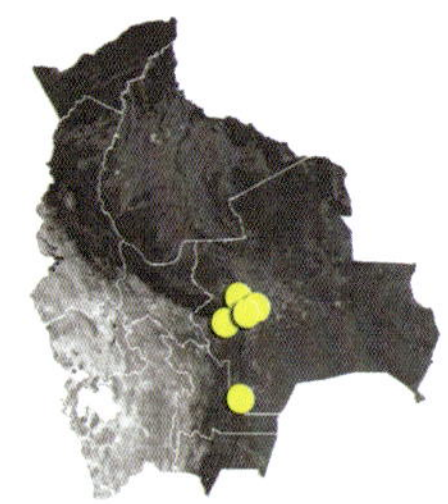

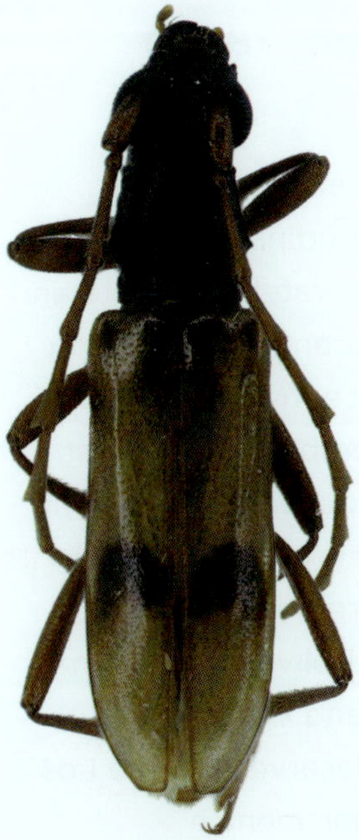

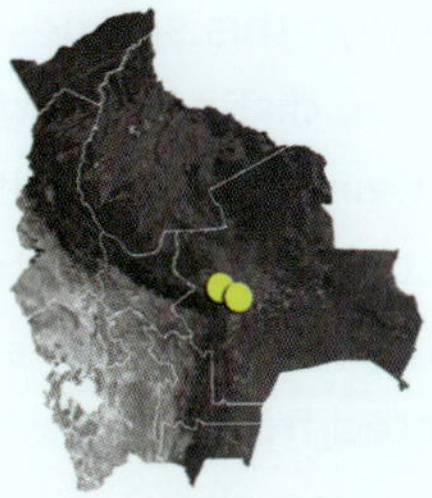

Plocaederus fasciatus (Martins & Monné, 1975)

Cerambycinae: Cerambycini

This is a small (6–9 mm), tan-colored species with dark reddish-brown pronotum and a narrow, dark fascia at the apical third of the elytra. The pronotum is minutely, closely punctate and widened laterally at the basal third into an obtuse tubercle. This is the smallest of the Bolivian *Plocaederus* and also the earliest to emerge, doing so at the very onset of the first seasonal rain. Following emergence, it is locally abundant at lights for just a few days, then is quickly gone until next season's emergence. The species is recorded from eastern Ecuador, French Guiana (where many of the adults lack the elytral fascia), and Bolivia, where it is known from only a couple localities in Santa Cruz Department, including the Flora and Fauna Hotel near Buena Vista and the Potrerillo del Guenda Reserve.

Plocaederus plicatus (Olivier, 1790)

Cerambycinae: Cerambycini

This is a medium-sized (13–15 mm), light tan, brightly shining species with a darker red to dark reddish-brown head and pronotum. Unlike some other species in the genus, it lacks elytral fasciae. The femoral spines are darker in color as are the small external spines on the elytral apices. This nocturnal species is an early season visitor to lights and locally abundant. It is a widespread lowland species found from Brazil, French Guiana, Surinamee, Venezuela, and Bolivia, where it is recorded from Cochabamba Department at Cristal Mayu and Santa Cruz Department, from the vicinity of the Flora and Fauna Hotel near Buena Vista and the Potrerillo del Guenda Reserve.

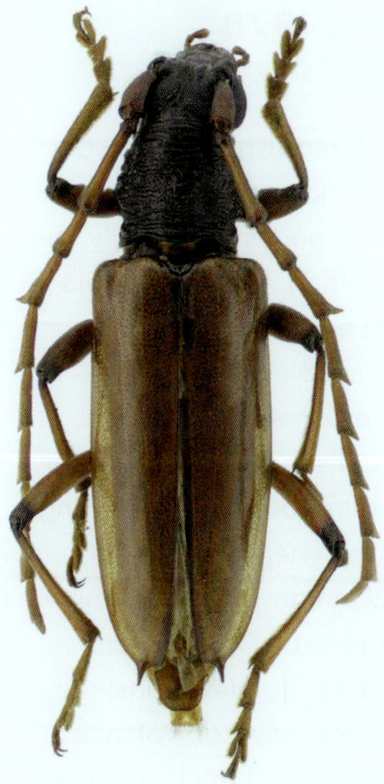

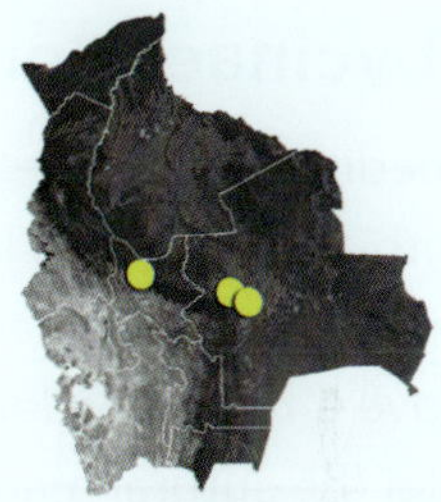

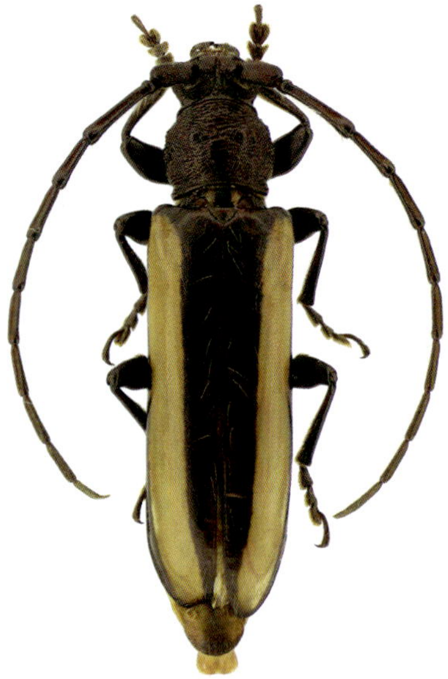

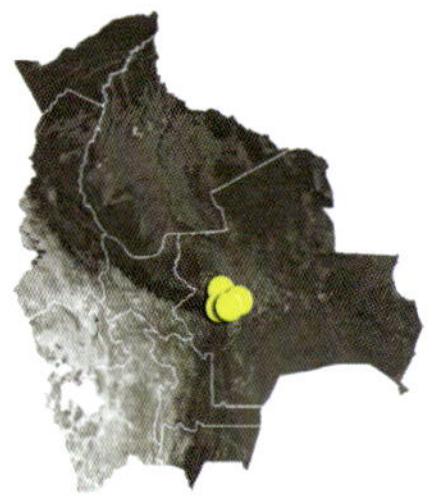

Poeciloxestia lanei Fragoso, 1978

Cerambycinae: Cerambycini

This is a medium-sized (12–22 mm), dark reddish-brown to black species with a tan-colored longitudinal stripe on each elytron. The elytra are minutely, sparsely punctate and, interestingly, the females have a row of widely separated, long, thin flying hairs that parallel the sutural margin and that are absent in males. The pronotal surface and disk is quite variable and complex, with shallow, widely separated punctures, shallow rugosities, and a variety of low tubercles. It can be distinguished from the seven other *Poeciloxestia* species found in Bolivia by its shining integument and parallel elytral fasciae with nearly straight edges. It is not common but occasional at lights. It has a limited distribution from central Brazil and Santa Cruz Department, Bolivia, where it has been collected at differing elevations: the Vicoquin area above Achira at 2,000 m, Refugio Los Volcanes at 1,200 m, and the lowlands of the Flora and Fauna Hotel near Buena Vista and the Potrerillo del Guenda Reserve.

Poeciloxestia lateralis (Erichson, 1847)

Cerambycinae: Cerambycini

This is a large (26–32 mm), slightly shining species with dark reddish-brown integument and a lateral longitudinal tan stripe on each elytron. The elytral surface is finely, moderately punctate and basally alutaceous on some specimens. The pronotum is shining and deeply, coarsely pitted with an irregular central glabrous ridge. Although this species superficially resembles *Poeciloxestia lanei*, it can be distinguished from it by its larger size, less shining integument, and coarsely pitted pronotum. It is occasional at lights but never common. It is distributed in northern South America and recorded from Peru, French Guiana, and Bolivia, where it has been collected from the vicinity of the Flora and Fauna Hotel near Buena Vista, the Potrerillo del Guenda Reserve, Refugio Los Volcanes, and near Cuevas in Santa Cruz Department.

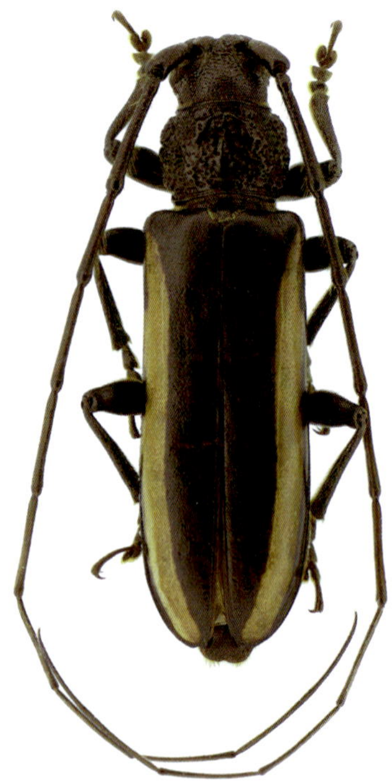

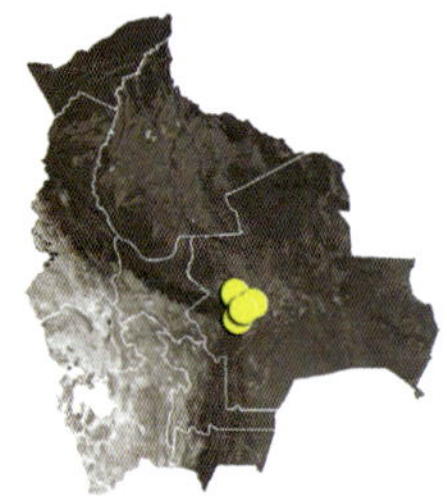

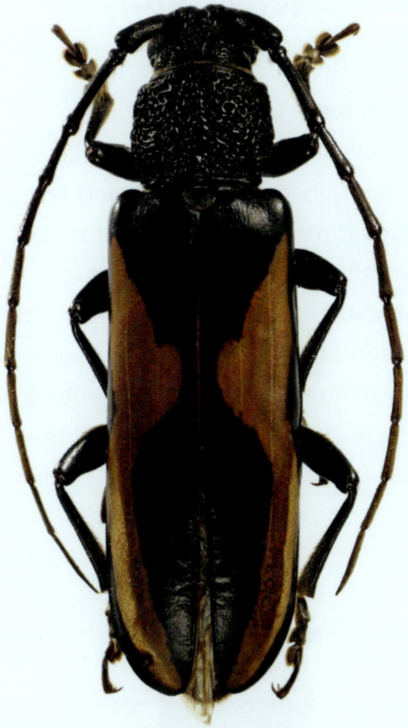

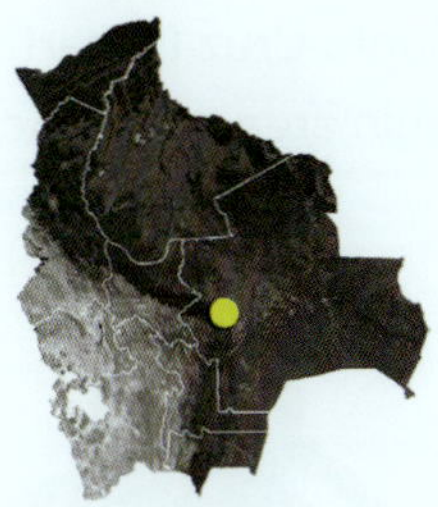

Poeciloxestia melzeri Lane, 1965

Cerambycinae: Cerambycini

This is a moderate to large cerambycine (18–24 mm), very robust and boldly patterned. The integument is dark brown except for an elongate, orange macula on the outer margin of the elytra, broadening toward the suture just anterior of the middle. The quadrate pronotum is completely rugose-punctate. This species is known from Peru, Brazil, and Bolivia, where it has been collected only in Refugio Los Volcanes, adjacent to Amboró National Park, at ultraviolet lights.

Sphallotrichus bidens (Fabricius, 1801)

Cerambycinae: Cerambycini

This is a large (26–30 mm), dark reddish-brown species with a slightly shining dorsal surface. The femora are dark reddish-orange and black tipped apically. The elytra are alutaceous and mostly finely, shallowly punctate with stout, double-spined apices. The pronotum is deeply coarsely punctate. This species is rarely encountered but occasional at lights. It is a northern South American species known from Brazil, the Guianas, and Bolivia, where it is recorded from Santa Cruz Department near Camiri and the Flora and Fauna Hotel near Buena Vista.

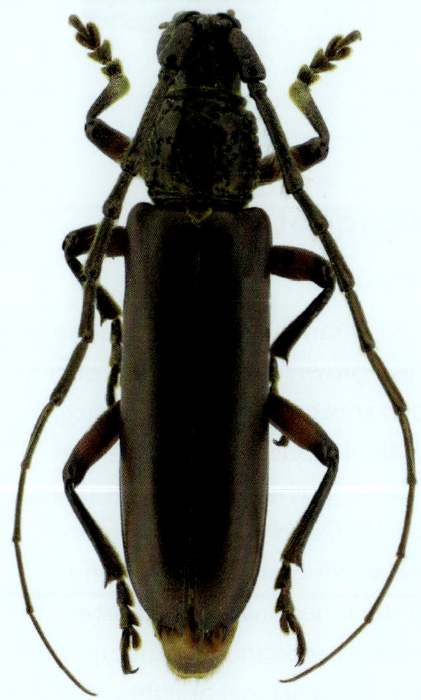

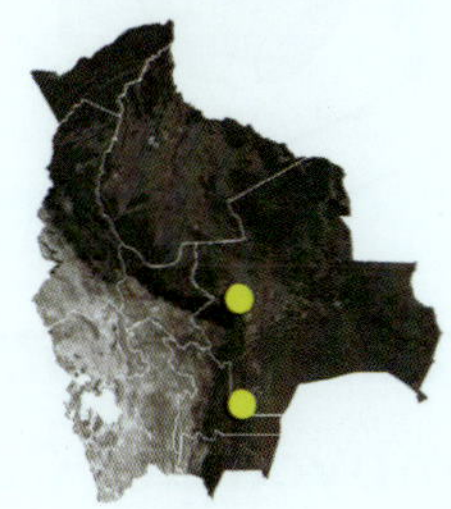

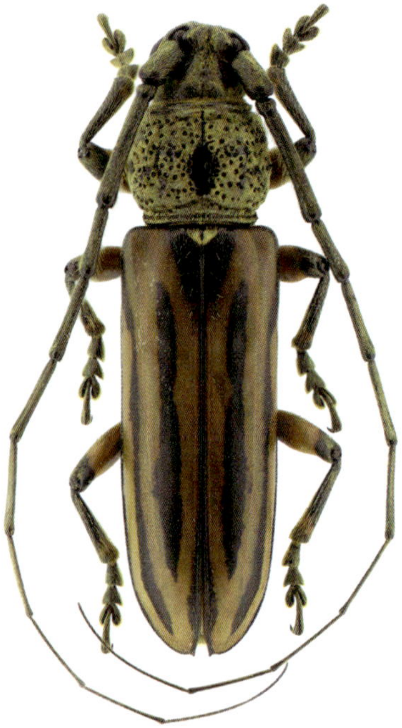

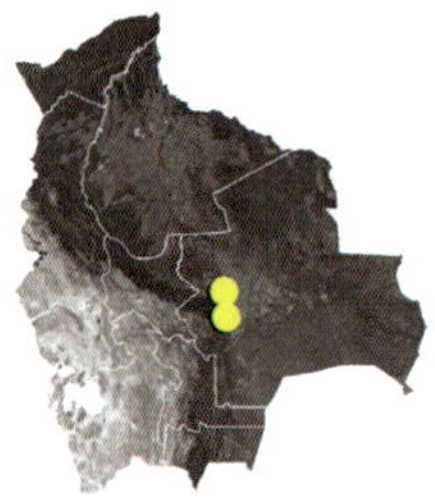

Xestiodion annulipes (Buquet, 1844)

Cerambycinae: Cerambycini

One of the more beautifully colored cerambycines, this moderately large species (26–28 mm) is reddish brown and tan with a deeply pitted, rugose pronotum covered in short, golden pubescence. The scutellum is also fringed with similar pubescence. The tan-colored elytra are minutely punctured and have a dark basal fascia surrounding the scutellum and two thin, dark, parallel stripes on each elytron with the outer one starting at the humeri and extending to near the apices with the inner one much shorter. This distinctive-looking species is easily recognizable among the Cerambycini. Rarely encountered, it is occasional at lights. It is known from Brazil, Paraguay, and Bolivia, where it is has been collected at the Flora and Fauna Hotel near Buena Vista and Refugio Los Volcanes in Santa Cruz Department.

Amyipunga armaticollis (Zajciw, 1964)

Cerambycinae: Clytini

This is a small- to medium-sized (8–18 mm), light rust-colored, diurnal species. The scutellum is apically fringed in whitish-yellow pubescence, and the elytra have three thin inverted V-shaped fasciae of the same pubescence. This species is most often encountered at high elevations (>1,500 m) in agricultural *chaqueos*, where it congregates on larger tree limbs stacked for posts or firewood. On warm, sunny mornings it can be observed actively buzzing about in this environment. It is known only from northwestern Argentina and Santa Cruz Department, Bolivia, northeast of Mairana and the Vicoquin area above Achira at 2,000 m.

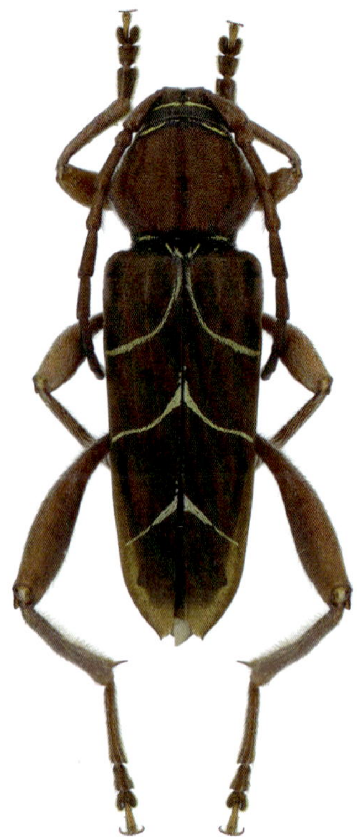

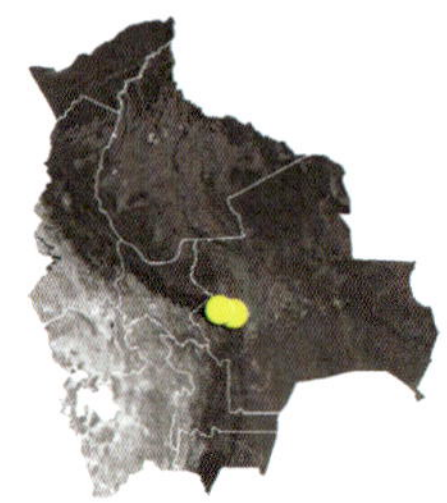

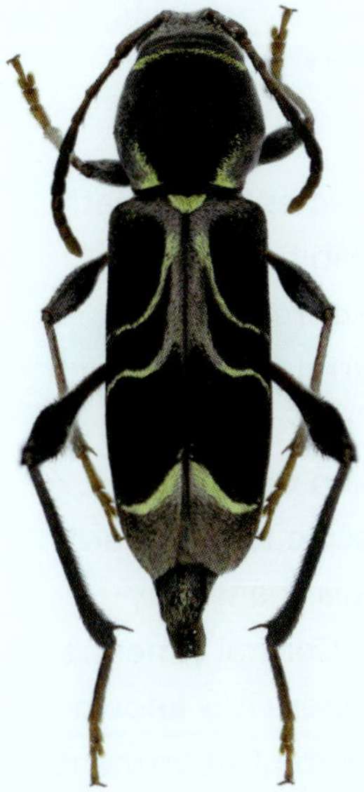

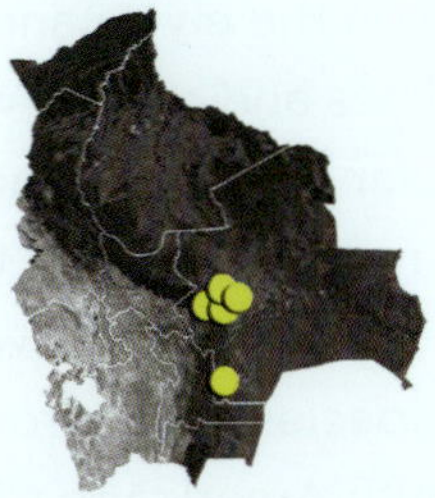

Cotyclytus sobrinus (Laporte & Gory, 1836)

Cerambycinae: Clytini

This is a small (6–10 mm), reddish-brown, diurnal species covered with short, recumbent pubescence that obscures the surface. Three contrasting cream- to yellow-colored pubescent fasciae are present on the elytra: two oblique ones before the middle and another one before the apices. The scutellum is densely covered with similar yellow pubescence, and the pronotum has a pair of small yellow pubescent fasciae on the basal margin. Typical of this genus, the pronotum is minutely punctate and has a raised central ridge with small, scattered tubercles. This species is diurnal and can be found flying around lowland agricultural *chaqueos* for much of the day. It is widely distributed in southern South America and known from Brazil, northern Argentina, Paraguay, Uruguay, and Bolivia. In Bolivia it is known from the vicinity of the Flora and Fauna Hotel near Buena Vista, the Potrerillo del Guenda Reserve, Refugio Los Volcanes, adjacent to Amboró National Park, north of Pampagrande, and the Camiri environs in Santa Cruz Department.

Cotyclytus stillatus (Aurivillius, 1908)

Cerambycinae: Clytini

This is a small (10–16 mm), reddish-brown to piceous, diurnal species covered in short, recumbent, dark pubescence that obscures the dorsal surface. The elytra are marked with bright yellow pubescent spots: two before the middle, three in a row at the middle, and another at the apical fourth. This yellow pubescent color scheme carries onto the densely pubescent scutellum and pronotum, where there are two round yellow basal spots, two more laterally at the middle, and linear fasciae at the apical margin. In this species the appendages are reddish orange. It is commonly found in mid- to high-elevation slashes where, like many of the tribe, it is active on warm, sunny days flying around or congregating on the larger limbs and trunks of recently cut or fallen trees. It is known from northern Argentina, Paraguay, and Bolivia, where it is recorded from Santa Cruz Department northeast of Mairana, the Vicoquin area above Achira, and Refugio Los Volcanes, adjacent to Amboró National Park.

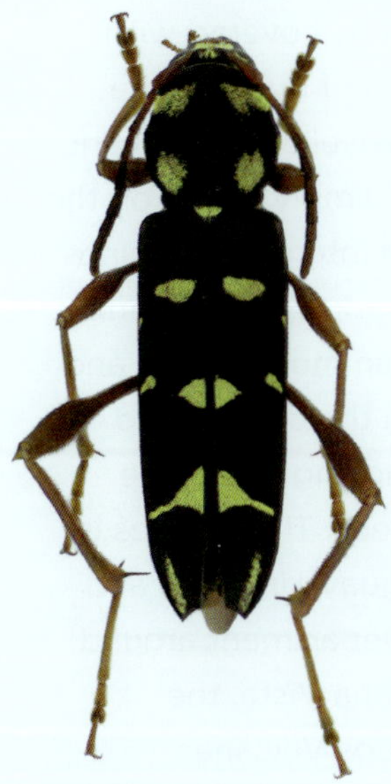

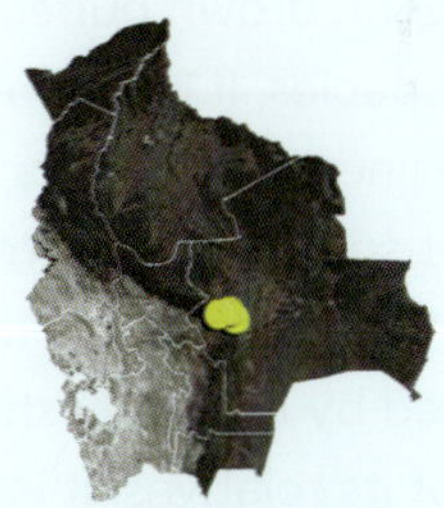

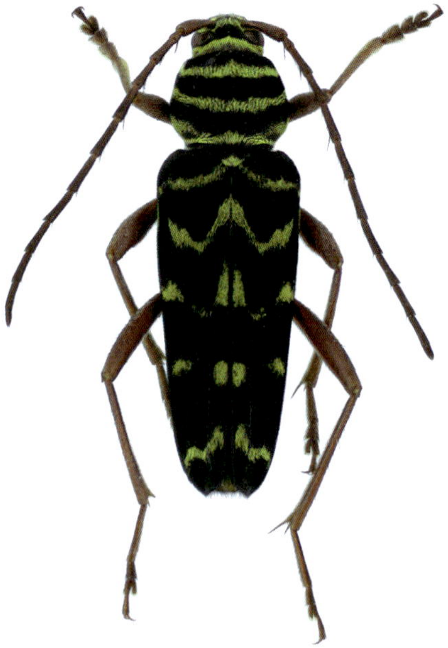

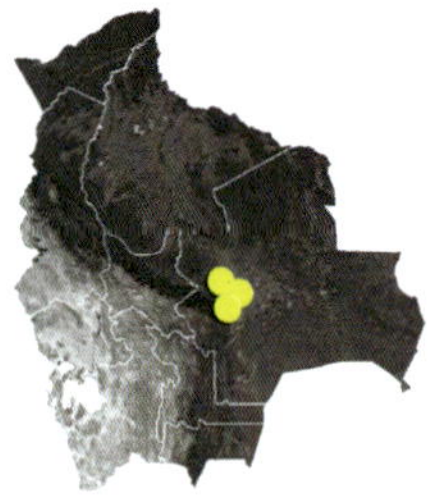

Megacyllene acuta (Germar, 1821)
Cerambycinae: Clytini

This is a small- to medium-sized (10–15 mm) species with reddish-brown to black integument moderately covered with short, recumbent, brown pubescence. Typical of many of the *Megacyllene* species, it has numerous narrow yellow pubescent fasciae, some linear and others V-shaped or simply spots, on the elytra. In most specimens the reddish-brown integument shines through, and the appendages are similarly dark red. This genus is characterized by relatively shorter legs than most Clytini, and unlike many of its relatives, it is active late in the day, toward evening, when it can be found flying to or moving about the remaining or partially burned trees in *chaqueos*. This species is widely distributed in Brazil, Argentina, Paraguay, Uruguay, and Bolivia, where it is known from Santa Cruz Department around Bermejo, the Flora and Fauna Hotel near Buena Vista, the Potrerillo del Guenda Reserve, and Refugio Los Volcanes.

Megacyllene angulata (Fabricius, 1775)
Cerambycinae: Clytini

This is a small- to medium-sized (10–15 mm) species that greatly resembles *Megacyllene acuta* and has similar habits. It is best distinguished from that species by having darker recumbent pubescence on the elytra and by its central elytral fasciae, which are narrow lines and extend well up the elytral margin in contrast to the irregular shape and short sutural extension of those of *M. acuta*. In most specimens the pubescent fasciae tend toward being more white than yellow as in *M. acuta*. This is a widely distributed species found from Mexico through Central America, northern South America to Chile, and Bolivia, where it is known from Beni Department near Riberalta, Cochabamba Department near Villa Tunari, and Santa Cruz Department from the vicinity of the Flora and Fauna Hotel near Buena Vista, the Potrerillo del Guenda Reserve, and Refugio Los Volcanes.

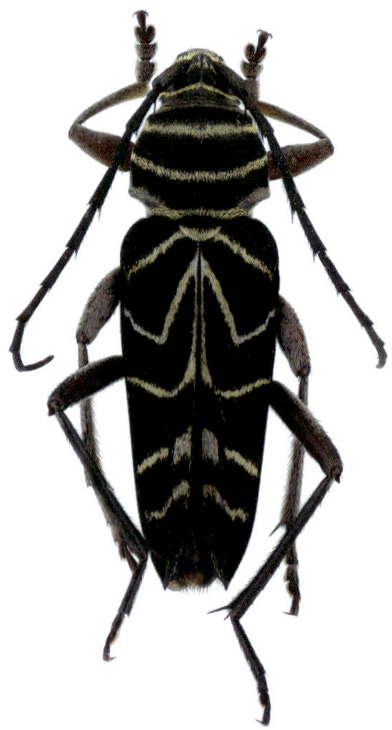

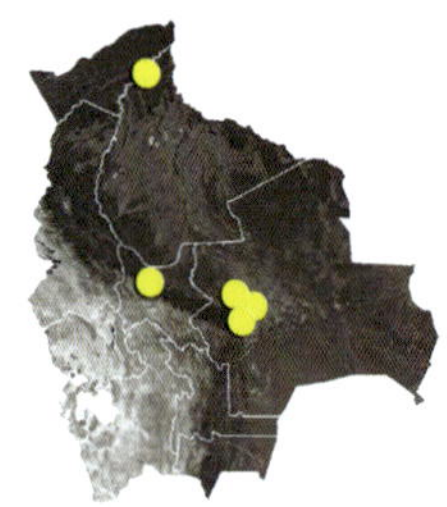

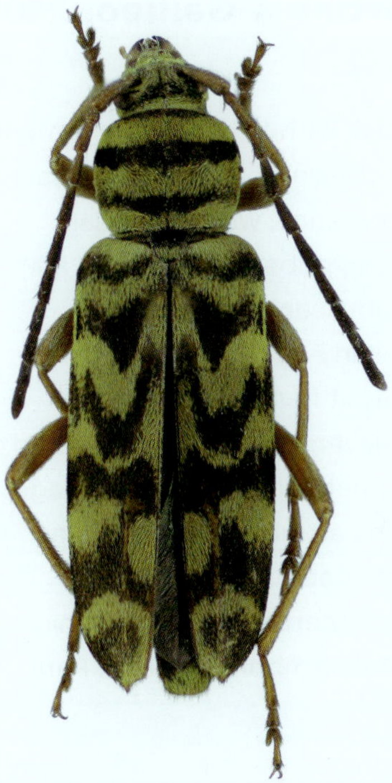

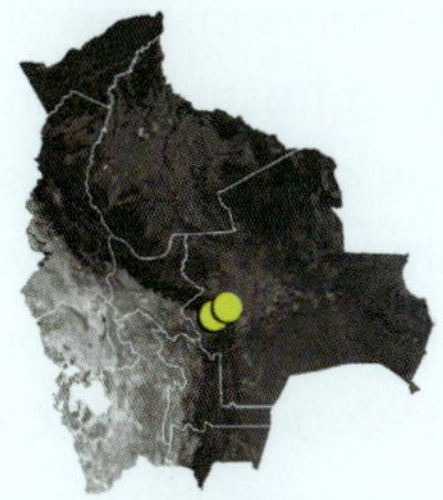

Megacyllene neblinosa Di Iorio, 1995

Cerambycinae: Clytini

This is a small (11–13 mm), reddish-orange to reddish-brown species with wide, dark yellow pubescent bands and fasciae on the pronotum and elytra, making its appearance more yellow versus its darker integumental color. This species is rarely encountered and primarily found at high elevations in Bolivia (above 1,500 m), where it has been collected by beating *chaqueo* debris or forest edges. It is known only from northern Argentina, Brazil, and Santa Cruz Department, Bolivia, from the Vicoquin area above Achira at 2,000 m and near Samaipata.

Megacyllene nebulosa (Laporte & Gory, 1836)

Cerambycinae: Clytini

This is a medium-sized (12–16 mm) diurnal species with brown integument and densely pubescent dorsum. The pronotum and elytra in freshly emerged specimens are covered in short, recumbent brown pubescence intermixed with areas of lighter-colored gray pubescence. Between them they form an intricate design of broad and narrow fasciae on the dorsal surface. In Bolivia collections of this species have been limited to a couple humid *chaqueo* areas, where they were found crawling on the foliage, limbs, and trunks of very recently felled leguminose trees. In the same environment they came to lights. *Megacyllene nebulosa* is recorded from Argentina, Chile, Paraguay, Uruguay, and Santa Cruz Department, Bolivia, near Pampagrande and north of Camiri.

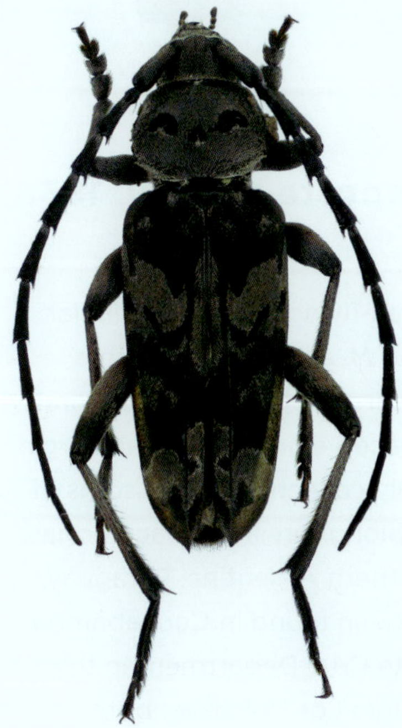

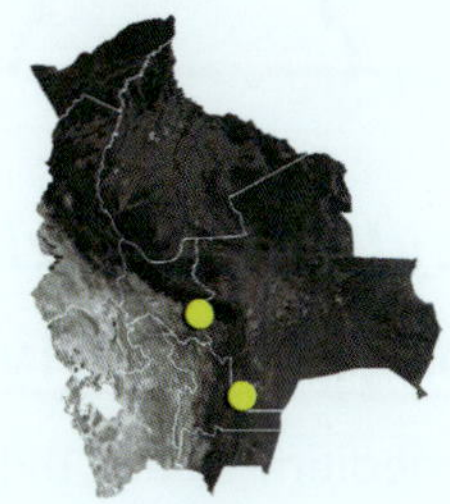

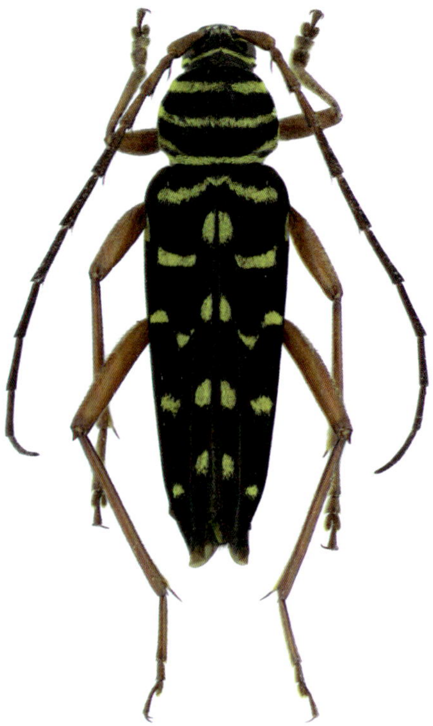

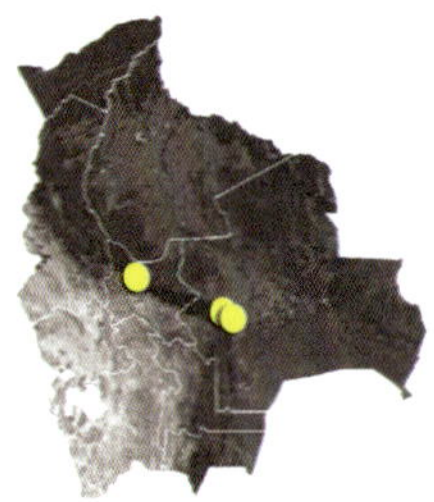

Megacyllene proxima (Laporte & Gory, 1836)

Cerambycinae: Clytini

This is a medium-sized (11–17 mm), day-flying species of similar appearance to *Megacyllene acuta* and *M. angulata* but with a more elongate body form and four pairs of elytral spots separated from the other fasciae and evenly spaced along the sutral margin. This species has been most commonly collected on blossoms of woody plants or perched on foliage along forest or chaco edges. It has been recorded from Brazil, northern Argentina, Paraguay, Venezuela, and Bolivia, where it has been found in Cochabamba Department at Cristal Mayu and Santa Cruz Department in the Vicoquin area above Achira and Refugio Los Volcanes, both adjacent to Amboró National Park.

Miriclytus triangularis Martins & Galileo, 2011

Cerambycinae: Clytini

This is a small (6–8 mm), diurnal species with light tan integument that is mostly obscured by dense, contrasting pubescent fasciae. Its scutellum is covered in yellow pubescence, with the basal elytral area around it covered in light tan pubescence. There are three pairs of bright yellow, narrow pubescent fasciae on the elytra, with the apical area having similarly colored, recumbent vestiture at the basal area. Interestingly, it has only eight antennomeres instead of the typical eleven. It has been collected by beating freshly cut small branches that still bear green foliage along the margins of slashed areas. This attractive little species was only recently discovered and described and is a Bolivian endemic known only from Santa Cruz Department above Achira and at Refugio Los Volcanes, adjacent to Amboró National Park.

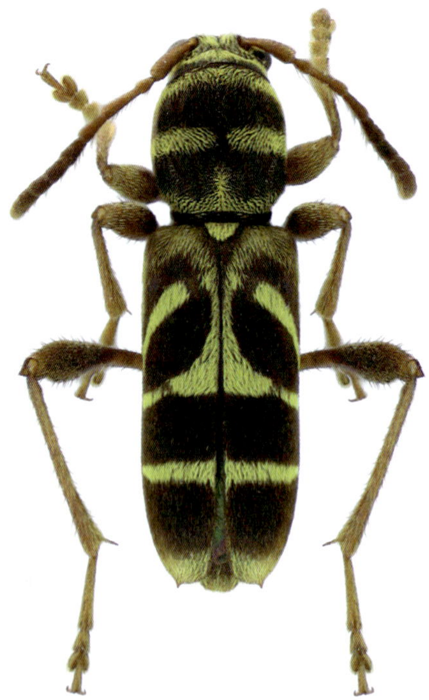

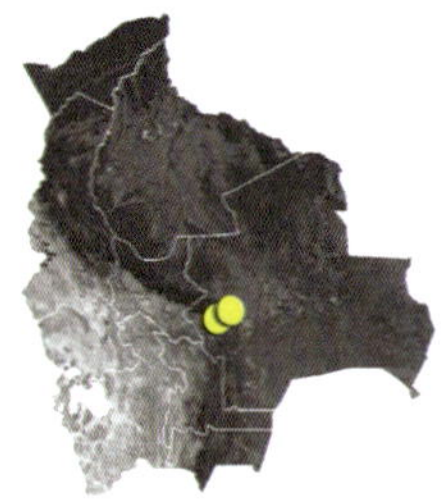

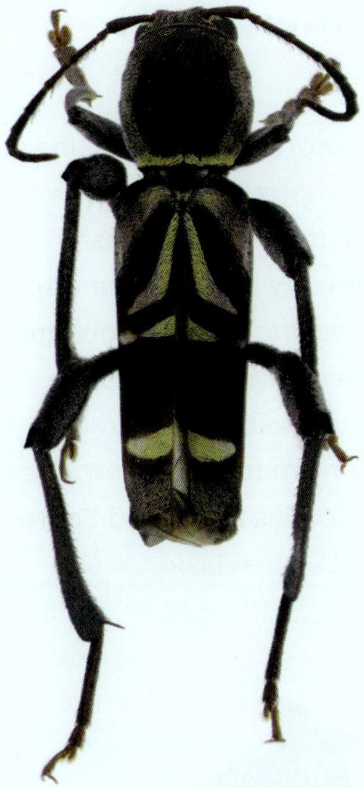

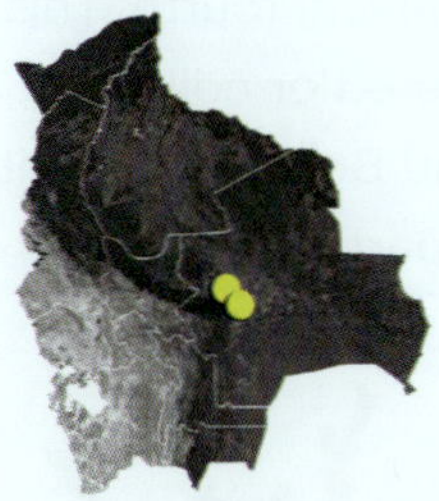

Neoclytus pusillus (Laporte & Gory, 1836)

Cerambycinae: Clytini

This is a small- to medium-sized (7–14 mm), dark brown cerambycid with a lighter reddish-brown pronotum. There are yellow to tan fasciae on the elytra and base of the pronotum. In this species the males are often less than half the size of the females, making for a truly odd-looking couple. This lowland species is common in agricultural *chaqueos*, preferring partially burned tree trunks on which their dark coloration makes them difficult to see unless they are moving. *Neoclytus pusillus* is widely distributed in South America and is known from Argentina, Brazil, Colombia, Guyana, Paraguay, Peru, Suriname, Uruguay, and Bolivia, where it is recorded from Santa Cruz Department from the vicinity of the Flora and Fauna Hotel near Buena Vista and the Potrerillo del Guenda Reserve.

Neoclytus rufus (Olivier, 1795)

Cerambycinae: Clytini

This is a small (8–12 mm), dark reddish species with narrow, irregular, white pubescent fasciae on the elytra. It is rarely encountered in Bolivia, with only one record from the humid Chaco in southern Santa Cruz (83 km north of Camiri), where it was collected on dead wood. It is more commonly found in the northern part of its range as evidenced by long series of specimens from Panama that have been examined. It ranges from there into Argentina, Paraguay, and Bolivia in South America.

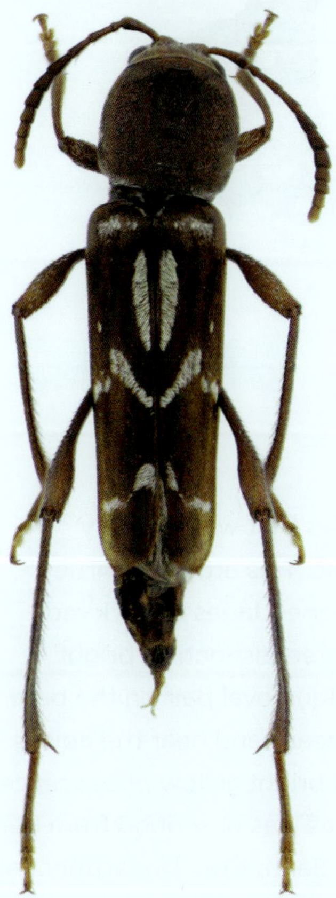

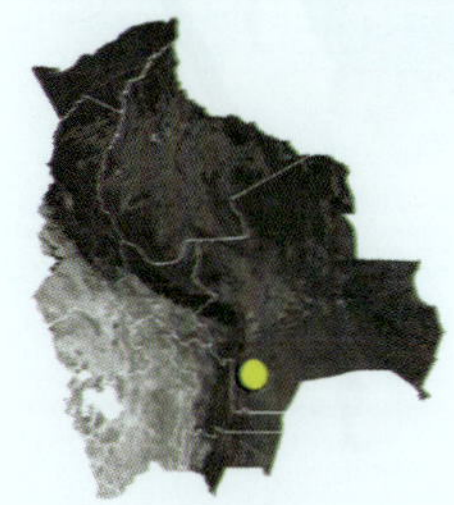

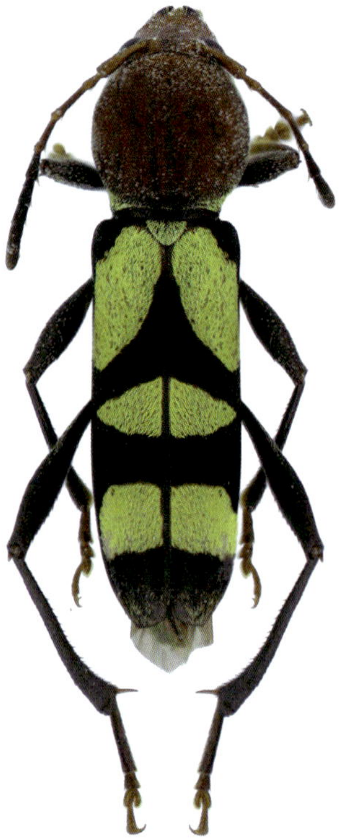

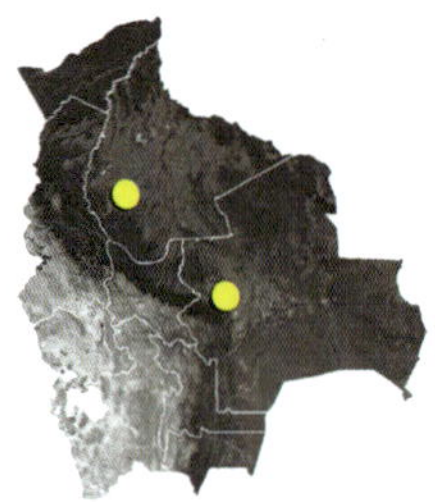

Pirangoclytus fraternus Martins & Galileo, 2011

Cerambycinae: Clytini

This is a small (8–9 mm), tricolored species that was recently described and is another Bolivian endemic. This attractive little cerambycid has a head, pronotum, and appendages of dark red and elytra of dark brown to black, with three distinctive, bright yellow pubescent fasciae. There is an oblique oval pair on the base, a linear oval spot centrally, and a wide, linear band near the apices. In addition, the scutellum is covered with bright yellow pubescence. Nothing is known of its habits. This species was described from Beni Department and now is also known from Santa Cruz Department at the Flora and Fauna Hotel near Buena Vista.

Pirangoclytus latecinctus (Bates, 1870)

Cerambycinae: Clytini

This is another small (9–10 mm), brightly colored, diurnal clytine that is densely covered, both dorsally and ventrally, with yellow pubescence that obscures the integument. Its long legs, basal antennal segments, and pronotum are primarily reddish orange. The elytra have several narrow, black fasciae, including an oblique one before the middle, a linear band at the middle, and another band at apical fourth. This species is a flower visitor but rarely encountered or collected. It is known only from northwestern Brazil and Bolivia, where it is recorded in Cochabamba and Santa Cruz Departments.

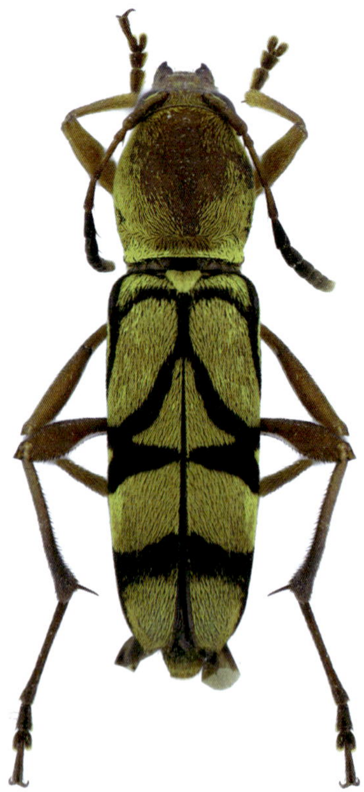

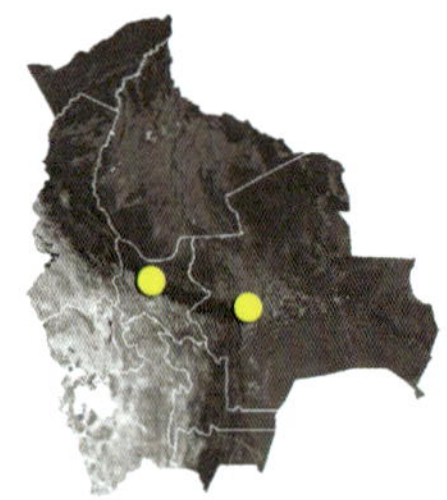

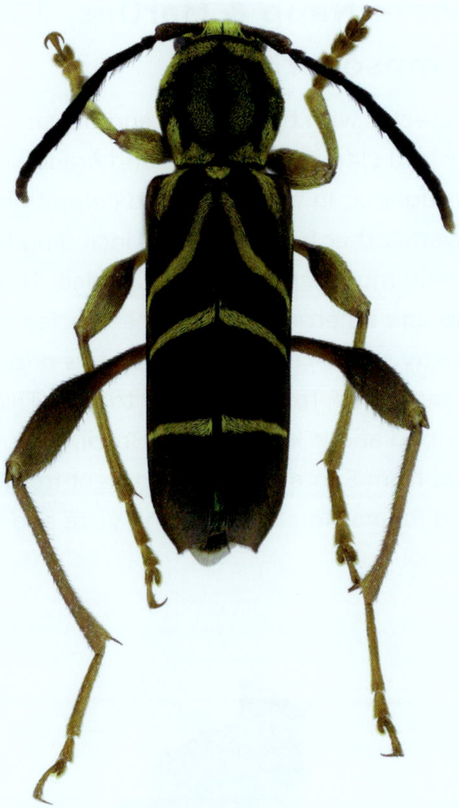

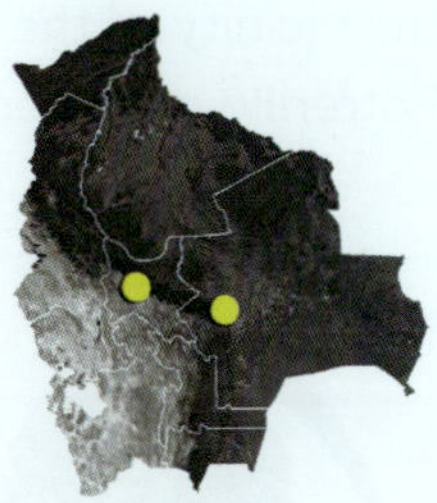

Unaiuba pinima (Galileo & Martins, 2007)

Cerambycinae: Clytini

This is a small- to medium-sized (11–12 mm), dark brown species with numerous yellow pubescent fasciae on the dorsum. The antennomeres are pubescent and black. This species is very rarely encountered and known from only a few specimens. It is a high-elevation inhabitant, which may help to explain its rarity, and has been recorded from Peru and Bolivia, where it has been collected in Cochabamba Department at San Jacinto at 2,800 m and Santa Cruz Department in the Vicoquin area above Achira at 2,000 m.

Aglaoschema ventrale (Germar, 1824)

Cerambycinae: Compsocerini

This species (14–17 mm) is very similar to *Aglaoschema haemorrhoidale*; however, the dorsal pubescence in that species is usually metallic blue, whereas in *A. ventrale* it is metallic green. The legs and antennae are black. This species is known from Brazil, Argentina, Paraguay, and Bolivia in South America. It is known from Refugio Los Volcanes and Potrerillo del Guenda in Santa Cruz Department in Bolivia.

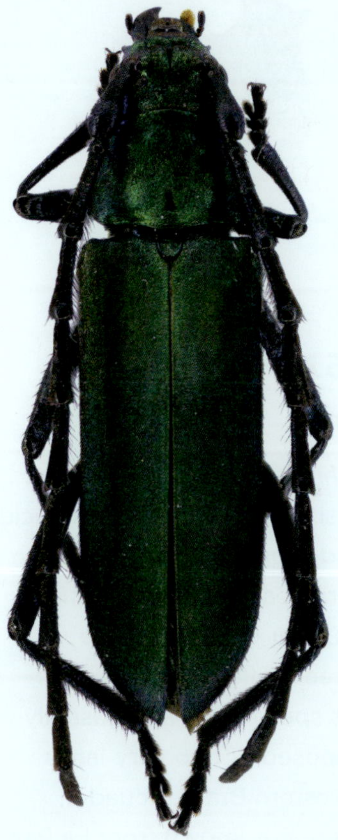

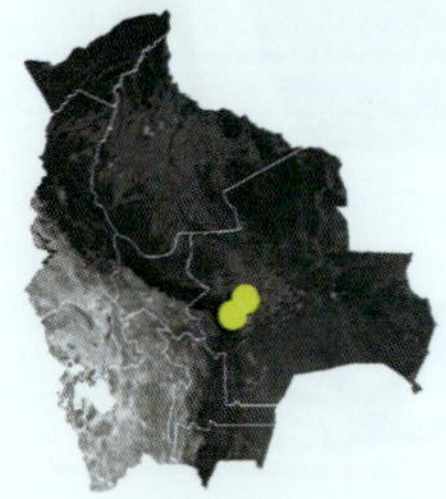

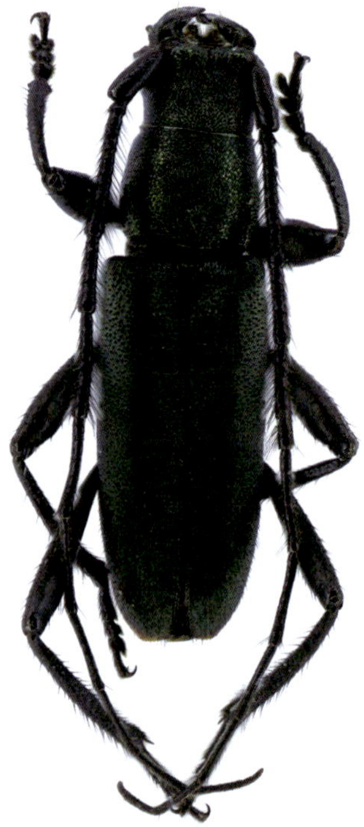

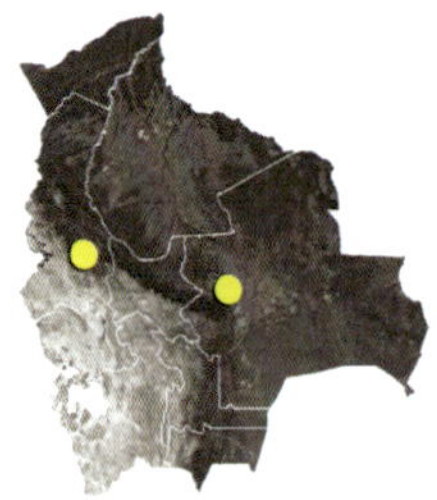

Chlorethe ingae Bates, 1867
Cerambycinae: Compsocerini

Superficially similar in appearance to *Chrysoprasis* species of the Heteropsini, this is a small- to medium-sized (7–11 mm) cerambycid with shining, densely punctate head and pronotum and a dark green, minutely punctate (each puncture bearing a short, stiff, semirecumbent, dark hair), dull elytra. Under direct light the head and pronotum provide metallic green reflections. The legs and antennae are black to piceous. This species has an unusually short face with most of its surface composed of two very large mandibles. It is distributed from northwestern Brazil, Ecuador, Peru, and French Guiana to Bolivia, where it is known from La Paz and Santa Cruz Departments.

Cosmoplatidius abare Napp & Martins, 2006
Cerambycinae: Compsocerini

This is a beautiful diurnal species with bold colors and dense pubescence. It is medium sized (14–18 mm), black and orange, and heavily pubescent throughout. In both form and coloration it resembles members of several other beetle families including the Cantharidae, Lampyridae, and most especially the Lycidae. The *Cosmoplatidius* species are late emergers, with some species only starting to appear in late rainy season and others, like this one, typically afield as the rain tapers off, from February to May. This species was recently described and is known from Brazil, Peru, and Bolivia, where it is recorded from Santa Cruz Department from the vicinity of the Flora and Fauna Hotel near Buena Vista and the Potrerillo del Guenda Reserve.

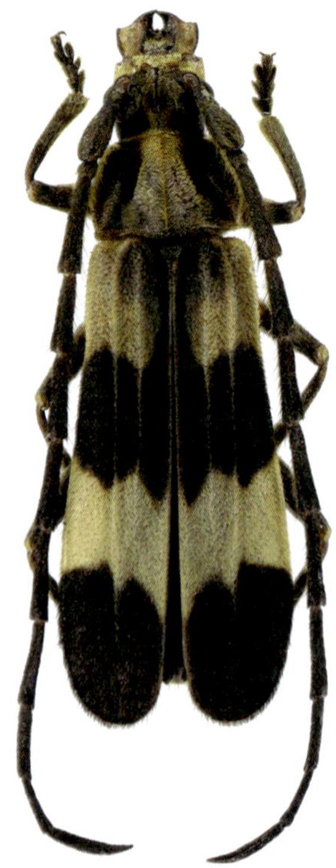

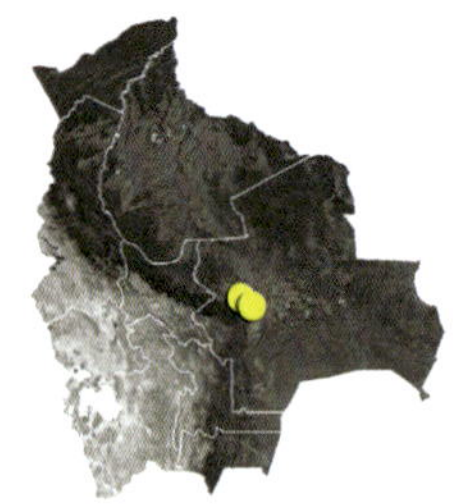

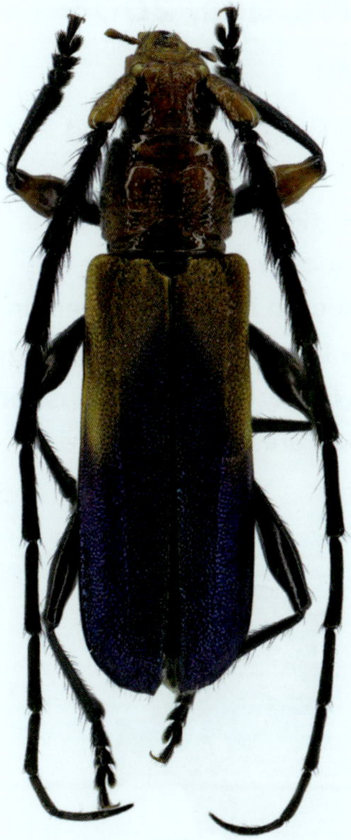

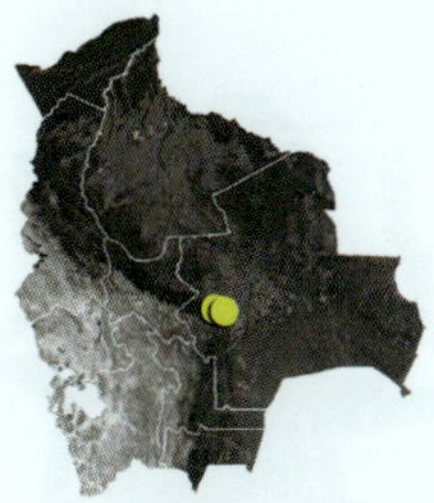

Goatacara boliviana Napp & Martins, 2006

Cerambycinae: Compsocerini

This is a small (11–12 mm), tricolored species with a shining orange head, pronotum, scape, and clavate portion of the profemora. The remaining antennomeres and legs are black. The elytra have a beautiful metallic blue or purple fascia (oblique at the front), which covers the apical two-thirds (the humeri and base are orange). This is an extremely rare species found only at high elevations. It was recently described from Bolivia, where it is endemic. It has been collected along the road above Achira Campo and northeast of Mairana in Santa Cruz Department.

Orthostoma abdominale (Gyllenhal, 1817)

Cerambycinae: Compsocerini

This is a medium-sized (10–13 mm), reflective green species. The integument is minutely punctate on both the dorsal and ventral surface, except for the dull orange abdomen. The legs are black and the antennae black to piceous, with the male antennae almost three times their body length. It is diurnal and often encountered perching on foliage along forest edges. Emergence of this common species is spread over a long period (even at the same locality), as evidenced by it having been collected from mid-November to May in the Buena Vista, Bolivia, area. *Orthostoma abdominale* is widely distributed and recorded from Ecuador, Colombia, Brazil, Paraguay, and Bolivia. In Bolivia it has been collected in Tarija Department (near Villamontes) and Santa Cruz Department (from the vicinity of the Flora and Fauna Hotel near Buena Vista, the Potrerillo del Guenda Reserve, and Refugio Los Volcanes, adjacent to Amboró National Park).

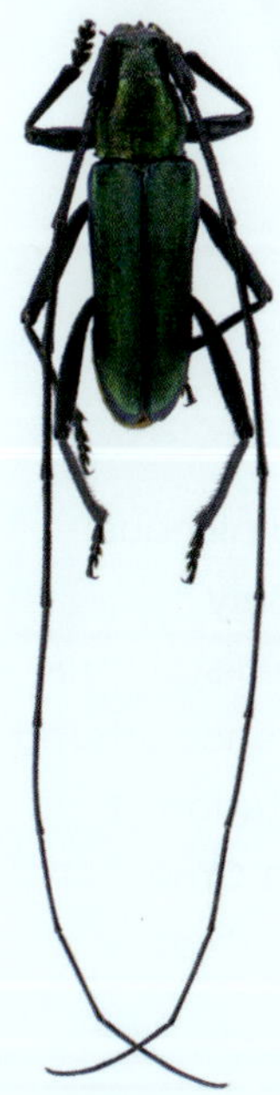

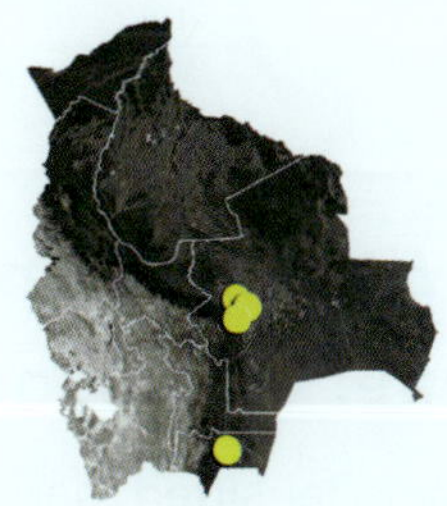

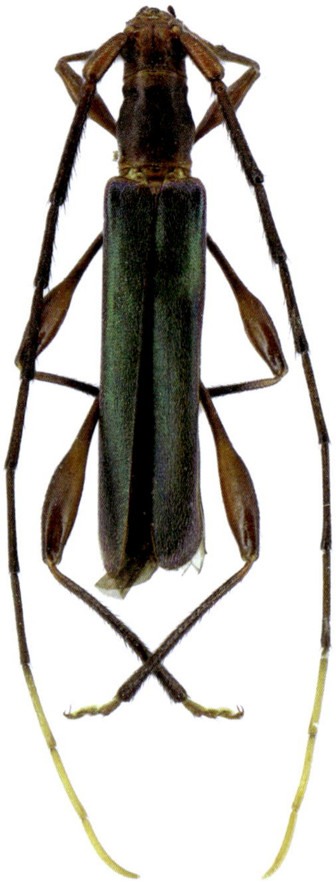

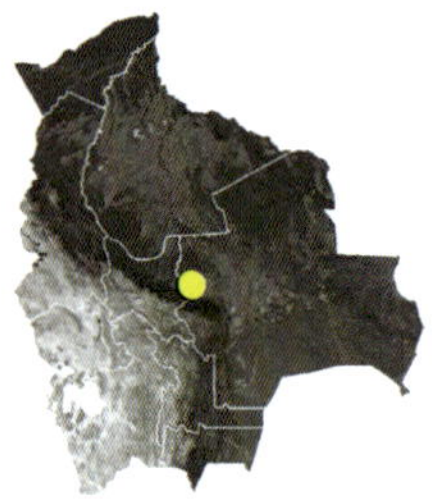

Rierguscha bicolor Viana, 1970

Cerambycinae: Compsocerini

This is a narrow, parallel-sided, dark reddish-brown to piceous species (11–13 mm) with dark green, minutely and closely punctate elytra. Its otherwise dark antennae have the distal three antennomeres light tan in color. This species is rarely encountered but has been collected by sweeping understory vegetation. It is known only from northwestern Argentina and Santa Cruz Department in Bolivia, where it has been collected north of Comarapa at Laguna Verde.

Rierguscha florida Napp & Martins, 2006

Cerambycinae: Compsocerini

This Bolivian endemic species is similar in size (11–15 mm) and shape to its congener *R. bicolor*, but differs by being dark brown to piceous with tan elytra. It is a moderately high-altitude cerambycid (above 1,500 m) that can be locally abundant at the right localities. Specimens have been collected by netting them in flight, from cut firewood stacks, and from foliage along forest edge habitat. It is known only from Santa Cruz Department northeast of Mairana (near Samaipata) and the Vicoquin area above Achira.

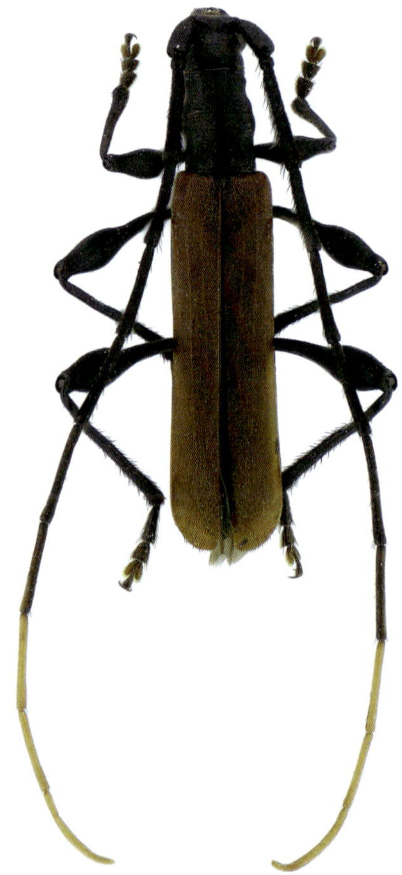

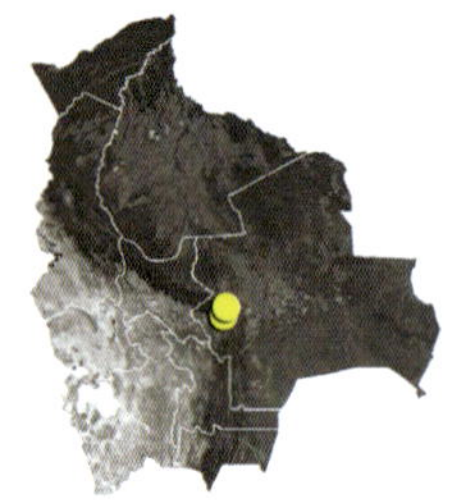

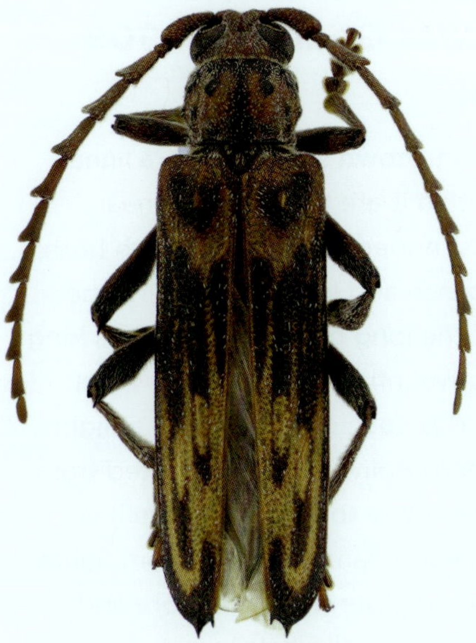

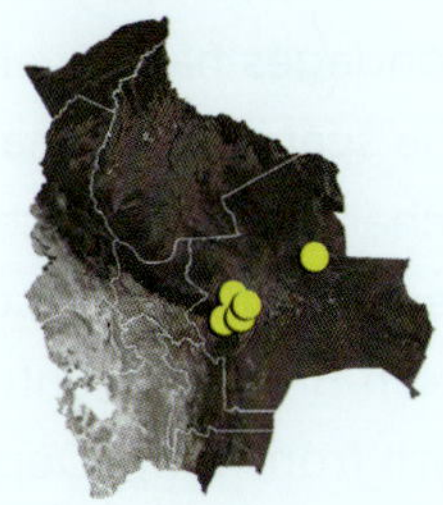

Diorus biapiculatus White, 1853

Cerambycinae: Diorini

This is a large (28–32 mm), nocturnal species easily recognizable by its tuberculate pronotal disk and ridged elytral base, flat femora (and in the males, flat middle tibiae also), short antennae with laterally expanded segments, and elytral apices each with a large spine externally and a smaller one suturally. Its dorsal surface is heavily punctate with scattered recumbent white setae intermixed. Its glabrous integument is a blend of tan to reddish brown without distinct, clear markings. The only species in the tribe Diorini, it is uncommon but occasionally collected at lights. It is recorded from French Guiana, Brazil, and Bolivia, where it is known from Santa Cruz Department in Concepción, the vicinity of the Flora and Fauna Hotel near Buena Vista, the Potrerillo del Guenda Reserve, Refugio Los Volcanes, adjacent to Amboró National Park, and Samaipata.

Beraba cheilaria (Martins, 1967)

Cerambycinae: Eburiini

This is a small (10–12 mm), beige species with three shining eburneous spots on each of its heavily punctate elytra (one basally and a contiguous pair behind the middle) and a pair of glabrous black tubercles on the pronotal disk. The sides of the pronotum are also distinctly spined in black. The mesal and metafemora are tipped with black, as is a long, acute spine at their apex. This nocturnal species is locally abundant at light but known from only a small geographic area in southern Brazil (Matto Grosso do Sul) and Bolivia, where it is recorded from Cochabamba Department and from Santa Cruz Department, at the Flora and Fauna Hotel near Buena Vista and the Potrerillo del Guenda Reserve.

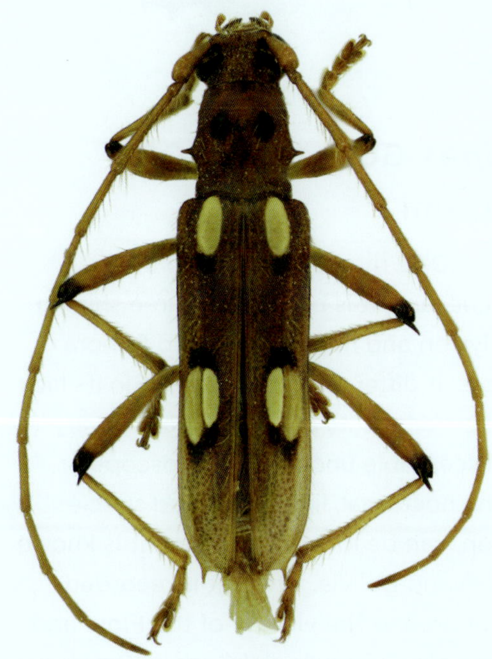

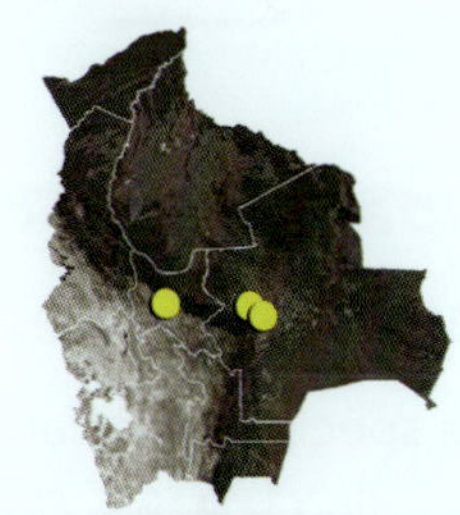

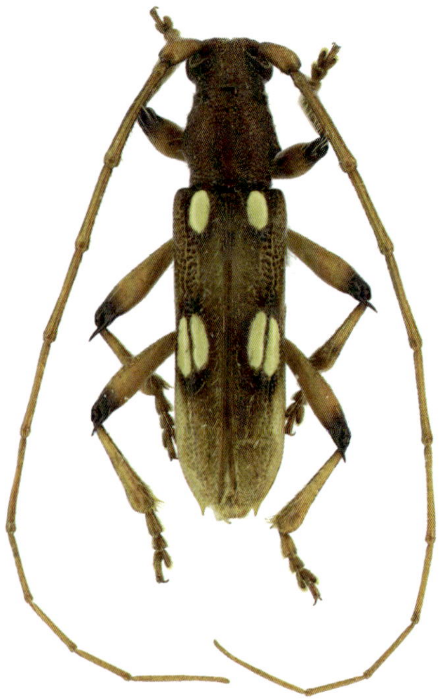

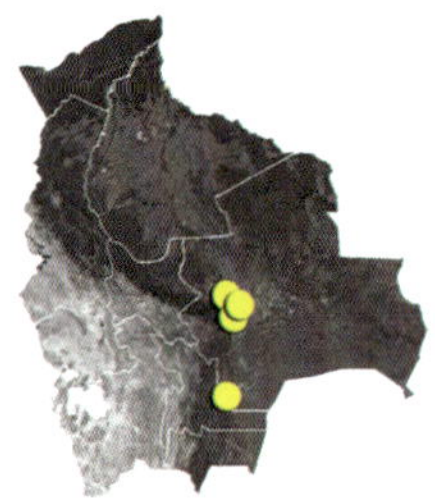

Cupanoscelis clavipes Gounelle, 1909

Cerambycinae: Eburiini

This species is similar in size (10–12 mm) and shape to the preceding species, *Beraba cheilaria*; it is beige with three shining eburneous spots on each elytron and has black-tipped femora with long spines at the apices. It differs from *B. cheilaria* in its lack of pronotal tubercles and the sides of its pronotum having very small spines that are only discernable under the microscope or hand lens. It is another of the nocturnal, light-attracted species that, early in the rainy season, can be locally abundant. It is known only from Brazil (Goias State) and Bolivia, where it is recorded from Santa Cruz Department around the vicinity of the Flora and Fauna Hotel near Buena Vista, the Potrerillo del Guenda Reserve, Refugio Los Volcanes, adjacent to Amboró National Park, and the Camiri environs.

Cupanoscelis heteroclita Gounelle, 1909

Cerambycinae: Eburiini

This is a small (10–12 mm), reddish-brown species with a linear, black fascia on each elytron. Within it are three small, linear eburneous marks. The dorsum is moderately covered with both short, white, recumbent pubescence and long flying hairs; the appendages have even more of the long hairs, with most as long as the scape. Males have distinctive inflated hind tibiae, which are characteristic of the genus. This species is common at lights, especially in the Chaco of southern Bolivia. It is distributed from eastern Brazil, Argentina, and Paraguay to Bolivia, where it is known from Tarija Department near Villamontes and from Santa Cruz Department near Pampagrande, Bermejo, the Flora and Fauna Hotel near Buena Vista, the Potrerillo del Guenda Reserve, and around Camiri.

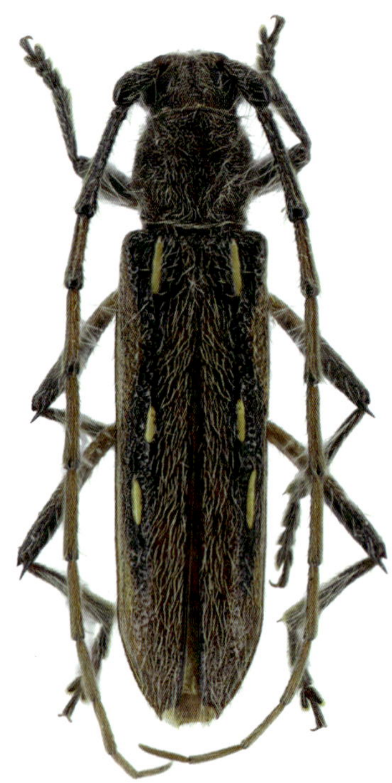

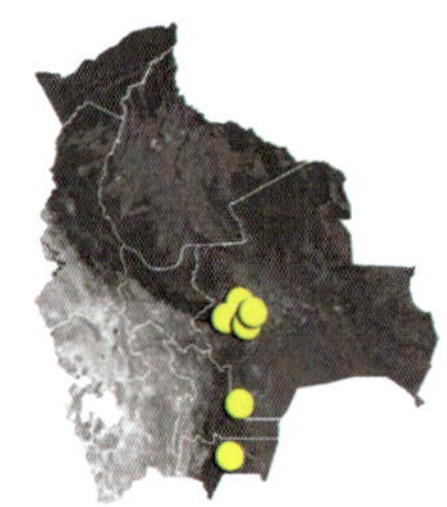

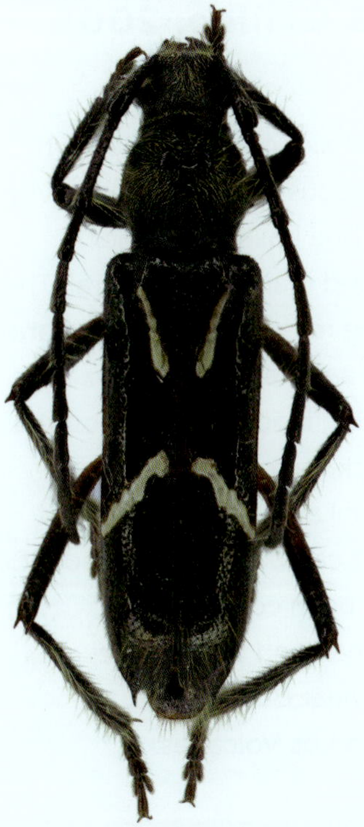

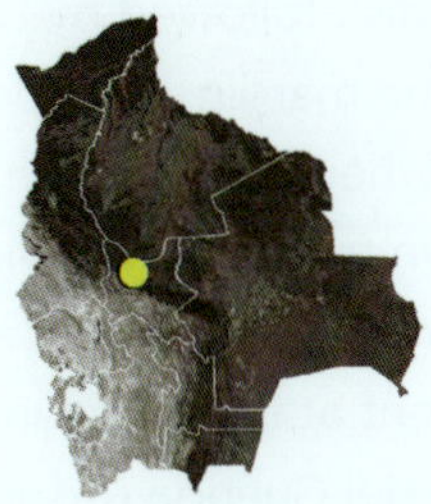

Dioridium borgmeieri (Lane, 1972)

Cerambycinae: Eburiini

This is a very unusual and distinct eburiine. It is relatively small (10–13 mm), with shining, dark integument. The elytra are minutely punctate and each has two narrow, oblique eburneous fasciae angled in different directions. It also has a mixture of fine, white hairs on the surface, some short and recumbent with others longer and sparser. The elytra have long spines on the external margin of the apices. The pronotum is densely pubescent, obscuring the closely punctate surface except for a pair of small, shining tubercles each side of the irregular, glabrous ridge. This species is rarely encountered but very occasional at lights. It is known only from northern Brazil, Ecuador, and Bolivia, where it is recorded from Cochabamba Department near Villa Tunari.

Eburella pinima Martins, 1997

Cerambycinae: Eburiini

This is a narrow, somewhat shining light-reddish species (10–11 mm) with three eburneous spots on each elytron, an oval one basally and two contiguous ones coalescing into a larger circular spot postmedially. The pronotum is dull, unmarked, and with small scattered punctures. Its antennomeres are slim and elongate with those of the males surpassing the elytral apices by four or more segments. It has been collected at lights and beaten from freshly cut, upright, smaller woody plants. It is known only from Peru and at middle to higher elevations in Santa Cruz Department, including the Vicoquin area along the Amboró Ridge above Achira at 2,000 m, northeast of Mairana, and Refugio Los Volcanes.

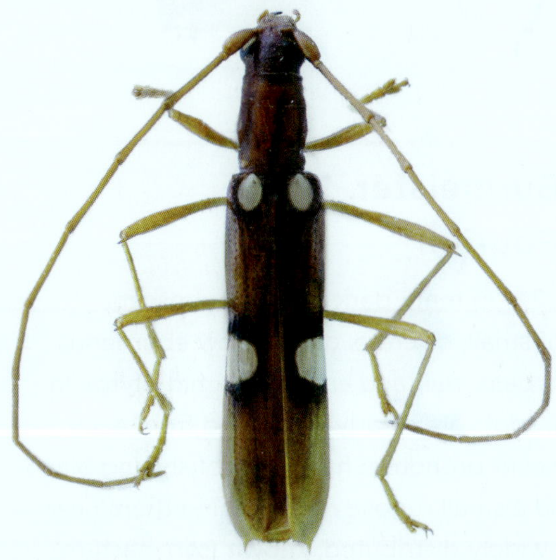

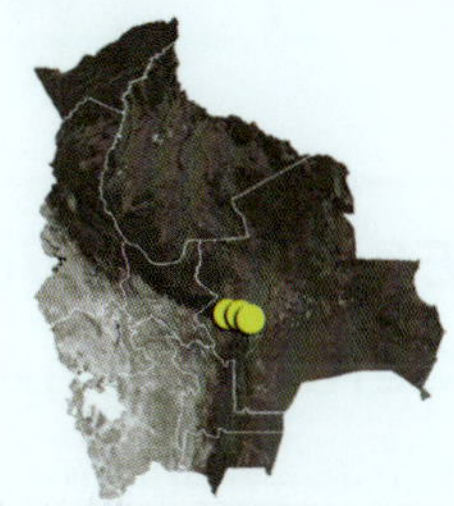

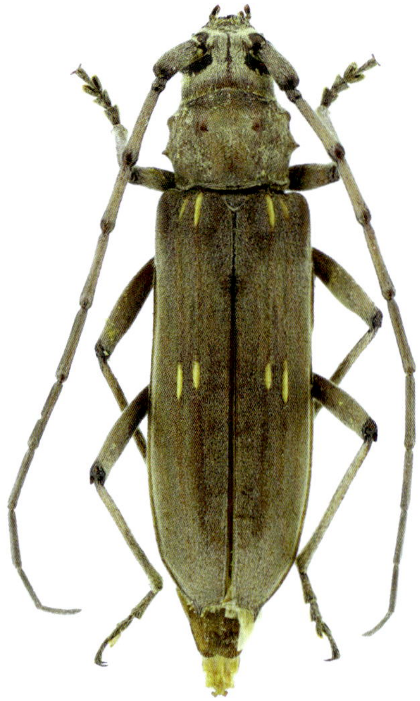

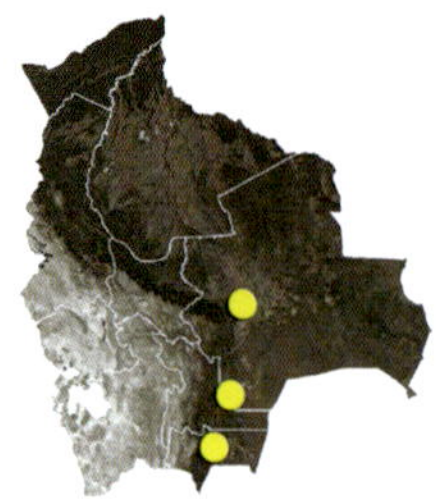

Eburia sordida Burmeister, 1865

Cerambycinae: Eburiini

This is a medium-sized (24–26 mm), dark brown, pubescent species with two pairs of small, discrete, dark yellow eburneous spots on each elytron—a basal pair and another at mid-elytra. In both pairs the individual spots are clearly separated from each other. The heavily pubescent pronotum has a pair of shining black tubercles on the disk and a small rugose ridge behind them near the base. The species is widely distributed, known from eastern Brazil, Argentina, Paraguay, Uruguay, and Bolivia, where it is recorded from Santa Cruz Department around Camiri and Potrerillo del Guenda Reserve and Tarija Department near Villamontes.

Eburodacrys apua Martins & Galileo, 2005

Cerambycinae: Eburiini

This is a medium-sized (13–18 mm), parallel-sided, tan species with three widely separated oval, shining eburneous spots on the heavily punctate elytra. They are positioned with one basally, another medially near the suture, and the last behind it and more lateral. The species has long, acute black spines on the middle and hind femora and at the external elytral apices. There is a rectangular, central black fascia behind the eyes and two shining black tubercles on the pronotal disk followed by two smaller black, rugose callosities in front of the margin. The species is nocturnal and readily attracted to lights where, early in the rainy season, it can be locally abundant. It is known from only Bolivia, where it has been collected in Cochabamba Department around Villa Tunari and Santa Cruz Department at the Flora and Fauna Hotel near Buena Vista, the Potrerillo del Guenda Reserve, and Refugio Los Volcanes.

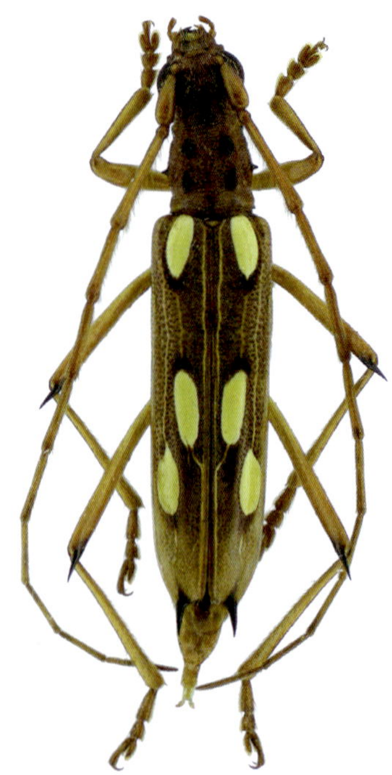

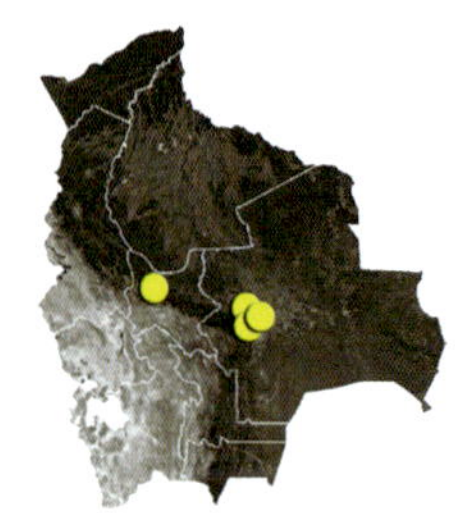

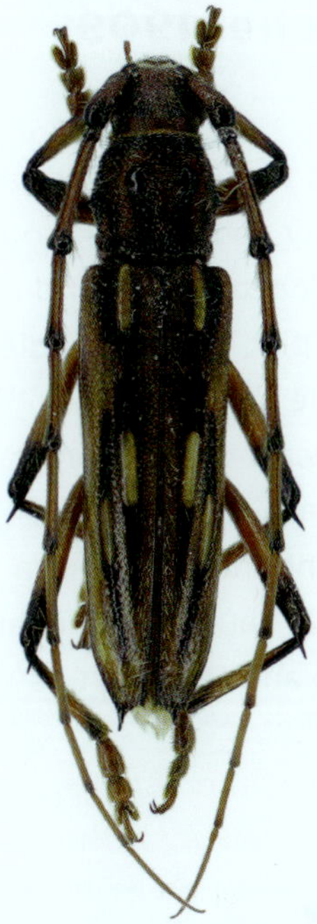

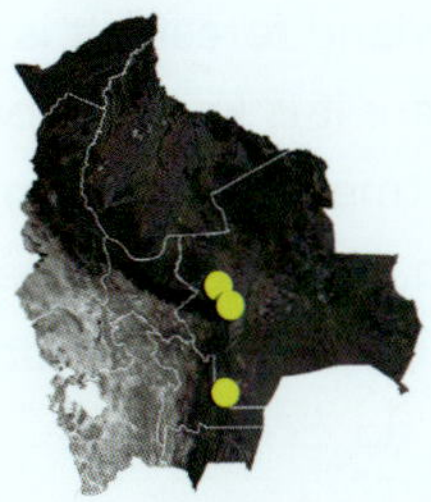

Eburodacrys crassimana Gounelle, 1909

Cerambycinae: Eburiini

This is a medium-sized (12–14 mm), tan species with dark reddish pronotum and a narrow, longitudinal black fascia on the elytra and pronotum. In addition there are three linear eburneous ridges, all positioned within the black fasciae, on each elytron: one basally, one centrally, and the last just behind and outside it. The dorsal surface is heavily punctate, but still shining, and has a mixture of small, fine, semirecumbent hair and sparser, long flying hairs. The pronotum has a pair of small glabrous tubercles at apical third. This species is most often encountered at lights but is never abundant. It is known from Brazil, Argentina, Paraguay, and Bolivia, where it has been collected at the Flora and Fauna Hotel grounds, the Potrerillo del Guenda Reserve, and the Camiri environs in Santa Cruz Department.

Eburodacrys errata Galileo & Martins, 2010

Cerambycinae: Eburiini

With its bold, elongate ivory markings on the elytra and long, black spines in the inner apex of the middle and hind femora and outer elytral apex, this 21–26 mm long eburiine is quite striking. The legs and elytra are otherwise yellowish, while the head and pronotum are darker reddish brown. This species, which is very similar in appearance to another South American species, *Eburodacrys lepida*, is endemic to Bolivia. It has been collected in the vicinity of the Flora and Fauna Hotel near Buena Vista in Santa Cruz Department.

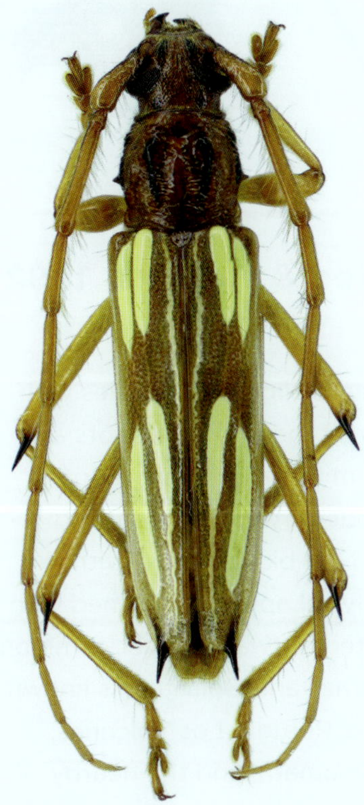

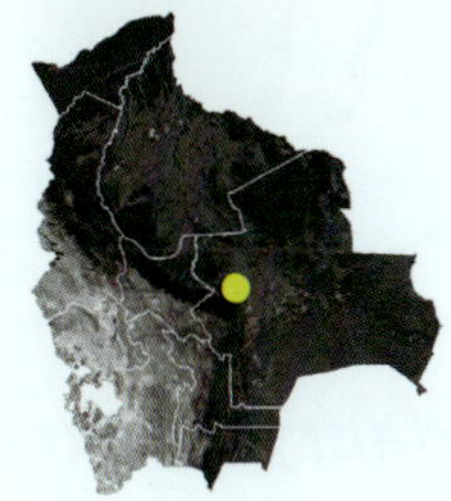

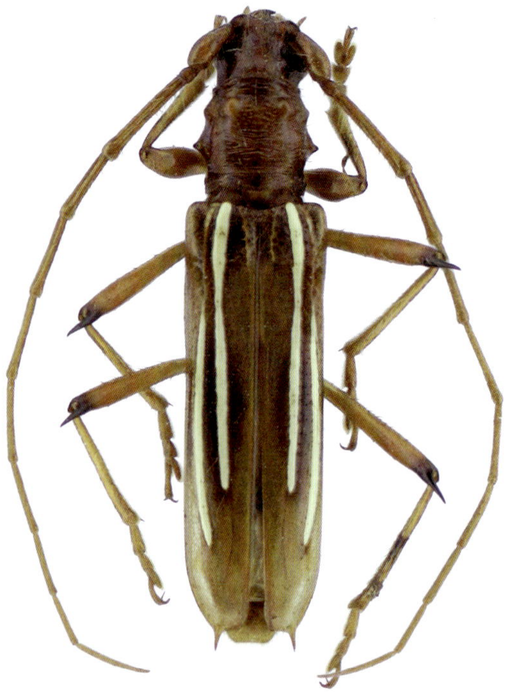

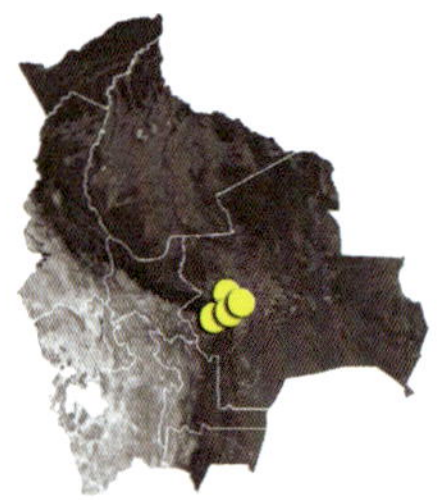

Eburodacrys gaucha Galileo & Martins, 1992

Cerambycinae: Eburiini

This is a medium-sized (12–15 mm), shining light tan species with two to four elongate eburneous spots (sometimes the spots are connected) on the elytra. The pronotum is transversely rugose on the basal two-thirds, apically preceded by two low, shining tubercles. The middle and hind femora are black tipped with the apices strongly spined. The external elytral apices are armed with a moderately long spine slightly directed outward. Never common, this species is known from southern Brazil and Bolivia. It is known from the Potrerillo del Guenda Reserve, Refugio Los Volcanes, adjacent to Amboró National Park, Samaipata, and the nearby Vicoquin area above Achira at 2,000 m in Santa Cruz Department in Bolivia.

Eburodacrys granipennis Gounelle, 1909

Cerambycinae: Eburiini

This is a medium-sized (12–18 mm), tan species with three elongate eburneous maculae on each elytron: one basally and a contiguous pair just past the middle. The dorsum has black accents behind the basal macula, before and after the contiguous pair, on the middle and hind femoral apices and their spines, on the apical elytral spines, and on the pronotal tubercles. This species is common at lights in lowland forests. It is recorded from Brazil, Ecuador, and Bolivia, where it is known from Beni Department and from Santa Cruz Department in the vicinity of the Flora and Fauna Hotel near Buena Vista, the Potrerillo del Guenda Reserve and Refugio Los Volcanes, adjacent to Amboró National Park.

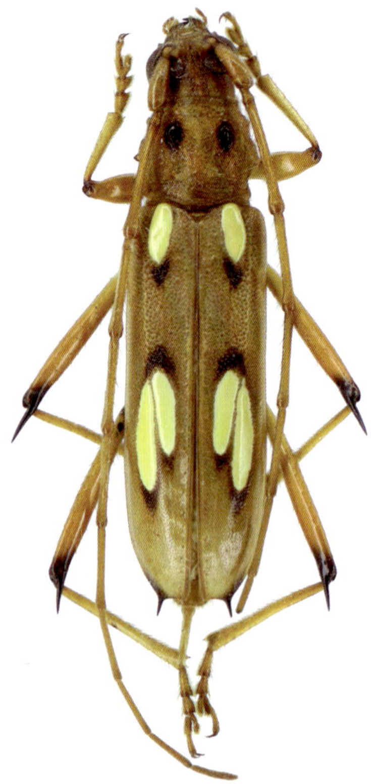

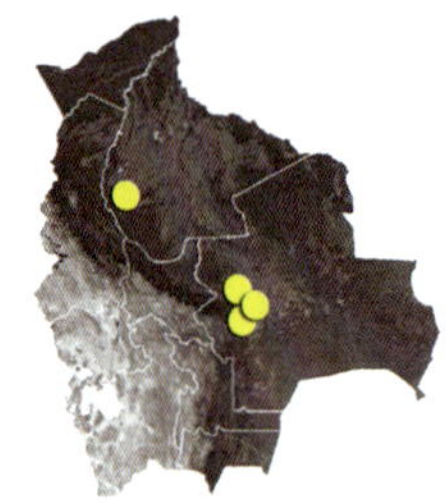

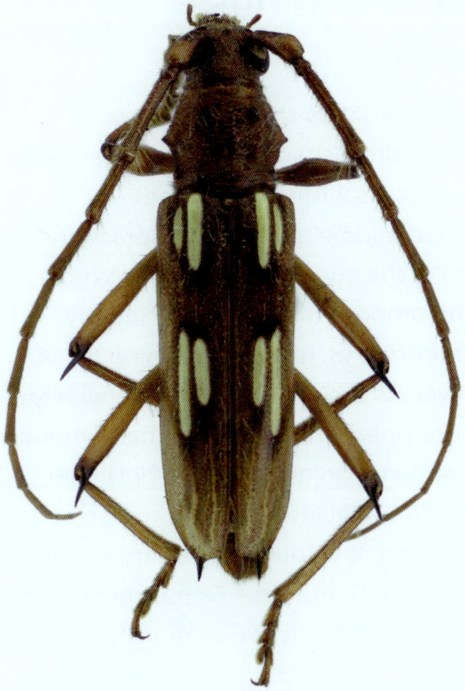

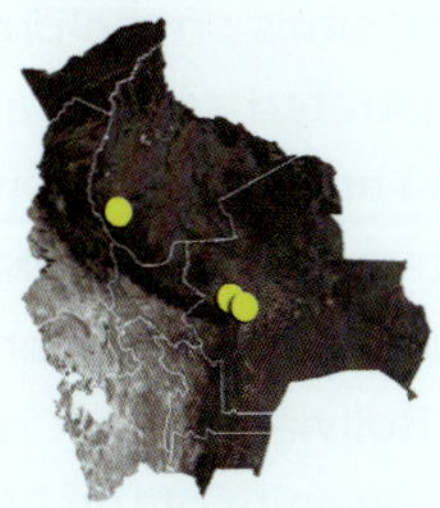

Eburodacrys havanensis Chevrolat, 1862

Cerambycinae: Eburiini

This is a medium-sized (12–18 mm), dark reddish-orange cerambycid with four linear eburneous fasciae on each elytron. The pronotum is transversely rugose with a pair of shining black tubercles and small lateral spines. The middle and hind femora are apically armed with long black spines, as are the external margins of the elytra. The appendages (legs and antennae) have long golden hairs equally spaced on their shining surface. This species is widely distributed from Mexico to Panama and Cuba and over much of South America including Bolivia. In Bolivia it is recorded from Beni Department and Santa Cruz Department, where it was collected at the Flora and Fauna Hotel near Buena Vista and the Potrerillo del Guenda Reserve.

Eburodacrys lanei Zajciw, 1958

Cerambycinae: Eburiini

This is a strikingly beautiful, medium-sized (18–20 mm) cerambycid, reddish orange with cream-colored eburneous markings nearly the length of the elytra. The antennomeres beginning with the second segment, tarsi, basal half of the tibia, and the apical spine of the femora are black. The pronotum is unicolorous except for the tip of the small lateral tubercles and is completely transversely rugose with two low tubercles near the apical margin and a low central ridge on the disk. It is rarely seen except at lights, where it is a frequent visitor. It is distributed from Peru, west-central Brazil, and Bolivia, where it is recorded from Santa Cruz Department from the vicinity of the Flora and Fauna Hotel near Buena Vista, the Potrerillo del Guenda Reserve, Refugio Los Volcanes, and the Vicoquin area above Achira at 2,000 m.

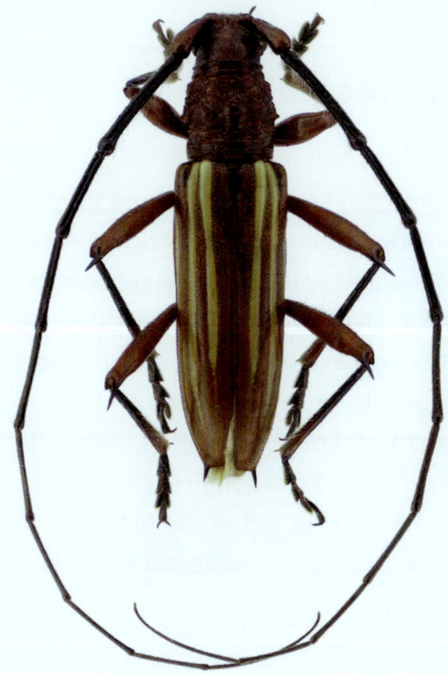

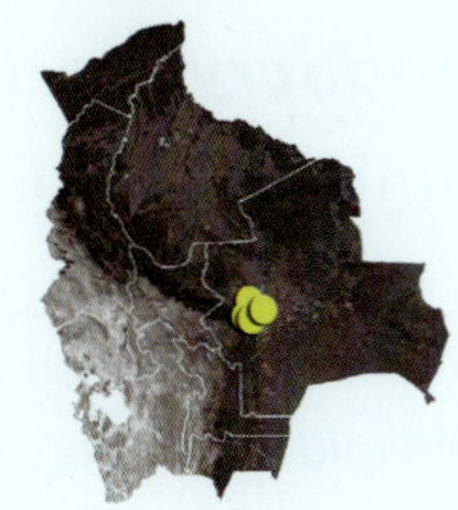

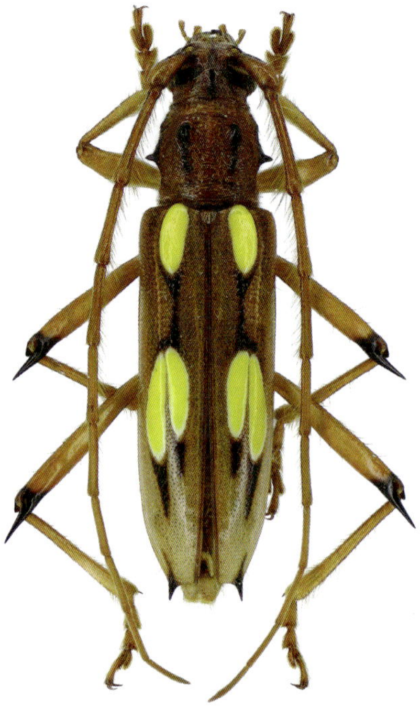

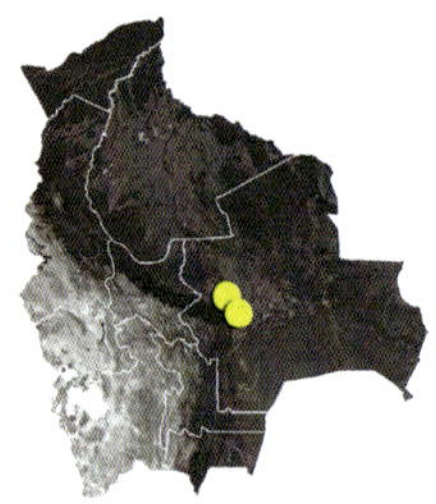

Eburodacrys raripila Bates, 1870

Cerambycinae: Eburiini

This is a medium-sized (17–20 mm), tan species with three bright yellow maculae on each elytron, one basally and a contiguous pair near the middle that are interconnected with the basal macula by a black fascia. The basal half of the elytra is densely punctate with the pronotum rugosely punctate and having a pair of shining black tubercles on the disk along with black, lateral tubercles on the sides. This species is rarely encountered except at lights, where it is an uncommon visitor. It is known from Ecuador, French Guiana, Peru, and Bolivia, where it is has been collected at the Flora and Fauna Hotel near Buena Vista and the Potrerillo del Guenda Reserve in Santa Cruz Department.

Eburodacrys truncata Fuchs, 1956

Cerambycinae: Eburiini

A medium-sized species (15–17 mm), this shining light tan longhorned beetle has three teardrop-shaped, widely separated eburneous spots on each elytron. As with many of the *Eburodacrys*, it has black-tipped middle and hind femora with long black spines. It differs though in the elytral apices being simply truncated and not spined. The pronotum is densely, minutely punctate and, unlike most *Eburodacrys*, has three shining black callosities on the pronotum (versus the typical two), two widely separated apically and a central linear ridge near the basal margin. The antennomeres have dense, long, golden hairs on both the internal and external margins. This species is fairly common at lights during the early rainy season. It is found in eastern Brazil and in Bolivia, where it is recorded from Santa Cruz Department from the vicinity of the Flora and Fauna Hotel near Buena Vista and the Potrerillo del Guenda Reserve.

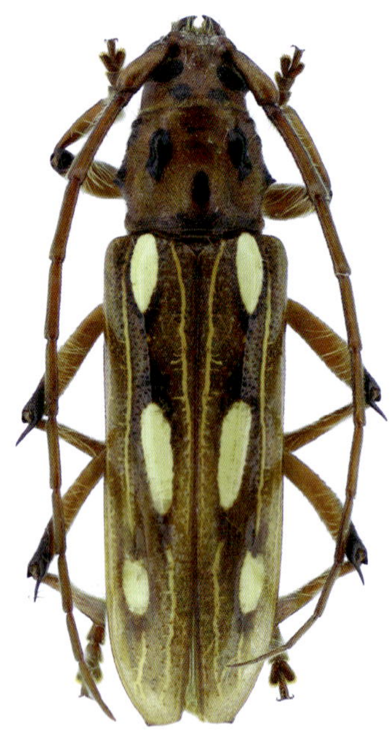

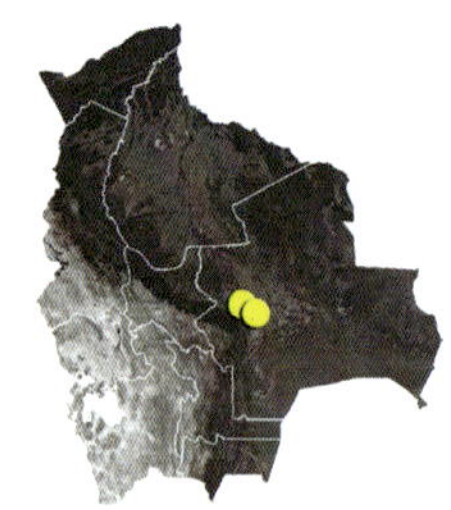

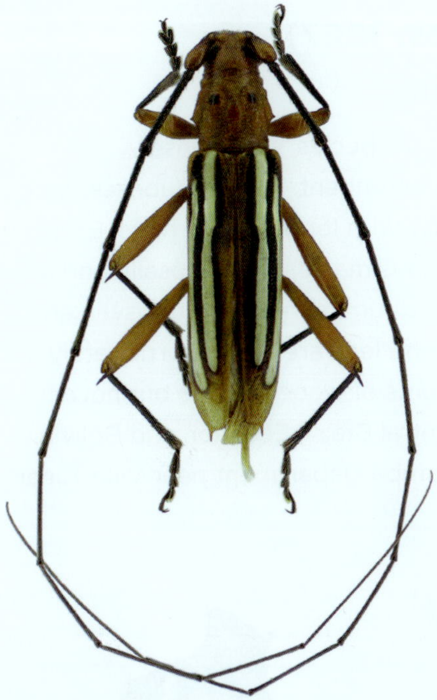

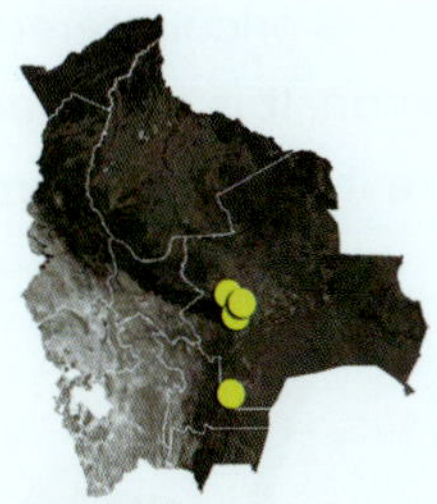

Erosida formosa (Blanchard, 1846)

Cerambycinae: Eburiini

This is a medium-sized (14–15 mm), orange-red species with black tibia, femoral spines, and antennomeres (beginning with the second). Each elytron is decorated with two long, shining eburneous fasciae, which begin at the base and end short of but near the apices. The fasciae are margined in a thin black line and separated similarly. The elytral surface is densely, minutely punctate as is the pronotum except for its shallowly rugose disk, which is impunctate and shining. This glamorous looking cerambycid is common at lights during the early rainy season. It is known from Brazil and Bolivia, where it is recorded from many lowland sites in Santa Cruz Department including Camiri, Cuevas, the Flora and Fauna Hotel near Buena Vista, and the Potrerillo del Guenda Reserve.

Pantomallus rugosus Martins & Galileo, 2005

Cerambycinae: Eburiini

This is a medium-sized (14–16 mm), dark brown species with the dorsum covered in short, recumbent, whitish-gray pubescence. Each elytron has two pairs of small yellowish eburneous maculae, one laterally and the other medially. The pronotum has a pair of slightly elevated, shining tubercles on the apical portion of the disk. This species was recently discovered and is a Bolivian endemic. To date it is known from only the Chaco near Pampagrande and the Potrerillo del Guenda Reserve in Santa Cruz Department.

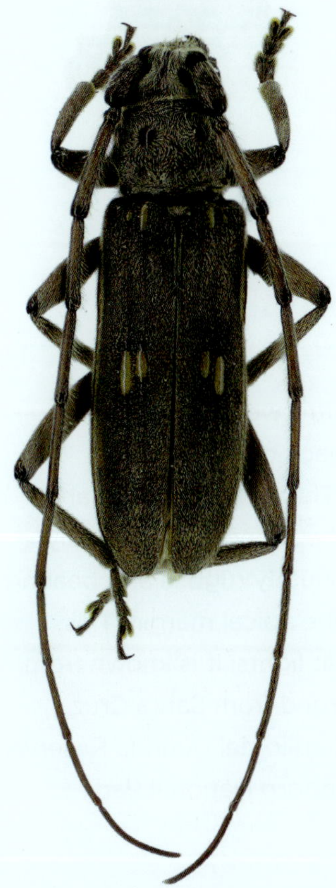

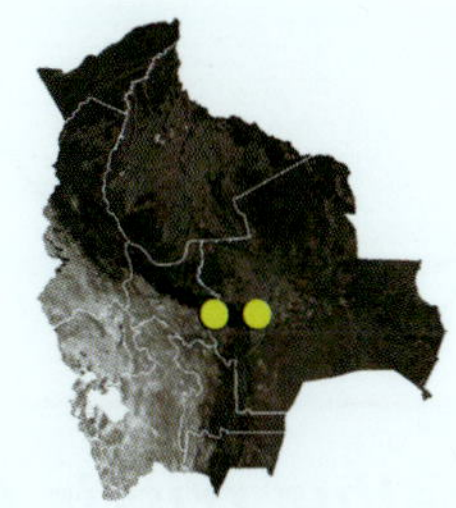

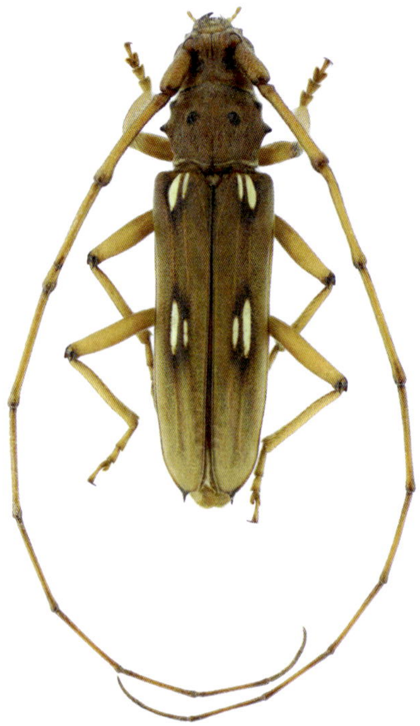

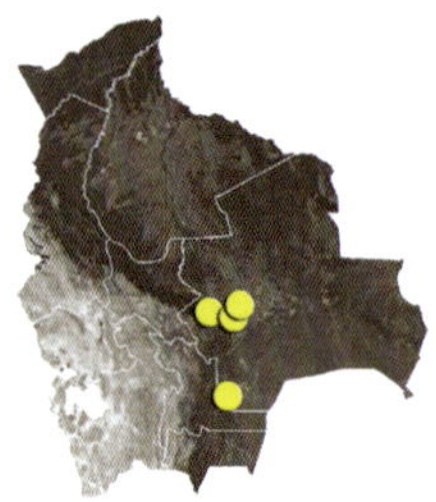

Pantomallus tristis (Blanchard, 1846)

Cerambycinae: Eburiini

This is a large (20–26 mm), tan species with two pairs of small, nearly contiguous eburneous fasciae on each of its minutely and sparsely punctate, pubescent elytra. The elytral margins are armed with a small, singular spine while the pronotum is densely pubescent with the underlying surface minutely rugose and bearing two shining black tubercles just short of the apical margin. This species is nocturnal and locally common at lights. It is known from central Brazil and Bolivia, where it is recorded from Santa Cruz Department in Pampagrande, Camiri, Potrerillo del Guenda Reserve, and Refugio Los Volcanes, adjacent to Amboró National Park.

Pronuba dorilis (Bates, 1867)

Cerambycinae: Eburiini

This is a small (9–10 mm), light tan species, with densely punctate elytra covered in semirecumbent, golden pubescence and intermixed at the elytral base with longer, erect hairs. Each elytron has three shining eburneous maculae, one basally and a contiguous pair centrally. Its pronotum is shallowly, transversely rugose, and in this species the hind legs are disproportionately long. It is encountered on blossoms of woody plants but never common. It is recorded from central Brazil, Ecuador, and Bolivia, where it is known from Cochabamba Department near Villa Tunari.

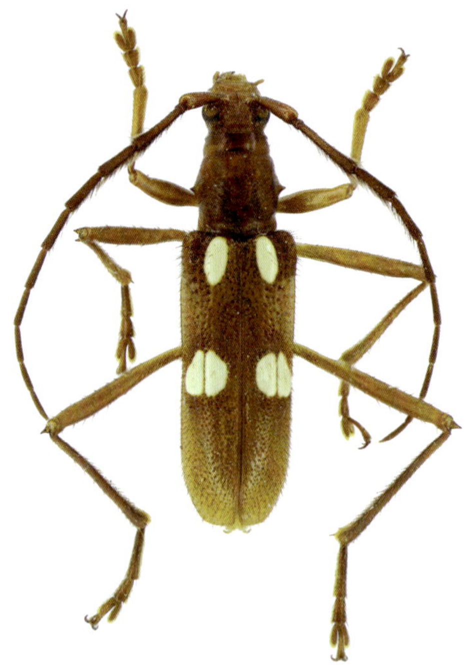

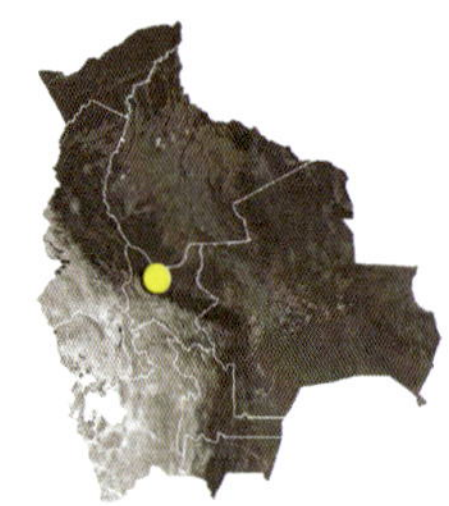

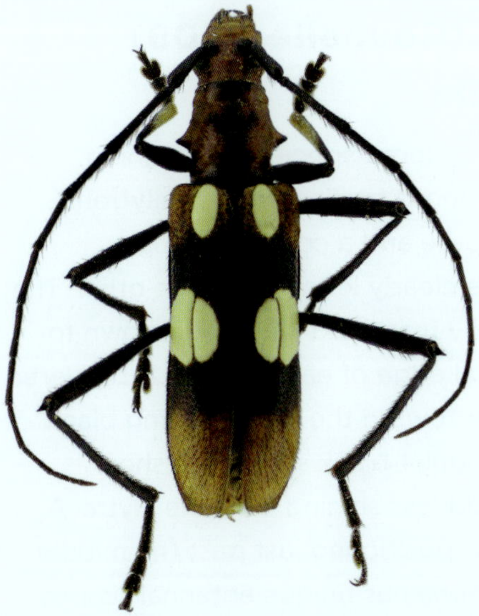

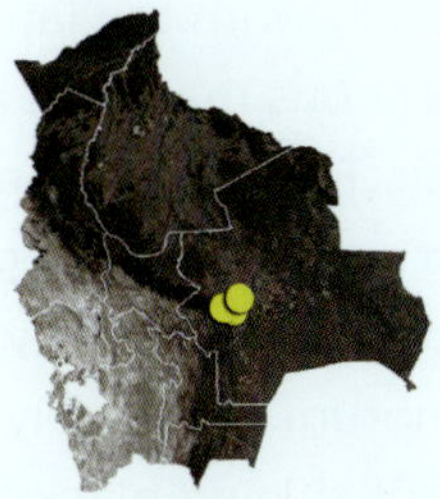

Pronuba lenkoi Monné & Martins, 1974
Cerambycinae: Eburiini

This is a small (10–11 mm), flower-visiting eburiine, orange to dark red in color with dark antennae and legs. The elytra are deeply, closely punctate at the base, with the punctation becoming sparser and shallower apically. Each elytron has three eburneous fasciae, a large oval one basally and two larger contiguous ones centrally. The elytra are orange on the basal fourth and apical third but, centrally, surrounding the eburneous fasciae, they are dusky black. The pronotum is unicolorous and moderately rugose basally and on the disk and has a transverse, concave collar bordering the apical margin followed by two low tubercles. This species has been collected on the blossoms of woody plants and also singularly at lights. It has a small distribution, being known only from Matto Grosso, Brazil, and Bolivia, where it is recorded from Santa Cruz Department on the road to Bella Vista (near Cuevas), the Vicoquin area above Achira at 2,000 m, and Refugio Los Volcanes, adjacent to Amboró National Park.

Styliceps sericatus (Pascoe, 1859)
Cerambycinae: Eburiini

This striking species is moderately large (22–25 mm), light tan to orange red, with a densely pubescent tan elytra, resulting in a velvety appearance. It is marked by a shining golden pubescent stripe the length of the elytra where, near the apices, it joins with another shorter, similar stripe. The elytral apices are armed with a singular black spine. Unlike the rest of the body of this beautiful species, the pronotum is glabrous, with the surface of the disk covered with round, bulbous-looking, contiguous tubercles of varying sizes. Apically, on each side of the midline, there are shining black tubercles. Interestingly, this species does not have any of the namesake eburneous fasciae or spots that are typical for members of the tribe Eburiini. Rarely encountered, it has been collected at lights and by beating forest edge foliage. It is rare but widely distributed from western Mexico to Panama, French Guiana, Suriname, northeastern Brazil, eastern Ecuador, and Bolivia. In Bolivia it has been collected from Santa Cruz Department at the Flora and Fauna Hotel near Buena Vista, Potrerillo del Guenda Reserve, and Refugio Los Volcanes.

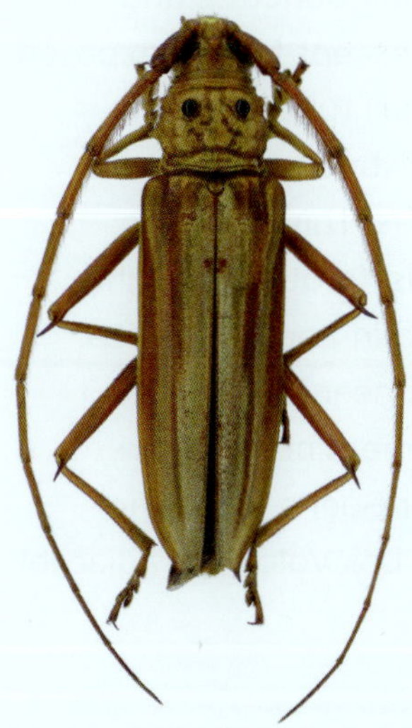

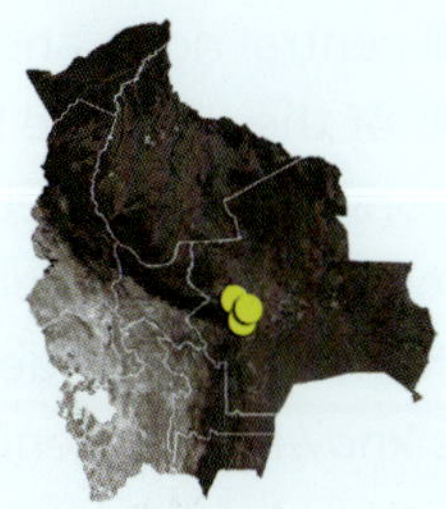

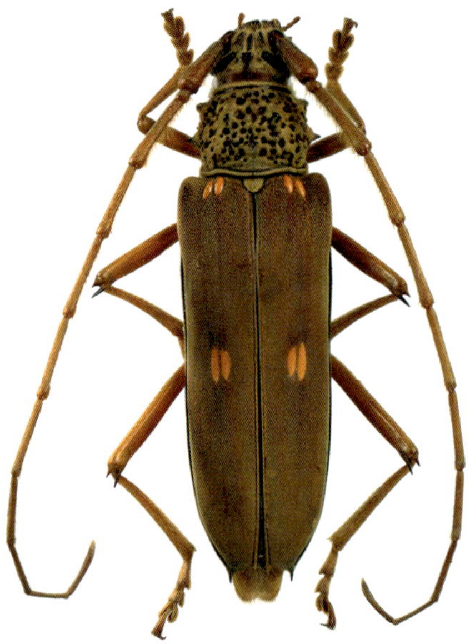

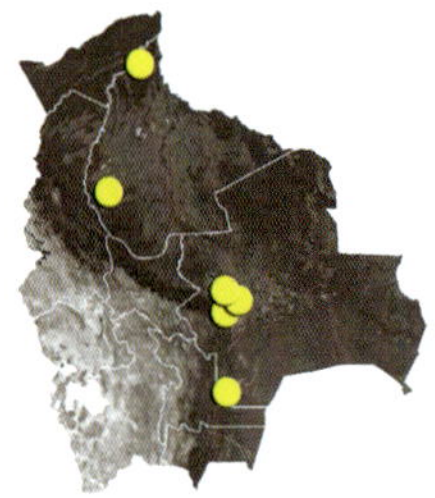

Susuacanga octoguttata (Germar, 1821)

Cerambycinae: Eburiini

One of the largest Bolivian eburiines at 22–32 mm, this species is light tan and has elytra densely covered in very short, recumbent pubescence. The pronotum is deeply, coarsely punctate and densely pubescent between the punctures except for the exposed top of a small central ridge and a pair of small tubercles on the apical half of the disk. The head has a distinctive square to rectangular dorsal tubercle between the eyes. This is among the most common of the larger cerambycids, frequent at lights and locally abundant. It is widely distributed in South America. In Bolivia it is known from Beni Department near Riberalta and Rurrenabaque and from Santa Cruz Department near Camiri, Concepción, the Flora and Fauna Hotel near Buena Vista, the Potrerillo del Guenda Reserve, and Refugio Los Volcanes, adjacent to Amboró National Park.

Uncieburia nigricans (Gounelle, 1909)

Cerambycinae: Eburiini

This is a large (20–25 mm), tan-colored species with shining elytra bearing three linear eburneous fasciae on each elytron: one moderate in length at the base and a contiguous pair medially, with the lateral fasciae clearly longer than the other. The eburneous markings are typically margined with dark brown to black. The sutural line and lateral edge of each elytron is bordered with a dark line that gets darker toward the apices, being black there, where it joins with the singular black spine. The short pronotum is minutely rugose and less shining than the elytra. A pair of shining black tubercles is positioned just past the middle on each side of the disk. *Uncieburia* has unique antennal scapes that are short and bulbous with a deep concave carina at the base. This species is one of the most common eburiines at light. It is found in eastern Brazil and Bolivia, where it is widely known from most surveyed localities in Santa Cruz Department including Camiri, Concepción, the road to Bella Vista near Cuevas, the Flora and Fauna Hotel near Buena Vista, Potrerillo del Guenda Reserve, and Refugio Los Volcanes, adjacent to Amboró National Park. It is also known from Tarija Department near Villamontes.

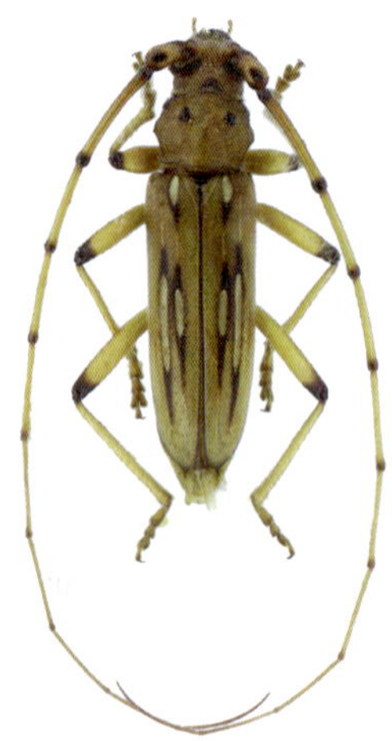

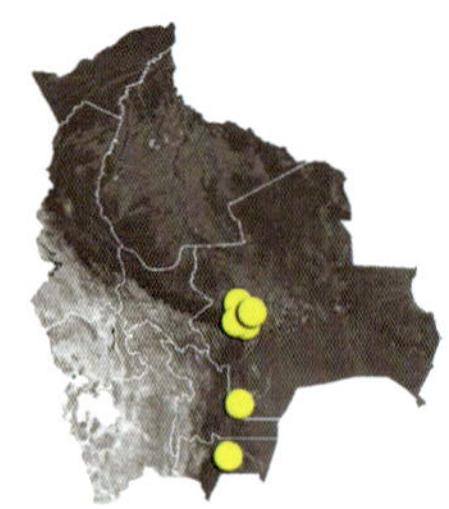

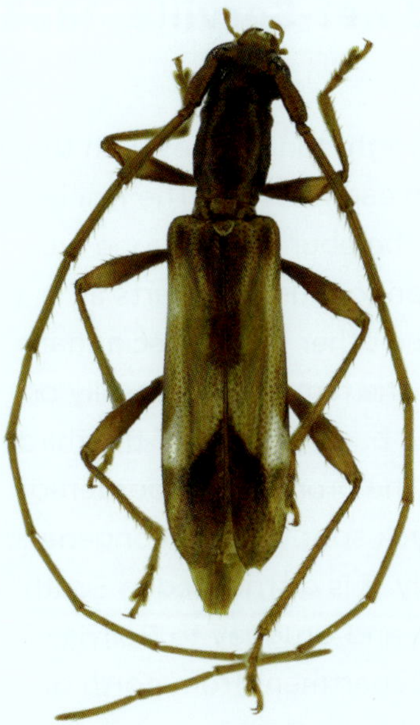

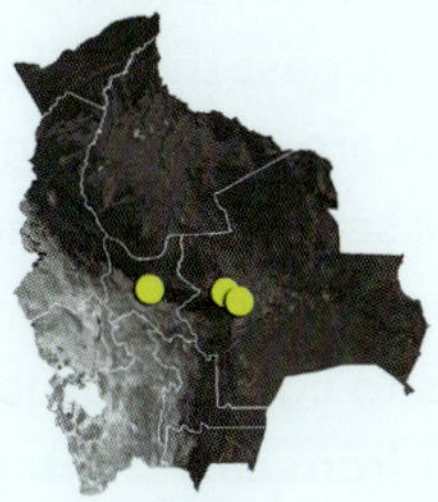

Bomarion carenatum Martins, 1962

Cerambycinae: Ectenessini

The genus *Bomarion* is characterized by its elongate, rounded pronotum and by its lack of any spines. This is a small (7–9 mm), dark reddish-brown species with lighter tan fasciae on the elytra. The elytral apices are merely emarginate to truncate in design. The elytra have a clearly defined lighter integumental spot followed by an oblique fascia on each elytron. The pronotum is darker reddish brown than the elytra, minutely punctate and uniform, without obvious vestiture or structures. Both its antennae and legs are long, slim, and without spines or color accents. In some specimens the scutellum is shining pubescent. Occasionally attracted to lights, this species is never abundant. This is another of the Bolivian endemics and is known from Cochabamba Department and Santa Cruz Department from the Flora and Fauna Hotel near Buena Vista and the Potrerillo del Guenda Reserve.

Ectenessidia nigriventris (Belon, 1902)

Cerambycinae: Ectenessini

This is a small (7–9 mm), parallel-sided species with black appendages. The dorsum is red except for the apical fifth of the elytra, which is shining dark blue-black. The elytra are closely, evenly punctate with long golden hairs on the apical surface and margin. The pronotum is minutely alveolate giving it its matte finish and smooth-appearing surface. This species is occasional at light but can be locally abundant. It is known only from eastern Brazil and Bolivia, where it is recorded from Santa Cruz Department near Camiri, the Flora and Fauna Hotel near Buena Vista, and the Potrerillo del Guenda Reserve.

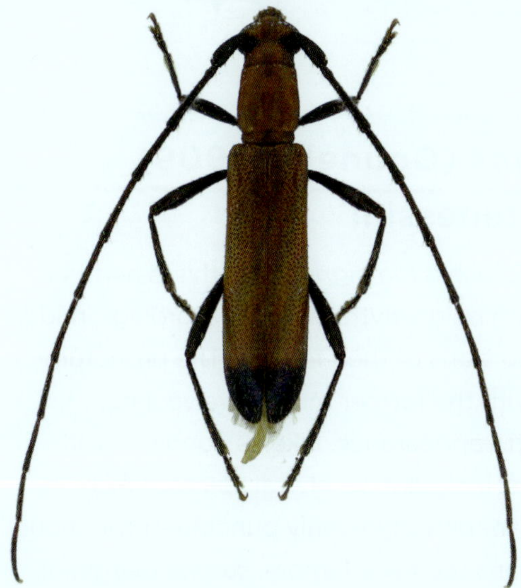

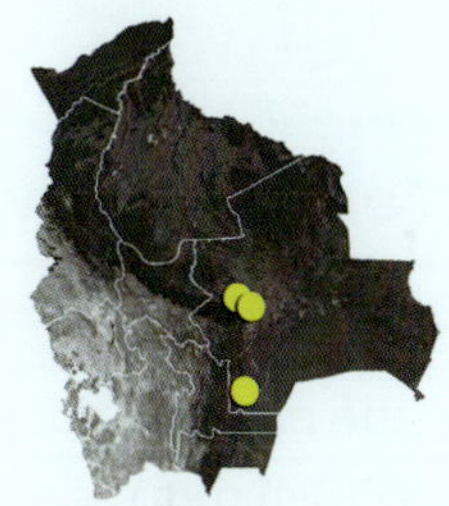

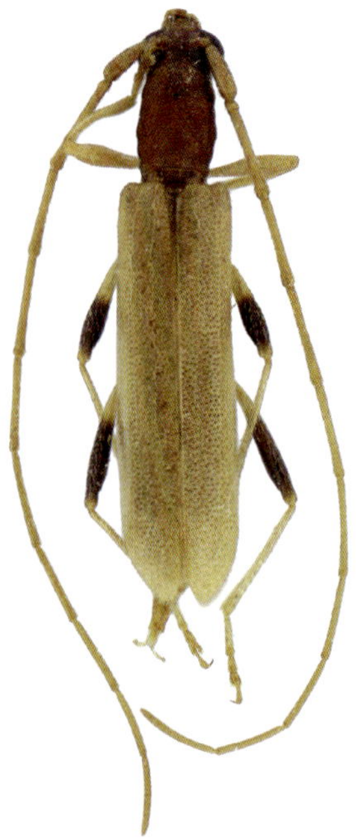

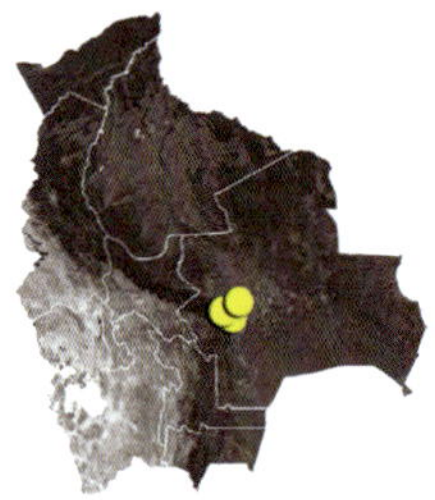

Ectenessidia varians (Gounelle, 1909)

Cerambycinae: Ectenessini

Like the previous species, *Ectenessidia nigriventris*, this is another small (10–11 mm) species with beige elytra, antennae, prolegs, middle and hind tibiae, and basal portions of their femora. The pronotum and head are reddish tan, with the former minutely alveolate, contributing to the dull matte appearance. Like its congeners, it lacks any spines or armature, including its elytral apices, which are evenly rounded. The elytra are densely, evenly punctate throughout, and in some female specimens the front femora, sutural margin of the elytra, and a small slash-like fascia above the elytral apices are also dark brown. It is never abundant but a regular visitor to lights. It is found in eastern Brazil and Bolivia, where it has been recorded from Santa Cruz Department from the Vicoquin area above Achira at 2,000 m, Potrerillo del Guenda Reserve, and Refugio Los Volcanes, adjacent to Amboró National Park.

Eurymerus eburioides Audinet-Serville, 1833

Cerambycinae: Ectenessini

This is a moderate to large (15–22 mm), light tan species with three typically ivory-colored, linear fasciae on each elytron. The fasciae are positioned very differently than on the eburiines as none are extremely basal in location. Rather, a contiguous pair starts at the basal fourth and slightly overlaps the other single fascia that begins just past the middle. The tips of the femora and basally on the tibiae are accented in black as is the basal portion of the third antennal segment in some specimens. The pronotum is bordered, both basally and apically, by a small black spot. Like its congeners, it comes to light but usually infrequently. It is distributed in South America from eastern Brazil, Argentina, and Uruguay to Bolivia. In Bolivia it is known from Santa Cruz Department from north of Pampagrande, around Camiri, above Achira at 2,000 m, at Refugio Los Volcanes, adjacent to Amboró National Park, and the road to Bella Vista near Cuevas.

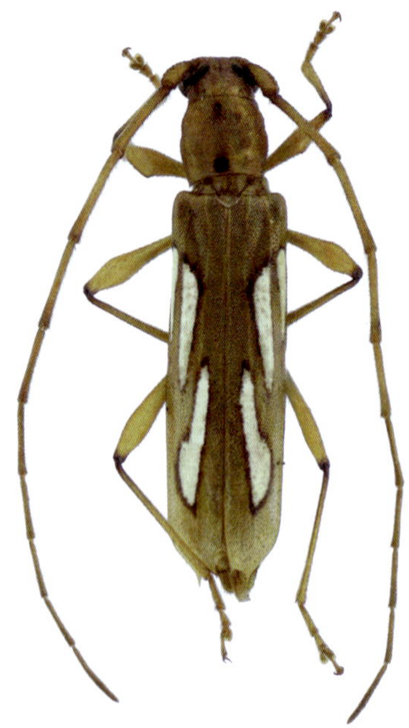

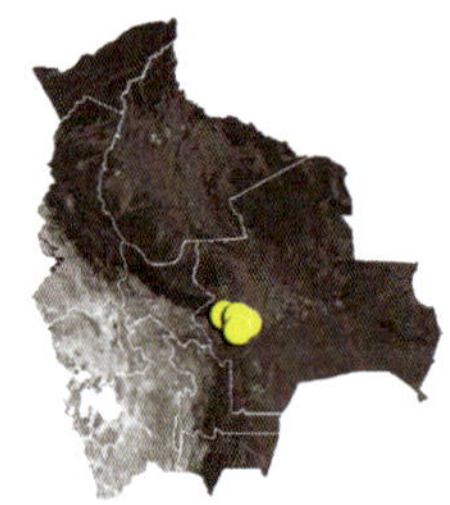

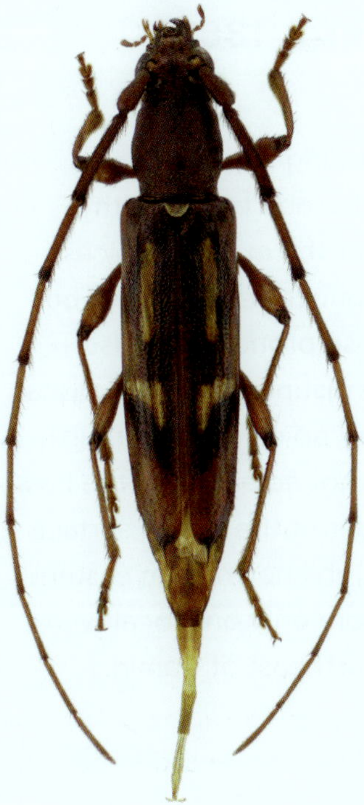

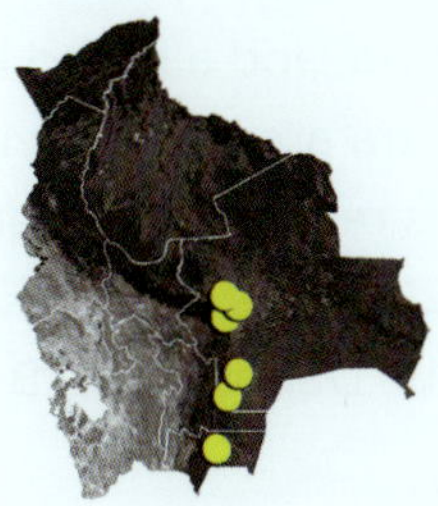

Tricheurymerus quadristigma (Gounelle, 1909)
Cerambycinae: Ectenessini

This is a medium-sized (13–16 mm), reddish-brown species with densely punctate elytra and even more densely, microscopically punctate pronotum. There are two tan fasciae on each elytron, a linear one paralleling the suture on the basal half and at the apical third a band that curves backward at the sutural margin. This species is commonly encountered at lights and distributed from French Guiana to Brazil, Argentina, Paraguay, and Bolivia, where it is known from Santa Cruz Department north of Camiri, the Flora and Fauna Hotel near Buena Vista, and the Potrerillo del Guenda Reserve and also from Tarija Department near Villamontes.

Ambonus distinctus (Newman, 1840)
Cerambycinae: Elaphidiini

This is a large (18–23 mm), mahogany-brown species with a densely pubescent scutellum and elongate whitish-yellow fasciae (three to four) on the elytra and on the lateral edges of the pronotum. The elytra are evenly, minutely punctate with short, fine, whitish pubescence scattered on the surface. The pronotum is densely punctured and finely rugose while the antennae are carinate dorsally and bear long, whitish-golden hairs along both sides of the antennomeres. This attractive species is a common visitor to lights during the early rainy season, when it can be collected in numbers. It is known from French Guiana, eastern Brazil, Argentina, Paraguay, and Bolivia, where it has been recorded from throughout Santa Cruz Department (including the vicinity of the Flora and Fauna Hotel near Buena Vista, the Potrerillo del Guenda Reserve, Refugio Los Volcanes, adjacent to Amboró National Park, Cuevas, Bermejo, and the Camiri environs) as well as Tarija Department near Villamontes.

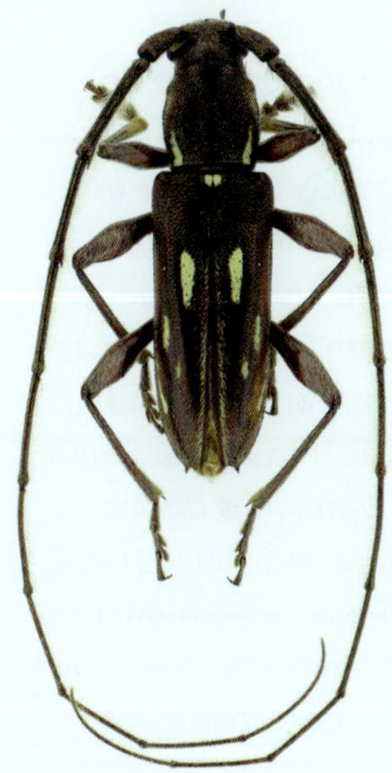

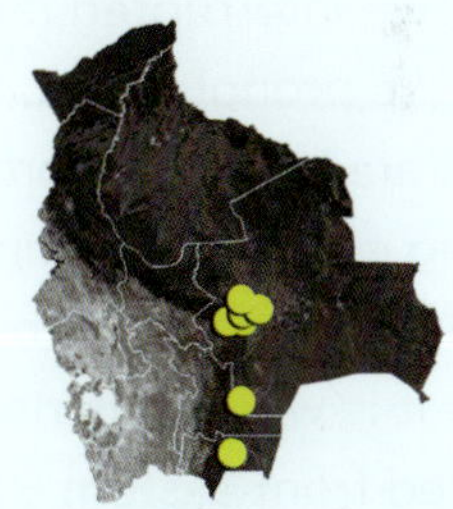

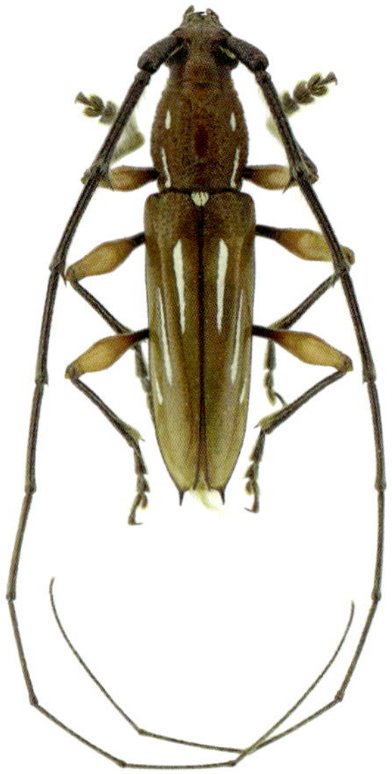

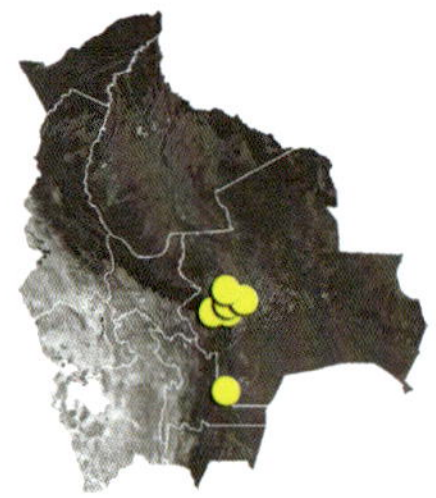

Ambonus lippus (Germar, 1824)

Cerambycinae: Elaphidiini

This is another of the large (18–20 mm) elaphidiines and is a distinctive reddish-orange species with white pubescent scutellum, three elongate, white pubescent fasciae on each elytron, and two interrupted fasciae on the pronotum, one on each side of the pronotal disk. The antennae, tibia, tarsi, and apical elytral spines are black. The pronotum is rugosely punctate except for the shining central disk. This species is rarely encountered except at lights, where it can be very common during the early rainy season. It is widely distributed in southern South America, being recorded from eastern Brazil, Argentina, Paraguay, and Bolivia, where it is known from lowland forests throughout Santa Cruz Department. Like the previous species, *Ambonus distinctus*, it has been collected in the vicinity of the Flora and Fauna Hotel near Buena Vista, the Potrerillo del Guenda Reserve, Refugio Los Volcanes, Samaipata, near Cuevas, and the Camiri environs. Unlike *A. distinctus*, however, it is not known from Tarija Department.

Amorupi fulvoterminata (Berg, 1889)

Cerambycinae: Elaphidiini

This is a moderate-sized (14–18 mm, excluding the everted genitalia), shining orange and black longhorned beetle with closely punctate elytra and short spines at the external apices of each elytron. The pronotum is widely punctate in females but minutely, densely punctate in males, whose pronotum is also much more compressed and without the distinct, elevated, discal tubercles of the females. The appendages, apical third to fourth of the elytra, and a broad band near, but not adjacent to, the base of the elytra are orange with the remainder of the dorsal surface black. It is rarely encountered, and although known from eastern Brazil and Argentina it has been found from only one locality in Santa Cruz Department, Bolivia, 22 km northeast of Camiri.

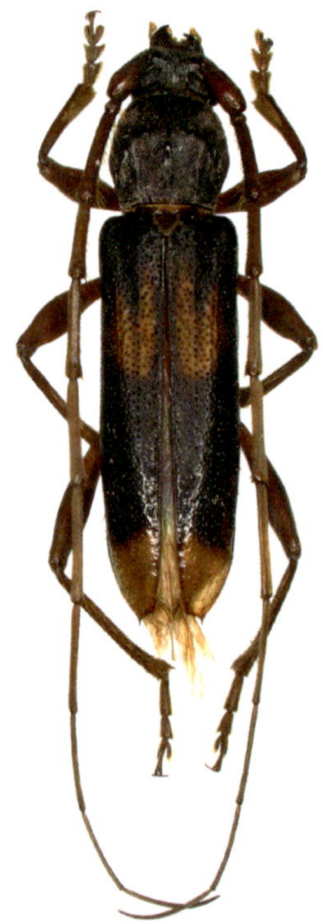

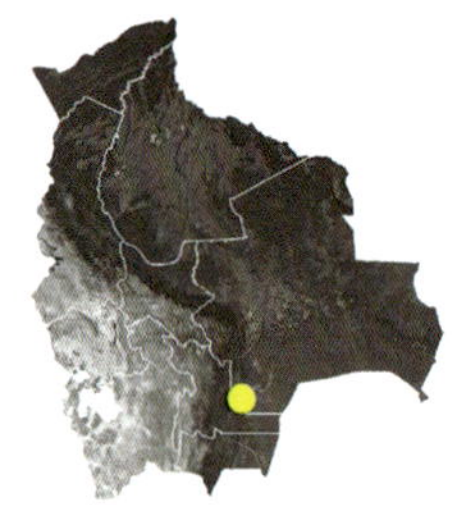

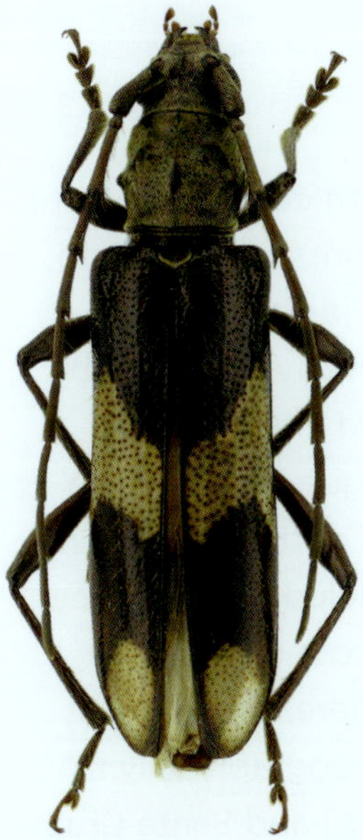

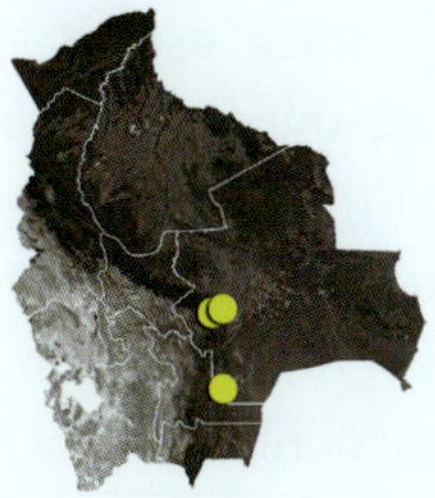

Amorupi hudepohli (Martins, 1974)

Cerambycinae: Elaphidiini

This is a slightly larger species than its congener *A. fulvoterminata*, with large specimens nearly 20 mm in length. It is dark brown with shining, evenly punctate elytra that have light tan, oval-shaped spots at the apices and an oblique pair of tan fascia near the middle. Females of this species have short antennae that do not attain the elytral apices unlike *A. fulvoterminata*, where they exceed the apices by one or more segments. This species is also easily separated from its congener by the truncate nonspinose elytral apices that, in *A. fulvoterminata*, are distinctly spined on the external apices. It is also more commonly encountered, being occasional at lights. This species is not yet known outside Bolivia. It has been collected in Santa Cruz Department from Samaipata, north of Camiri, and Refugio Los Volcanes, adjacent to Amboró National Park.

Anelaphus souzai (Zajciw, 1964)

Cerambycinae: Elaphidiini

This is a moderate-sized (12–14 mm), brown species whose dorsum has scattered, irregular patches of light tan pubescence on its heavily punctate surface. The genus *Anelaphus* is large, having more than 60 species distributed from the United States, Mexico, all of Central America, the Caribbean, and down through South America to Argentina and Brazil. In Bolivia the two known species are this one and *A. cerussatus* (Newman). They are very similar in form and coloration except the latter differs by its whitish patches of pubescence (versus tan in *souzai*) and is much less common. This species can be locally abundant at lights very early in the rainy season but is rarely encountered after that. It is known from central Brazil, French Guiana, Argentina, and Bolivia, where it has been recorded in Santa Cruz Department from Pampagrande, Santa Cruz Botanical Gardens, Potrerillo del Guenda Reserve, and Refugio Los Volcanes, adjacent to Amboró National Park.

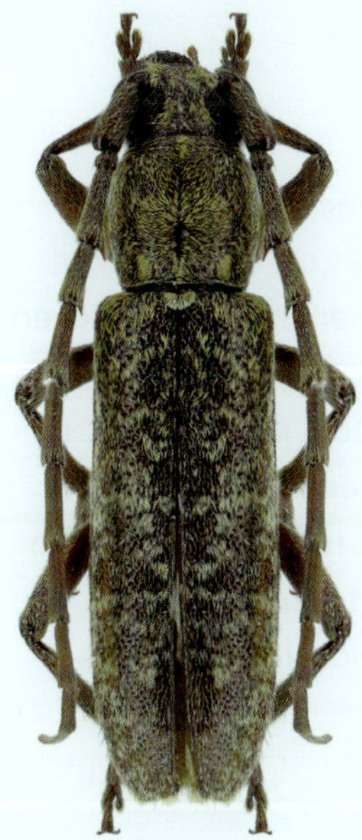

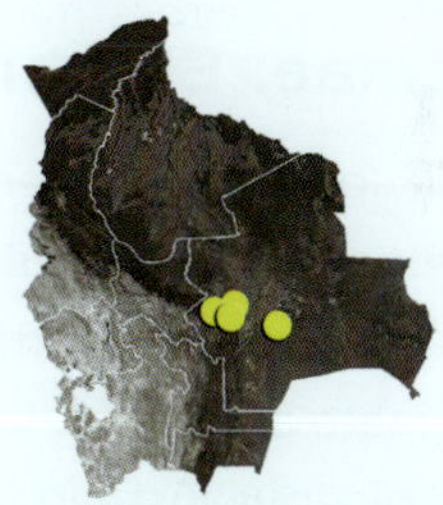

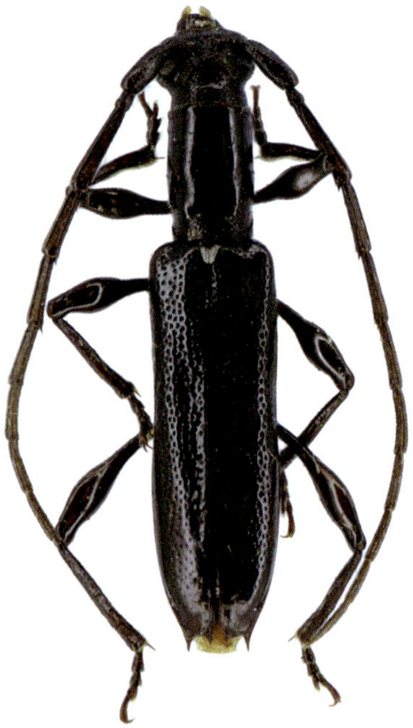

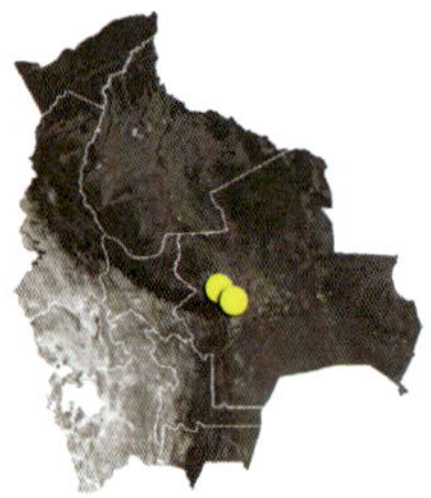

Aposphaerion nigritum Galileo & Martins, 2010

Cerambycinae: Elaphidiini

A shining black species with densely, finely punctate elytra, impunctate pronotum, and double apical spines on each elytron, this species is very small (6–8 mm). Antennomeres three to eleven are doubly carinate and much less shining than the rest of the dorsum, being piceous instead of black. This Bolivian endemic is most commonly collected early in the rainy season at lights but occasionally also collected by beating along forest edges. Known only from lowland forests in Santa Cruz Department, it has been collected at and around the Flora and Fauna Hotel near Buena Vista and the Potrerillo del Guenda Reserve.

Appula aliena Martins, 1981

Cerambycinae: Elaphidiini

This is a small (9–10 mm), black elaphidiine with closely punctate elytra and antennae that have the basal segments piceous then progressively lighter in color apically, with the last four or five segments a light tan. The dorsal surface is partially obscured by silvery pubescence, especially a wedge-shaped area covering the apical two-thirds of the pronotum, an oblique fascia on the basal third of the elytra, and an irregular band at the apical third covering much of the elytra toward, but not including, the apices. This species is the least common of its genus found in Bolivia and rarely encountered other than at lights where, even then, it is uncommon at best. Its distribution is small geographically as only Matto Grosso and Rondônia States in Brazil and Santa Cruz Department in Bolivia (Refugio Los Volcanes, the Flora and Fauna Hotel near Buena Vista, and the Potrerillo del Guenda Reserve) have recorded this species.

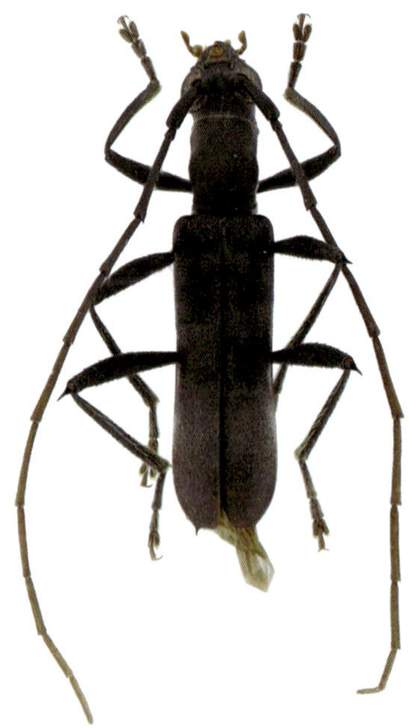

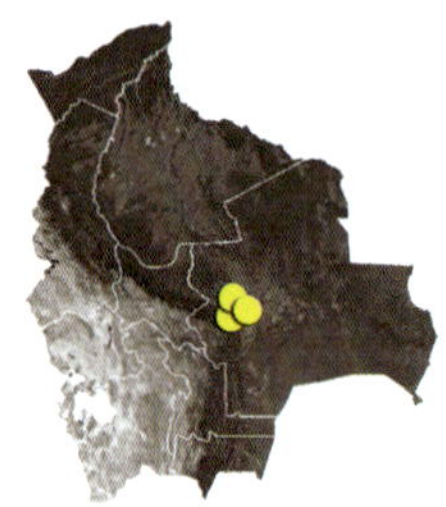

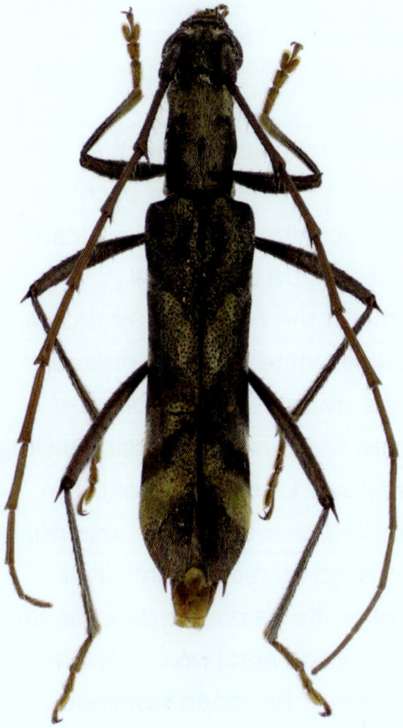

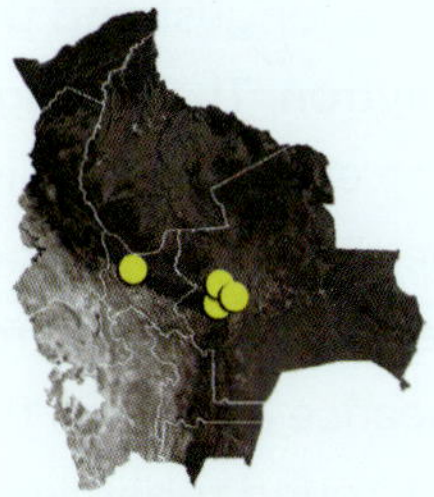

Appula melancholica Gounelle, 1909
Cerambycinae: Elaphidiini

This is a moderate-sized (13–15 mm), narrow, parallel-sided, reddish-brown species with shining gold and silvery pubescent fascia. It has noticeably long, thin legs with an acute spine on the femoral apices of all six legs and the external margin of the elytral apices. The pronotal disk has five low tubercles, and the apical half is densely gold pubescent. This species can be regularly taken at lights or by beating dead leaf clumps hanging in living trees in lowland forests where it can be locally common. It is recorded from east-central Brazil, Paraguay, and Bolivia, where it is has been collected in Cochabamba Department and Santa Cruz Department from the Flora and Fauna Hotel near Buena Vista, the Potrerillo del Guenda Reserve, and Refugio Los Volcanes, adjacent to Amboró National Park.

Appula nigripes Bates, 1870
Cerambycinae: Elaphidiini

This is the largest (16–22 mm) of the five *Appula* species found in Bolivia. It has dark reddish-brown integument covered with gold and silver pubescent bands. The pronotum is densely pubescent with the apical half covered in bright golden vestiture and the disk with two denuded, impunctate, shining tubercles positioned centrally followed by a low, shining elongate ridge. This species is easily distinguished from its congeners by the much shorter, stouter antennal segments with large spines on segments three to six or seven. Less common than *Appula melancholica*, it can still be routinely collected at lights early in the rainy season. It has been recorded from northern Brazil, French Guiana, and Bolivia, where it is known from La Paz and Santa Cruz Departments. It has been collected recently north of Camiri, at the Flora and Fauna Hotel near Buena Vista, and in and around the Potrerillo del Guenda Reserve.

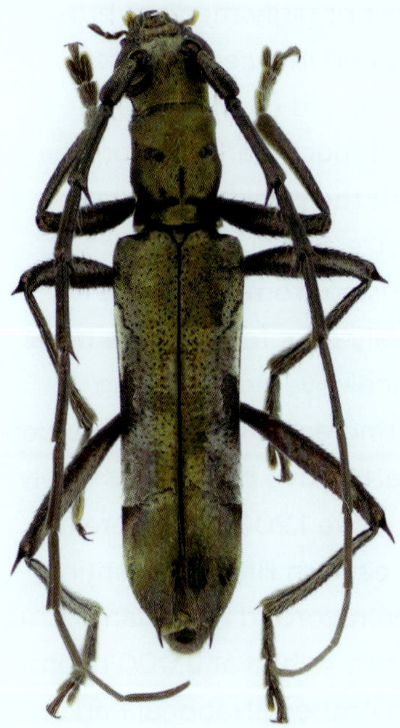

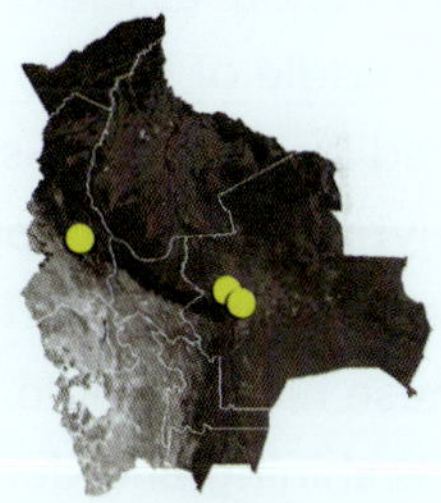

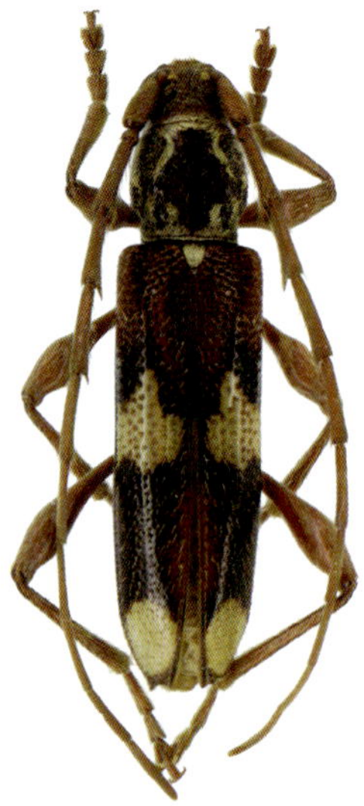

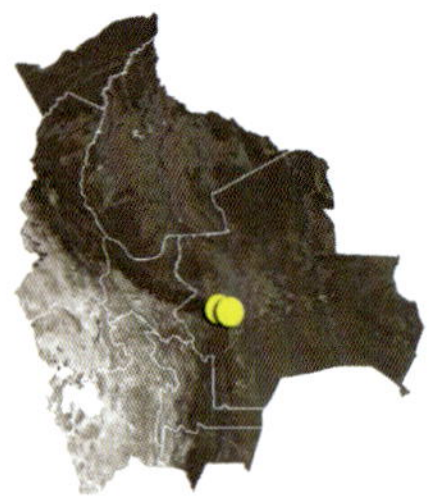

Centrocerum elegans (Chevrolat, 1861)

Cerambycinae: Elaphidiini

This is a small- to medium-sized (8–15 mm), distinctive, multicolored species. The integument is primarily medium brown with light tan spots at the elytral apices and in a central fascia across the middle of the elytra. The apical and medium fasciae are bordered by a variable-sized band of much darker color. The glabrous elytra are mostly punctate with the punctures largest at the base then gradually reducing in size and number apically. Many of these have a single, fine hair arising from them. The pronotum is shining and densely, rugosely punctate with its basal and apical margins and part of its surface covered in shining golden pubescence. The nicely contrasting earth tones and golden pubescence patterns make for a most attractive beetle. It is rarely encountered except at high altitudes (above 1,200 m), where it is occasional at lights. It is known from eastern Brazil, Argentina, Paraguay, and Bolivia, where it has been recorded from Santa Cruz Department from the Vicoquin area above Achira at 2,000 m and Refugio Los Volcanes, both adjacent to Amboró National Park.

Eurysthea nicolai (Aurivillius, 1908)

Cerambycinae: Elaphidiini

This is a large (19–22 mm), shining dark brown species with a tan fascia, roughly chevron shaped, across the middle of each elytron. Members of this genus are characterized by their intricate and variable vestiture. In this species the elytra have small, semirecumbent hairs arising from some of the elytral punctation and longer, scattered, stout, almost spine-like, white hairs intermixed, which are longest and thickest near the double-spined apices of each elytron. These long, stout hairs are scattered, but clearly visible, over the entire dorsum including the appendages. The pronotum has the disk heavily punctate, surrounding an impunctate, shining, elongate tubercle, and moderately covered by recumbent silver pubescence as well as the longer stout white hairs. The disk has four shining, round tubercles on the apical portion and two larger ones basally. This species is occasional at lights. It has been recorded from only Argentina and Bolivia. In Bolivia it is known from the type locality of Mojos in La Paz Department and from many localities in Santa Cruz Department, including Samaipata, the Vicoquin area above Achira at 2,000 m, Refugio Los Volcanes, the vicinity of the Flora and Fauna Hotel near Buena Vista, and Camiri.

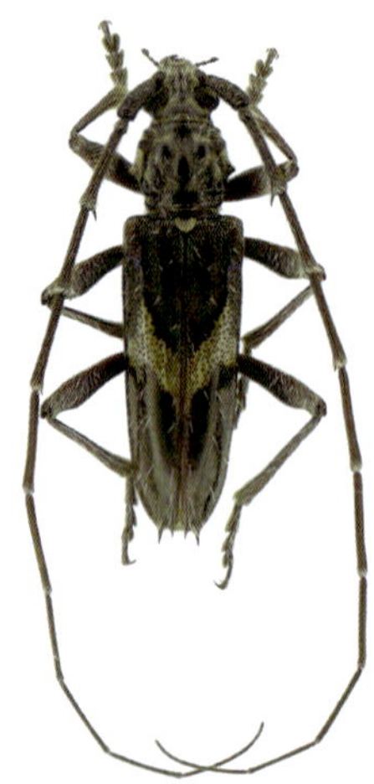

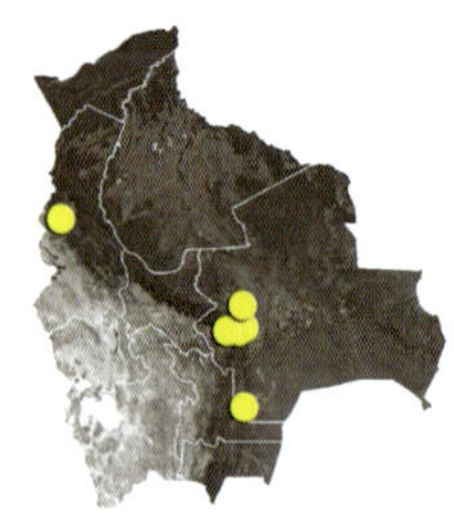

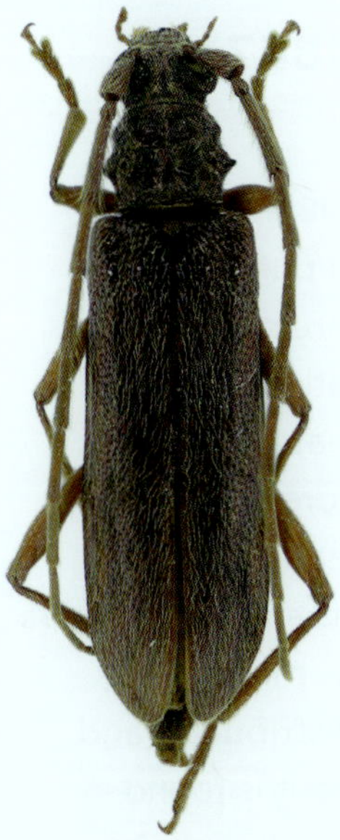

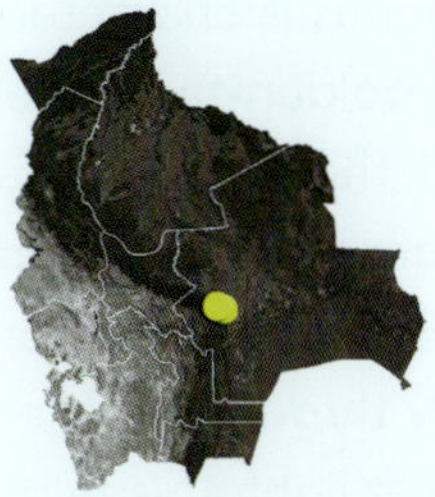

Eurysthea sordida (Erichson, 1847)
Cerambycinae: Elaphidiini

This is a medium-sized (14–16 mm), medium to dark brown species with heavily punctate, shining elytra. In most specimens the legs and antennae are reddish brown, appearing noticeably lighter in color than the elytra. The head and pronotum are typically darker than the elytra and more coarsely punctate except for five glabrous impunctate tubercles on the pronotal disk. Males have antennae that surpass the elytral apices by slightly less than one segment whereas the antennae of females fall short of the apices by less than a segment. Males also have a tendency toward longer, denser vestiture on the elytra and pronotum. This species is typically collected at lights above 1,000 m and is most abundant at elevations of 1,800–2,100 m. It is widely distributed in northern South America, being recorded from Colombia, Ecuador, Peru, Venezuela, and Bolivia, where it is known from only Santa Cruz Department at the Vicoquin area above Achira at 2,000 m and Refugio Los Volcanes, both adjacent to Amboró National Park.

Ironeus pulcher Bates, 1880
Cerambycinae: Elaphidiini

This is a moderate-sized species (13–15 mm), reddish orange to black in color. The pronotum is reddish orange as is a small, basal, central area on the elytral disk. Laterally and apically the elytral integument is darker and much of it obscured by three elongate, whitish-gray, recumbent, pubescent fasciae on each elytron, producing a striped appearance. The head, scape, pronotum, and femora are bright orange red with the tibiae black and remaining antennomeres and tarsi piceous. The distinctive striped elytra make this one of the most recognizable cerambycid species found in Bolivia. *Ironeus pulcher* is common at lights and has a wide distribution from Mexico to Panama and Venezuela to Bolivia, where it is a lowland forest species in Santa Cruz Department. It has been collected there at the Flora and Fauna Hotel near Buena Vista and the Potrerillo del Guenda Reserve.

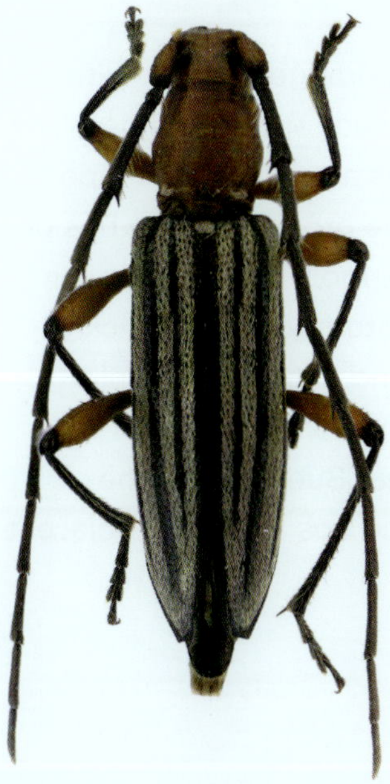

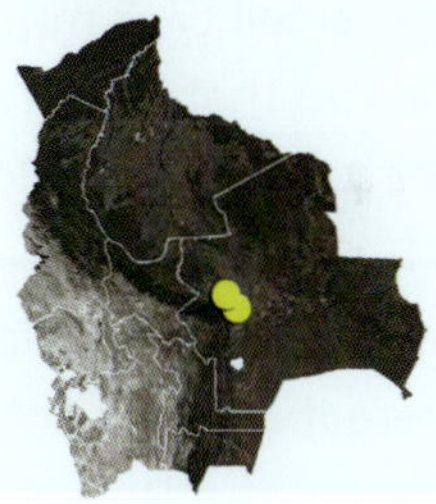

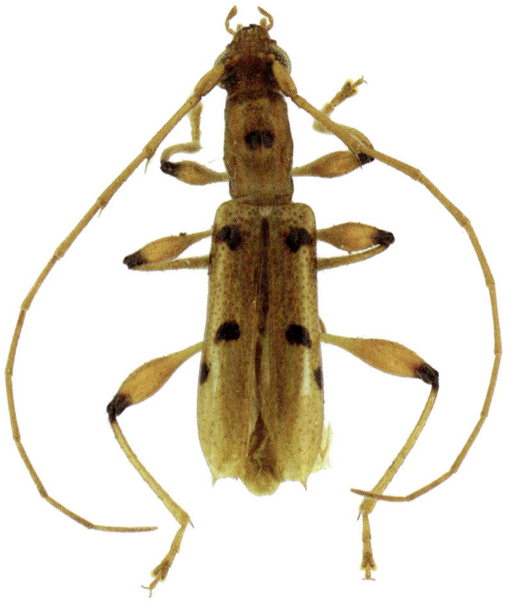

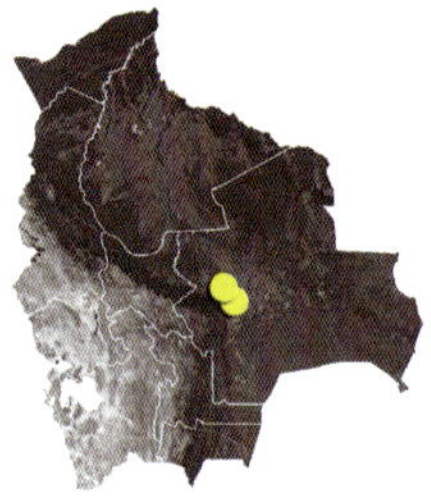

Iuaca nigromaculata Galileo & Martins, 2000
Cerambycinae: Elaphidiini

This species is small (7–8 mm) and bright orange in color, with translucent elytra that have two pairs of black spots, one basally and the other across the middle, as well as a moustache-shaped black fascia fronting the discal ridge of the pronotum. The tips of the femora are similarly shining black. It is most commonly seen at lights, where it remains active and if disturbed drops quickly and disappears. It is a Bolivian endemic and known from only the lowland forests of Santa Cruz Department, where it has been collected at the Flora and Fauna Hotel near Buena Vista, the Potrerillo del Guenda Reserve, and Refugio Los Volcanes, adjacent to Amboró National Park.

Lanephus njumanii (Haldeman, 1847)
Cerambycinae: Elaphidiini

This is a strikingly patterned, medium-sized (16–18 mm), stout cerambycid with reddish-brown integument and pubescent maculae. Although exposed areas of the elytra are shining, this monotypic genus is characterized by the alternating bands of white and golden-tan pubescence that dominate its dorsal surface. In all, there are five, rather distinctly colored, alternating bands of pubescence on the dorsum. The appendages are uniformly covered with golden-tan vestiture. This species appears intermittently throughout the rainy season and is most often encountered at light. It is fairly restricted in its distribution and known from only central Brazil and Bolivia, where it is found in the lowland forests of Santa Cruz Department. It is known from Bermejo, Refugio Los Volcanes, adjacent to Amboró National Park, the vicinity of the Flora and Fauna Hotel near Buena Vista, and the Potrerillo del Guenda Reserve.

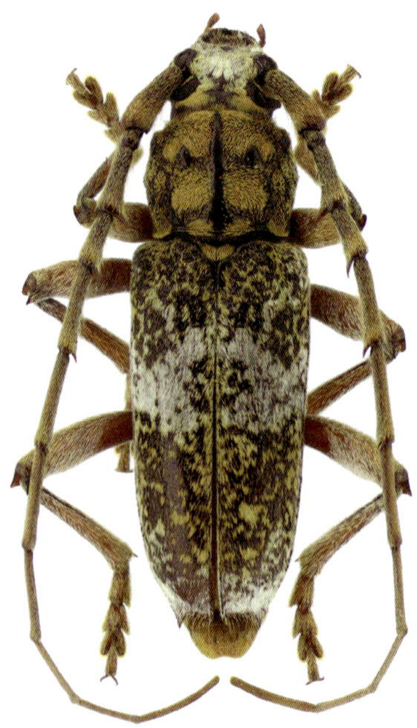

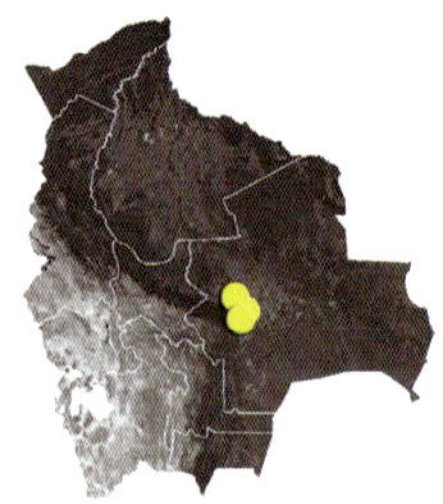

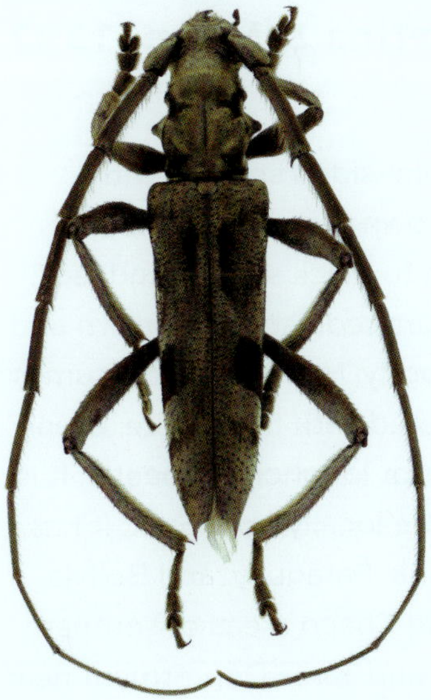

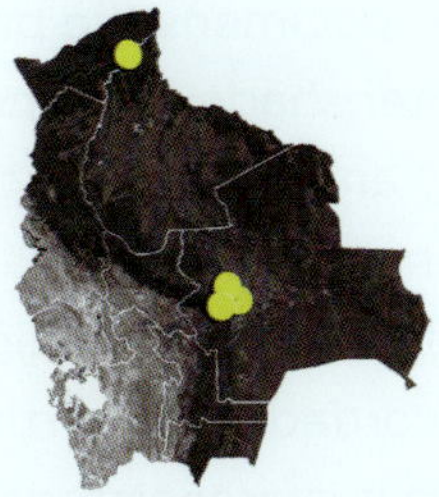

Mallocera amazonica Bates, 1870

Cerambycinae: Elaphidiini

This is a large (16–24 mm), reddish-brown, parallel-sided species with shining pubescent bands on the elytra. It is the largest of the four *Mallocera* species recorded from Bolivia. All are rather similar in appearance, having the same general form and shining pubescent bands. The elytra are minutely punctate with each puncture bearing a single backward-directed hair. Its dull surface is accented by shining silvery-white, pubescent fasciae, one basally, an oblique one at the basal third, another behind the middle, and a fourth in front of the apices and laterally including them. This species is rarely encountered except at lights. It is a widely distributed species known from central Brazil, Ecuador, Peru, Trinidad, French Guiana, and Bolivia. In Bolivia it is known from Pando Department, the Riberalta area of Beni Department, and from Santa Cruz Department at the specific localities of the Vicoquin area above Achira at 2,000 m, the vicinity of the Flora and Fauna Hotel near Buena Vista, and the Potrerillo del Guenda Reserve.

Mephritus blandus (Newman, 1841)

Cerambycinae: Elaphidiini

This is a medium-sized (17–20 mm), orange species with distinctive elytral maculae in the form of a pair of large, oval, ivory-colored spots covering much of the basal half. They are completely encircled by a narrow black ring that makes them appear as large eyespots. The rest of the dorsum, with the exception of the piceous to black antennomeres (not including the scape or pedicel), tibiae, and tarsi is a medium orange with a matte finish. Although this species is widely distributed it is rarely encountered, typically showing up singularly at lights. It is known from the Guianas, Brazil, Paraguay, and Bolivia, where it is recorded from La Paz Department and Santa Cruz Department, where it has been collected at the Flora and Fauna Hotel near Buena Vista and the Potrerillo del Guenda Reserve.

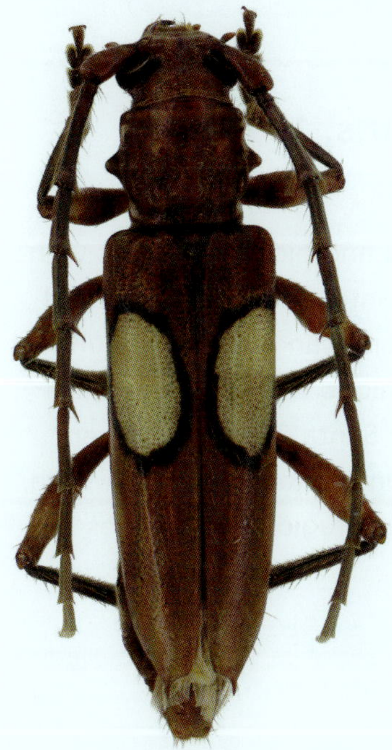

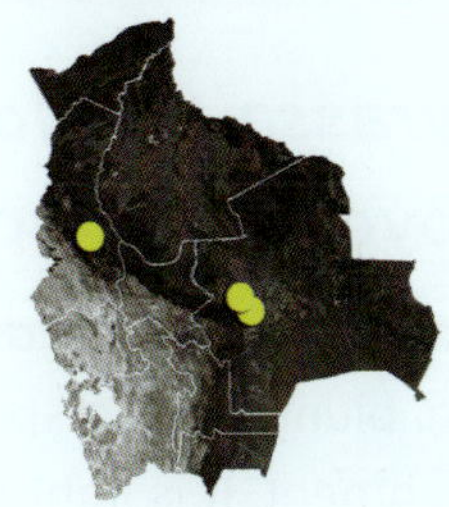

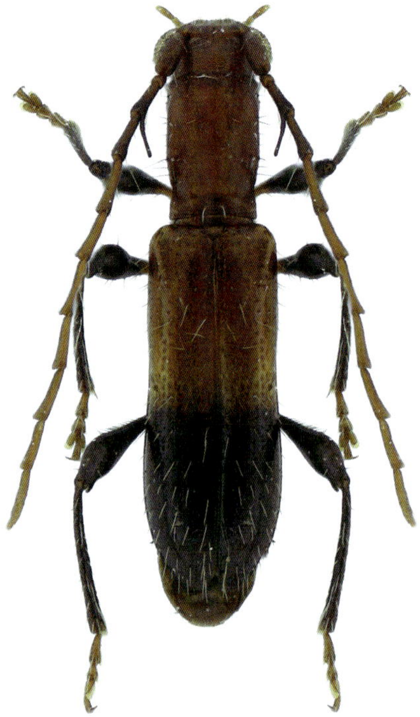

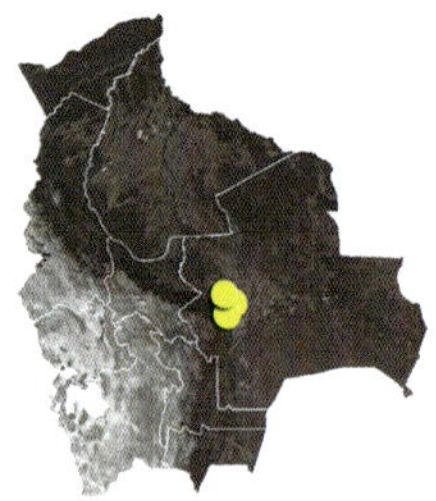

Minipsyrassa bicolor Martins, 1974

Cerambycinae: Elaphidiini

This is a tiny (5–8 mm), reddish-orange and piceous species with a very long, blunt, curved spine on the internal apex of the third antennal segment. It is light attracted and can be locally abundant early in the rainy season. It is an endemic to Bolivia, where it is known from La Paz Department and in Santa Cruz Department from the localities of the Flora and Fauna Hotel near Buena Vista, the Potrerillo del Guenda Reserve, and Refugio Los Volcanes, adjacent to Amboró National Park.

Morphaneflus prolixus Martins & Napp, 1992

Cerambycinae: Elaphidiini

This is a small (11–12 mm), narrow, parallel-sided species with reddish to dark brown elytra, lighter brown tarsi and most antennomeres, and shining dark brown to black scapes and legs. In some specimens the pronotum is more reddish than brown and of a lighter shade than the rest of the body. Most of the dorsum is minutely and evenly punctate and covered with fine, white, evenly spaced hairs. Overall, this is a nondescript longhorned beetle. It is most often collected at lights and can be locally abundant. It has been recorded from northwest Argentina, Paraguay, and Bolivia. In Bolivia it is associated with the humid Chaco areas of southern Santa Cruz Department around Camiri and Tarija Department near Villamontes.

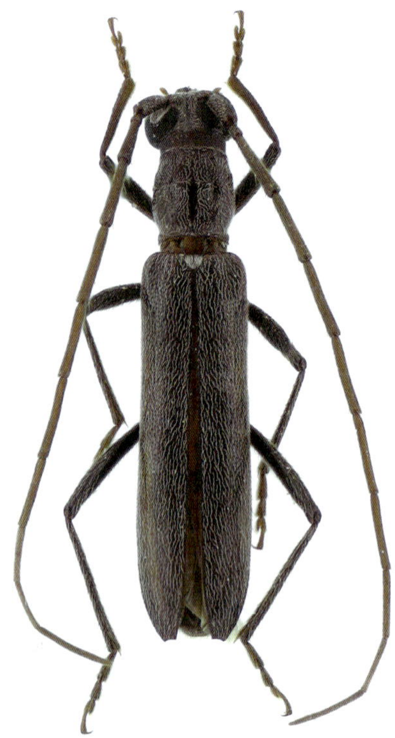

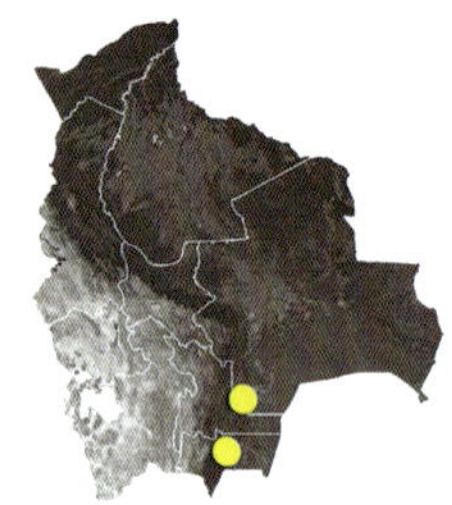

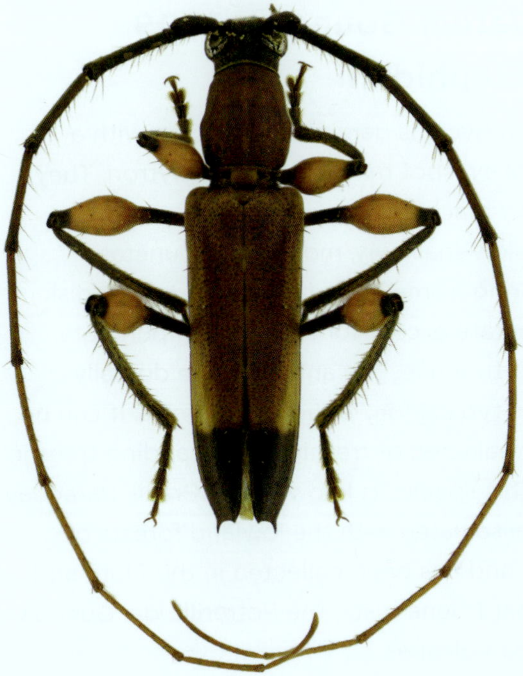

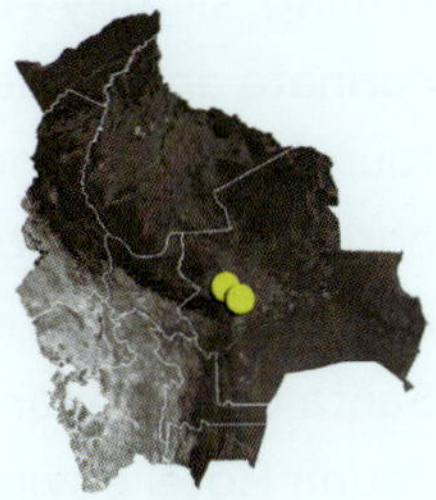

Nyssicostylus paraba Martins, 2005

Cerambycinae: Elaphidiini

This is a small (10–12 mm), orange and black species with long legs and antennae. Although primarily of just two colors (orange and black), it provides a complex and beautiful presentation of them. The head, basal antennal segments, tibia, femoral tips, bases of the mid and hind femora, and the apical fourth of the elytra are black. The rest of the dorsum and the femoral clubs are orange. The remaining antennomeres transition from the prior black segments to piceous, to tan, and then to light tan distally. This Bolivian endemic species is known from Santa Cruz Department from the vicinity of the Flora and Fauna Hotel near Buena Vista and the Potrerillo del Guenda Reserve, where it is most often encountered at lights.

Pantonyssus bitinctus Gounelle, 1909

Cerambycinae: Elaphidiini

This is another small (9–11 mm), orange and black cerambycid that strongly resembles *Nyssicostylus paraba* Martins in form and color pattern when compared with the naked eye. However, when examined under the microscope or with the aid of a good hand lens the differences become readily apparent. Quite simply, this species is brightly shining (*N. paraba* is a dull luster) and has long, stout black spines on the elytral apices, on the middle and hind femoral apices, and internally on the third to fifth antennal segment whereas *N. paraba* has short spines on the elytral apices, none on the femora, and only minuscule spines internally on the antennal segments. This species can be very common at lights in forest environments and is also encountered occasionally by beating agricultural chaco or forest edges. It is known from Brazil and Bolivia, where it is recorded in Santa Cruz Department from the vicinity of the Flora and Fauna Hotel near Buena Vista, the Potrerillo del Guenda Reserve, and Refugio Los Volcanes, adjacent to Amboró National Park.

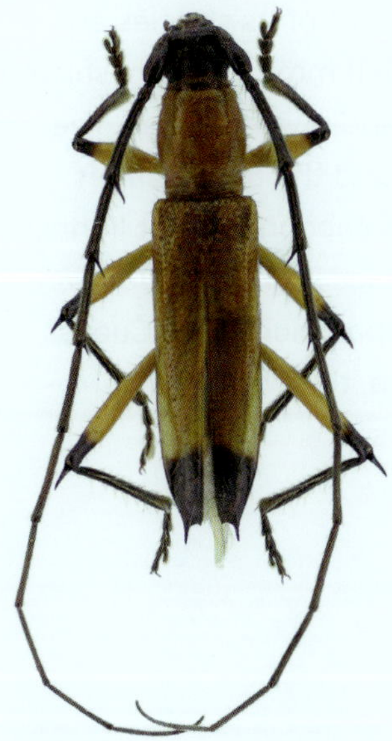

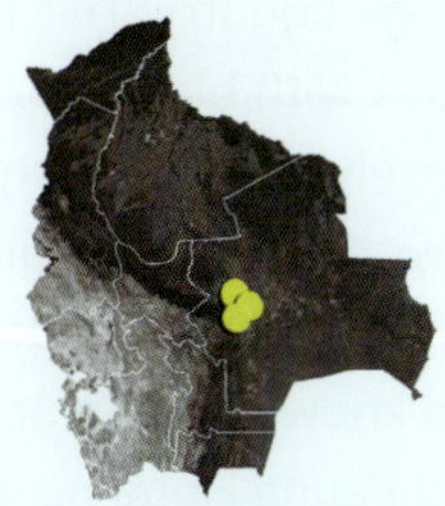

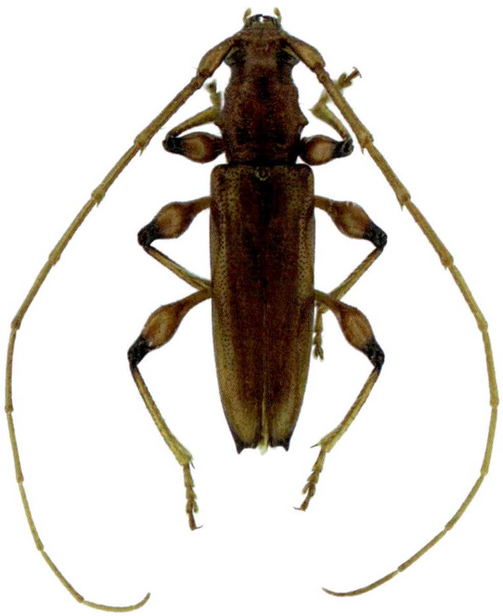

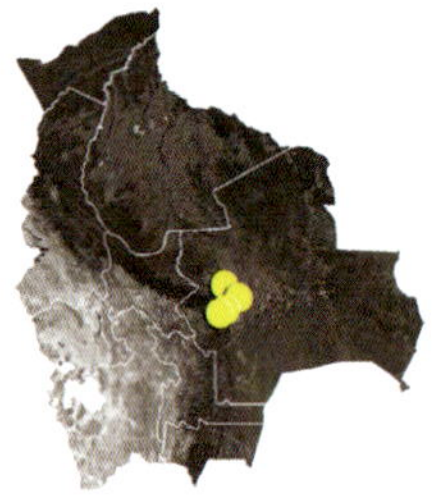

Periboeum bolivianum Martins & Monné, 1975
Cerambycinae: Elaphidiini

Periboeum is a common genus in Bolivia as 10 of its species are known from the country. This is a small (10–11 mm), strongly shining orange species with black-tipped femora and elytral apices. The elytra are closely, but shallowly, punctate, and the appendages are clothed with long, silky hairs. It is often encountered at lights, where it can be locally abundant. This species is another Bolivian endemic and known only in Santa Cruz Department near Cuevas, the Flora and Fauna Hotel near Buena Vista, the Potrerillo del Guenda Reserve, and Refugio Los Volcanes.

Periboeum ocellatum Gounelle, 1909
Cerambycinae: Elaphidiini

This 16–18 mm long cerambycid is dark reddish brown with a distinctive, dark tan, oval eyespot macula on each elytron. They are encircled by a narrow black ring making them even more apparent. The elytral disk is shallowly, moderately punctate with each puncture bearing a small, fine hair. The pronotal disk is densely, rugosely punctate except for a central, impunctate, shining, slightly elevated tubercle. The antennae are dorsally carinate and spined as is typical for *Periboeum* species. It can be found at lights and on weakened or freshly dead standing trees in agricultural *chaqueos*. This species is known from Brazil, Paraguay, and Bolivia, where it is associated with the lowland forests of Santa Cruz Department and has been collected in the Flora and Fauna Hotel environs near Buena Vista, the Potrerillo del Guenda Reserve, and Refugio Los Volcanes.

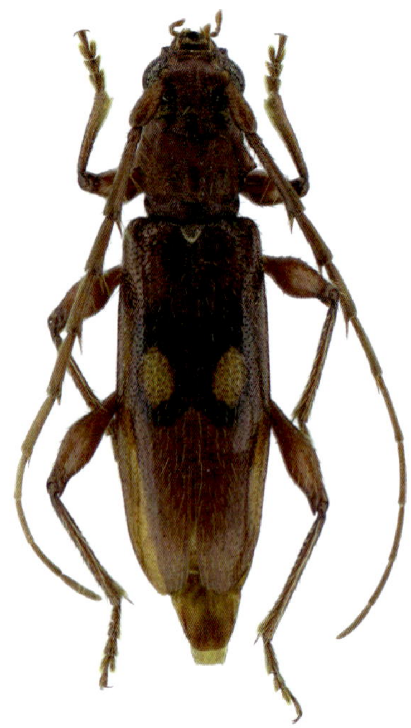

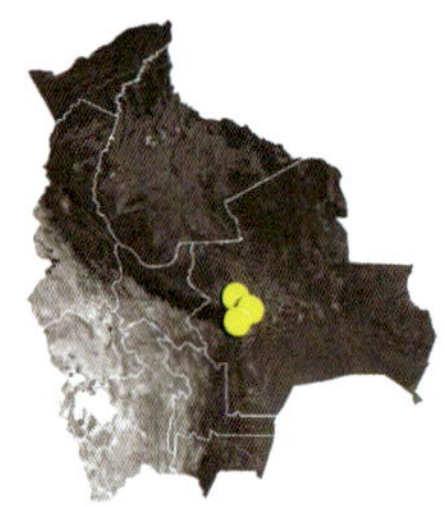

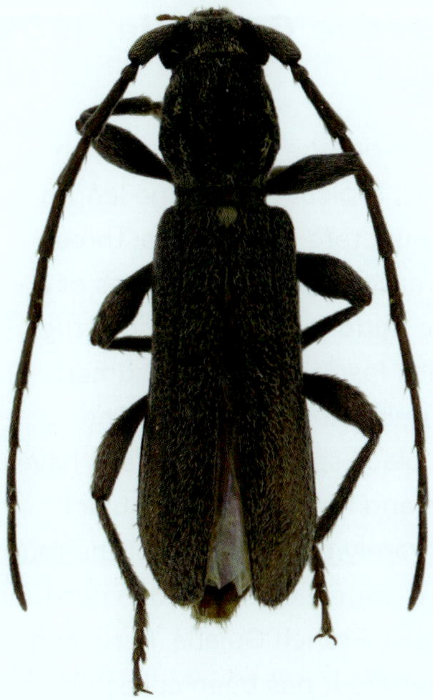

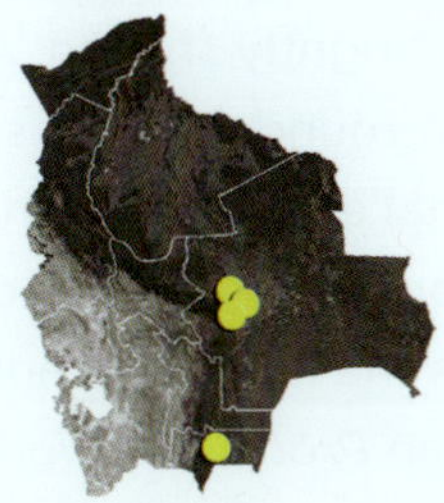

Piezophidion intricatum Galileo & Martins, 1992

Cerambycinae: Elaphidiini

This is a small (8–10 mm), dark brown, rather nondescript species with a densely punctate dorsum irregularly covered with short, fine, silver-gray, recumbent pubescence that partially obscure the surface. A few longer, golden hairs are also evident on the legs and apical third of the elytra. The most noteworthy attribute of this species may be its inflated pronotum being widest at middle, to near the apical margin, and strongly restricted at the base. Although rarely found, when it is, it can be abundant. It has been recorded from Brazil, Paraguay, and Bolivia, where it is known from Santa Cruz Department at the Flora and Fauna Hotel near Buena Vista, the Potrerillo del Guenda Reserve, Refugio Los Volcanes, adjacent to Amboró National Park, and from Tarija Department near Villamontes.

Protomallocera hilairei (Gounelle, 1909)

Cerambycinae: Elaphidiini

This is a moderate-sized (13–15 mm), brown species with a densely pubescent pronotum and shining pubescent fasciae on the elytra. This species strongly resembles the *Mallocera* species found in Bolivia but can be distinguished by the lack of lateral pronotal tubercles. The elytra in this species are reddish brown, and its shining pubescent bands are not distinct or well defined. Beginning with the third antennomere, the segments are dorsally carinate and lightly fringed internally with long, silky hairs. It is fairly common at lights early in the rainy season and is known from Brazil, Paraguay, Suriname, and Bolivia, where it has been collected in the vicinity of the Flora and Fauna Hotel near Buena Vista, the Potrerillo del Guenda Reserve, and Refugio Los Volcanes in Santa Cruz Department and Villamontes in Tarija Department.

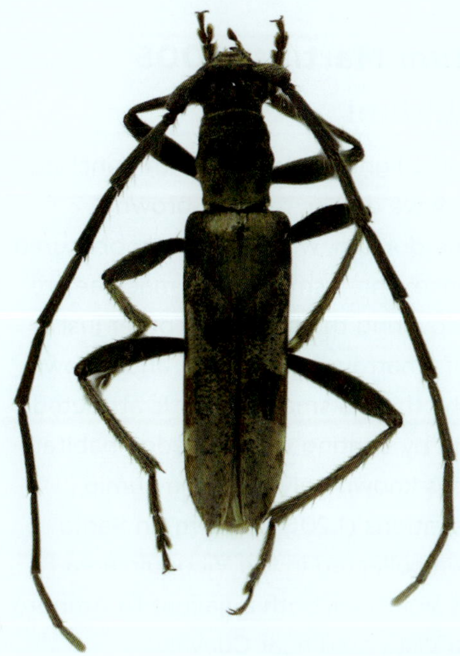

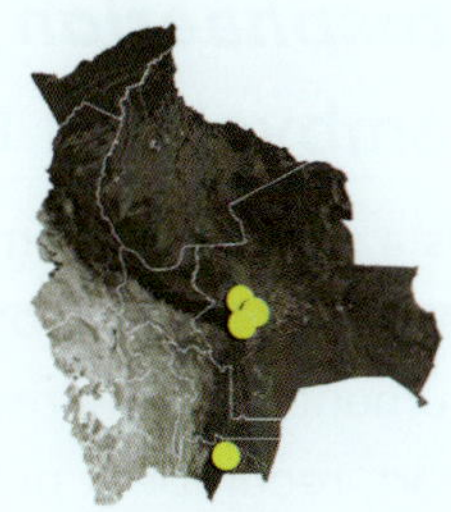

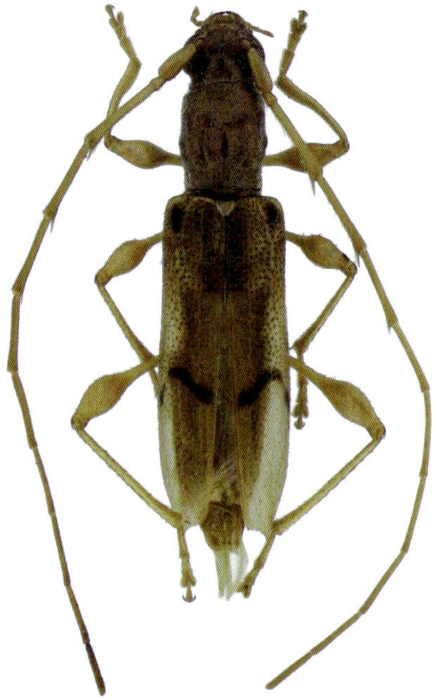

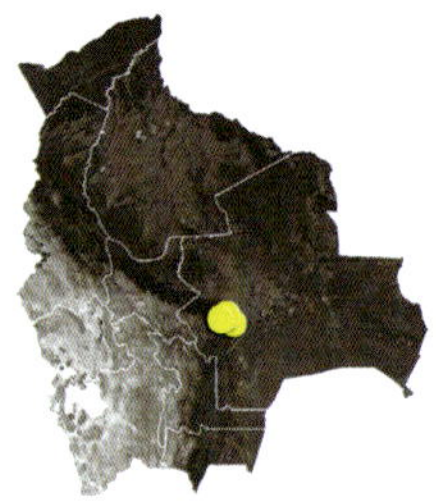

Protosphaerion pictum Martins, 2005
Cerambycinae: Elaphidiini

This is a small (11–12 mm), shining light tan species with punctate elytra. Its head and pronotum have darker, reddish-brown integument than the rest of the dorsum, which is mostly obscured by short, recumbent pubescence. Three small, black maculae are located on each elytron, one covering the humeri, another just past the middle, in the form of a narrow oblique line, and below it, near the lateral margin, is the third, a small spot. It is attracted to lights and can also be found by beating wooded edge habitat. This Bolivian endemic species is known only from the humid forests at moderately high elevations (1,200–2,100 m) in Santa Cruz Department and has been collected in the Vicoquin area above Achira and Refugio Los Volcanes, both adjacent to Amboró National Park, and along Bella Vista road near Cuevas.

Protosphaerion signatipenne Gounelle, 1909
Cerambycinae: Elaphidiini

This is a large (18–20 mm), distinctive orange-tan species with the males having very long antennae, typically twice the length of the body. The elytra are heavily punctate on the basal three-fifths; the punctures markedly decrease within the confines of a black M-shaped fascia, just past the middle, with the rest of the elytra brightly shining. Another black fascia covers the humeri and surrounds the densely golden pubescent scutellum. The external elytral apices have a large black spine, and the legs have the connecting tip of both the tibia and femora similarly black. This is the least common and most rarely encountered of the four Bolivian *Protosphaerion* species. It is widely distributed in northern South America and known from Brazil, French Guiana, Peru, and Santa Cruz Department in Bolivia, where it has been collected at Flora and Fauna Hotel near Buena Vista.

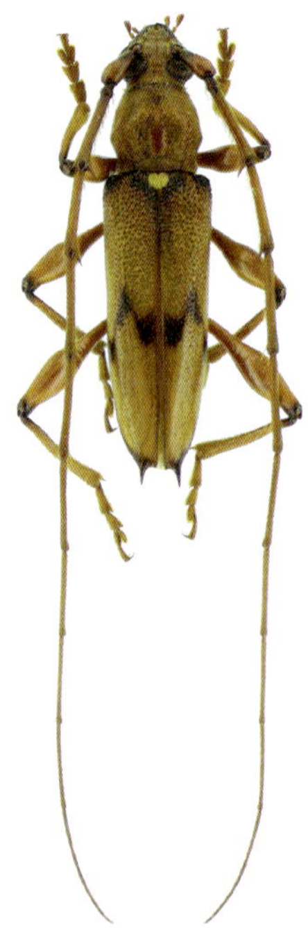

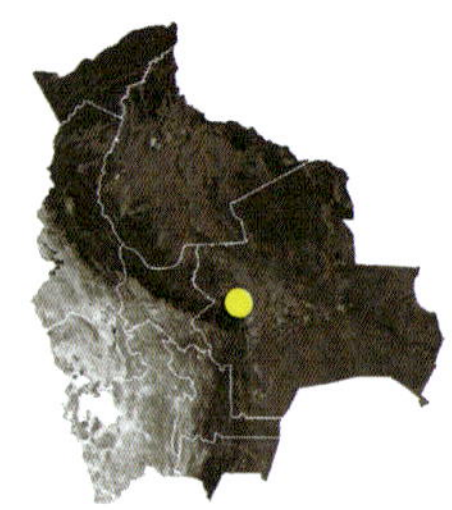

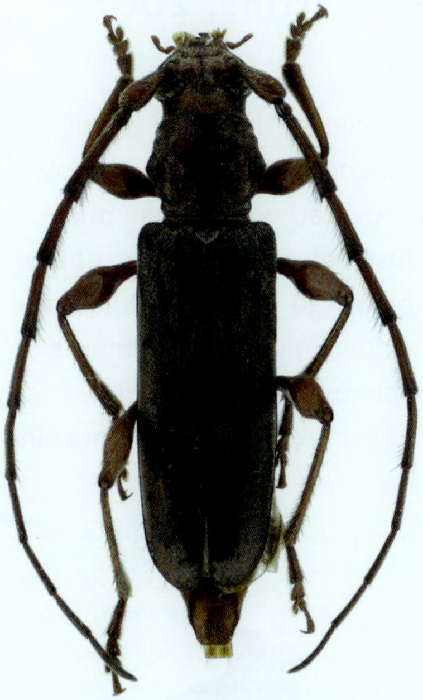

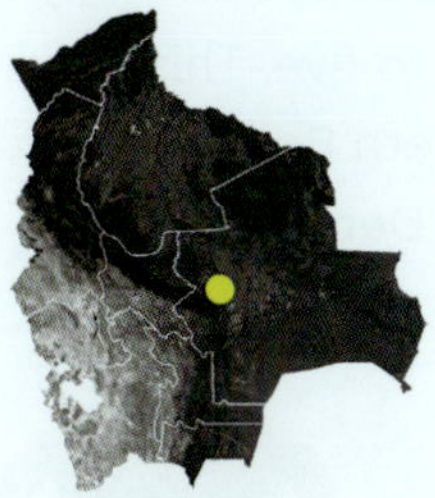

Sphaerion cyanipenne Audinet-Serville, 1834

Cerambycinae: Elaphidiini

This is one of the smaller members (10–12 mm) of this genus. It is reddish orange and clothed with short, dense, recumbent pubescence, resulting in a velvety appearance. The elytra are closely punctate at the base but much less so apically, where a few widely scattered seta-bearing punctures are present. Like the elytra, the pronotum is densely pubescent, more so laterally, and closely punctate on the disk except the impunctate, shining midline and to a lesser extent a pair of separated low tubercles fronting it. This species is uncommon and only rarely encountered at lights. It is known from Brazil, Paraguay, and Bolivia, where it has been recorded from the Flora and Fauna Hotel near Buena Vista in Santa Cruz Department.

Sphaerion exutum (Newman, 1841)

Cerambycinae: Elaphidiini

This is a moderate-sized (16–18 mm), dark reddish-orange species with piceous to black antennomeres and tibiae. Much of its dorsum is closely, finely punctate and clothed with very short, fine, recumbent pubescence, giving it a velvety appearance. This species, like its congener *S. sladeni* Gahan, lacks contrasting fasciae or maculae but is still easily recognizable by its form and coloration. The pronotum differs between the sexes, with the males having a rounded, more compressed form, being less tuberculate on the disk, and lacking the lateral tubercles that are present in the females. It is one of if not the most common and abundant species at lights—seemingly ever present. This species has been recorded from the southern United States through northern Argentina and Uruguay. In Bolivia it has been recorded from almost every locality sampled in Santa Cruz Department. It is also known from Beni and Tarija Departments near Riberalta and Villamontes, respectively.

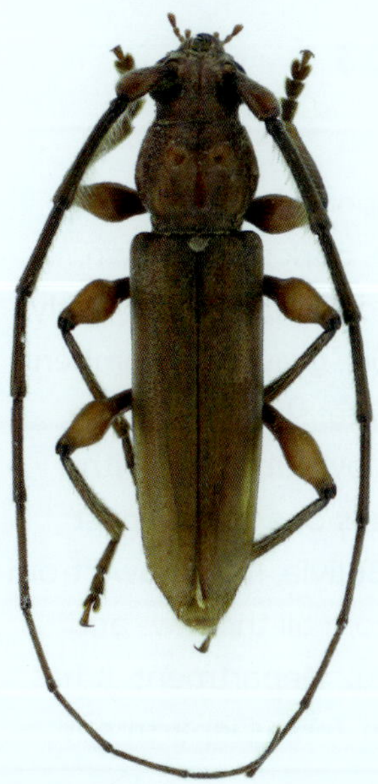

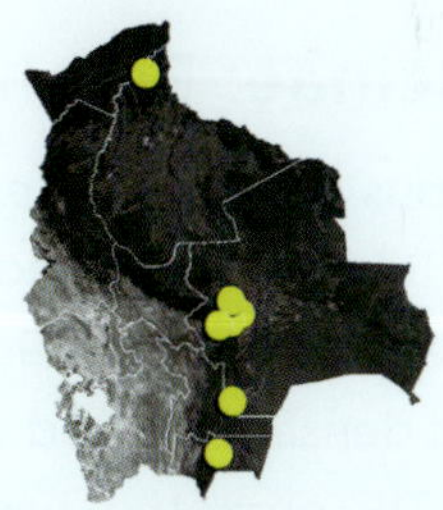

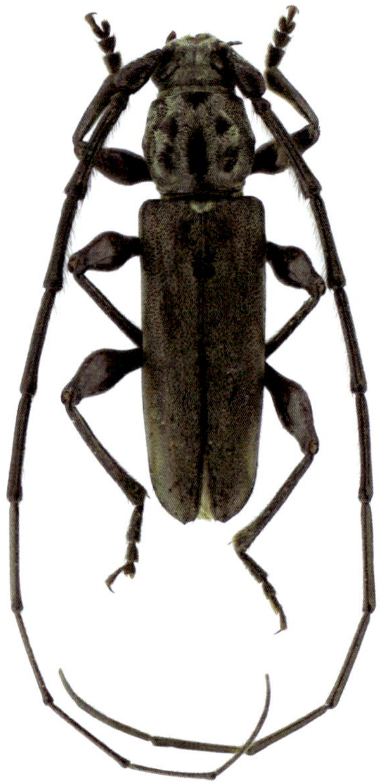

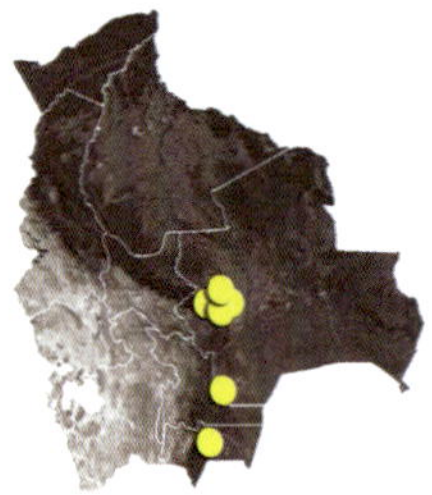

Sphaerion sladeni Gahan, 1903

Cerambycinae: Elaphidiini

Like its previous congener, *S. exutum*, this species is rather nondescript. It is medium sized (16–20 mm), dark slate brown, and completely lacking contrasting fasciae or maculae. It is finely punctate and densely clothed with fine, short, grayish, recumbent pubescence. The pronotum has an impunctate, shining midline on the disk, and in the males there are four low tubercles with elevated, weakly shining apices as well. This is one of the most abundant species encountered at lights in Bolivia. It is known from Brazil and Bolivia. In Bolivia it is recorded from all the low- and midelevation localities surveyed in Santa Cruz Department. It has also been recorded from near Villamontes in Tarija Department.

Stizocera asyka Galileo & Martins, 2004

Cerambycinae: Elaphidiini

This is one of the smaller and rarer species of the large genus *Stizocera*, with specimens 8–12 mm in length. The body is shiny, lustrous, and mostly dark bluish and covered with sparse but long and conspicuous white setae. Just before the middle of the elytra are angulate, transverse pale brown to yellow maculae. The antennae are strongly spined on the inner margin of segments three to five. This species is known only from Paraguay and southern Bolivia, where it has been collected near Villamontes in Tarija Department.

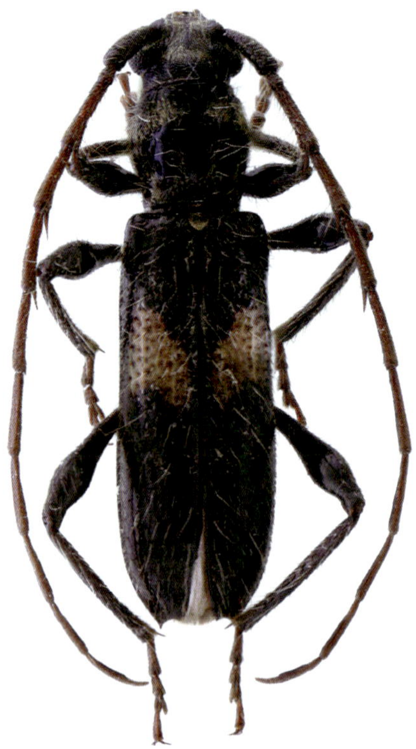

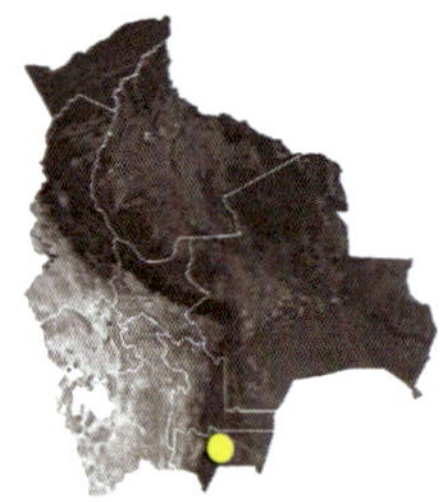

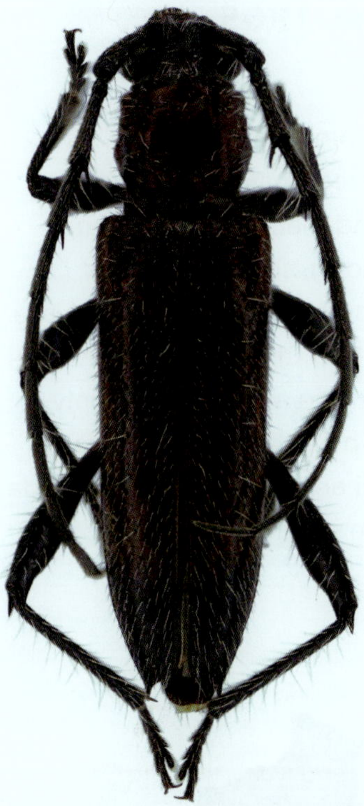

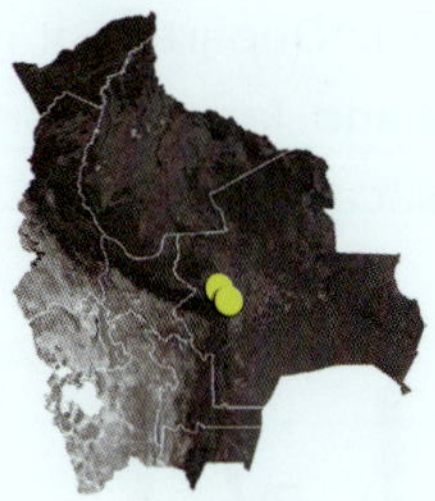

Stizocera ichilo Lingafelter, 2004

Cerambycinae: Elaphidiini

This is a small (8–10 mm), shining red species with the head, apical fourth of the elytra, and all appendages black. The elytra and pronotum are moderately punctate with most of the punctures bearing short, fine, semirecumbent white hairs. Intermixed on the elytra are much longer, widely spaced, silky white hairs. These longer hairs are also present, but much denser and closer together, over the rest of the dorsum and the appendages. This endemic Bolivian species is known from only the lowland forests of Santa Cruz Department from the vicinity of the Flora and Fauna Hotel near Buena Vista and the Potrerillo del Guenda Reserve and is rarely encountered.

Stizocera lissonota (Bates, 1870)

Cerambycinae: Elaphidiini

This is a small- to medium-sized (13–14 mm), somewhat nondescript species. It is somewhat similar in color and shape to *Stizocera ichilo*; however, it is larger with a longer and narrower body. The black integument is restricted to the antennomeres beyond the scape, tibiae, the femora (except for the clavate areas), and the apical spine of each elytron. This species is the most common and widespread of the genus and is known from Panama, French Guiana, Brazil, and Bolivia, where it is recorded from Riberalta in Beni Department and in Santa Cruz Department from Cuevas, Refugio Los Volcanes, adjacent to Amboró National Park, the vicinity of the Flora and Fauna Hotel near Buena Vista, and the Potrerillo del Guenda Reserve.

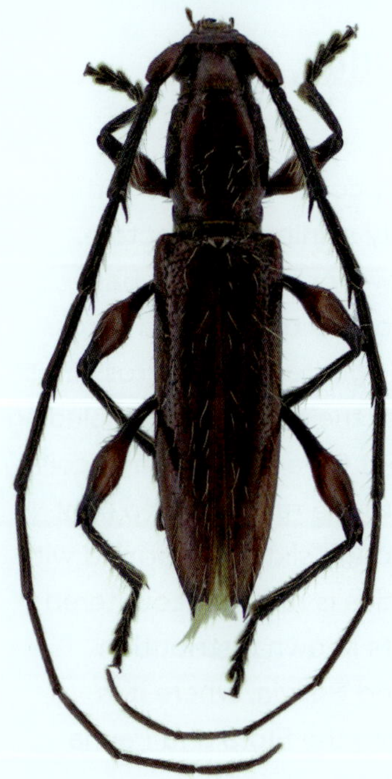

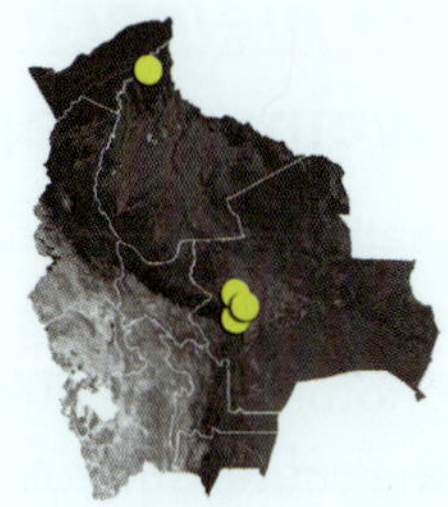

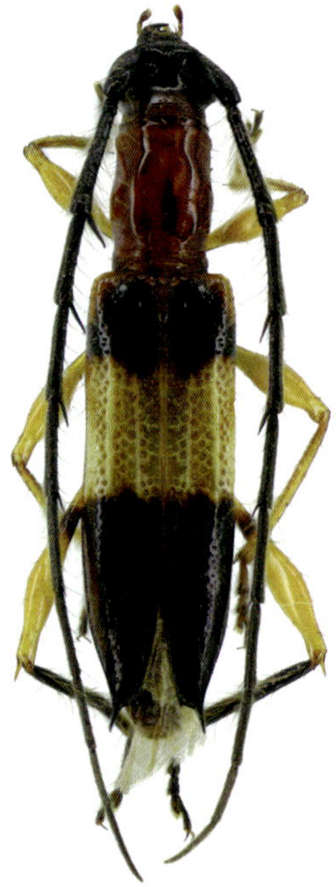

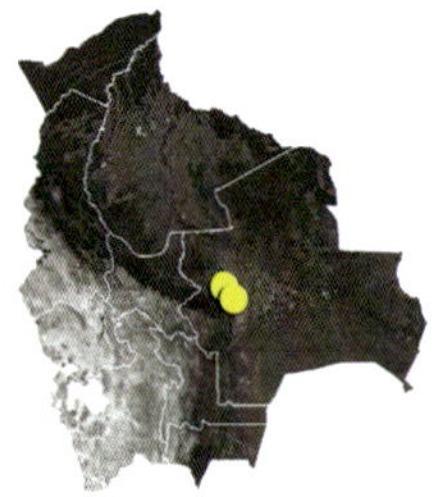

Stizocera meinerti (Aurivillius, 1900)

Cerambycinae: Elaphidiini

This is a small (10–12 mm), handsome, tricolored species with a striate-punctate elytral disk and brightly shining, impunctate, orange pronotum. The elytra are narrowly orange at the base, including the scutellum, followed by a broad black fascia across the basal fourth, then apically a broad, light tan band across most of the central half with the remainder of the elytra black, including the large apical spines. The head (except for a basal orange collar), antennae, and hind tibiae are black, with the rest of legs a light tan. The basal antennomeres (third through sixth) are armed with a stout, slightly curved spine. This species is rarely encountered and has been collected only at lights. Its known distribution includes Brazil, Venezuela, Paraguay, and Bolivia, where it is recorded in Santa Cruz Department from the Flora and Fauna Hotel near Buena Vista and Potrerillo del Guenda Reserve.

Stizocera nigroflava Zajciw, 1965

Cerambycinae: Elaphidiini

This is a small- to medium-sized (9–14 mm), dark reddish-black, shining cerambycid with moderately punctate elytra. The appendages are alternating light tan and black and, as is most of the dorsum, moderately covered in long, silky flying hairs, resulting in a most interesting and attractive appearance. It is a locally abundant species at lights in humid Chaco forests. This species has a very small area of distribution and is known only from northern Argentina and Bolivia, where it is recorded from Santa Cruz Department north of Camiri and Potrerillo del Guenda Reserve and Tarija Department near the southwest border of Chuquisaca Department.

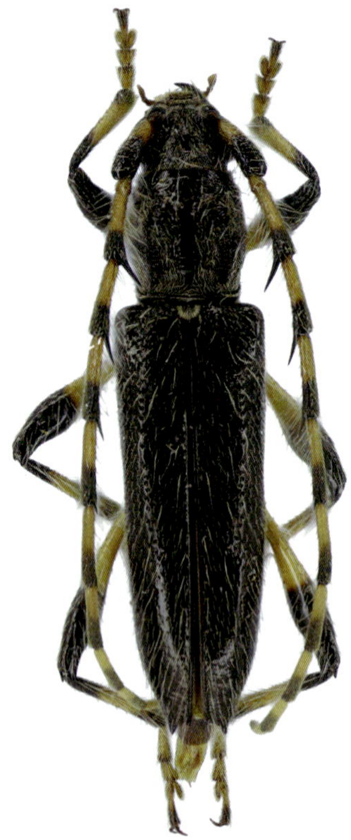

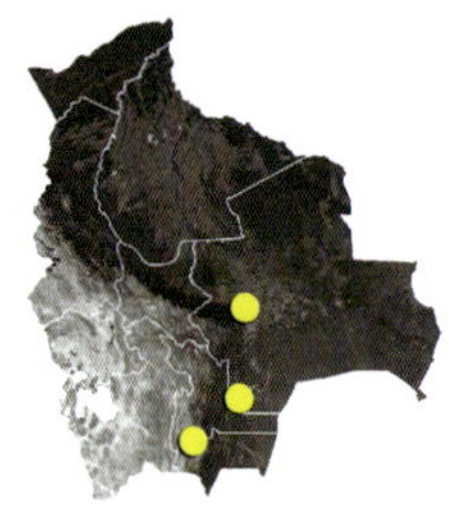

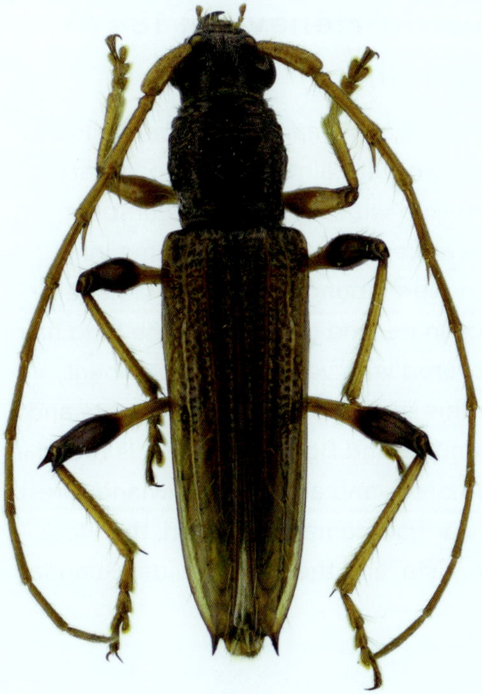

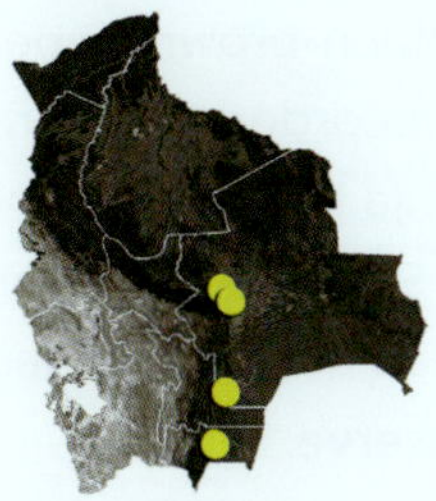

Stizocera plicicollis (Germar, 1824)

Cerambycinae: Elaphidiini

This medium-sized (16–20 mm) species is tan and dark red with coarsely punctate elytra basally that become impunctate and brightly shining distally to the apices. The pronotum is brightly shining dark red and has a moderately, transversely rugose disk. The appendages are a light tan with the apical third of the femora a darker reddish brown. The hind femoral apices are distinctive, being strongly armed with stout, double spines. This is a common species at light and widely distributed from Mexico to central South America, including Argentina, Uruguay, and Bolivia, where it is recorded from virtually all the lowland localities surveyed in Santa Cruz and Tarija Departments.

Stizocera poeyi (Guérin-Méneville, 1838)

Cerambycinae: Elaphidiini

This is a medium-sized (13–15 mm), reddish-orange species with an elongated, oval, tan macula on each elytron. The macula is positioned centrally on the disk and encircled by a narrow black ring. The tibiae and antennomeres (except the scape) are piceous to black with long, silky hairs scattered on them and to a lesser extent across the dorsum. This widespread species is never common but encountered occasionally at lights. It was described from Cuba and is known also from Panama, northern South America (Colombia, Venezuela, and French Guiana), Brazil, and Bolivia. In Bolivia it has been collected in Santa Cruz Department from the vicinity of the Flora and Fauna Hotel near Buena Vista, the Potrerillo del Guenda Reserve, and Refugio Los Volcanes, adjacent to Amboró National Park.

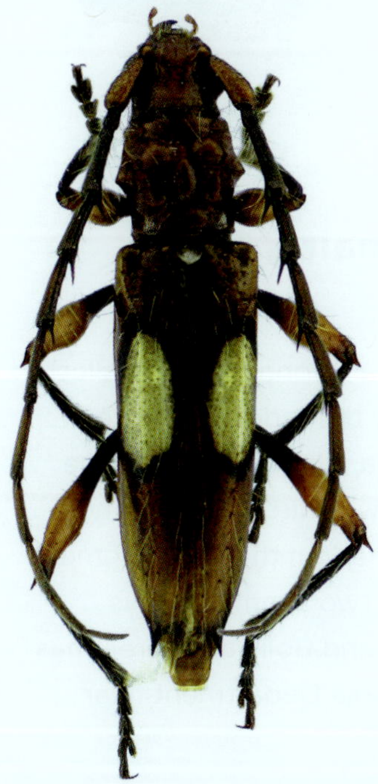

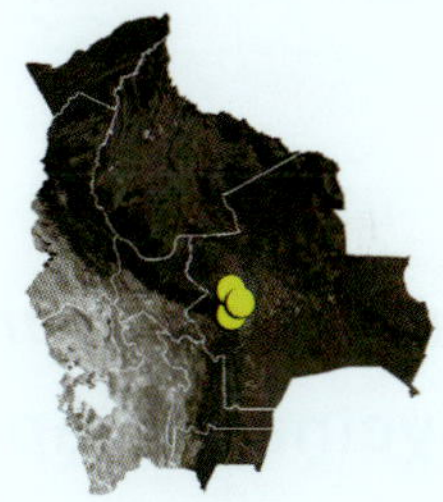

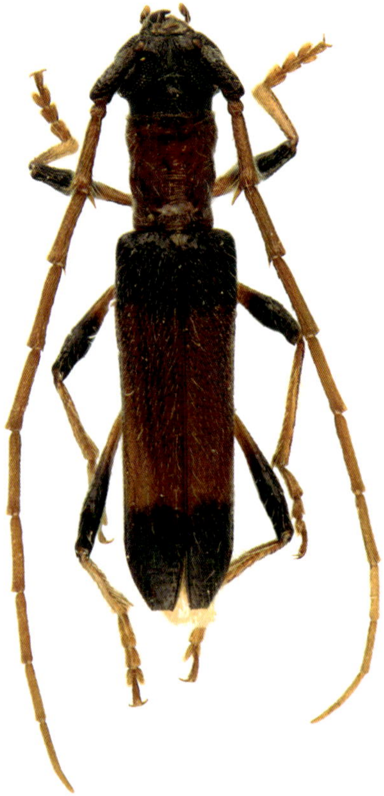

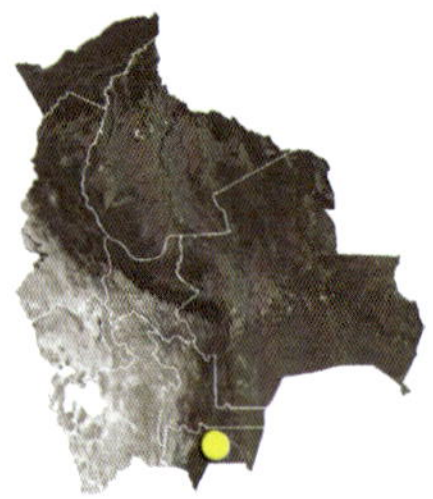

Stizocera spinicornis (Fairmaire, 1864)
Cerambycinae: Elaphidiini

This very distinctive *Stizocera* species is also one of the smallest (8–10 mm) and rarest. It is recognized by the reddish-orange pronotum, middle two-thirds of the elytra, tibiae, and tarsi. The head, elytral base and apex, and apical two-thirds of the femora are black. The antennae are strongly spined on the third segment but weakly so on the subsequent one to two segments. It is known from Argentina, Brazil, Paraguay, and Bolivia, where it has been collected only once, at a light, in Tarija Department near Villamontes.

Stizocera tristis (Guérin-Méneville, 1844)
Cerambycinae: Elaphidiini

This is one of the most pubescent and smallest *Stizocera* species. It is 8–12 mm, dark reddish brown, with a moderately punctate dorsum, and long hairs intermixed with short, semirecumbent, white hairs on the elytra. The appendages have the same long, silky hairs, but they are much more abundant there. The flagellar antennomeres are dorsally carinate, and underneath the long hairs the surface is moderately covered with very short, recumbent, golden-brown pubescence. This species is common at lights and is recorded from Brazil, Argentina, and Bolivia, where it is found in Tarija Department near Villamontes and also in the lowland forests of Santa Cruz Department near Pampagrande, Camiri, the Flora and Fauna Hotel near Buena Vista, and the Potrerillo del Guenda Reserve.

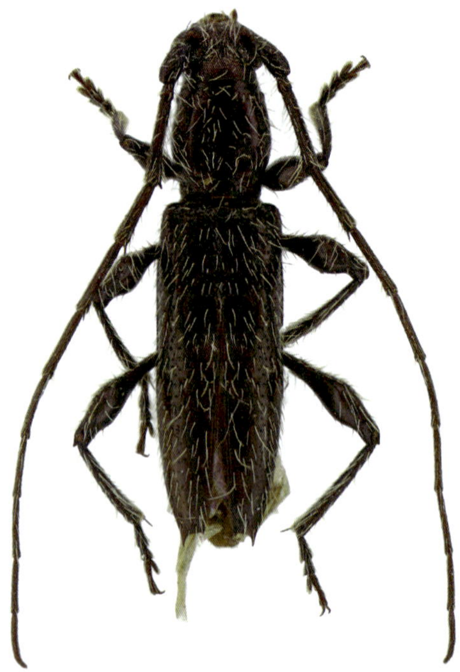

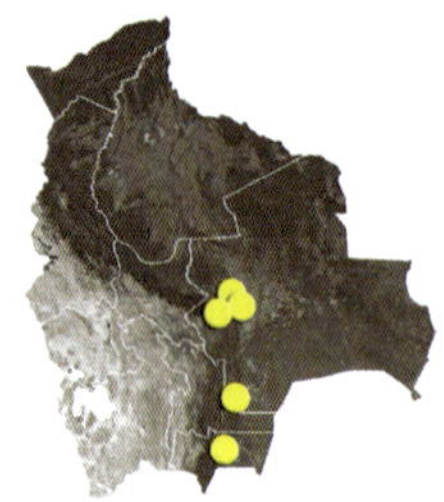

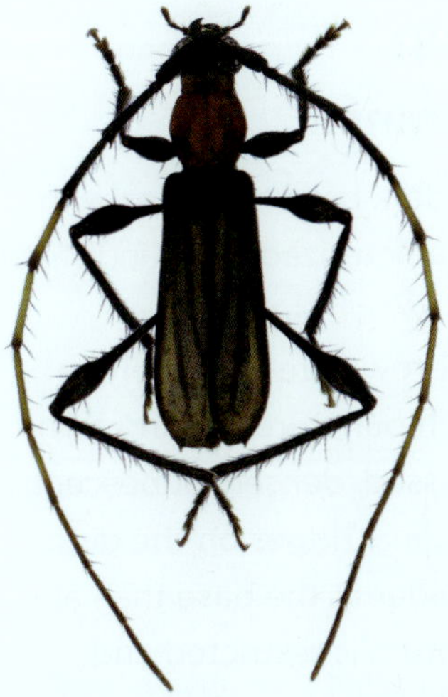

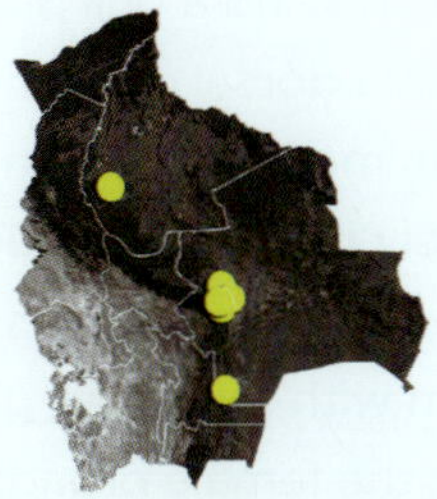

Terpnissa listropterina Bates, 1867

Cerambycinae: Elaphidiini

A very unusual elaphidiine, this small longhorn (9–13 mm) is easily recognized by its black color except for the base of the head and pronotum, which are orange. The elytral disk is closely, shallowly punctate and clothed with short, silvery-gray pubescence that obscures the surface and is lighter in color apically. Each elytron has a pair of dark, elongate, parallel carinae. These begin at the base and extend for much of its length, with the inner carina much shorter than the outer. The fourth through eleventh antennal segments are dull light tan, unlike the shining black basal segments. This species is attracted to lights where it often remains active and bounces around making it difficult to capture. It is a widely distributed species that is found from southern Mexico through Central America to Ecuador, French Guiana, Brazil, Argentina, and Bolivia. In Bolivia it has been recorded from Beni and La Paz Departments and most surveyed localities in Santa Cruz Department, including the Vicoquin area above Achira at 2,000 m, the vicinity of the Flora and Fauna Hotel near Buena Vista, the Potrerillo del Guenda Reserve, Refugio Los Volcanes, and the Camiri environs.

Erlandia inopinata Aurivillius, 1904

Cerambycinae: Erlandiini

Highly unusual in its morphology, this medium-sized (13–20 mm), shining brown, flattened species has a moderately punctate dorsum, large head, and very large, forward-pointing mandibles. It is the only South American member of the tribe Erlandiini, which has only one other species. This uncommon species is encountered above 1,000 m at lights, showing up late into the night. It is known from only northern Argentina, Paraguay, and Bolivia, where it has been recorded in Santa Cruz Department from Achira Campo, north of Camiri, Pampagrande, and Refugio Los Volcanes.

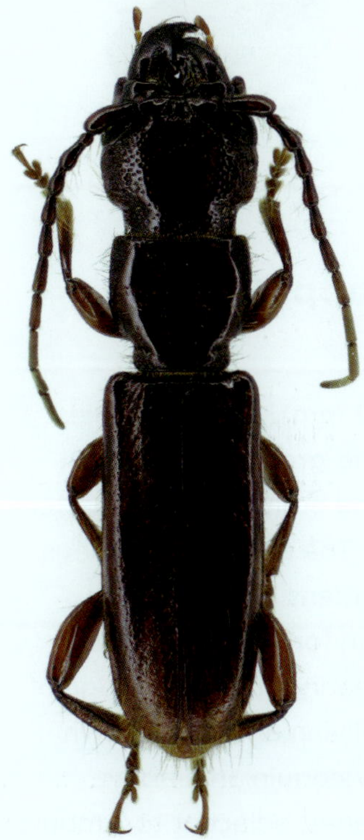

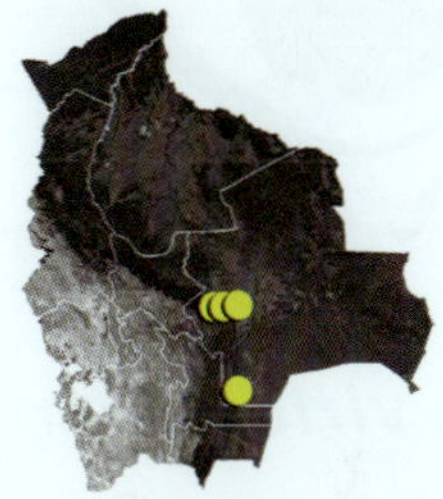

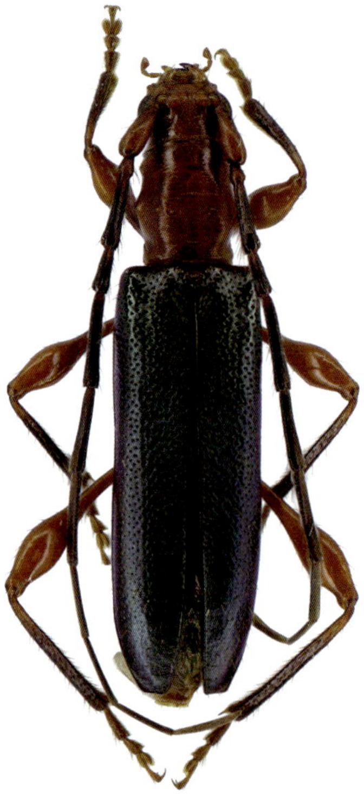

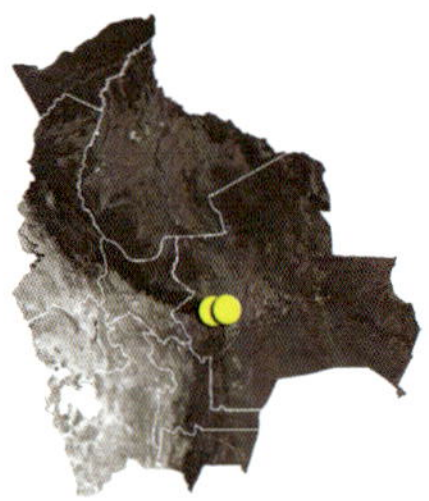

Aruama viridis Martins & Napp, 2007

Cerambycinae: Graciliini

This is an especially attractive small (7–12 mm), orange species with punctate, but velvety appearing, blue-green to blue-gray elytra. The head and pronotum are shining orange with the latter shallowly and finely, transversely rugose. The antennae and legs are typically orange but with some specimens having darker tibiae. This species inhabits high-elevation forests (>1,000 m) and is occasional at lights or found by beating the forest edge of agricultural *chaqueos*. This is another Bolivian endemic known in the Santa Cruz Department from the Vicoquin area above Achira at 2,000 m and Refugio Los Volcanes, adjacent to Amboró National Park.

Alastos batesi (Pascoe, 1888)

Cerambycinae: Hesperophanini

This is a medium-sized (10–16 mm) tan to dark brown species with shining elytra and a dull pronotum. It is characterized by its indistinct and variable dorsal coloration and pattern, which is a blend of lighter tan markings intermixed with dark brown areas. The elytra are densely punctate throughout, with each puncture bearing a fine, recumbent hair. Both sexes have a compressed, densely pubescent pronotum with three low, glabrous longitudinal ridges on the disk. However, the female pronotum is clearly wider at the base than at the apical margin, whereas the male pronotum is restricted and much narrower at the base, with a distinctly broader apical margin. *Alastos batesi* is rarely encountered except, very occasionally, at lights. It is a widely distributed species known from Costa Rica to Ecuador, Venezuela, and Bolivia, where it is has been collected in only the Vicoquin area above Achira Campo, Santa Cruz Department.

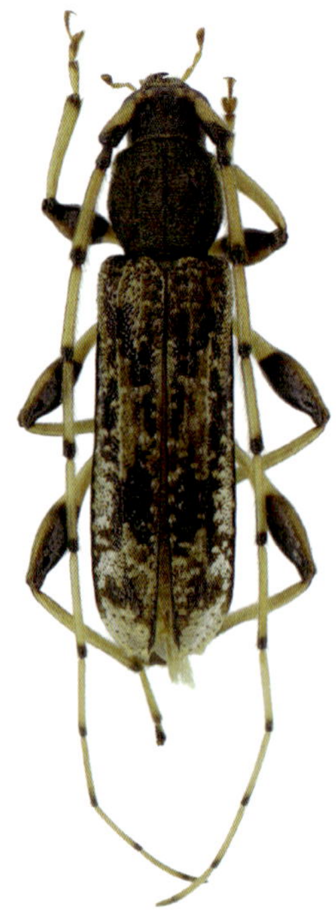

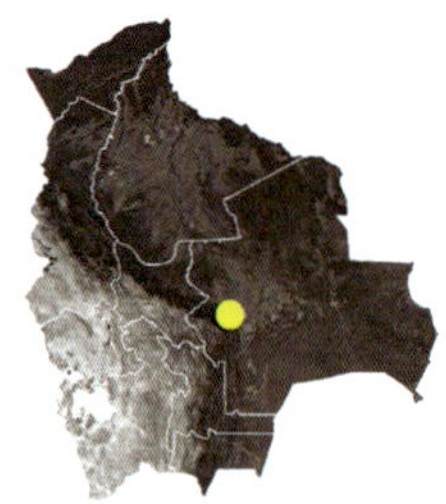

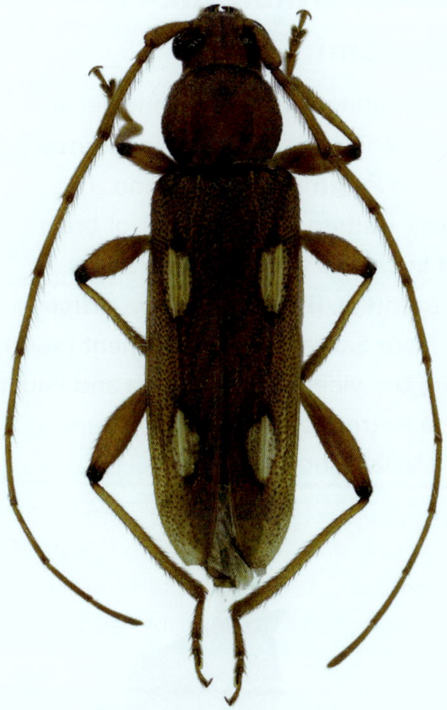

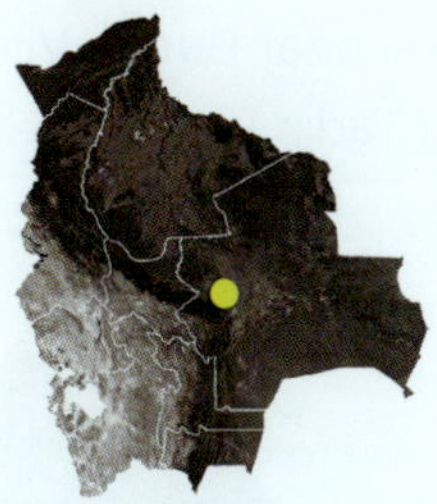

Anoplomerus rotundicollis Guérin-Méneville, 1844

Cerambycinae: Hesperophanini

This medium-sized (11–17 mm), brown hesperophanine looks superficially like members of Eburiini because of the raised eburneous markings on the elytra. The pronotum is rounded laterally and the elytral apices are truncate. The uniform light brown color is present throughout except on the extremities of the femora and antennal segments, along with the perimeter of the eburneous ridges, which are dark brown to black. It is known from Brazil, Paraguay, Argentina, and Bolivia, where it has been collected rarely at the Flora and Fauna Hotel near Buena Vista.

Eusapia amazonica (White, 1855)

Cerambycinae: Hesperophanini

This is a medium-sized (12–20 mm), slightly shining brown species with evenly punctate elytra. Each puncture bears a short, semirecumbent hair. The pronotum is uneven, shining, and more coarsely punctate with mostly longer, erect hairs. This species is devoid of any fasciae or spines and has evenly rounded elytral apices. It is typically encountered at lights in lowland forests and is locally abundant. It is widely distributed from French Guiana and Suriname to Brazil, Paraguay, and Bolivia, where it is recorded from La Paz Department and from Santa Cruz Department in the localities of the Flora and Fauna Hotel near Buena Vista, the Potrerillo del Guenda Reserve, and Refugio Los Volcanes, adjacent to Amboró National Park.

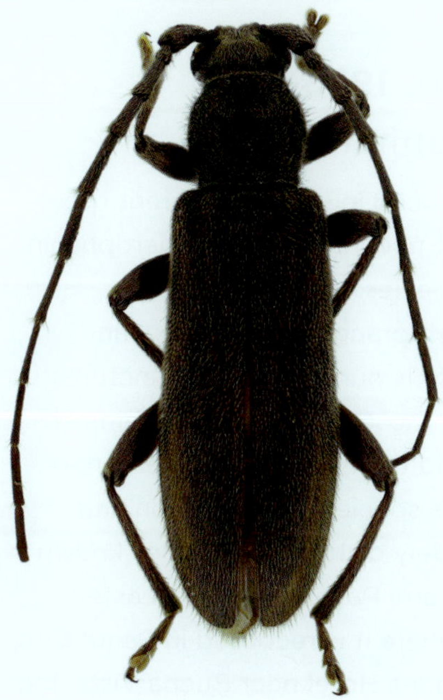

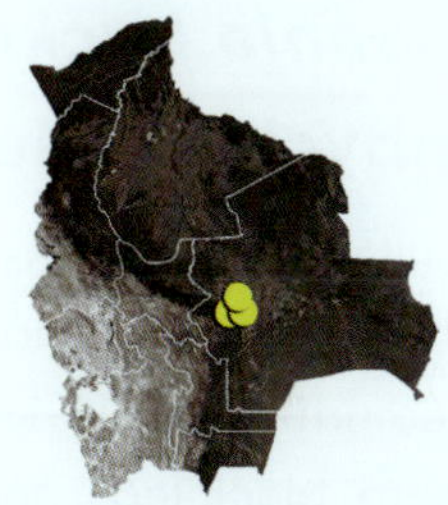

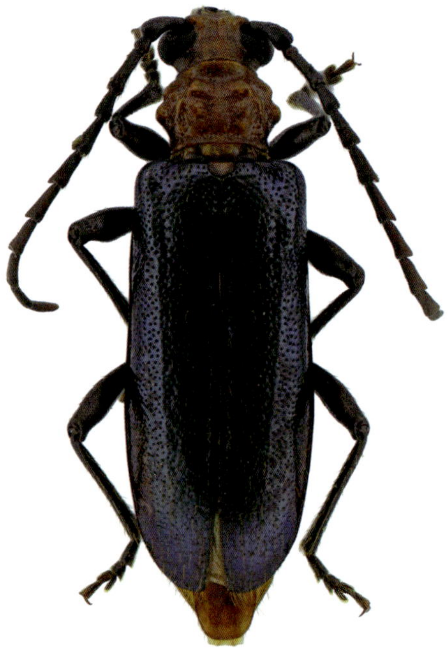

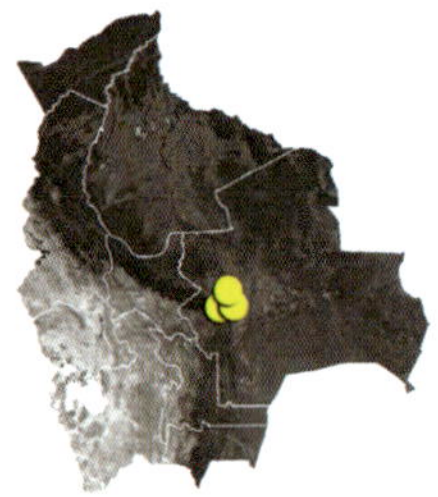

Tippmannia bucki (Lane, 1973)

Cerambycinae: Hesperophanini

This peculiar longhorn has been placed in several different tribes due to its unusual morphology but now resides in Hesperophanini. It is a medium-sized (12–18 mm) species with shining blue-black elytra, black legs, and reddish-orange venter, head, and pronotum. The entire upper surface is punctate, with punctures on the head very fine and shallow, those on the pronotum much larger and more widely spaced, and the elytral punctures larger still, dense, but evenly spaced. This species is uncommon and only occasional at lights. It is a widely distributed species known from Central America (Costa Rica and Panama), Brazil, eastern Ecuador, Argentina, and Bolivia, where it is recorded in Santa Cruz Department from the Flora and Fauna Hotel near Buena Vista, the Potrerillo del Guenda Reserve, and Refugio Los Volcanes, adjacent to Amboró National Park.

Chrysoprasis aurigena (Germar, 1824)

Cerambycinae: Heteropsini

This is a small (7–9 mm), flower-visiting, metallic green species with black antennae, green to greenish-black legs, and iridescent red to cupreous lateral fasciae the length of each elytron. This species can be found in numbers visiting the blossoms of bush *Croton*. It is widely distributed in central and southern South America and recorded from Argentina, Brazil, Paraguay, Uruguay, and Bolivia, where it is known from Santa Cruz Department (south of Cabezas and north of Camiri, the vicinity of the Flora and Fauna Hotel near Buena Vista, and the Potrerillo del Guenda Reserve) and Tarija Department (around Villamontes).

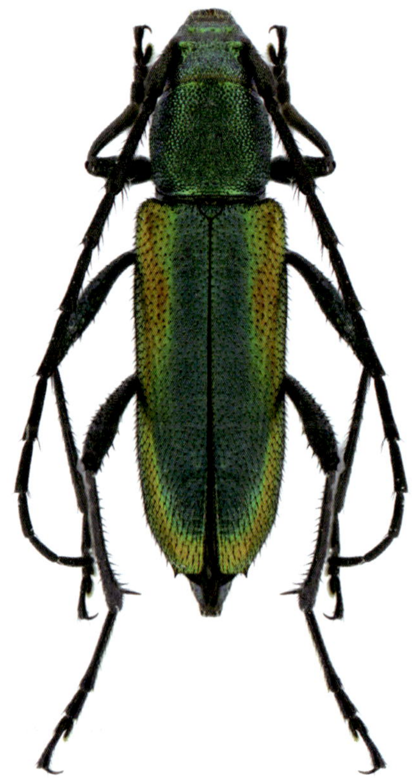

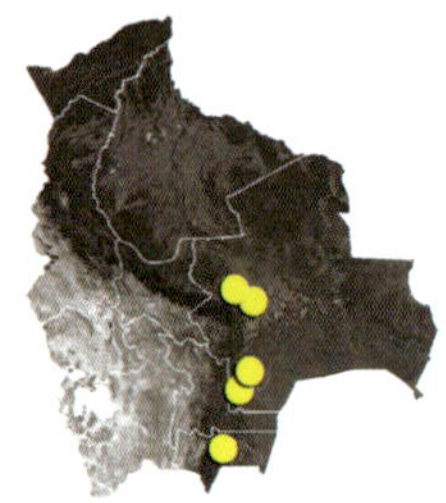

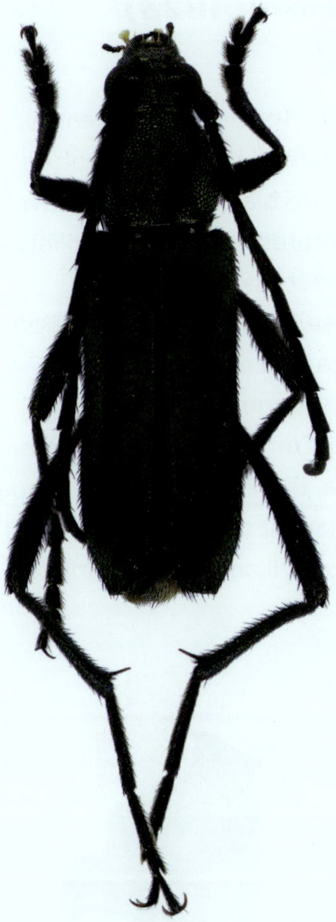

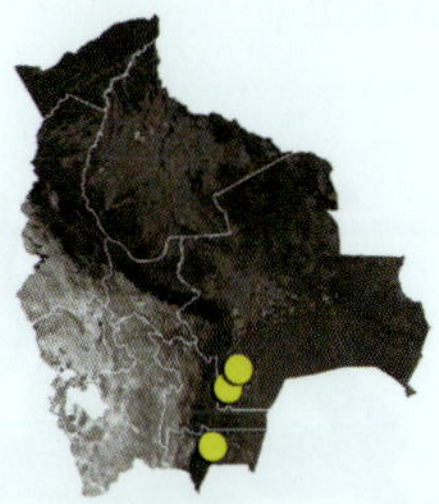

Chrysoprasis concolor Redtenbacher, 1867

Cerambycinae: Heteropsini

This is a small (6–8 mm), greenish-blue to green species with black antennae, greenish to bluish-black legs, and reflective, densely punctate pronotum. Typical of this genus, it is a frequent visitor to bush *Croton* during its bloom. It has been recorded from Argentina, Brazil, Paraguay, Uruguay, and Bolivia, where it is known from Santa Cruz Department (south of Cabezas and north of Camiri) and Tarija Department (north of Villamontes).

Chrysoprasis hypocrita Erichson, 1847

Cerambycinae: Heteropsini

This is a small (9–15 mm), shining blue-green species with black antennae and legs except for the shining metallic green femora. Ventrally, the minutely punctate, dark orange abdomen and metallic green mesosternum with widely spaced punctures will help to distinguish this species from other *Chrysoprasis*. It is a frequent visitor to the blossoms of many different woody plants. Widely distributed, this species is found from Mexico through Central America to southern South America. In Bolivia it is recorded from Cochabamba, La Paz (Madidi National Park), Santa Cruz (near Cabezas, Samaipata, from the vicinity of the Flora and Fauna Hotel near Buena Vista, the Potrerillo del Guenda Reserve, and Refugio Los Volcanes), and Tarija (west of Villamontes) Departments.

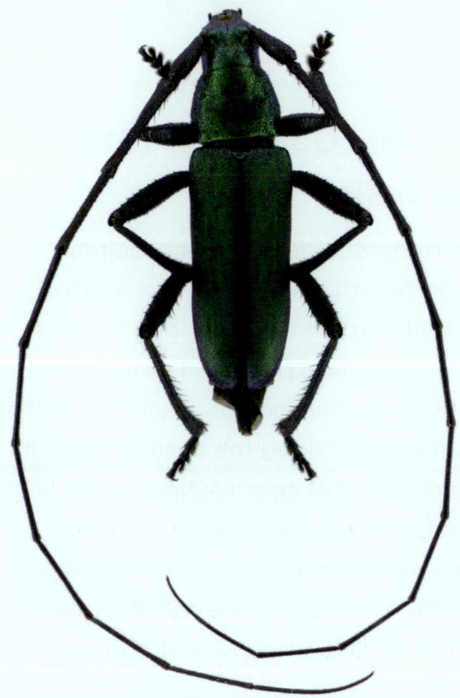

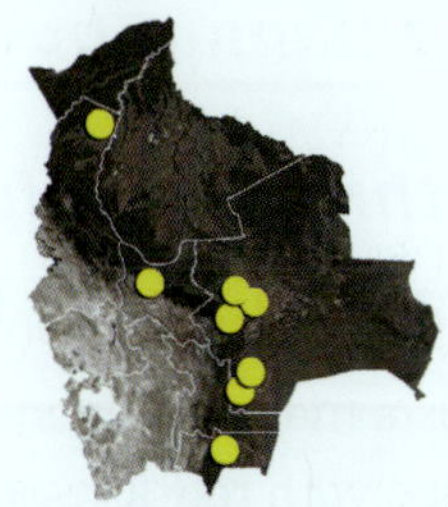

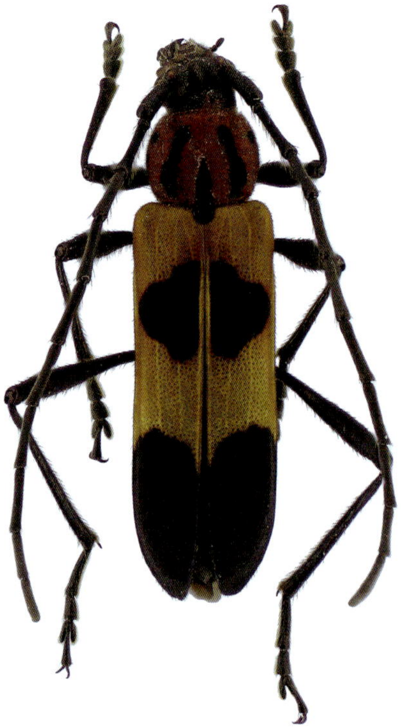

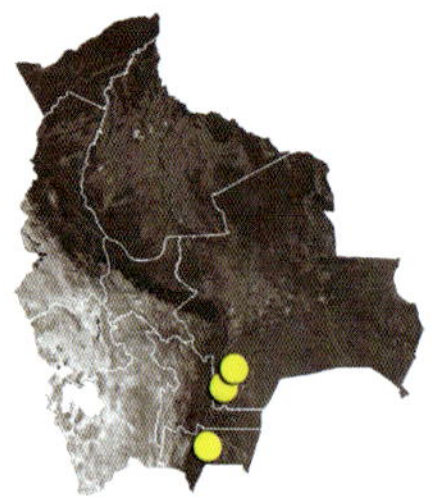

Erythrochiton jucundum (Gounelle, 1913)

Cerambycinae: Heteropsini

This is a medium-sized (10–14 mm), orange and black, diurnal species with reddish-orange pronotum. The elytra have a broad, light tan band covering the basal third and another beginning at the middle and covering half of the apical portion. The pronotum has three linear, longitudinal black fasciae, one on each side of the disk and a shorter one placed centrally along the basal margin. It has shining black antennae and legs. This species can be locally abundant on *Croton* blossoms in the humid Chaco (see also frontispiece). It is known from Argentina, Uruguay, and Bolivia, where it is recorded from southern Santa Cruz Department near Cabezas and Camiri and from Tarija Department west and north of Villamontes.

Mallosoma zonatum (Sahlberg, 1823)

Cerambycinae: Heteropsini

This is a medium-sized (14–16 mm), black species with a central, orange-tan band on the elytra and the lateral third of pronotum velvety orange. The antennae and legs are black. This distinctively marked, diurnal species is often encountered perching on foliage along forest trails and edge and is also occasional at lights (perhaps when they are perched in close proximity to the lights). It is widely distributed in South America and recorded from Argentina, Brazil, French Guiana, Peru, Paraguay, Uruguay, and Bolivia, where it is known from north of Camiri, the vicinity of the Flora and Fauna Hotel near Buena Vista, the Potrerillo del Guenda Reserve, and Refugio Los Volcanes, adjacent to Amboró National Park, in Santa Cruz Department and near Villamontes in Tarija Department.

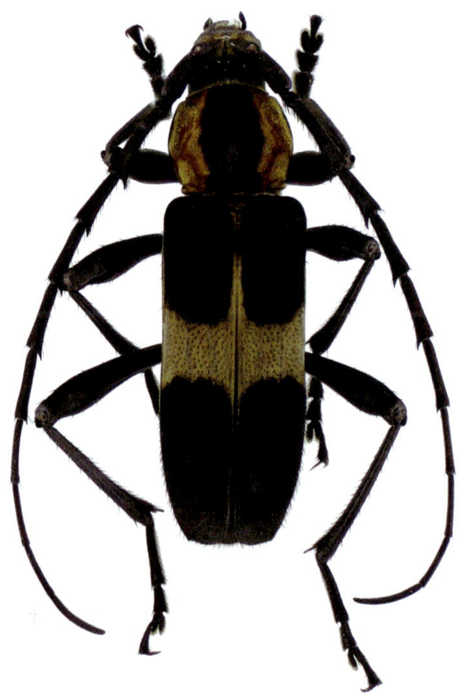

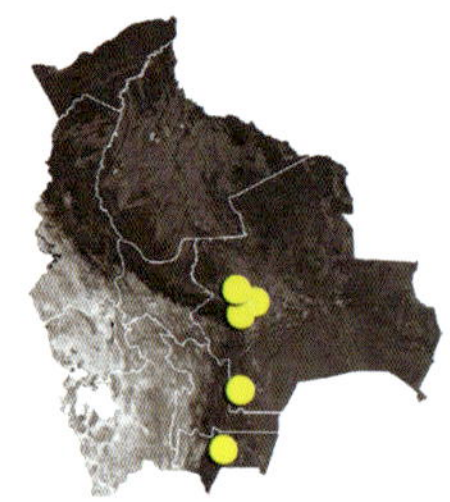

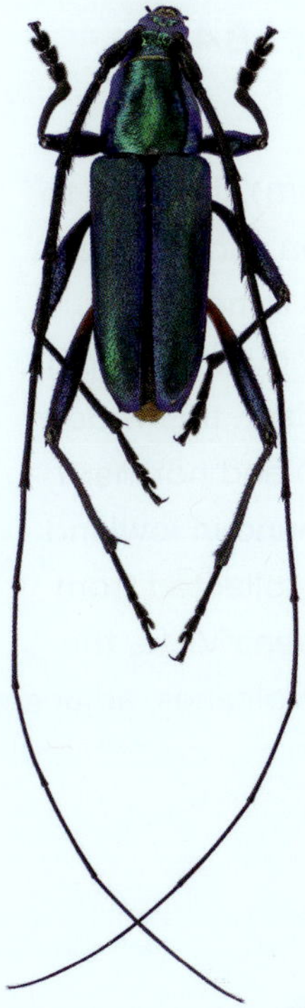

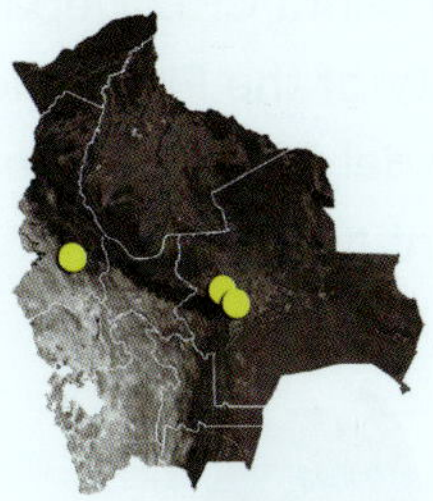

Tobipuranga ruficoxis (Bates, 1870)

Cerambycinae: Heteropsini

This is a small- to medium-sized (9–13 mm), shining green species with black antennae and greenish-black legs, except for the base of the metafemora, which are reddish orange where they connect with the coxae (hence the name *ruficoxis*). The pronotum is brightly shining iridescent green and minutely, shallowly rugose. The males, like the one pictured here, have very long antennae. This rare diurnal species has been found on freshly cut small trees, where it was congregating to mate and oviposit. It is known from central Brazil, eastern Ecuador, Peru, and Bolivia, where it is known from the vicinity of the Flora and Fauna Hotel near Buena Vista and the Potrerillo del Guenda Reserve in Santa Cruz Department and from La Paz Department.

Ctenoplon x-littera (Thomson, 1865)

Cerambycinae: Hexoplonini

This is a small (7–9 mm), dark, heavily punctate and pubescent species with shining elytra. There is a narrow, partially eburneous, X-shaped fascia on the elytra. The remaining integument is somewhat obscured by whitish-gray, recumbent pubescence with intermixed longer, stiff, whitish hairs arranged in five longitudinal rows per elytron. Typical of the Hexoplonini, in side profile the pronotum is apically humpbacked. This species is occasional at lights and recorded from Argentina, Brazil, Paraguay, and Bolivia. In Bolivia it is known in Santa Cruz Department from the vicinity of the Flora and Fauna Hotel near Buena Vista, the Potrerillo del Guenda Reserve, Refugio Los Volcanes, adjacent to Amboró National Park, and the road to Bella Vista near Cuevas.

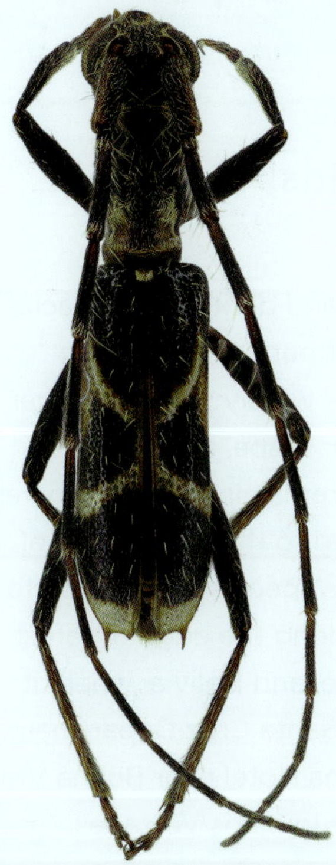

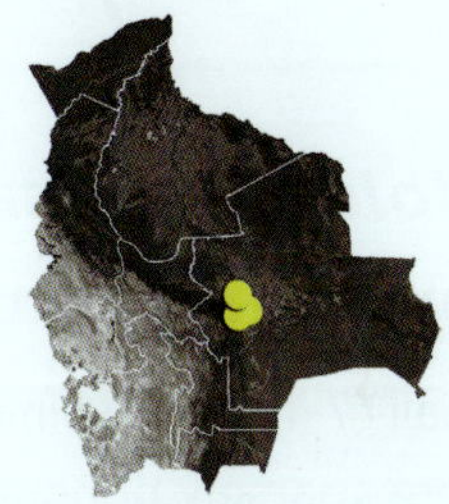

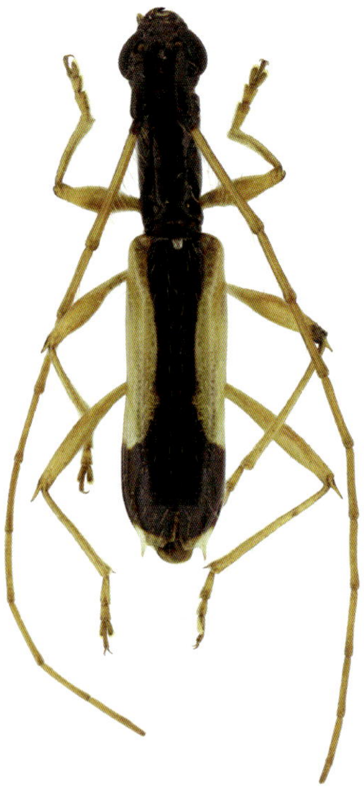

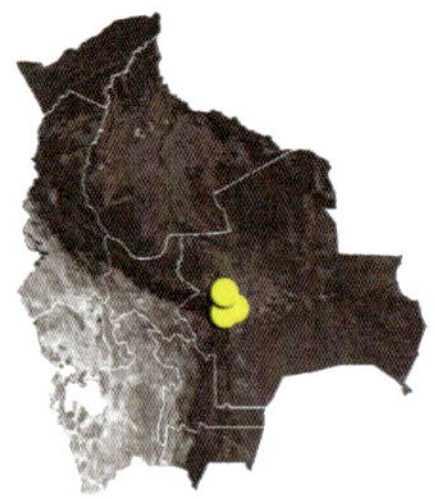

Gnomidolon grantsaui Martins, 1967

Cerambycinae: Hexoplonini

This is a small (7–11 mm), shining light tan and dark brown species. The elytra are sparsely punctate with most punctures bearing short, whitish setae. Intermixed are rows of widely spaced, longer, semirecumbent hairs. The head, pronotum, scape, and first antennomere are shining dark brown; the remaining antennomeres, legs, elongate wide fasciae from the humeri to the apical third of the elytra, and the apices are light tan. This species is attracted to lights and also found perched on foliage along the edge of forest trails. It is known from Paraguay, Venezuela, and Bolivia, where it is recorded from lowland forest locales in Santa Cruz Department, including the vicinity of the Flora and Fauna Hotel near Buena Vista, the Potrerillo del Guenda Reserve, and Refugio Los Volcanes.

Gnomidolon musivum (Erichson, 1847)

Cerambycinae: Hexoplonini

This shining *Gnomidolon* species is small (8–11 mm), light tan to dark brown, with two oblique, light tan fasciae on each elytron that together form an open X marking. The hind femora are double spined with the external elytral apices singularly spined. This species comes to lights and is locally abundant. It is widely distributed from Panama (a new country record) and northern South America to Peru and Bolivia, where it is found in lowland forests in Santa Cruz Department and has been collected from the vicinity of the Flora and Fauna Hotel near Buena Vista, the Potrerillo del Guenda Reserve, and Refugio Los Volcanes, adjacent to Amboró National Park.

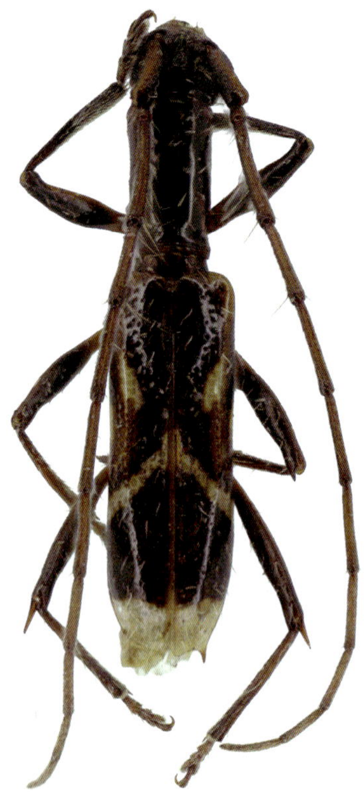

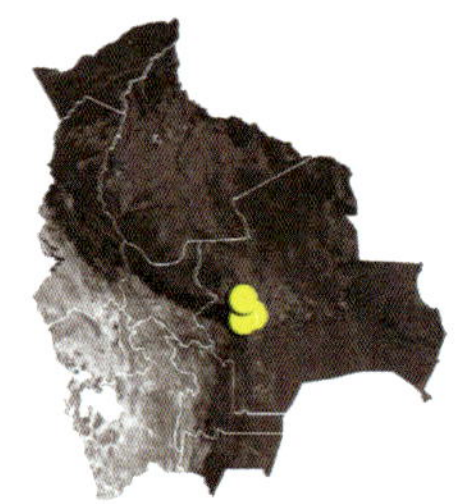

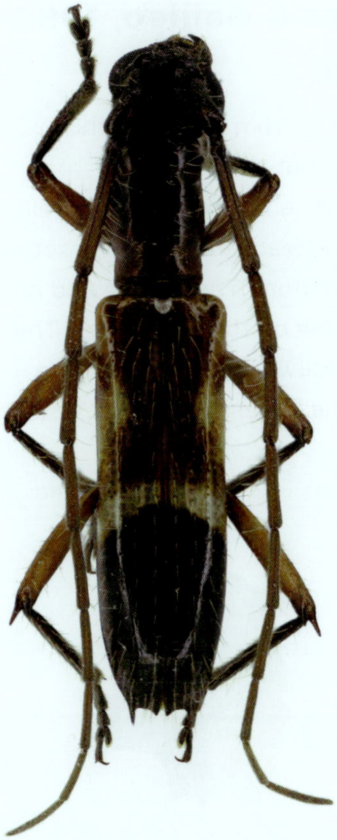

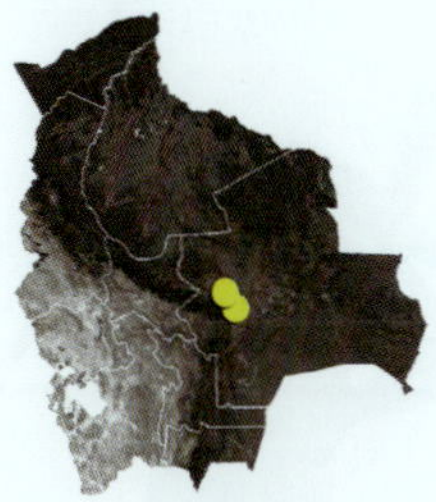

Gnomidolon nanum Martins, 1962

Cerambycinae: Hexoplonini

This is a midsized (12–14 mm), reddish-orange to dark brown species with two lighter-colored elytral fasciae. The head, apical portion of the pronotum, scape, and legs, except for the reddish-orange femora, are all dark brown. In addition to the fasciae there are five rows of long, silky, semirecumbent hairs the length of each elytron. This species is a Bolivian endemic, rarely encountered, and known only from a few specimens from the Flora and Fauna Hotel near Buena Vista and the Potrerillo del Guenda Reserve in Santa Cruz Department.

Hexoplon annulatum Martins & Galileo, 2011

Cerambycinae: Hexoplonini

Recently discovered and described, this small (8–10 mm) Bolivian endemic is light tan and black and known only from Refugio Los Volcanes, adjacent to Amboró National Park, in Santa Cruz Department. It has black annulations on the femora, apices of the scape, and third antennal segment. The elytra also have a narrow black band near, but in front of, the elytral apices. Interestingly, both the middle and hind femora have long spines, but on the middle legs they are on the inner margin and on the hind legs the external margin.

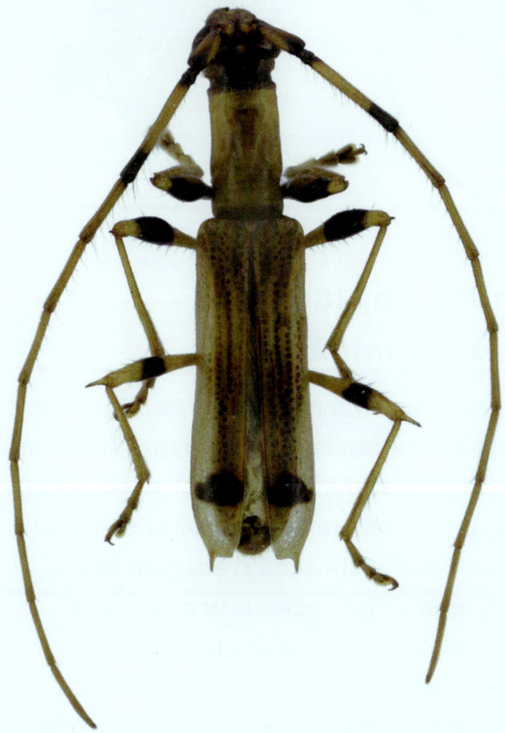

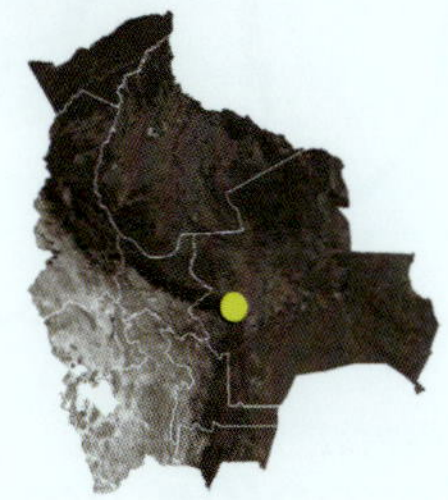

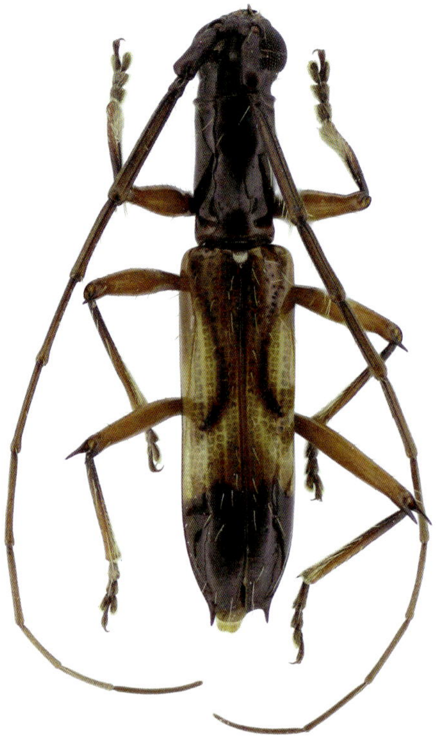

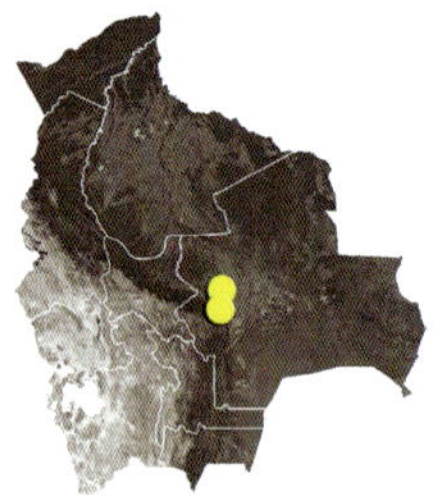

Hexoplon juno Thomson, 1865

Cerambycinae: Hexoplonini

This is a midsized (12–14 mm), shining tan and dark reddish-brown species with indistinct, light brown fasciae on the elytra. The head, pronotum, scape, and apical third of the elytra are all dark reddish brown. The legs, remaining antennomeres, and basal two-thirds of the elytra are light tan with an indistinct, curved, dark fascia, centered lengthwise, in this area. This species is occasional at lights but never common. It is recorded from Argentina, Brazil, Paraguay, Peru, and Bolivia, where it is known only from Santa Cruz Department from the vicinity of the Flora and Fauna Hotel near Buena Vista and Refugio Los Volcanes, adjacent to Amboró National Park.

Isostenygra monnei Martins & Galileo, 1999

Cerambycinae: Hexoplonini

One of the largest hexoplonines at over 20 mm in length, this beautiful species is glossy black throughout the body and appendages. It is mostly glabrous but has a few sparse, long yellow setae scattered on the elytra, pronotum, head, and legs. The setae form an indistinct row along the elytral suture. The middle and hind femora are armed with a long, acute spine on the outer apex. This species is known only from Brazil and Bolivia, where it has been collected at lights at the Flora and Fauna Hotel near Buena Vista in Santa Cruz Department.

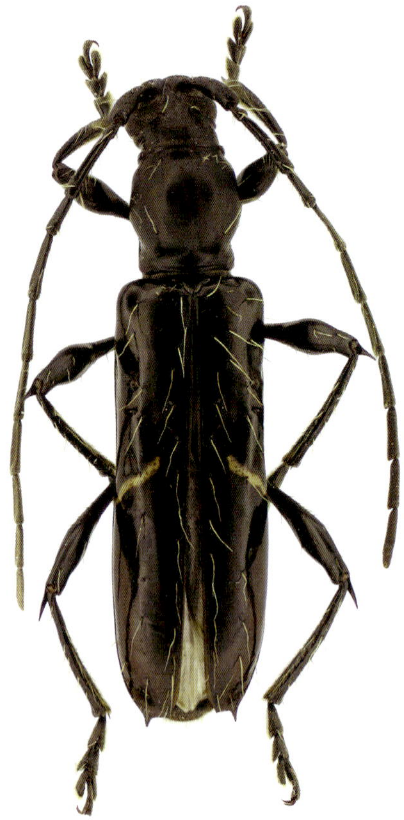

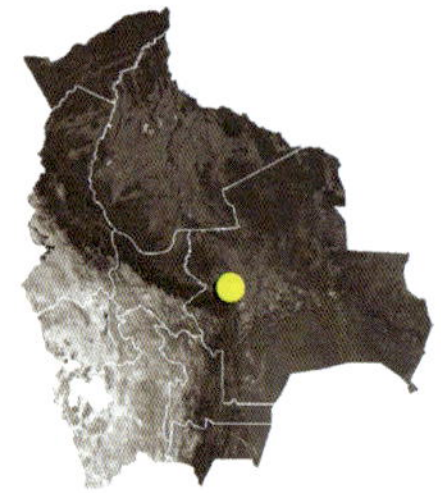

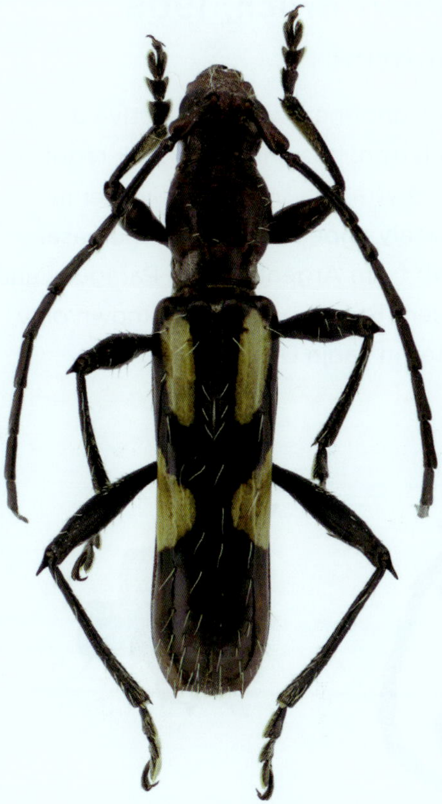

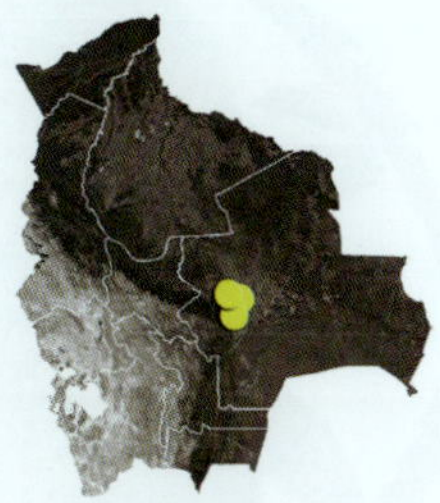

Stenygra cosmocera White, 1855

Cerambycinae: Hexoplonini

This is an elongate, midsized (12–15 mm), wine-colored species with four ivory-colored spots on its smooth, shining elytra. The pronotum is finely punctate, and its antennae are much shorter than the elytra. This is a lowland forest species rarely encountered but most often seen at lights. Its distribution includes the Amazon Basin of Brazil and Peru south to Argentina and Paraguay. In Bolivia it is known only from Santa Cruz Department at the Flora and Fauna Hotel near Buena Vista, the Potrerillo del Guenda Reserve, and Refugio Los Volcanes, adjacent to Amboró National Park.

Stenygra setigera (Germar, 1824)

Cerambycinae: Hexoplonini

This is a midsized (13–17 mm), tan to reddish-brown and black, shining species with dark legs and antennae. The inflated, globose, dark red pronotum, dark elytra with two large eburneous spots, and scattered, long, stiff, cream-colored setae will distinguish this species from its Bolivian congeners. It is rarely collected but has been collected by beating forest edge habitats. It is recorded from Argentina, Brazil, Paraguay, and Bolivia, where it is known from the vicinity of the Flora and Fauna Hotel near Buena Vista in Santa Cruz Department.

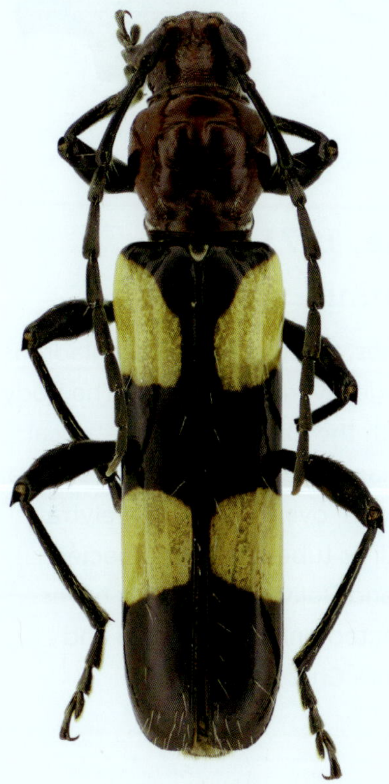

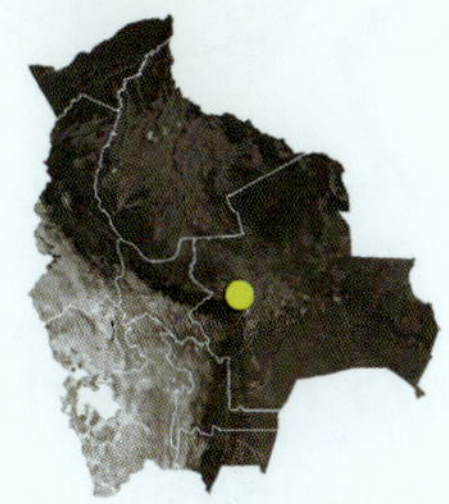

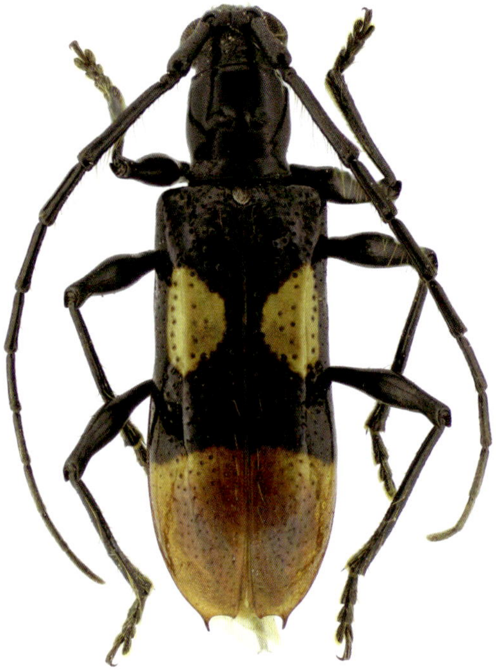

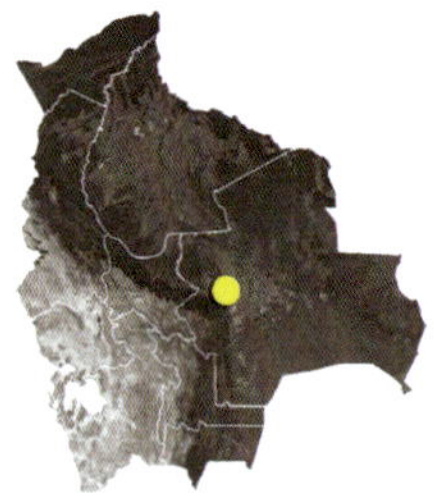

Tapuruia felisbertoi Lane, 1973

Cerambycinae: Hexoplonini

One of the most elegant hexoplonines, this glossy, mostly glabrous longhorned beetle is moderately large, averaging 16 mm. The body and appendages are dark brown with the exception of the apical third of the elytra and middle maculae, which are yellow to orange. There are punctures, somewhat in rows, over most of the elytra. The pronotum lacks any distinct calli or tubercles. This species ranges from French Guiana, Brazil, and Bolivia, where it is never common. In Bolivia it has been collected only at the Flora and Fauna Hotel near Buena Vista in Santa Cruz Department.

Lissonotus andalgalensis Bruch, 1908

Cerambycinae: Lissonotini

This small (11 mm) but robust lissonotine is very distinctive due to its basal depressions on each elytron. It is uniformly dark reddish brown, but the tibiae, tarsi, and elytra can be slightly lighter in color. The antennae are moderately appendiculate on the basal segments. This species is known from Argentina and Paraguay and recently was collected in southern Bolivia, where it is known only from an area west of Villamontes in Tarija Department.

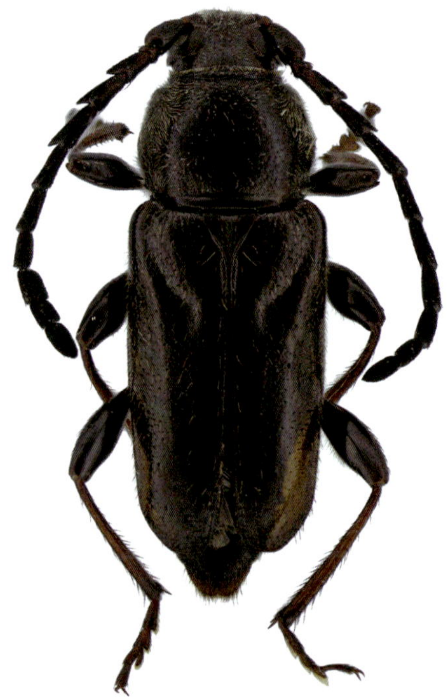

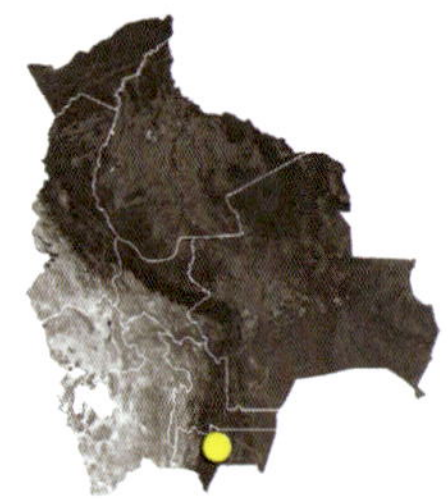

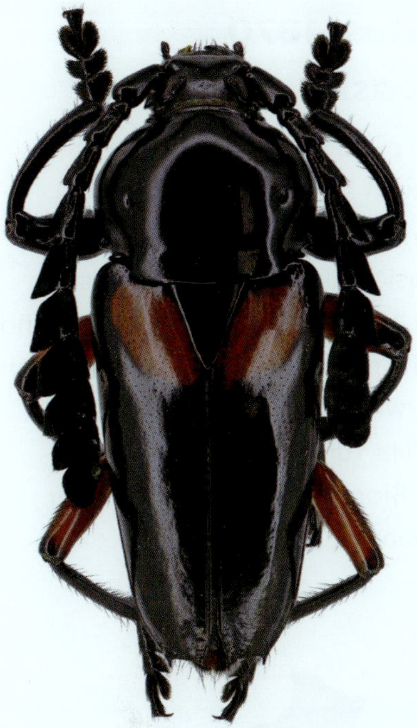

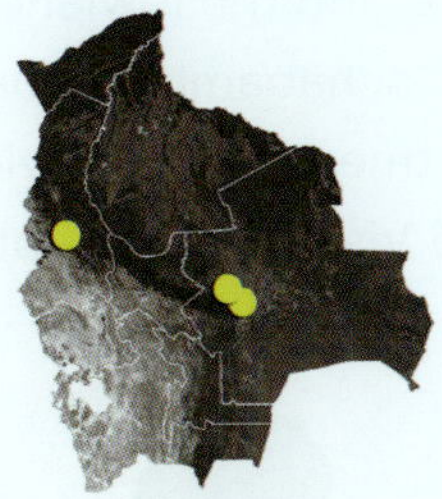

Lissonotus unifasciatus Gory, 1831

Cerambycinae: Lissonotini

This is a midsized (12–18 mm) but robust, shining black, diurnal species with a large, red, oval spot on the basal area of the elytra. This genus is characterized by its large, globose pronotum, typically as wide as the elytra, and its flat, serrate antennae. This species is minutely, sparsely punctate over most of the dorsum. It is often found flying about in agricultural cuts. It has a limited distribution and is recorded only from southeastern Brazil and Bolivia, where it is known from La Paz Department in Madidi National Park and from Santa Cruz Department in the vicinity of the Flora and Fauna Hotel near Buena Vista and the Potrerillo del Guenda Reserve.

Merionoedopsis brevipennis Melzer, 1934

Cerambycinae: Molorchini

This is an odd-looking, medium-sized (12–15 mm) cerambycid that is sexually dimorphic. Both sexes have short elytra that do not cover the abdomen and end in an acute point. The female is primarily black with the pronotum centrally black but laterally shining red and the basal portion of the hind tibia light tan. The antennae of females are very short. The males tend to be more dark brown than black, with the legs (except for the apices of the hind femora, the hind tibia, and tarsi, which are black) a light yellow orange and the elytral pads a dark tan to brown. The male also has an elongated pronotum. The antennae of males are extremely long, being two to three times the length of their body. The species is a visitor to woody plant blossoms but rarely encountered. In South America it is found in southern Brazil, Argentina, Paraguay, and Bolivia, where it is known from only Potrerillo del Guenda Reserve in Santa Cruz Department.

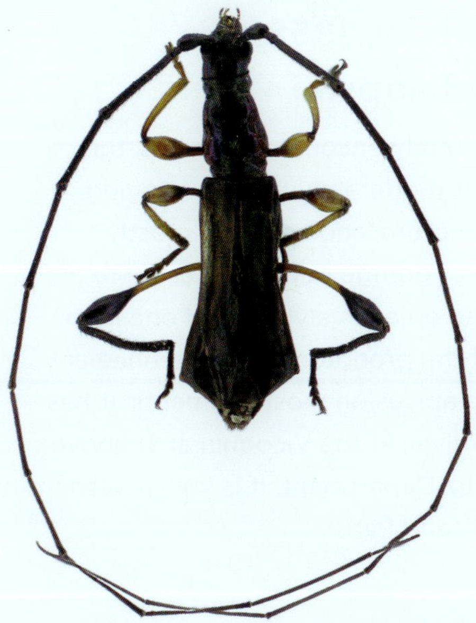

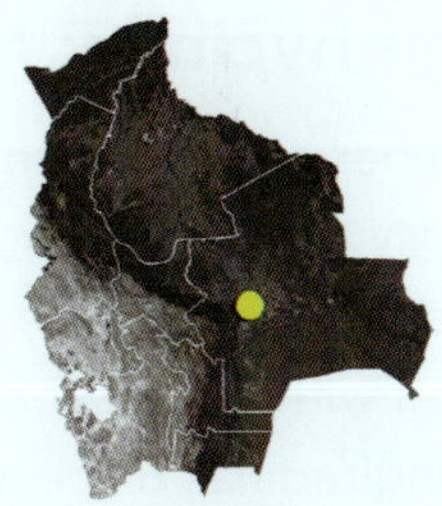

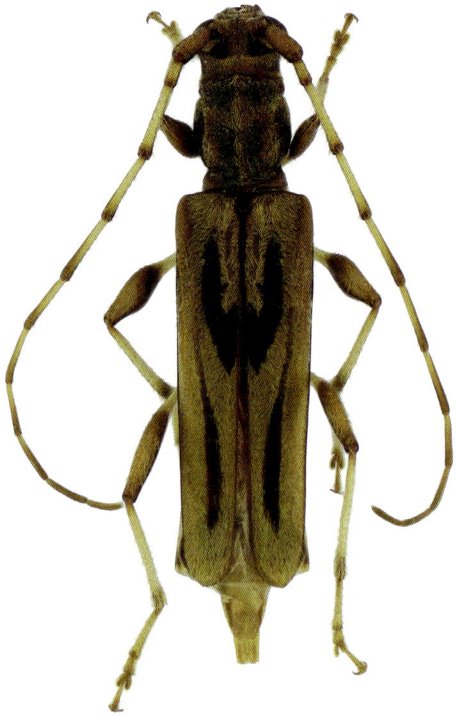

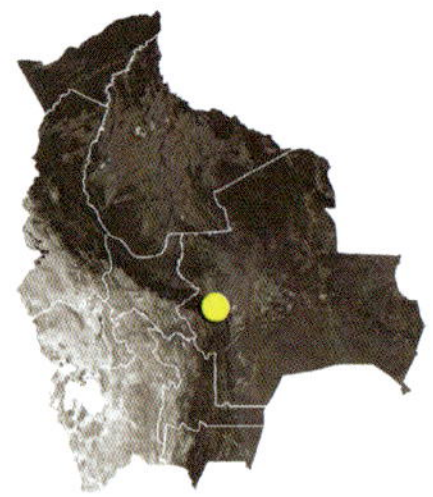

Neozodes signata Zajciw, 1958

Cerambycinae: Necydalopsini

A member of one of the more rarely encountered tribes, this strikingly patterned longhorn is moderately large, averaging around 20 mm. The elytra, pronotum, and head are mostly covered with dense, yellowish to orange pubescence. There are two longitudinal, black maculae on each elytron and one large, black macula on either side of the pronotum. The antennae are pale yellow with dark annulate apices on most segments. It has been collected only once in Bolivia, at the Vicoquin area above Achira at 2,000 m in Santa Cruz Department. It is known also from Ecuador, French Guiana, Brazil, and Peru.

Ozodes malthinoides Bates, 1870

Cerambycinae: Necydalopsini

This is a medium-sized (18–20 mm), parallel-sided, light tan species (except for the pronotum) with an intricate, golden, pubescent pattern on both the elytra and the pronotum. Rarely encountered other than occasionally at lights, it is easily distinguishable from similar species by the golden pubescent band crossing the pronotum. It is known from the Amazon Basin, French Guiana, and the lowlands of Bolivia, where it has been recorded from Cochabamba Department in Villa Tunari and Santa Cruz Department from the vicinity of the Flora and Fauna Hotel near Buena Vista and the Potrerillo del Guenda Reserve.

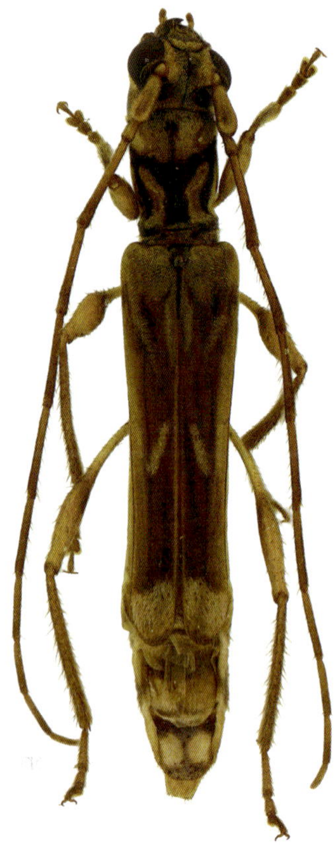

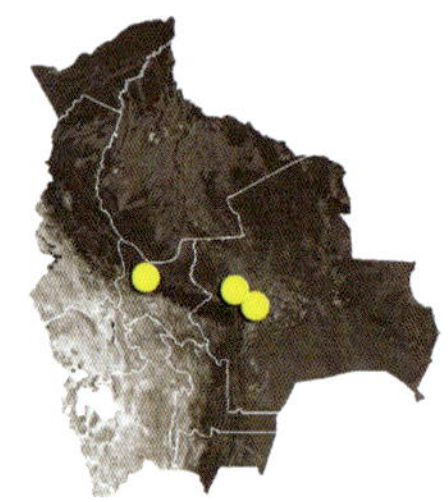

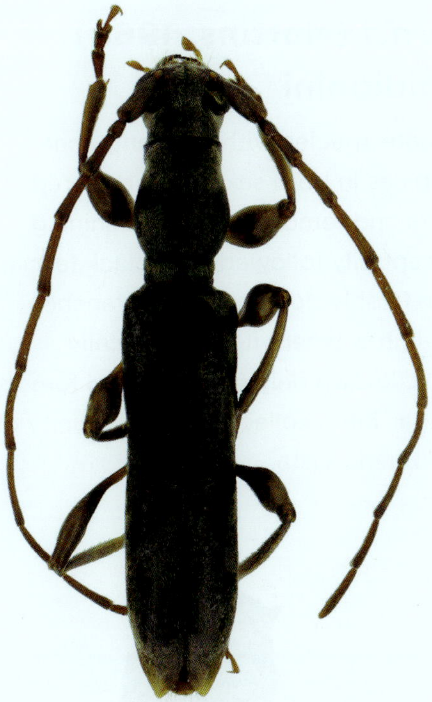

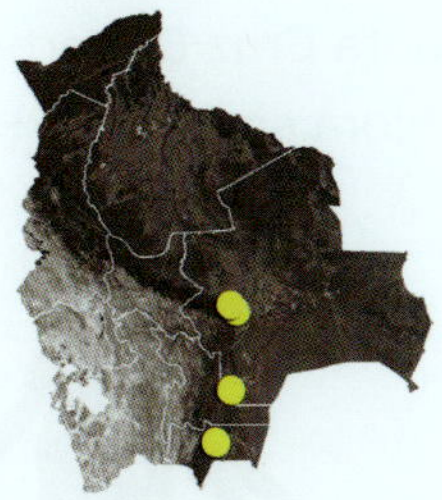

Neocorus ibidionoides (Audinet-Serville, 1834)

Cerambycinae: Neocorini

This is a small (10–11 mm), parallel-sided, dark brown to black species with inflated globular pronotum, constricted both basally and apically, and densely pubescent dorsum. It can be separated from its close relative, *Neocorus diversipennis* Belon, by the lack of long flying hairs on the legs and antennae. This is the more common of Bolivia's two species and can be collected at lights. It is widely distributed and recorded from Argentina, Brazil, French Guiana, Paraguay, and Bolivia, where it is known from Santa Cruz Department north of Camiri, the Vicoquin area above Achira at 2,000 m, Potrerillo del Guenda Reserve, Refugio Los Volcanes, adjacent to Amboró National Park, and in Tarija Department near Villamontes.

Compsa quadriguttata (White, 1855)

Cerambycinae: Neoibidionini

This is a small (7–9 mm), narrow, orange to reddish-brown species with adjacent light tan and black fasciae on the orange elytra and a reddish-brown pronotum. The males have the third to fifth antennal segments swollen. This species is locally abundant at lights and is also found by beating overhead cut vines in and along the edge of agricultural clearings. It is widely distributed in northern South America to Paraguay and Bolivia, where it is known from Beni (Rurrenabaque), Cochabamba, La Paz, and Santa Cruz (between Camiri and Santa Cruz, the Flora and Fauna Hotel near Buena Vista, and the Potrerillo del Guenda Reserve) Departments.

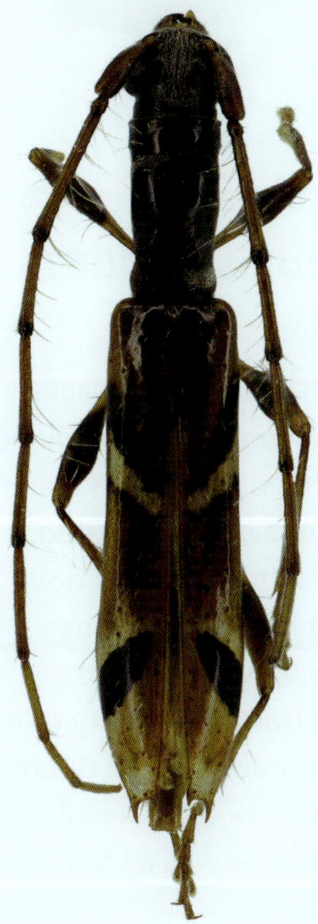

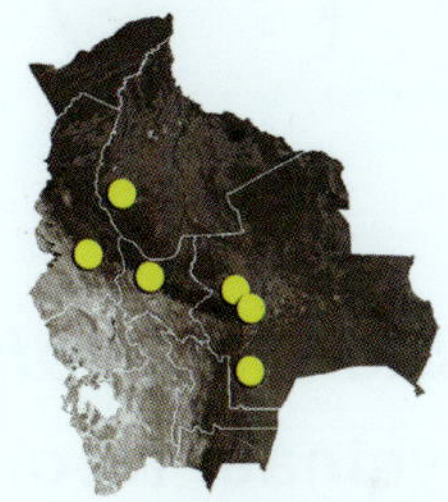

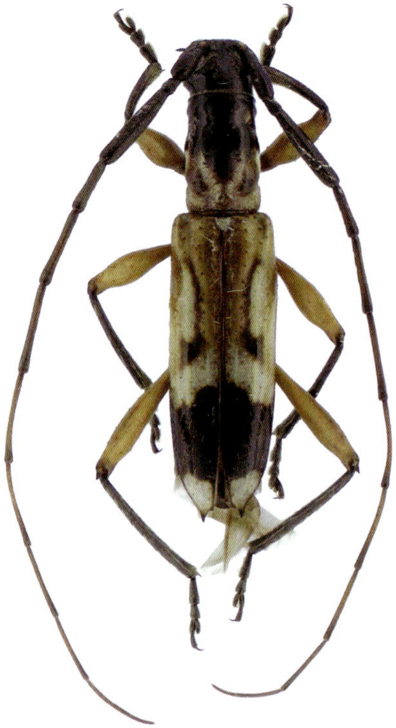

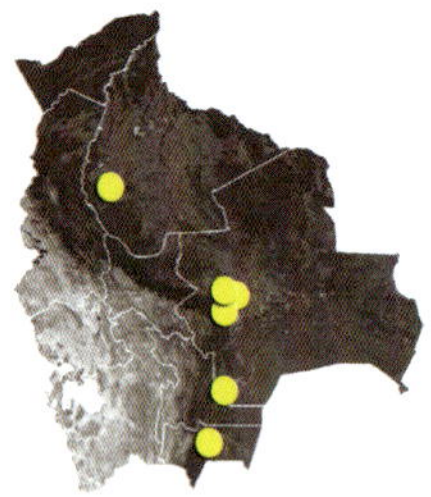

Compsibidion graphicum (Thomson, 1867)

Cerambycinae: Neoibidionini

This is a small (9–12 mm), orange to reddish-orange and black species with two light tan fasciae on each elytra and an extended U-shaped, densely pubescent marking on the pronotal disk. In this species males have third and fourth antennal segment inflated. It is common well into the rainy season and can be collected both at lights and by beating recently cut small branches and vines. It is widely distributed and is recorded from Argentina, Brazil, Paraguay, and Bolivia. In Bolivia it is known from the vicinity of the Flora and Fauna Hotel near Buena Vista, the Potrerillo del Guenda Reserve, Refugio Los Volcanes, adjacent to Amboró National Park, and north of Camiri in Santa Cruz Department as well as from Beni and Tarija Departments.

Compsibidion reichardti (Martins, 1962)

Cerambycinae: Neoibidionini

This is a small (11–12 mm), elongate species with long antennae greatly surpassing the elytral apices in both sexes. It is dark red with the apical third of the elytra and femora black. The shining elytra have a tan spot located centrally followed by a black fascia. Although occasionally found on freshly dead or dying branches it is most often encountered at lights, where it can be locally common. It is a Bolivian endemic known only from lowland forests in Santa Cruz Department and has been collected in the vicinity of the Flora and Fauna Hotel near Buena Vista.

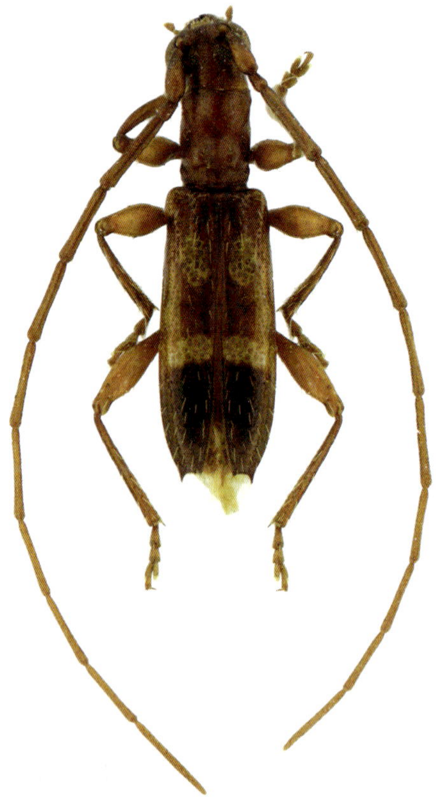

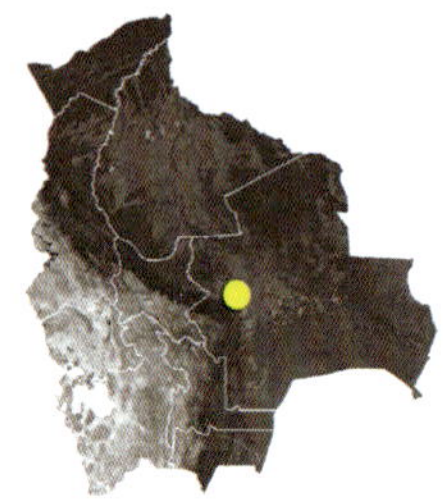

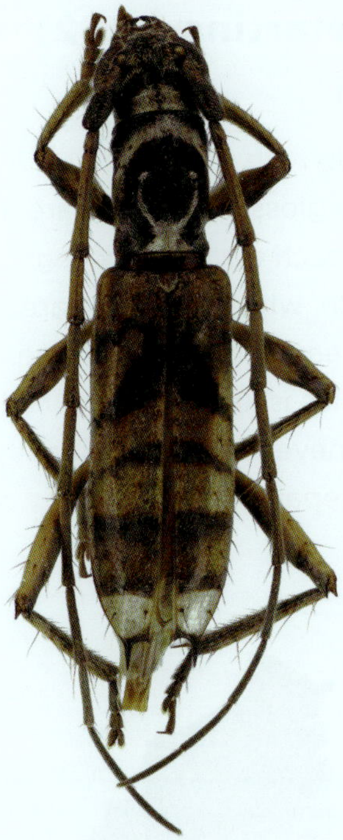

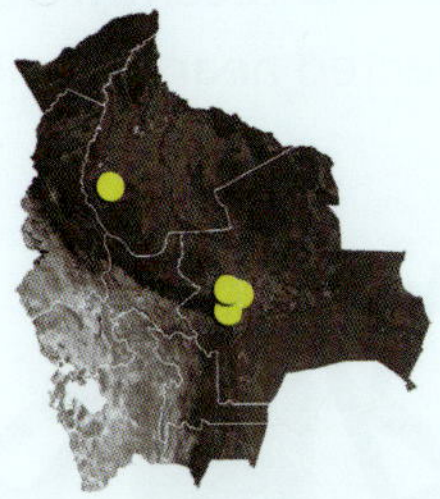

Compsibidion thoracicum (White, 1855)

Cerambycinae: Neoibidionini

This is a small (7–12 mm), tan and brown species with several, mostly transverse brown fasciae on the tan elytra and a distinctly marked pronotal disk. The disk is dark brown, minutely and densely punctate, with a small, triangular, light tan area basally and a golden pubescent, open U-shaped mark in front of it. The margins of the disk are also ringed in dense golden pubescence. This species is occasional at lights or by beating in agricultural cuts but never common. It is known from central Brazil, Peru, and Bolivia, where it is recorded from near Rurrenabaque in Beni Department and the lowland forest areas of Santa Cruz Department, from the vicinity of the Flora and Fauna Hotel near Buena Vista, the Potrerillo del Guenda Reserve, and Refugio Los Volcanes, adjacent to Amboró National Park.

Compsibidion ytu

Martins, Galileo, & Oliveira, 2011

Cerambycinae: Neoibidionini

This is a small (7–9 mm), shining reddish-orange species with black antennae and legs except for the orange clavate portion of the femora. It is rarely collected and most often encountered at light. Its known distribution is very limited and includes northern Brazil and Santa Cruz Department in Bolivia, where it has been collected at Potrerillo del Guenda Reserve and Refugio Los Volcanes.

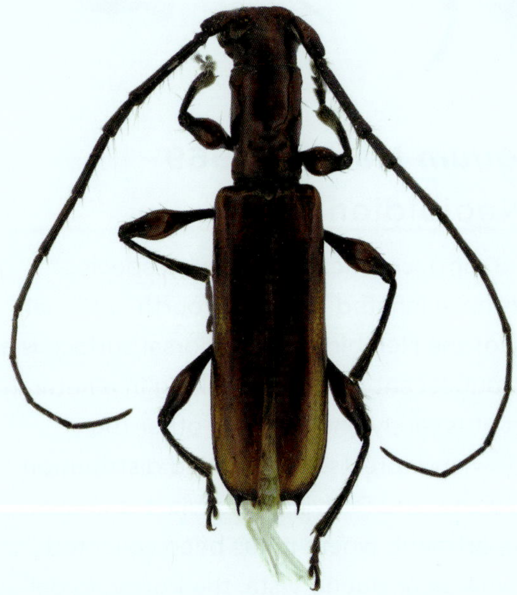

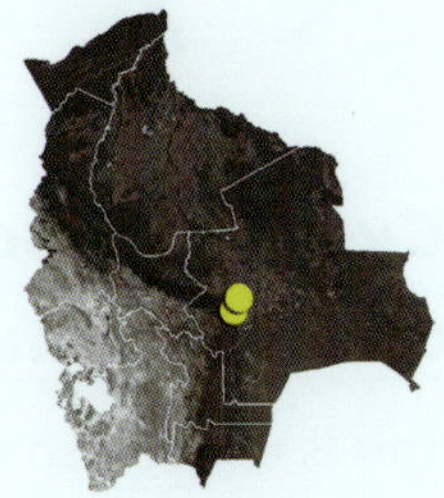

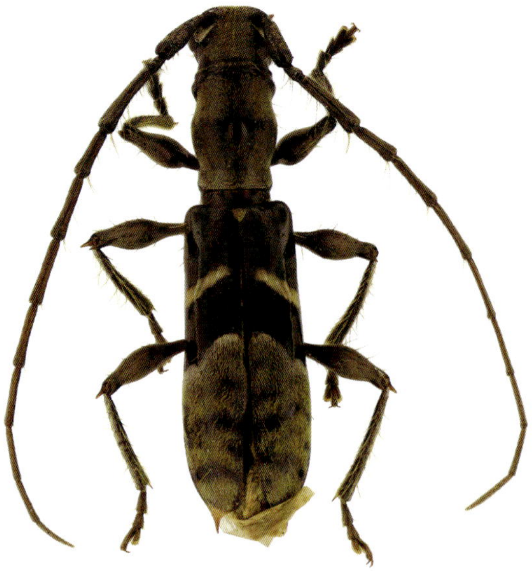

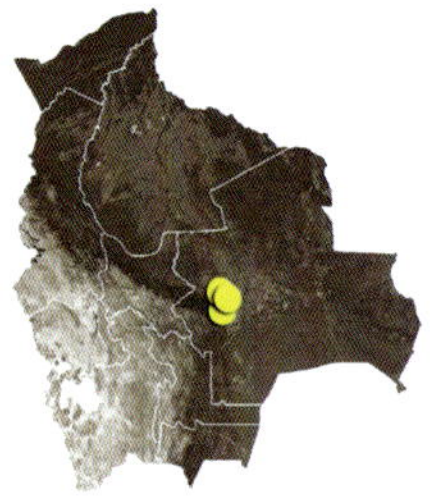

Cycnidolon obliquum Martins, 1969
Cerambycinae: Neoibidionini

This is another small (6–8 mm), cryptic, dark-colored species. The males of this species have inflated third and fourth antennal segments. As with many of the Neoibidionini its dorsal surface is a complex mix of intricate pubescent patterns resulting in a natural and beautiful work of art. It is rarely encountered other than occasionally at lights in lowland forests. Its recorded distribution includes Brazil, French Guiana, and Bolivia. In Bolivia it is known only from Santa Cruz Department, where it has been collected at the Flora and Fauna Hotel near Buena Vista, the Potrerillo del Guenda Reserve, and Refugio Los Volcanes, adjacent to Amboró National Park.

Diasporidion argentinense (Martins, 1962)
Cerambycinae: Neoibidionini

This is a very attractive, large neoibidionine, with specimens averaging 20 mm long. The body is mostly glossy, but there are a few widely separated rows of setae down each elytron. The legs, antennae, and pronotum are mostly reddish, while the elytra are darker blue. Just anterior of the middle of each elytron is a large, irregularly round, yellow macula. This species is known from Brazil, Argentina, and recently, Bolivia. Although never common, it has been collected near Villamontes in Tarija Department.

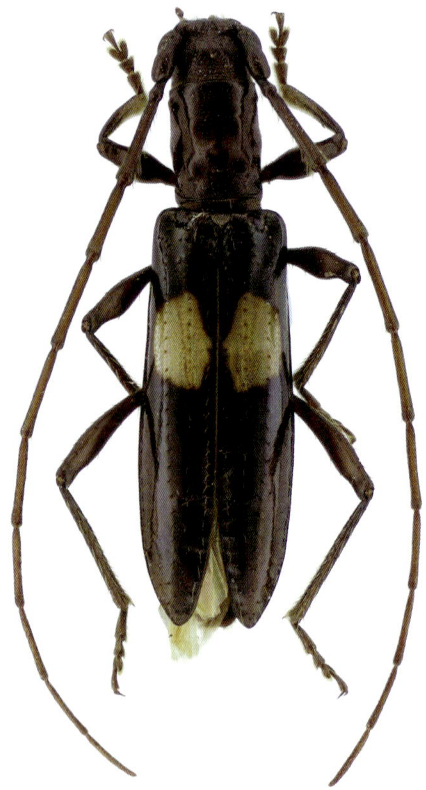

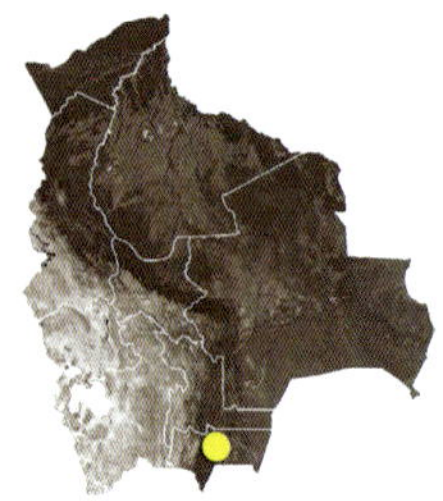

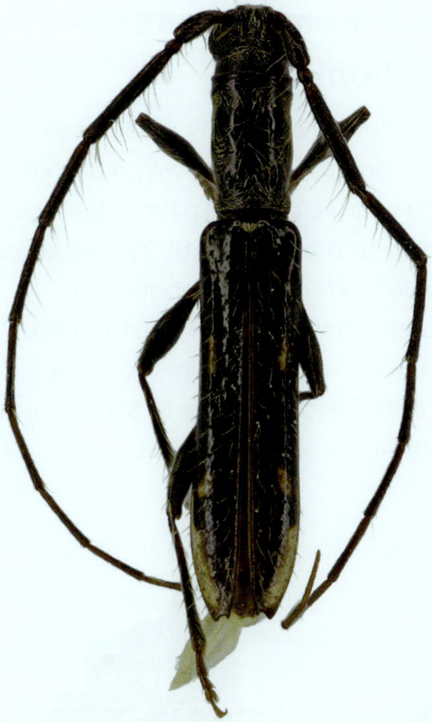

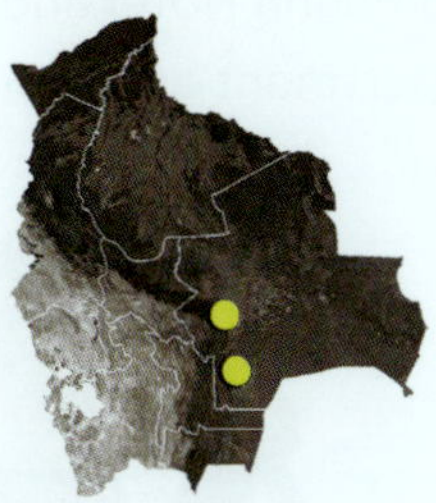

Dodecaibidion bolivianum Martins & Galileo, 2012

Cerambycinae: Neoibidionini

This is a small (7–9 mm), narrow, shining brown species with two small, light tan fasciae on each elytron, one linear and right before the middle and the other a spot at apical third. Rarely, in a few specimens, the fasciae are absent. This species is characterized by the very long setae on the antennal segments (longer than the width of the pronotum). This distinctive but plain-looking Bolivian endemic was only recently discovered and described from Refugio Los Volcanes in Santa Cruz Department. It has since also been found 83 km north of Camiri.

Engyum fasciatum Martins, 2009

Cerambycinae: Neoibidionini

This is a small (5–8 mm), shining yellow-orange species with two oblique black fasciae on the elytra: one before the middle facing forward and another after the middle facing backward. It has been collected at lights and beaten from fairly freshly cut hanging branches with the leaves still intact. This species is endemic to Bolivia, where it has been collected in Santa Cruz Department at Refugio Los Volcanes, adjacent to Amboró National Park, and the Vicoquin area above Achira at 2,000 m.

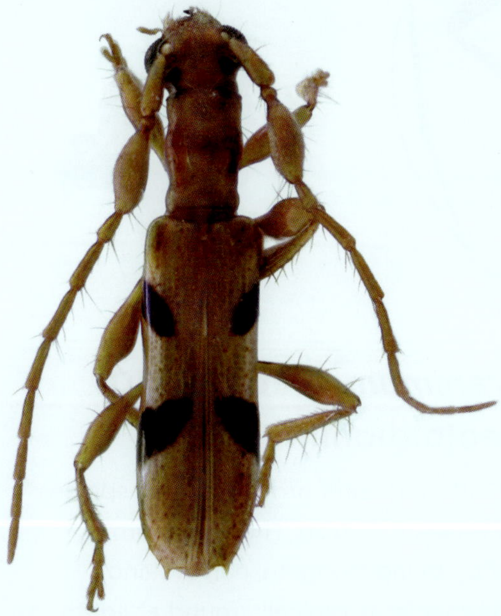

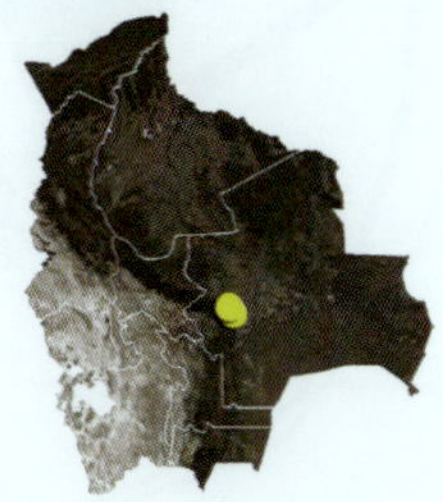

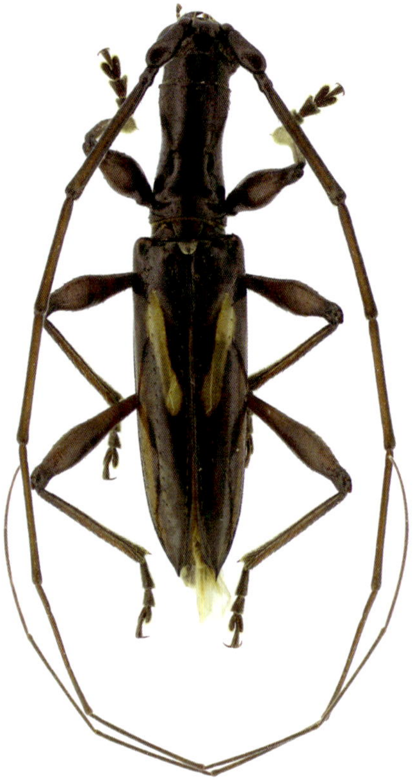

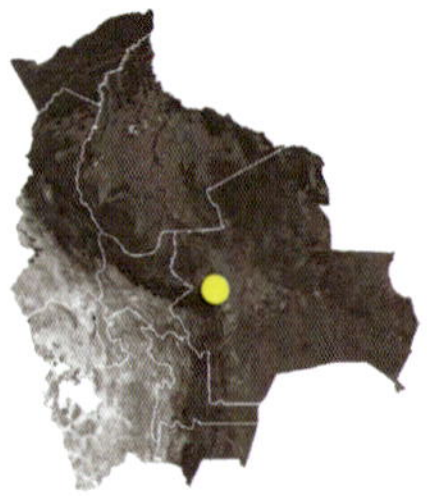

Gnomibidion digrammum (Bates, 1870)

Cerambycinae: Neoibidionini

This is a medium-sized (15–18 mm), dark brown, shining species with an oblique tan fascia on each elytron. The pronotum is virtually hourglass shaped by being elongate, narrow, and pinched in the middle. This species is typically found at lights and occasionally encountered at night coming to or crawling on medium to small freshly cut branches along the edges of agricultural *chaqueos*. It is distributed widely across the Amazon Basin of Brazil to French Guiana, Peru, and Bolivia, where it is known from only Santa Cruz Department in the vicinity of the Flora and Fauna Hotel near Buena Vista.

Gnomibidion fulvipes (Thomson, 1865)

Cerambycinae: Neoibidionini

This large, strikingly colored neoibidionine is easily recognized by the size (averaging about 18 mm) and the light brown color that is interrupted on each elytron by a large, circular, pale yellow spot extending from just behind middle to the basal sixth. The antennae are quite long, particularly in males, extending by about five segments beyond the elytral apices. This is a widely distributed species in South America. In Bolivia it is known from the Flora and Fauna Hotel and Potrerillo del Guenda Reserve in Santa Cruz Department and near Villamontes in Tarija Department.

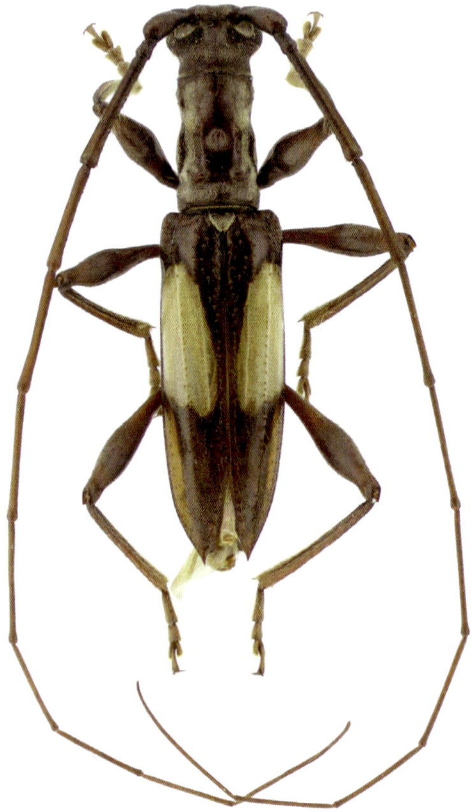

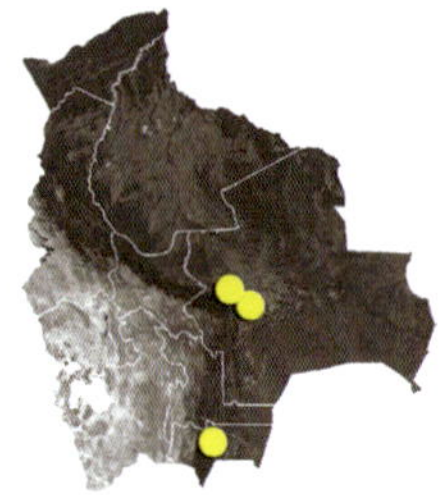

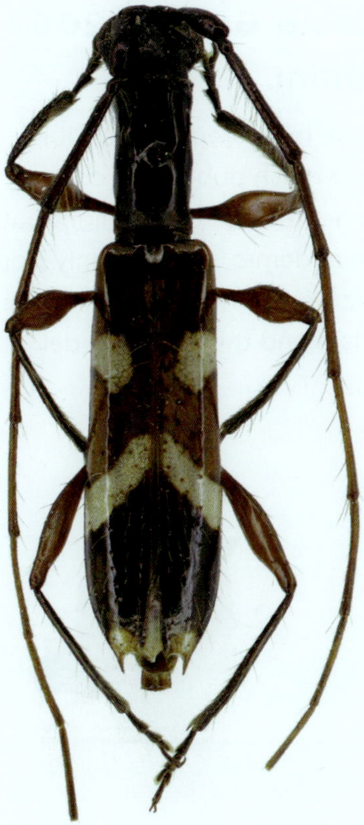

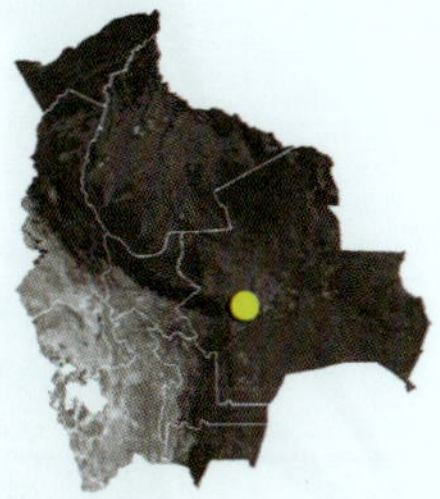

Heterachthes tysiphonis (Thomson, 1867)

Cerambycinae: Neoibidionini

This is a small (10–12 mm), narrow, parallel-sided, tricolored species with shining integument. Much of the dorsum is reddish orange with the head, femora, and most of the distal third of elytra piceous. Each elytron is decorated with three light beige fasciae: one small, circular spot basally, an oblique forward-directed band just past the middle, and the apices tipped in a semicircular spot. This species is attracted to lights but only rarely encountered. It is recorded from north-central Brazil, French Guiana, Peru, and Bolivia, where it is known from Santa Cruz Department in the vicinity of the Potrerillo del Guenda Reserve.

Heterocompsa eburata Martins, 1970

Cerambycinae: Neoibidionini

This is a small- to medium-sized (10–14 mm) cerambycid, shining dark orange in color and with black-tipped femora apically. Each elytron has two roughly oval, light tan spots with a dark-edged margin. This Bolivia endemic is known only from Santa Cruz Department, where it has been rarely collected at lights in the Refugio Los Volcanes reserve, adjacent to Amboró National Park, and Potrerillo del Guenda Reserve.

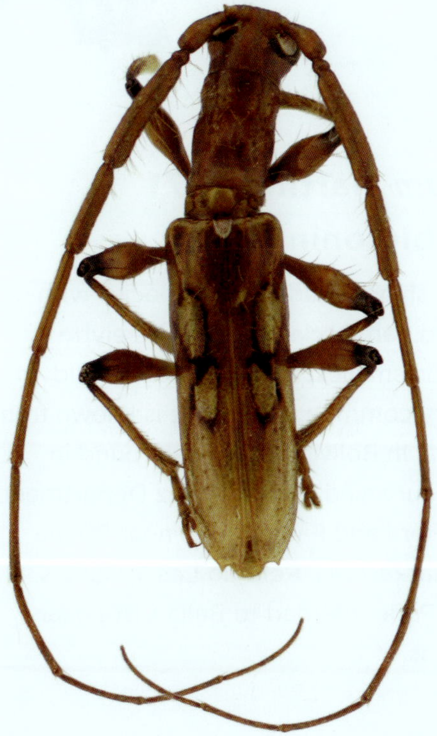

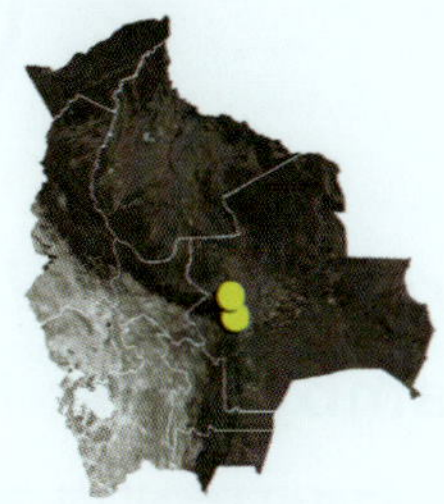

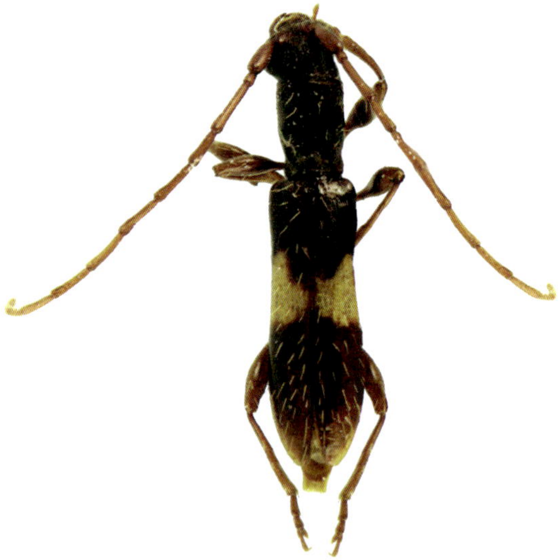

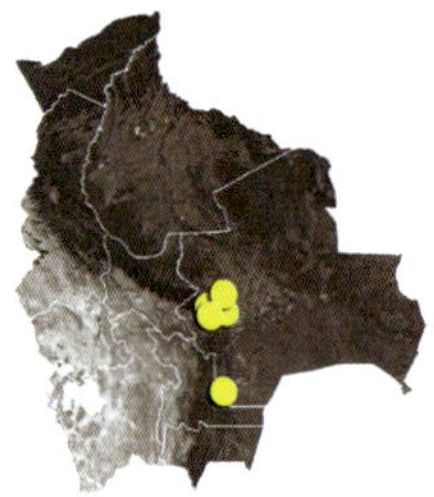

Microibidion mimicum Martins, 1971

Cerambycinae: Neoibidionini

This is a very small (4–6 mm), shining dark brown species with a transverse, broadly U-shaped, tan fascia crossing the elytra near the midline. It is one of the smallest Bolivian cerambycid species and can be locally very common at lights. It is known from northern Argentina and Bolivia. In Bolivia it has been found in most of the lowland localities surveyed in Santa Cruz Department, including Pampagrande, the Flora and Fauna Hotel near Buena Vista, the Potrerillo del Guenda Reserve, Refugio Los Volcanes, adjacent to Amboró National Park, the road to Bella Vista near Cuevas, and the Camiri environs.

Minibidion tricolor Martins & Galileo, 2007

Cerambycinae: Neoibidionini

This is a small (6–7 mm), parallel-sided, tricolored species. It is yellow to beige to dark brown with a whitish pubescent collar at the base of the pronotum. It is rarely seen other than at light but can be locally common. This Bolivian endemic is known only from lowland forest localities in Santa Cruz Department, including the Flora and Fauna Hotel near Buena Vista and the Potrerillo del Guenda Reserve.

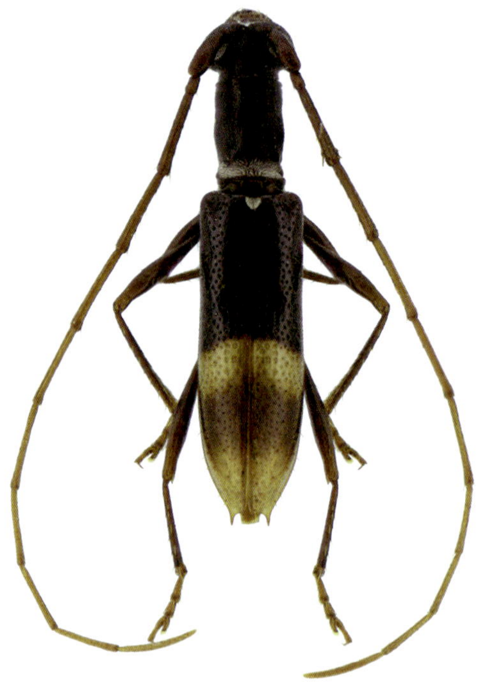

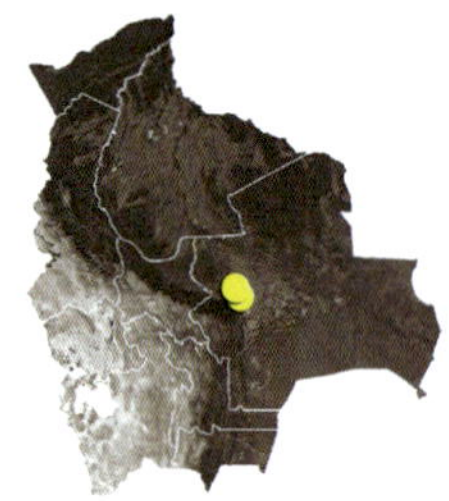

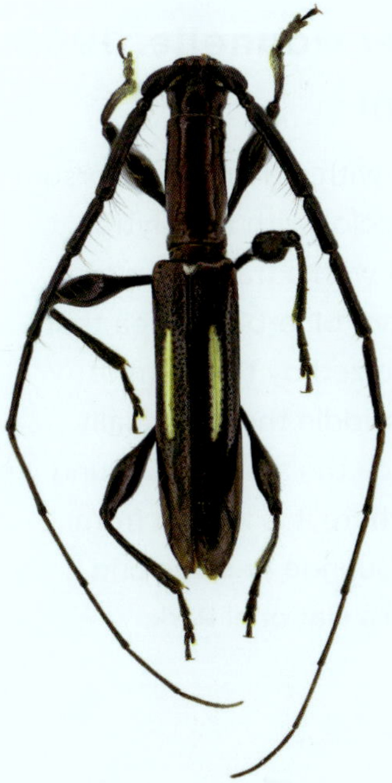

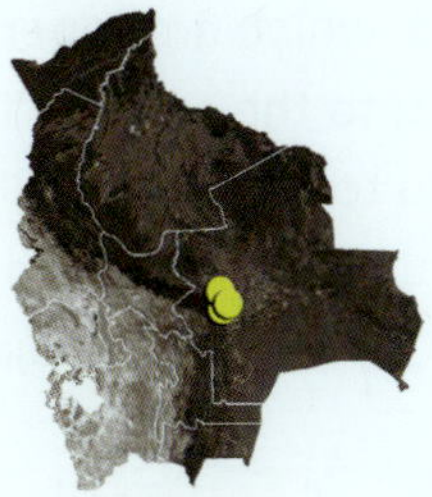

Neocompsa lineolata (Bates, 1870)

Cerambycinae: Neoibidionini

This is a narrow, parallel-sided species (9–11 mm) with shining dark reddish integument. Its antennae vary in color from medium brown to black and, along with the legs, have numerous long flying hairs. A pair of linear ivory fasciae, paralleling each other, are present on the elytra. Both males and females have the third through the fifth or sixth antennomeres expanded and compressed. This species has been collected by beating live leguminose trees in coffee plantations, although whether those trees are true larval hosts or simply gathering sites for the species is not known. It is also collected occasionally at lights. It is known from Central America south to Peru and Bolivia, where it has been found only in Santa Cruz Department in the vicinity of the Flora and Fauna Hotel near Buena Vista, the Potrerillo del Guenda Reserve, Refugio Los Volcanes, adjacent to Amboró National Park, and Bermejo.

Ophtalmoplon inerme Martins, 1965

Cerambycinae: Hexoplonini

This is a small (6–7 mm), compact, orange species with central beige markings on each elytron bordered behind by a narrow, dark fascia. It was recently transferred to the Hexoplonini, which was formerly in the Ibidionini (Neoibidionini). Although rarely encountered in Bolivia it is widely distributed in northern South America from the Guianas to central Brazil. In Bolivia it is known from just one location, the vicinity of the Flora and Fauna Hotel near Buena Vista in Santa Cruz Department.

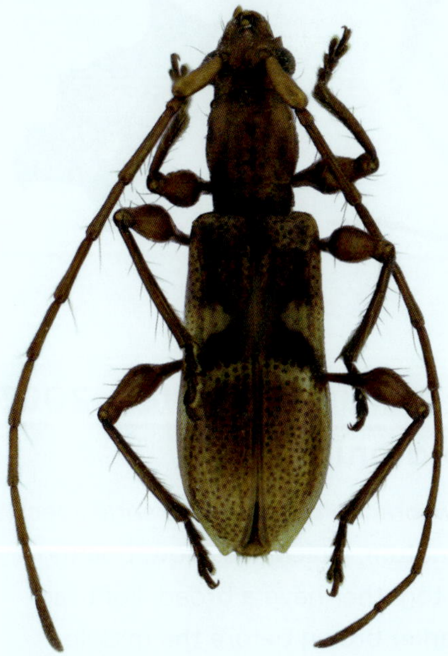

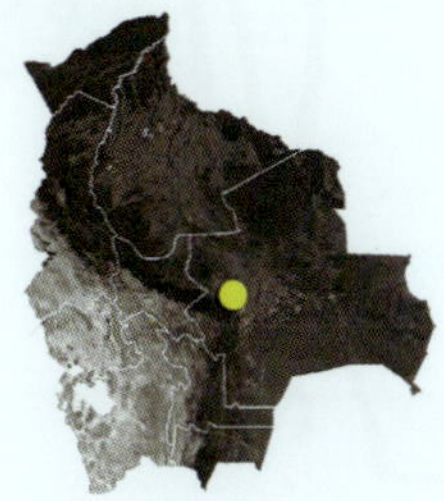

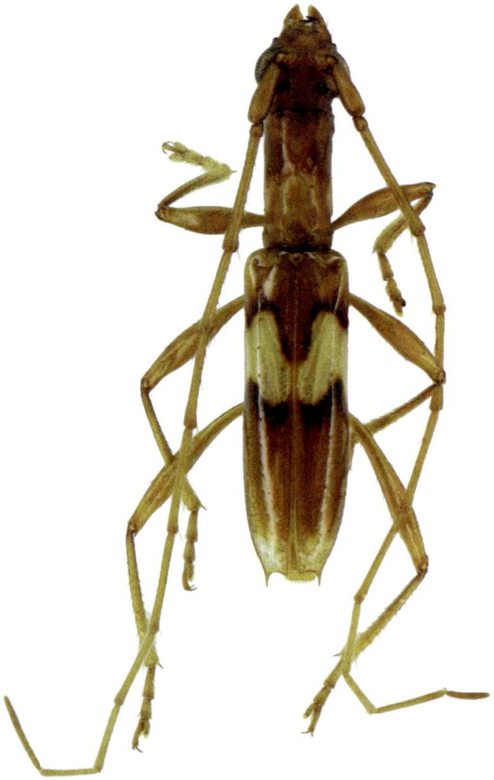

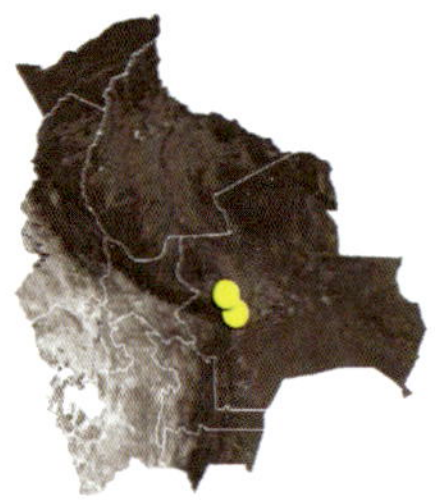

Palpibidion minimum Martins & Galileo, 2003

Cerambycinae: Neoibidionini

This is a small (6–9 mm), narrow, orange to reddish-orange species with an especially elongate pronotum, which has a low, pointed tubercle on the disk. The elytra together have a broad, light tan M-shaped fascia margined in darker brown before the middle. Both the elytral base and the apices are a noticeably lighter color. This species is rarely encountered, singularly attracted to lights, and never common. Its known distribution is limited to areas of Brazil adjacent to Bolivia and in Bolivia to Santa Cruz Department from the vicinity of the Flora and Fauna Hotel near Buena Vista and the Potrerillo del Guenda Reserve.

Perissomerus hilairei hilairei Gounelle, 1909

Cerambycinae: Neoibidionini

This is a small (9–11 mm), narrow species with a tricolored dorsum and shining integument. It is dark red in color with the elytra somewhat lighter orange red basally and with a transverse tan fascia across the middle followed by a dark blue-black area that continues to the apices. It is easily recognized by the form of the elytra, which are tapered (widest at middle then gradually narrowing to the apices). In South America the species is found from Peru to central Brazil and Bolivia, where it is known from Santa Cruz Department in Potrerillo del Guenda Reserve and Refugio Los Volcanes, adjacent to Amboró National Park.

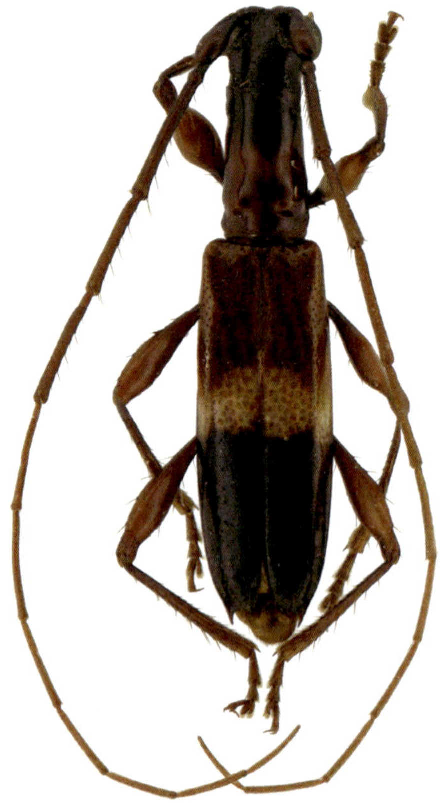

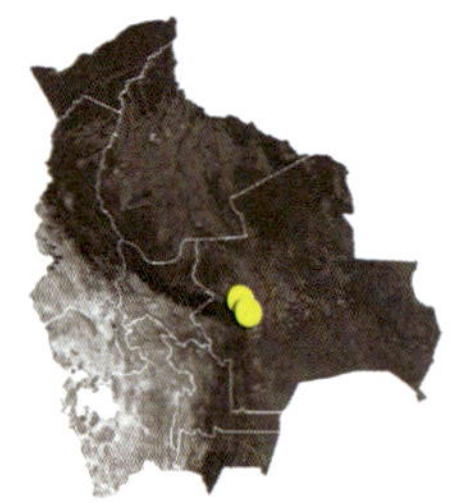

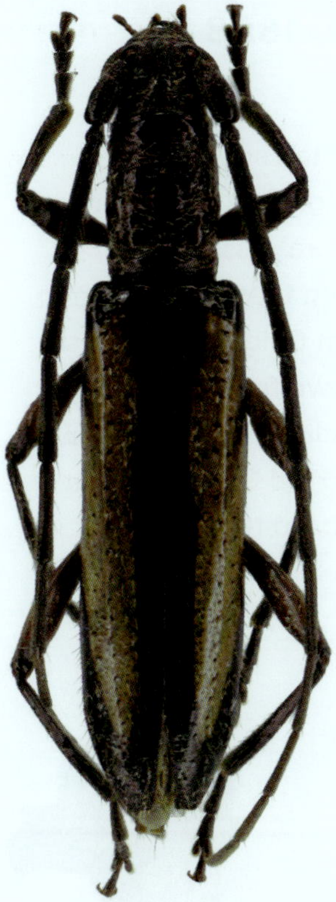

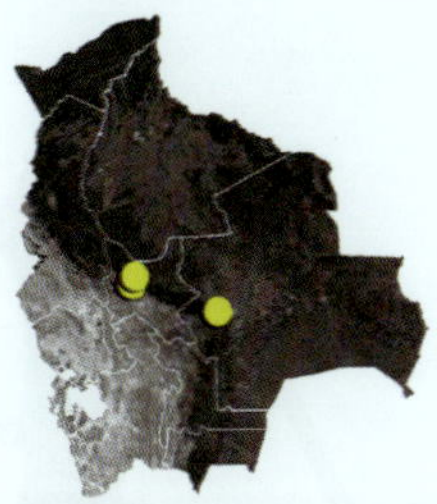

Prothoracibidion xanthopterum Martins, 1962

Cerambycinae: Neoibidionini

This is a midsized (10–14 mm), tan and dark red to dark brown species with sparse, shallowly punctate elytra and rugosely punctate pronotum with a small, shining central disk. The elytra have a broad tan fascia the length of each elytron that begins just after the humeri and ends just before the apices. This species is locally abundant at lights at higher elevations (higher than 1,500 m) but is also occasionally recorded from wet forests at lower altitudes. This Bolivian endemic is recorded from Cristal Mayu and Villa Tunari in Cochabamba Department and in Santa Cruz Department from the Vicoquin area above Achira at 2,000 m.

Psiloibidion leucogramma (Perty, 1832)

Cerambycinae: Neoibidionini

This is an elongate, medium-sized (15–22 mm), shining reddish-brown species with an elongate beige fascia at the middle of each elytron. Males have antennae nearly twice the length of their body whereas females' antennae are much shorter, typically exceeding the elytral apices by less than two segments. It is a fairly common species that comes to lights and can also be found at night coming to small- and medium-sized branches or trees along the edge of agricultural cuts. It is found in southern South America from central Brazil to Argentina and Paraguay. In Bolivia it is found in lowland forest localities in Santa Cruz Department, including the vicinity of the Flora and Fauna Hotel near Buena Vista, the Potrerillo del Guenda Reserve, and north of Camiri.

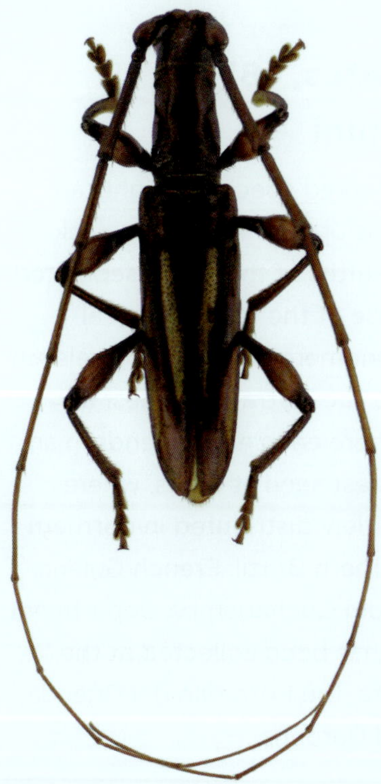

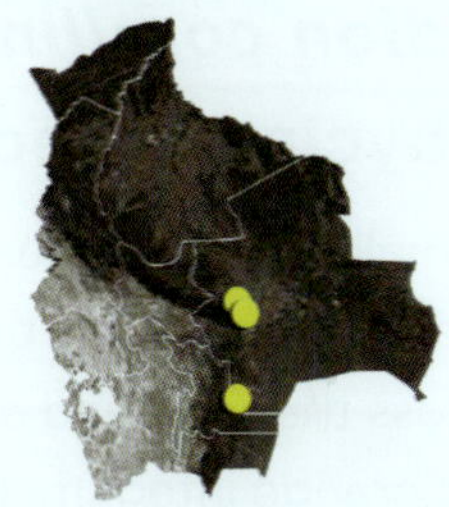

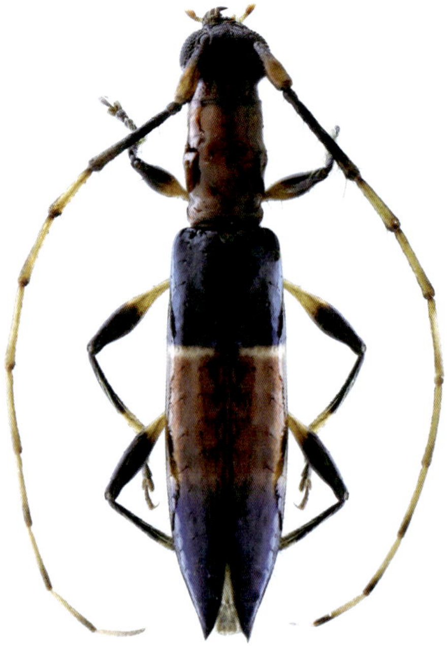

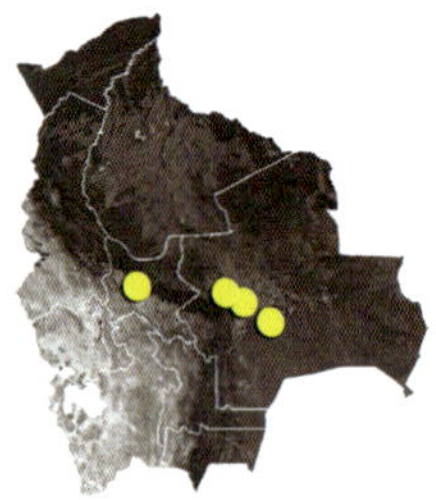

Stenoidion corallinum (Bates, 1870)

Cerambycinae: Neoibidionini

This is a small (7–9 mm), beautifully marked species of light tan, orange, and blue black. Its body color is orange with blue-black bands across the base and apical two-fifths of the elytra separated by shining orange integument. The base of the femora, apical portion of the tibiae, scape, and antennomeres four through eleven are light tan with the remainder black. The elytra are widest at the middle and taper to the apices, where each elytron ends in an acute spine. Rarely encountered, it is best seen at lights, where even then it is very uncommon. It is widely distributed in northern South America being known from northern Brazil, French Guiana, Peru, and Bolivia, where it is known from Cochabamba Department and Santa Cruz Department, where it has been collected at the Flora and Fauna Hotel near Buena Vista, the Potrerillo del Guenda Reserve, and the Santa Cruz Botanical Gardens.

Sydax stramineus Lacordaire, 1868

Cerambycinae: Sydacini

This is a delicate, very narrow-bodied, light tan species (13–15 mm) with a distinctive shape. Males have a greatly inflated third antennal segment and expanded front femora. It is seldom encountered other than at lights, where even there it is a rare visitor. Its known distribution stretches from Colombia to most of Brazil and into Bolivia. In Bolivia it is known from only a few localities in Santa Cruz Department that include the Flora and Fauna Hotel near Buena Vista, the Potrerillo del Guenda Reserve, and Refugio Los Volcanes, adjacent to Amboró National Park.

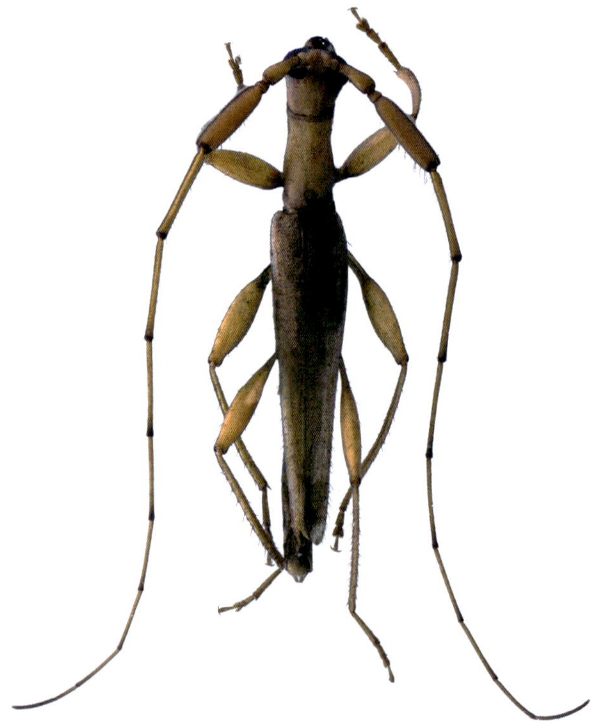

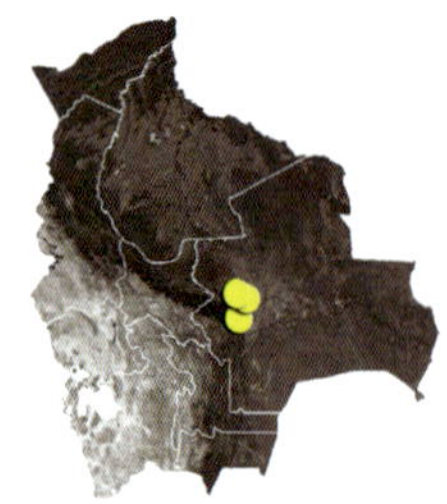

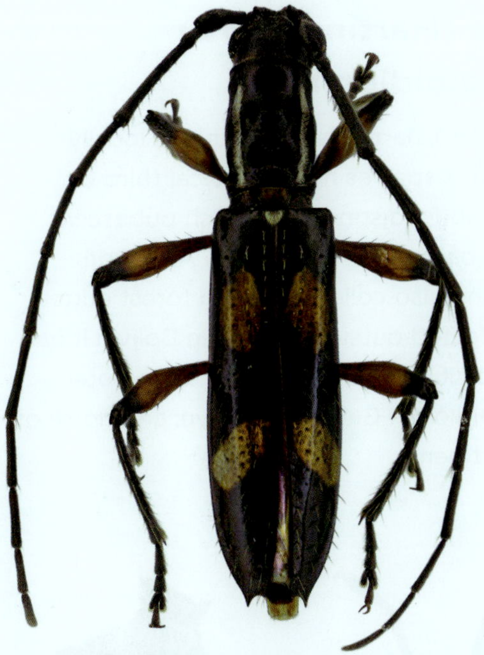

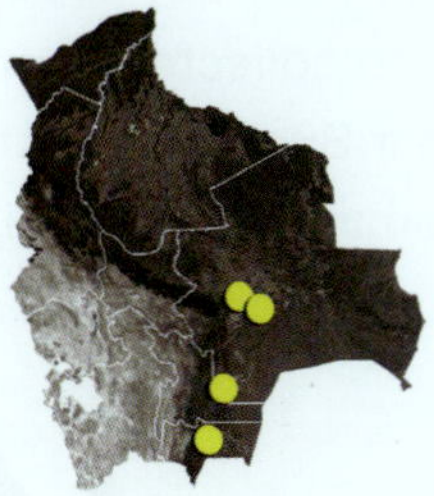

Thoracibidion flavopictum (Perty, 1832)

Cerambycinae: Neoibidionini

This is another of the small (12–13 mm), narrow, multicolored species that is attracted to lights but rarely otherwise encountered. Its dorsum is primarily black with two orange bands on the elytra: one behind the humeri, covering much of the basal half, the other narrower and placed midway across the apical half. It has a centrally rugose pronotum bordered laterally by whitish pubescence fascia. It is found throughout most of central and southern Brazil, Argentina, and Paraguay, but is never common. It is known from the Potrerillo del Guenda Reserve, Rio Selva near Warnes, and the Camiri environs in Santa Cruz Department in Bolivia. It has also been collected in Tarija Department west of Villamontes.

Tropidion contortum Martins, 1968

Cerambycinae: Neoibidionini

This is a small (10–13 mm), dark reddish-brown, black, and tan species with a heavily pubescent pronotum that obscures much of the surface. The elytra are black with a variable, tan-colored pattern typically composed of an elongate spot followed by an inverted V. Often these two markings are connected, appearing as a single, larger fascia. The head, legs, and antennae are primarily a dark reddish brown. This is probably the most common small species attracted to lights, where it can be locally abundant and extremely active on the collecting sheet. Surprisingly, this Bolivian endemic species has been collected at virtually all survey localities in the lowlands of Santa Cruz Department, including the vicinity of the Flora and Fauna Hotel near Buena Vista, the Potrerillo del Guenda Reserve, Refugio Los Volcanes, adjacent to Amboró National Park, and around Camiri. It has also been collected in Tarija Department west of Villamontes.

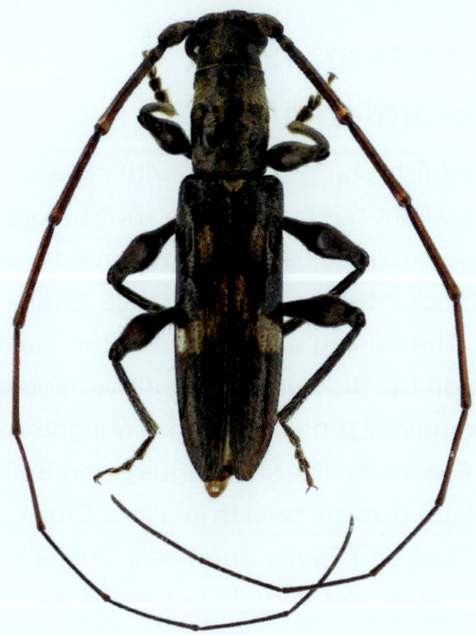

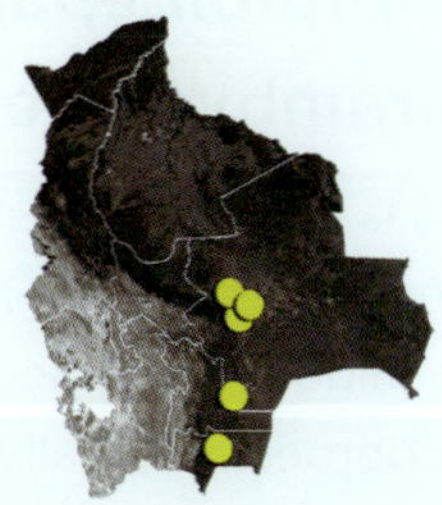

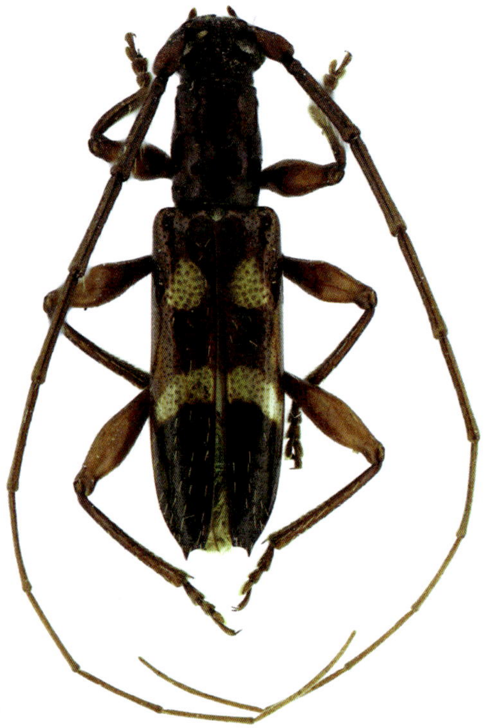

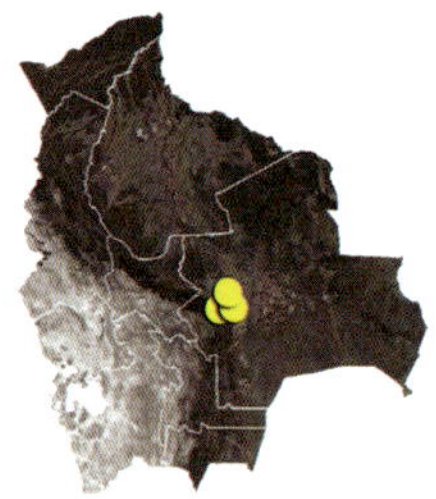

Tropidion epaphum (Berg, 1889)

Cerambycinae: Neoibidionini

This is a small (6–10 mm), reddish-orange species with closely punctate elytra that have two light tan fasciae, one spot before the middle, the other a slightly oblique band after the middle, and are bordered apically by a black integumental area that covers the remainder of the elytra. The shining dark red pronotum has a pair of low, round tubercles on the disk just apical of the middle. This species is one of the commonest neoibidionines at lights in Bolivia. It is recorded from Argentina, Brazil, Paraguay, and Bolivia, where it is known from lowland tropical forests in Santa Cruz Department, including the Flora and Fauna Hotel near Buena Vista, the Potrerillo del Guenda Reserve, and Refugio Los Volcanes.

Tropidion pusillum (Martins, 1960)

Cerambycinae: Neoibidionini

This is another small (6–9 mm), narrow species that is locally abundant. This primarily black species has the apical third or fourth of the elytra orange and a distinctive whitish pubescent band clearly evident at the base of the pronotum. Although locally abundant at lights and also collected along forest edges, thus far it has not been collected outside Bolivia. In Bolivia it has been collected from the vicinity of the Flora and Fauna Hotel near Buena Vista, the Potrerillo del Guenda Reserve, and north of Camiri in Santa Cruz Department.

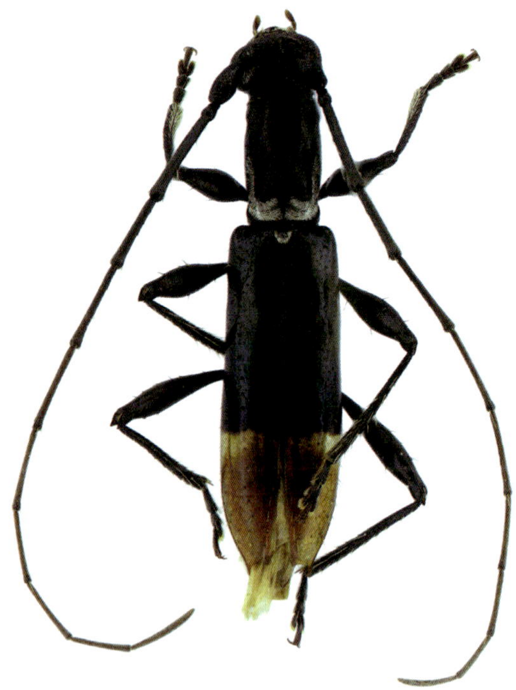

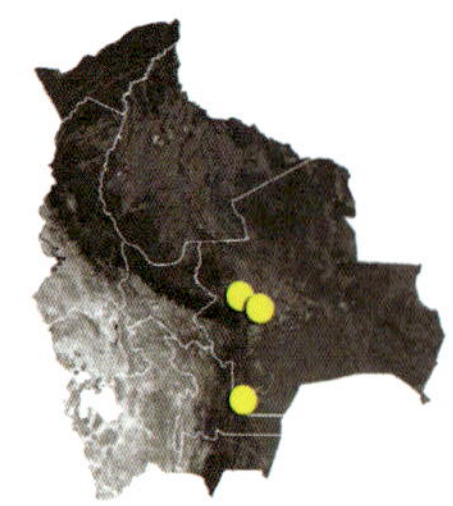

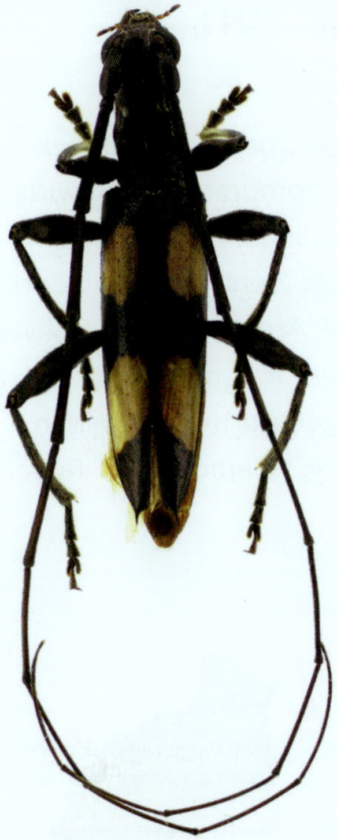

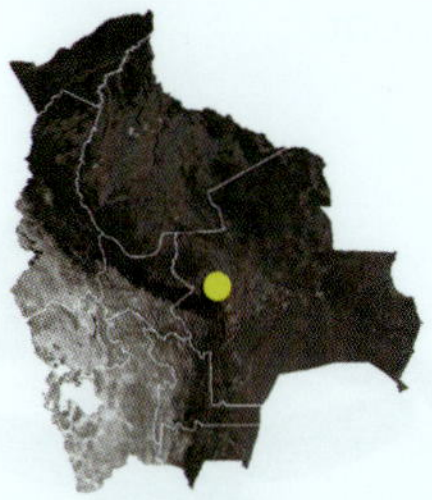

Tropidion subcruciatum (White, 1855)

Cerambycinae: Neoibidionini

This is a midsized to large (20–24 mm), dark reddish-brown species with two pairs of large tan fasciae on the elytra: one separated pair on the basal half and the other a contiguous pair centered apically. This is likely the largest *Tropidion* species and has tapered elytra narrowing apically. Its pronotum has a mix of low tubercles on the disk and very fine, short, whitish, recumbent pubescence that partially obscures the surface. This uncommon species is occasional at lights but never abundant. It is known from Brazil, Colombia, the Guianas, Venezuela, and Bolivia, where it has been recorded in Santa Cruz Department at the Flora and Fauna Hotel near Buena Vista.

Obrium cicatricosum Gounelle, 1909

Cerambycinae: Obriini

This is among the smallest (5–6 mm) of the Bolivian cerambycid species and one of four similar-sized and -colored *Obrium* species recorded from the country. All have a pattern of several fine, brown, narrow fasciae on the lighter tan elytra and pronotum. These patterns are diagnostic for each species, which are otherwise structurally very similar. Although occasionally found on dead twigs and vines along the edge of agricultural cuts, they are more typically collected at lights. This is a widely distributed species in southern South America and is known from southeastern Brazil, Argentina, Paraguay, and Bolivia, where it is recorded from Santa Cruz Department in Bermejo, near Warnes, the Flora and Fauna Hotel near Buena Vista, and near Camiri.

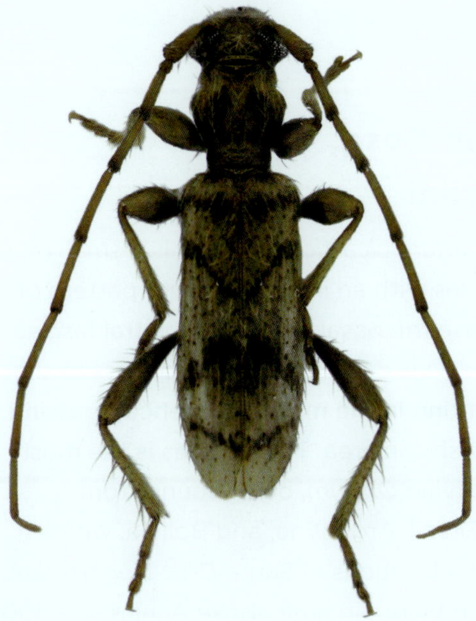

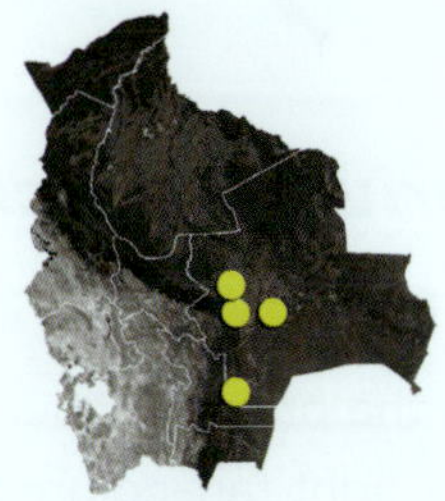

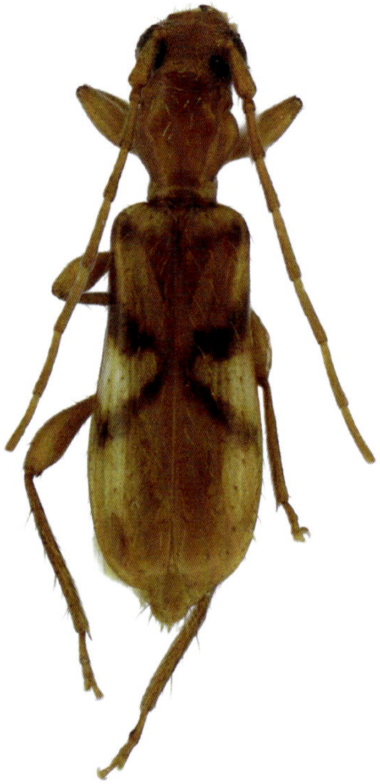

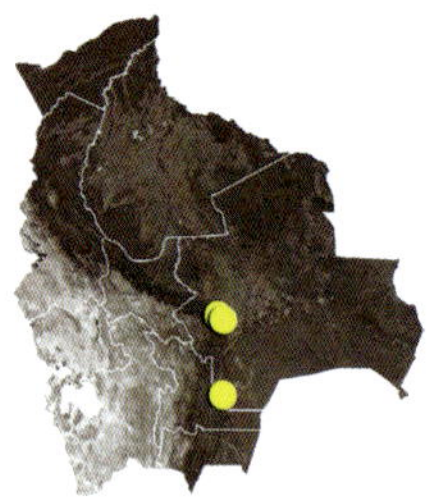

Obrium trifasciatum Bosq, 1951

Cerambycinae: Obriini

Like the previous species, *Obrium cicatricosum,* this is a small (4–6 mm), light tan-colored species with an intricate elytral pattern of fine brown fasciae. In this case the basal and apical elytral fasciae are subtle but the central fascia is dark brown and appears as the halves of two diamonds meeting in the middle and encompassing a much lighter, distinct, whitish-tan area. This species is the most widely distributed of the Bolivian *Obrium*, being found from French Guiana to southern Brazil, Argentina, and Bolivia, where it is recorded from mid to high elevations in Santa Cruz Department, including north of Camiri, the Vicoquin area above Achira at 2,000 m, Refugio Los Volcanes, adjacent to Amboró National Park, and the road to Bella Vista near Cuevas.

Argentinoeme schulzi Bruch, 1911

Cerambycinae: Oemini

This is a medium-sized (13–16 mm), orange species with black antennae, tibiae, and tarsi. Its femora are compressed and widened like a blade or paddle. Its pronotum has a linear raised ridge with minute tubercles that makes it readily identifiable. In South America it is recorded from central Brazil, Argentina, and Bolivia. In Bolivia it is known from the Chaco where it comes to lights but is never abundant. It has been collected north of Camiri in southern Santa Cruz Department and near Villamontes in Tarija Department.

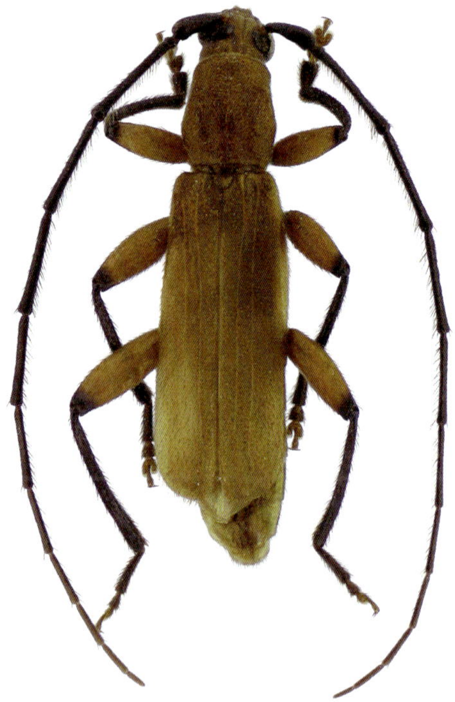

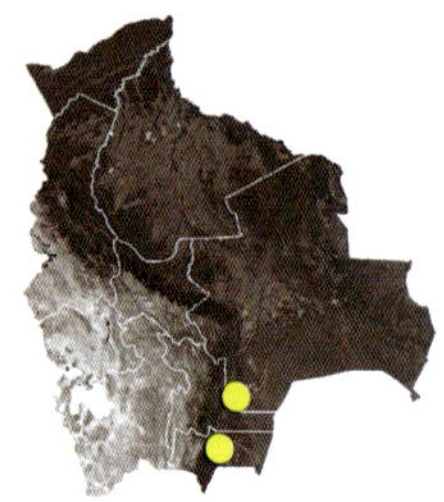

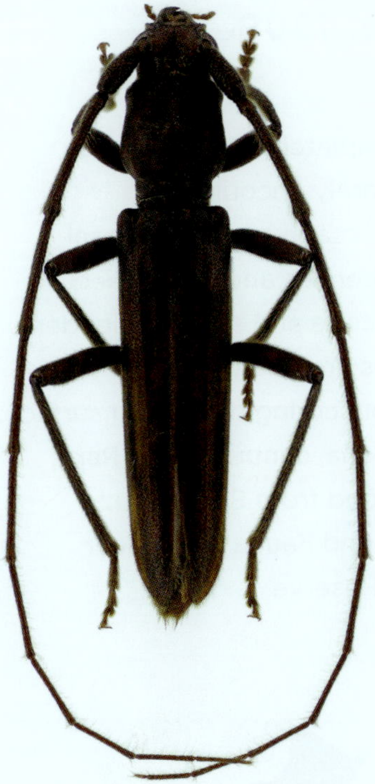

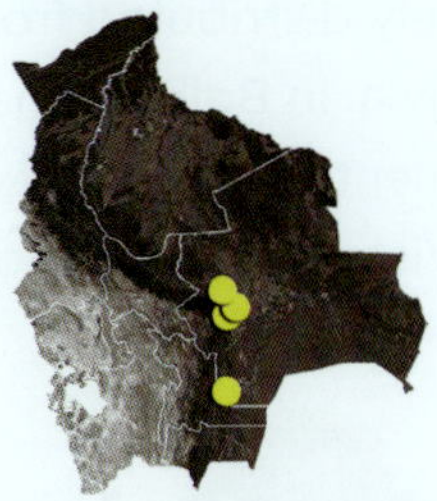

Macroeme priapica (Thomson, 1857)
Cerambycinae: Oemini

This is one of the largest oemines at 22–26 mm long. It is tan to medium brown with a densely punctate head, minutely, closely punctate pronotum, and moderately punctate elytra. Each elytron has a lateral, narrow, dark fascia and a dark, contiguous, sutural fascia. This species is very setose with variable length golden hairs on the appendages and elytra. It is a common species that is attracted to lights in numbers during the early rainy season. It is widely distributed in Brazil and also known from Argentina, Paraguay, Uruguay, and Bolivia, where it is recorded from Santa Cruz Department from north of Camiri, the vicinity of the Flora and Fauna Hotel near Buena Vista, the Potrerillo del Guenda Reserve, and Refugio Los Volcanes near Bermejo.

Malacopterus tenellus (Fabricius, 1801)
Cerambycinae: Oemini

This is one of the larger members of the tribe Oemini (20–24 mm) and is beige in color with slightly darker lines the length of each elytron. The femoral tips are black and third to fifth or sixth antennomeres are apically black as well. It is a widely distributed species in the New World, being found from the southwestern United States and the West Indies to central Brazil and Bolivia, where it is known from Santa Cruz Department along the Bella Vista road near Cuevas, the Flora and Fauna Hotel near Buena Vista, Potrerillo del Guenda Reserve, and Refugio Los Volcanes, adjacent to Amboró National Park.

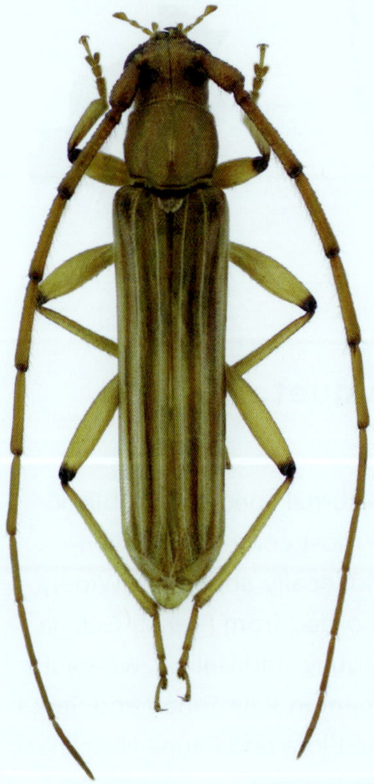

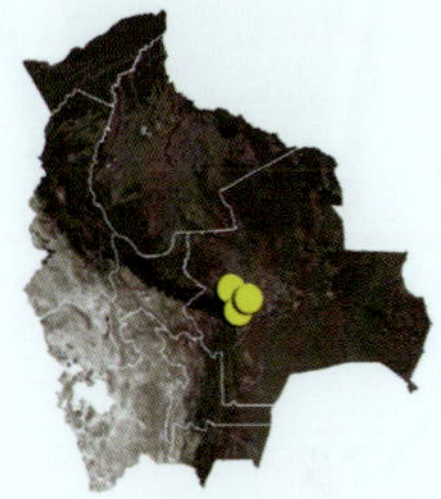

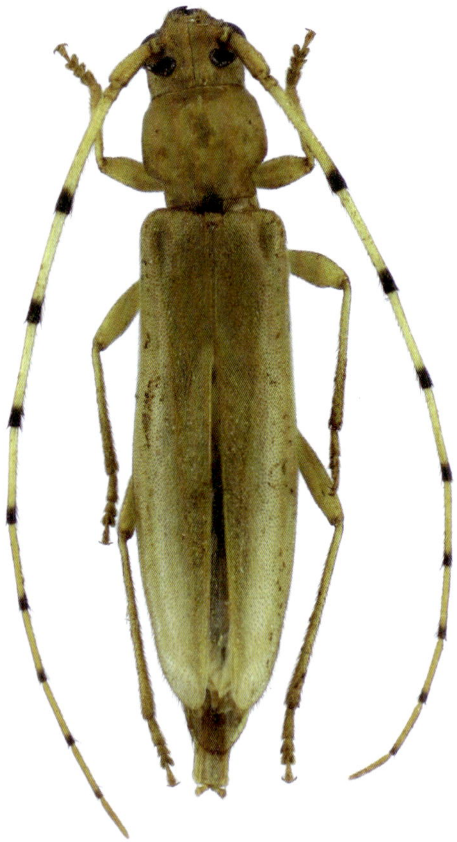

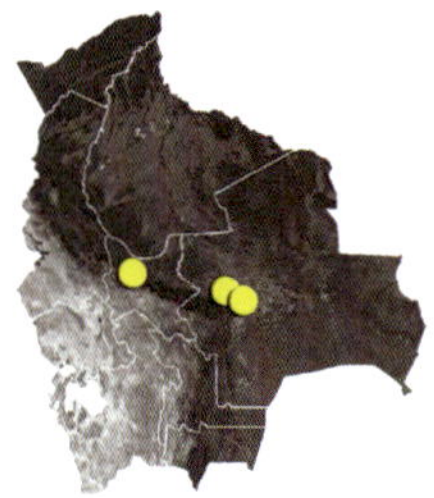

Neoeme annulicornis (Buquet, 1859)

Cerambycinae: Oemini

This is a large (17–20 mm), beige, nocturnal species with black annulate antennae. It is probably the most common Bolivian oemine, readily attracted to lights and locally abundant. Widely distributed in South America, it is recorded from French Guiana, eastern and southeastern Brazil, Paraguay, and Bolivia, where it is known from Cochabamba Department in Villa Tunari and Santa Cruz Department in the vicinity of the Flora and Fauna Hotel near Buena Vista and Potrerillo del Guenda Reserve.

Neoeme bouvieri Gounelle, 1909

Cerambycinae: Oemini

This is a medium-sized (14–16 mm), completely brown species, without fascia or other markings. It is rarely encountered except at lights. The males have a large scape and stout basal antennomeres, covered with minute tubercles and intermixed long hairs, unlike the female's smaller scape and smooth slender antennomeres (shown here). Both sexes are finely, densely pubescent with the vestiture partially obscuring the dorsal surface. It is widely distributed from French Guiana, central Brazil, Peru, and Bolivia. In Bolivia it has been recorded from Santa Cruz Department at the vicinity of the Flora and Fauna Hotel near Buena Vista and Potrerillo del Guenda Reserve.

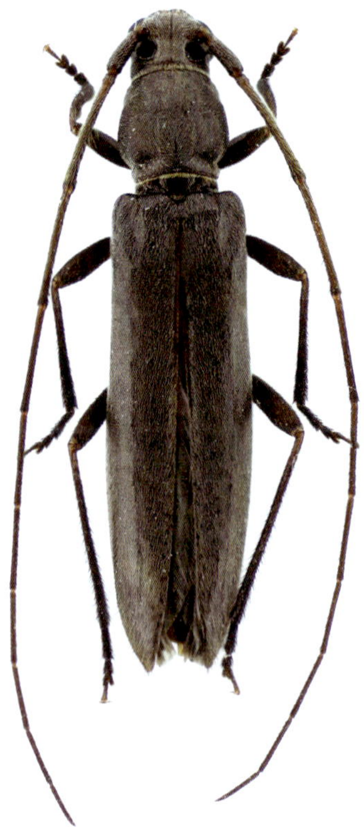

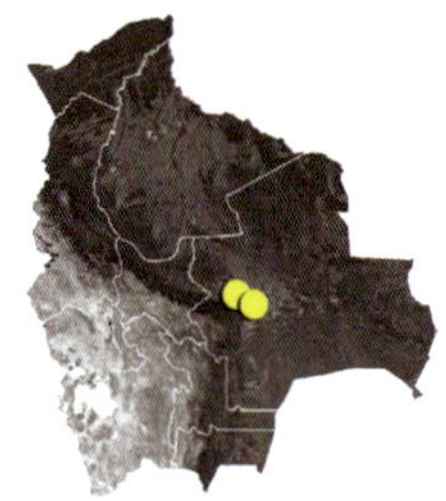

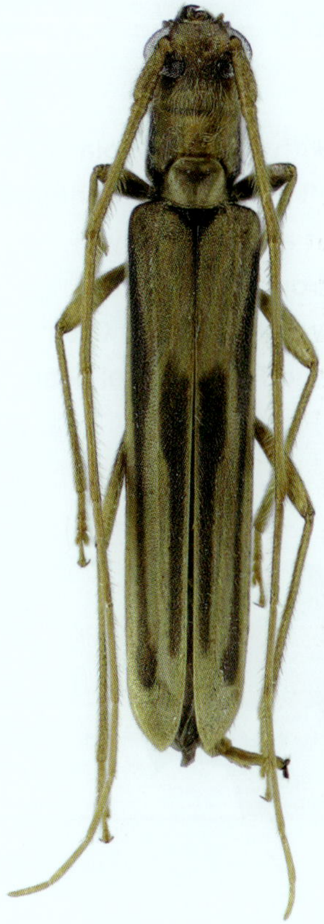

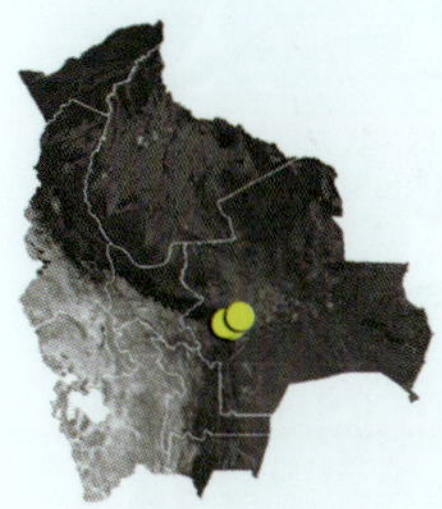

Placoeme wappesi Galileo & Martins, 2010

Cerambycinae: Oemini

This is a midsized (13–22 mm), light tan, elongate species with black fasciae on the sides of the pronotum and lateral margin of the elytra. There is a brown, linear fascia on each elytron that begins before the middle and extends nearly to the apices. This pale-looking, nocturnal species was only recently discovered and described and is endemic to Bolivia. It is known from only a few locations in Santa Cruz Department, including the Potrerillo del Guenda Reserve, Refugio Los Volcanes, adjacent to Amboró National Park, and the road to Bella Vista near Cuevas.

Sphagoeme sahlbergi Aurivillius, 1893

Cerambycinae: Oemini

This is a small (7–11 mm), light tan species with an orange pronotum, black-tipped elytra, and black, annulate antennomeres. Its dorsum is minutely, densely punctate. It is rarely encountered except at lights where, early in the rainy season, it can be locally abundant in lowland forest locales. It is known from Brazil and Bolivia, where it is recorded from Santa Cruz Department from the vicinity of the Flora and Fauna Hotel near Buena Vista and the Potrerillo del Guenda Reserve.

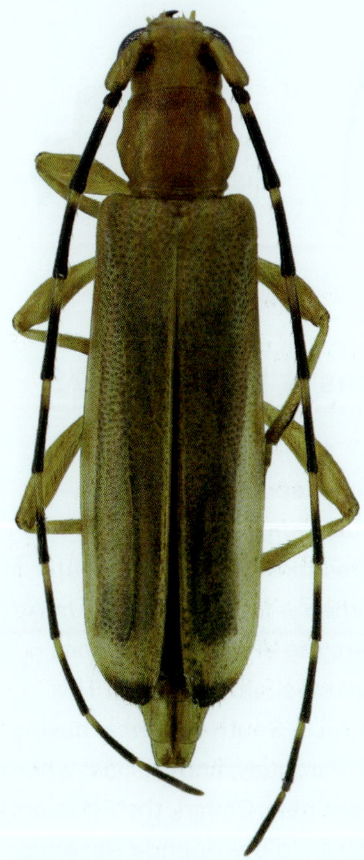

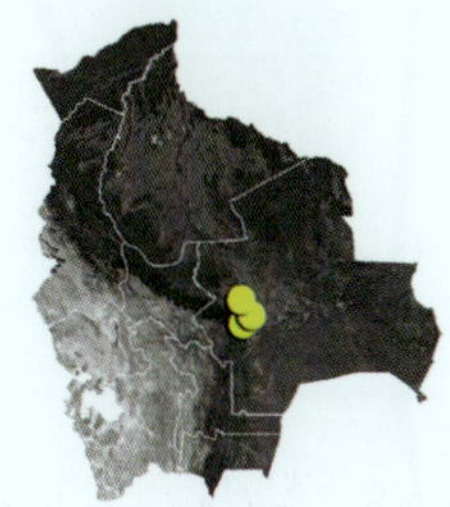

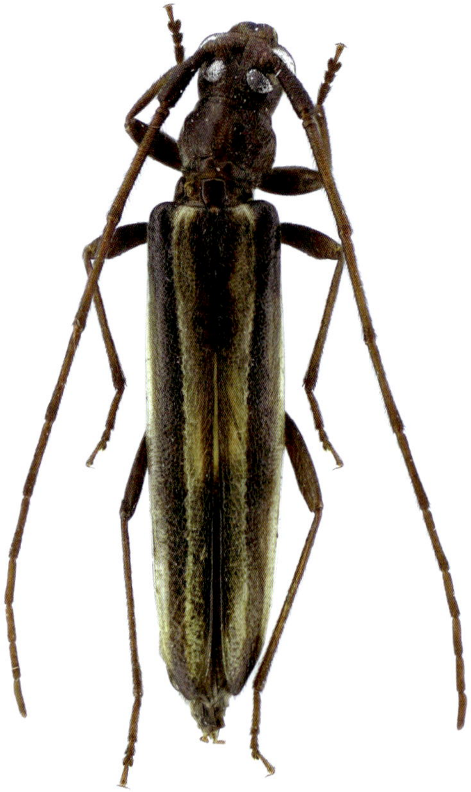

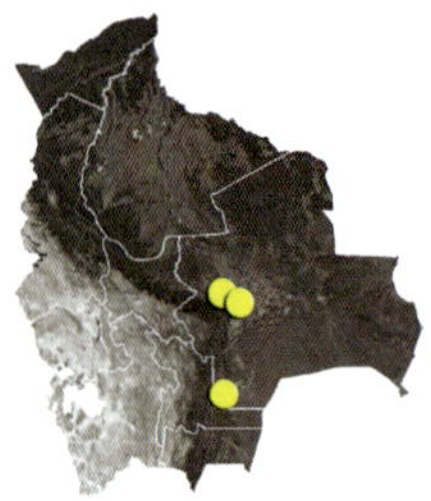

Temnopis latifascia Martins & Monné, 1975

Cerambycinae: Oemini

This is a medium-sized (9–14 mm), brown species with yellow-fasciate elytra along the suture and outer angle. It is sexually dimorphic: the scape and antennomeres three to five are stouter in males than in females (a female specimen is pictured here). Unlike on the closely related *Neoeme bouvieri*, the elytral pubescence is shorter and scattered, leaving the elytra slightly shining. This species has a more southern distribution in South America, having been recorded from Argentina, Brazil, Paraguay, and Bolivia, where it is known from Santa Cruz Department near Camiri, the Flora and Fauna Hotel near Buena Vista, and Potrerillo del Guenda Reserve.

Temnopis oculata Zajciw, 1960

Cerambycinae: Oemini

This is a small (8–15 mm), orange species with dark legs, black antennae, and costate black-tipped elytra. It is rarely encountered at lights in lowland tropical forests, where it appears early in the rainy season. This species is recorded from central Brazil, French Guiana, and Bolivia, where it is known from Santa Cruz Department from the vicinity of the Flora and Fauna Hotel near Buena Vista and the Potrerillo del Guenda Reserve.

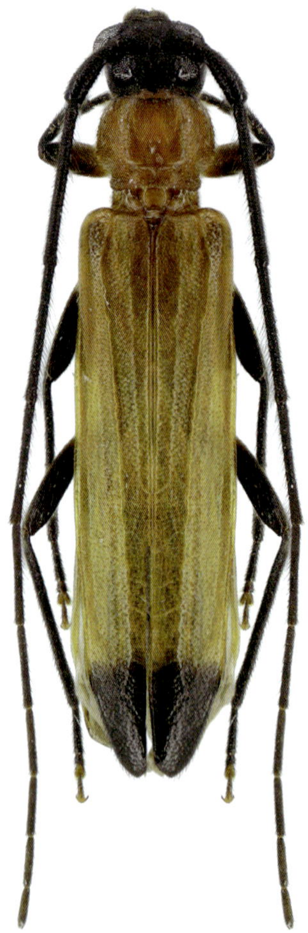

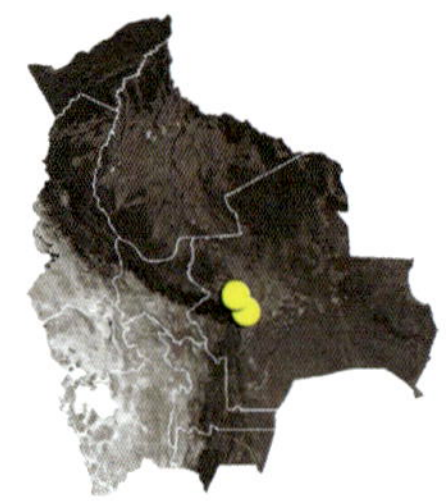

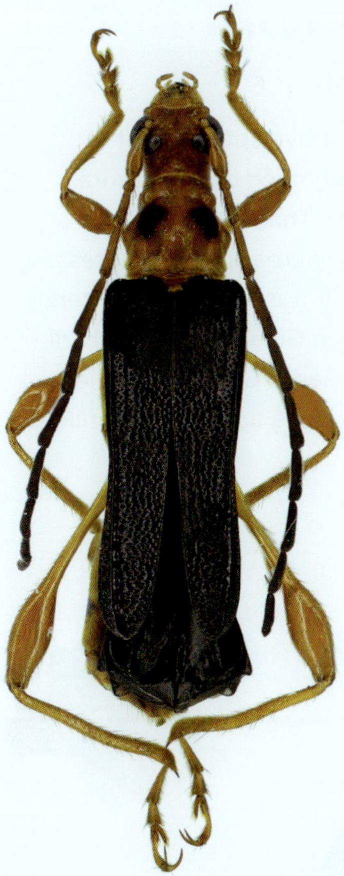

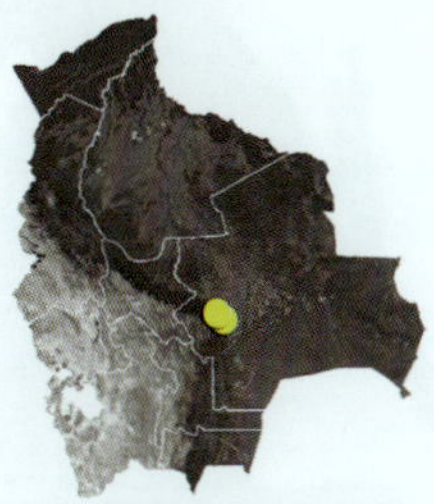

Oxycoleus flavipes Martins & Galileo, 2006

Cerambycinae: Oxycoleini

This is a small (9–10 mm), shining orange species with densely punctate, laterally costate, black elytra. The elytra are shorter than the abdomen, thus revealing the tips of the folded wings underneath. This diurnal cerambycid is a woody plant flower visitor and rarely encountered. It is a Bolivian endemic known only from a couple localities in Santa Cruz Department, including the road to Bella Vista (near Cuevas) and the Vicoquin area above Achira at 2,000 m.

Phoracantha semipunctata (Fabricius, 1775)

Cerambycinae: Phoracanthini

This is a midsized (20–25 mm), dark brown species with a pattern of light tan fasciae on the elytra. It is not a native species but appeared with its natural host, *Eucalyptus*, when numbers of that tree were imported from Australia to Bolivia years ago. It is very destructive to its host, killing many of the larger trees by continually reinfesting them. Fortunately, it is not known to attack any native plants. This species is found throughout the New World wherever *Eucalyptus* has been introduced. In Bolivia it is recorded from Santa Cruz around Samaipata but is very likely much more widespread.

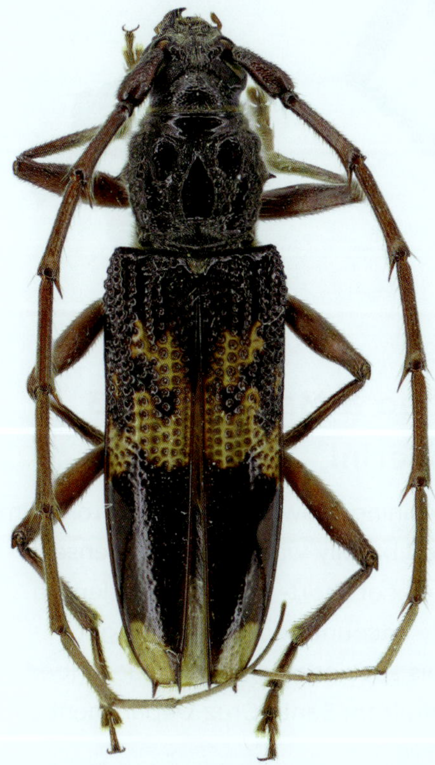

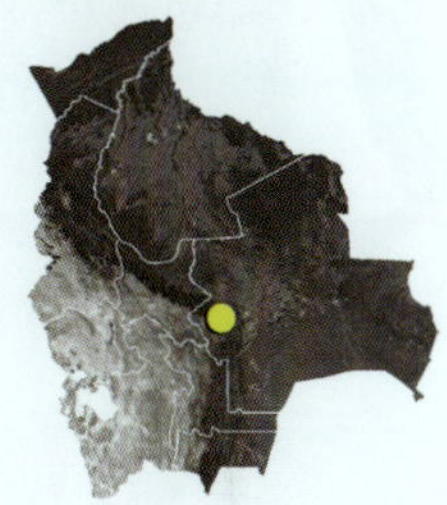

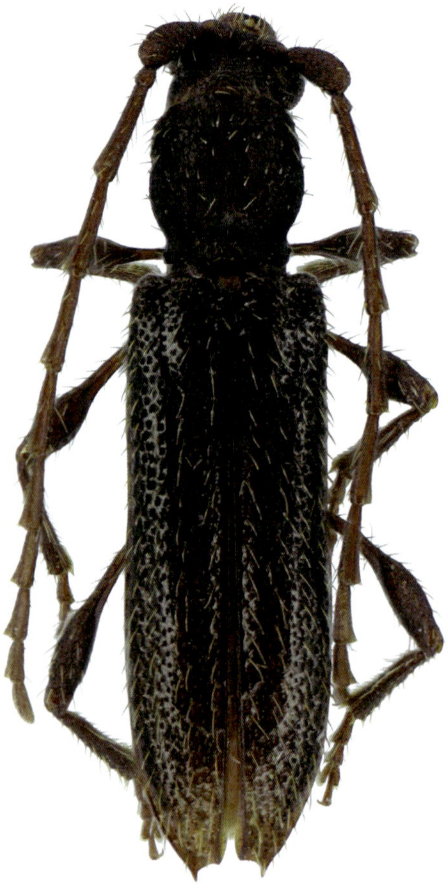

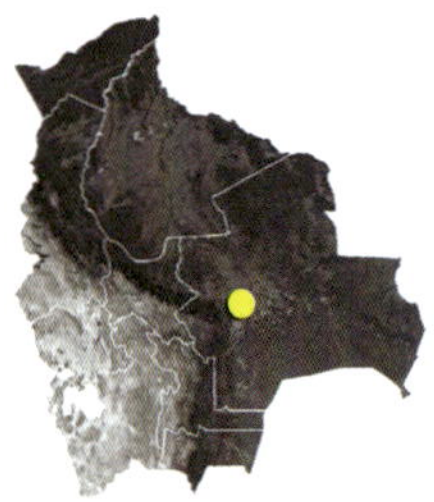

Gorybia guenda Galileo & Martins, 2013

Cerambycinae: Piezocerini

This is a very small (4–6 mm), shining brown species. Its pronotum is alveolate and tightly restricted basally with the elytra densely, closely punctate and having a mix of short- and medium-length semirecumbent hairs. It is rarely encountered except at lights, where it is locally abundant. This species was recently described and is a Bolivian endemic known from Santa Cruz Department at the Potrerillo del Guenda Reserve.

Gorybia suturella Martins, 1976

Cerambycinae: Piezocerini

This is a diminutive (5–8 mm) but attractive species that is shiny reddish brown with a distinct and much darker sutural stripe nearly the length of each elytron. This, along with its alveolate pronotal surface and distally, increasingly compressed antennomeres help to readily identify it. It is rarely encountered except at lights. It is known from French Guiana, central Brazil, and Bolivia, where it is recorded from Santa Cruz Department at the Flora and Fauna Hotel near Buena Vista and the Potrerillo del Guenda Reserve.

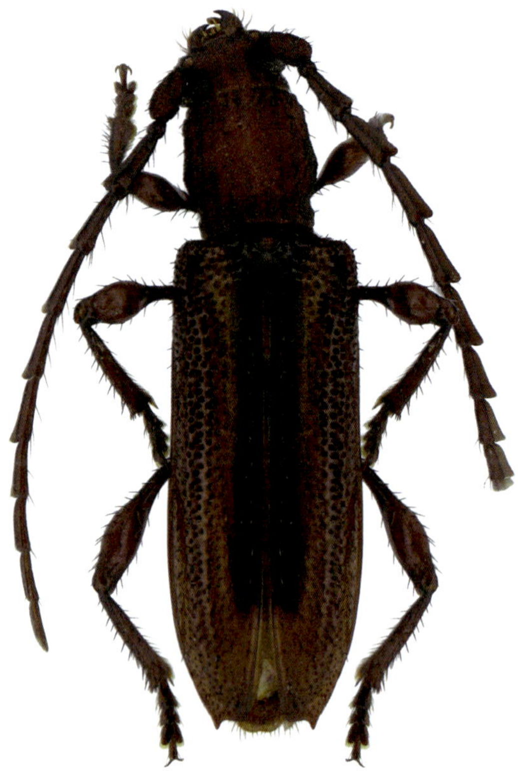

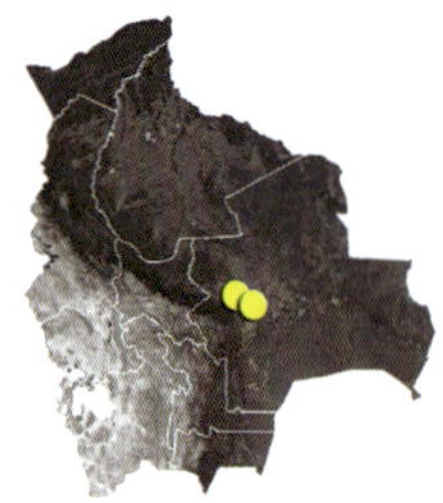

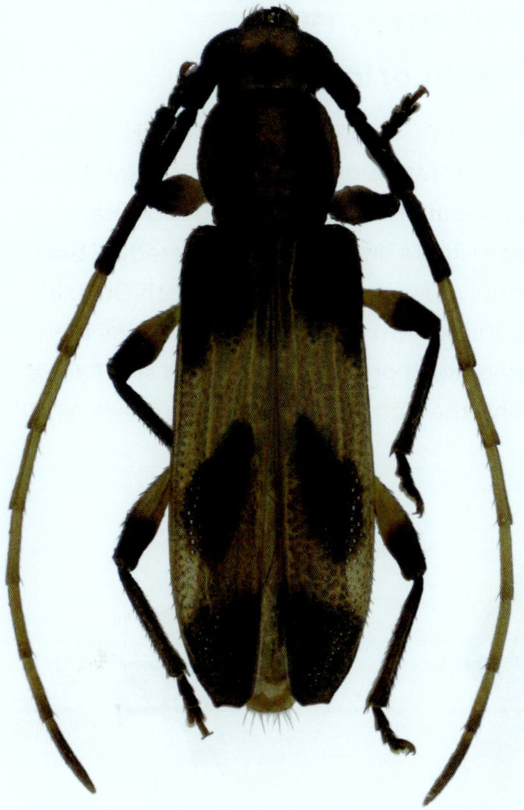

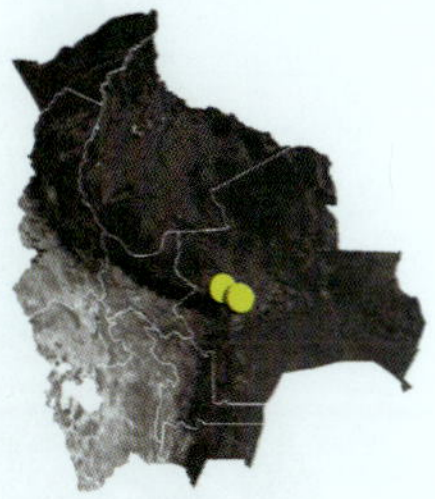

Haruspex bivittis (White, 1855)

Cerambycinae: Piezocerini

This is a small (7–10 mm), black and tan, heavily punctured species. It is locally common at lights and occasionally collected by beating freshly cut or dead wood on the edges of agricultural cuts. Widely distributed, known from Costa Rica and Panama in Central America to French Guiana, Brazil, Argentina, Peru, and Bolivia, where it is recorded from Santa Cruz Department in the vicinity of the Flora and Fauna Hotel near Buena Vista and the Potrerillo del Guenda Reserve.

Hemilissa quadrispinosa Gounelle, 1913

Cerambycinae: Piezocerini

This is a small (7–10 mm), dark brown, nocturnal species with alveolate pronotum, heavily punctate elytra, doubly carinate antennomeres, and a pair of strong spines on the apices of each elytron (hence its specific name). It is the commonest of the several dozen piezocerines found in Bolivia and especially abundant at light. It is widely distributed from northern Brazil, Peru, Argentina, and Bolivia. In Bolivia it is known from the vicinity of the Flora and Fauna Hotel near Buena Vista, the Potrerillo del Guenda Reserve, and Refugio Los Volcanes, adjacent to Amboró National Park, in Santa Cruz Department.

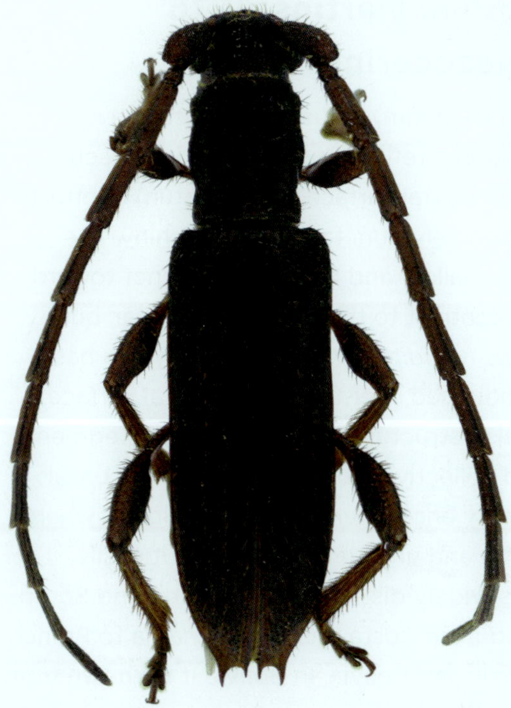

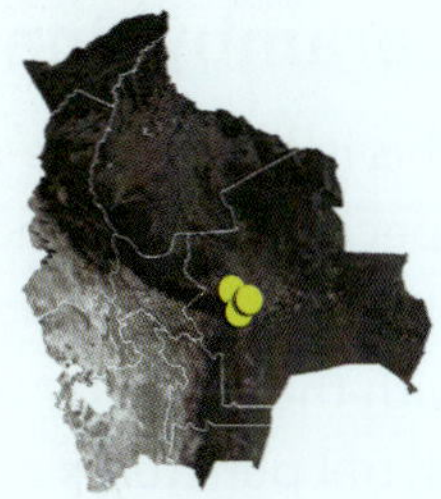

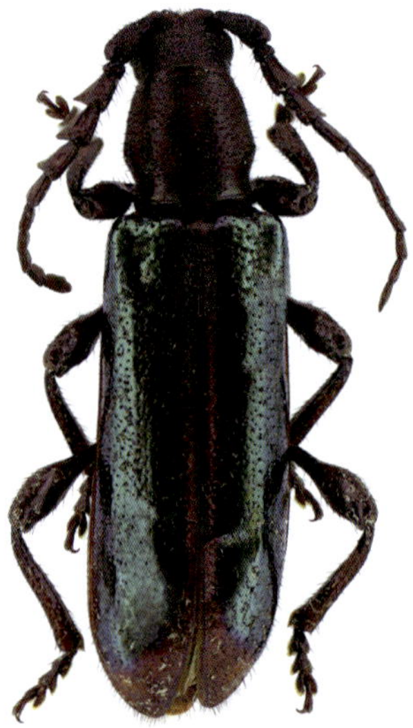

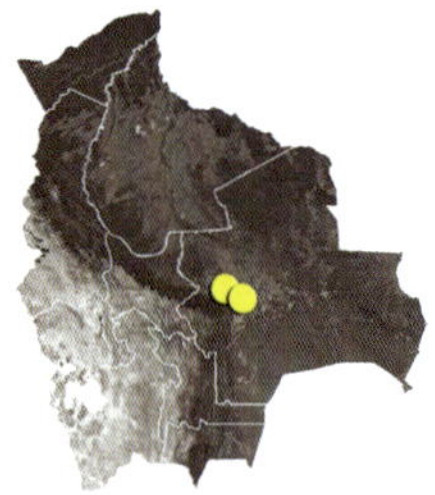

Piezarina smaragdina Martins, 1976

Cerambycinae: Piezocerini

This is a moderate-sized (17–19 mm), unusual, and attractive species that has metallic green reflections the length of each elytron on a shining brown elytral surface. Its pronotum contrasts with the elytra by being sparsely punctate and less shiny. The elytral punctures are very shallow and become indistinct toward the apex of the elytra, in contrast to the otherwise similar, but much smaller, *Thyellocerus fulgidipennis* (Gounelle), which has punctures deep and pronounced over the entire elytral surface. The antennae are unusual in structure and length. Their segments are spatulate in form and collectively barely reaching the basal third of the elytra. It is rarely encountered except at lights, where its reflective metallic elytra and unusual shape make it easily seen and recognized. It is widely distributed in Central and South America, where it is found from Costa Rica and Panama to French Guiana, north-central Brazil, and Bolivia. In Bolivia it is known from Santa Cruz Department in the vicinity of the Flora and Fauna Hotel near Buena Vista and the Potrerillo del Guenda Reserve.

Piezocera ataxia Martins, 1976

Cerambycinae: Piezocerini

This is a small (6–8 mm), reddish-orange species. The elytra are shining and have a dark, linear fascia on the disk and edges. Its antennomeres are greatly expanded laterally and doubly carinate. This species is occasional at lights and also encountered by beating the edge of agricultural cuts. It is recorded from French Guiana, Brazil, and Bolivia, where it is known from several locations in Santa Cruz Department, including north of Camiri, the Flora and Fauna Hotel near Buena Vista, and the Potrerillo del Guenda Reserve.

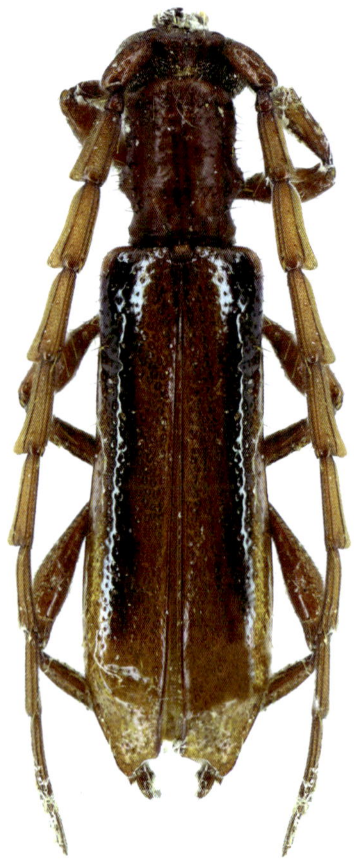

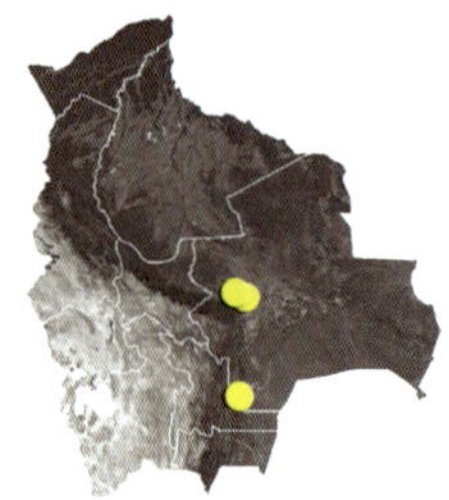

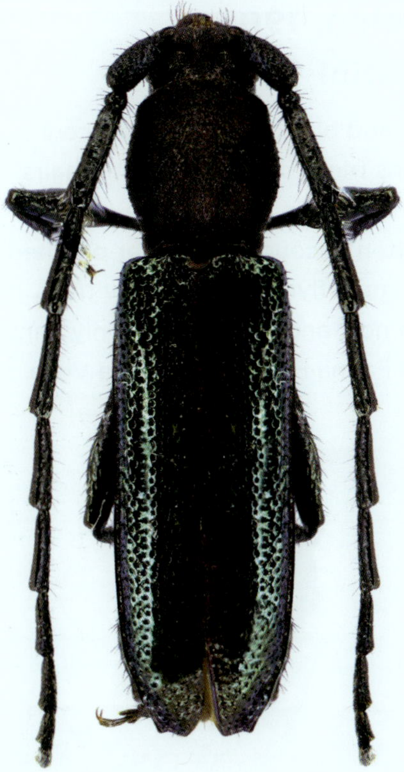

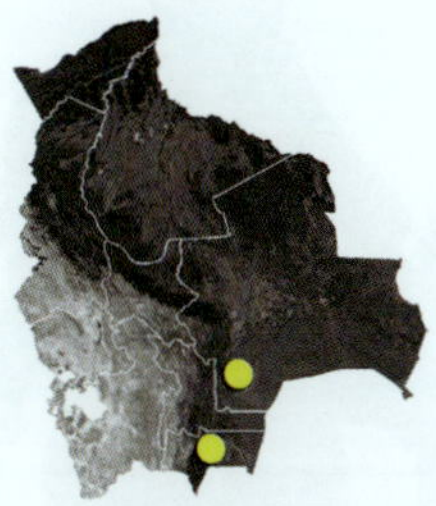

Thyellocerus fulgidipennis (Gounelle, 1909)

Cerambycinae: Piezocerini

This is a small species (7–8 mm), with dark metallic-green legs and antennae and dark reddish pronotum and sutural margin of the elytra. The remainder of the elytra is heavily punctate, dark metallic green. The deep elytral punctures and very dull pronotum (combined with much smaller size) distinguish this species from the similar *Piezarina smaragdina* Martins. This species is found in the humid Chaco of southern Bolivia, where it is rare at lights. It is known from Brazil, Peru, and Bolivia. In Bolivia it has been collected in Santa Cruz Department between Santa Cruz and Camiri and Villamontes in Tarija Department.

Pleiarthrocerus opacus Bruch, 1914

Cerambycinae: Pleiarthrocerini

This is a rather large (18–26 mm), densely pubescent, dark piceous-colored species with long pectinate antennae. It is so anatomically unique that it is placed in a tribe by itself. This species is nocturnal and has been collected only at lights, where it is very uncommon. It is recorded from northeastern Brazil, north-central Argentina, and Bolivia, where it is known from Santa Cruz Department at four localities: North of Pampagrande, 20 km north of Camiri, near Samaipata, and Refugio Los Volcanes.

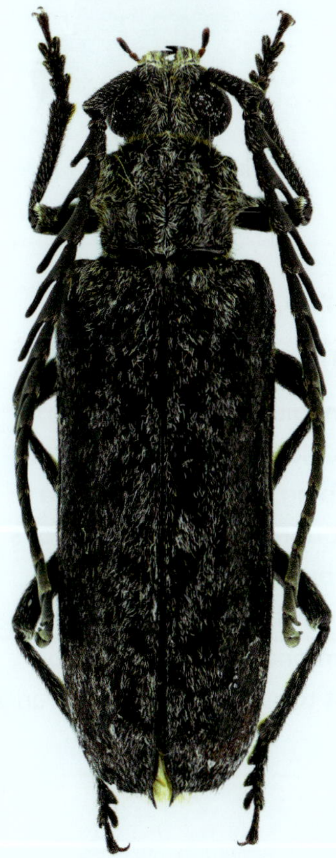

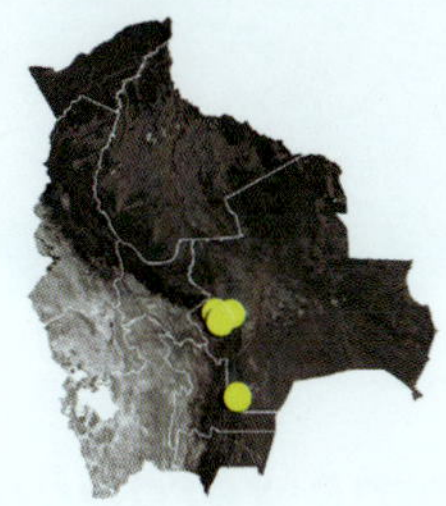

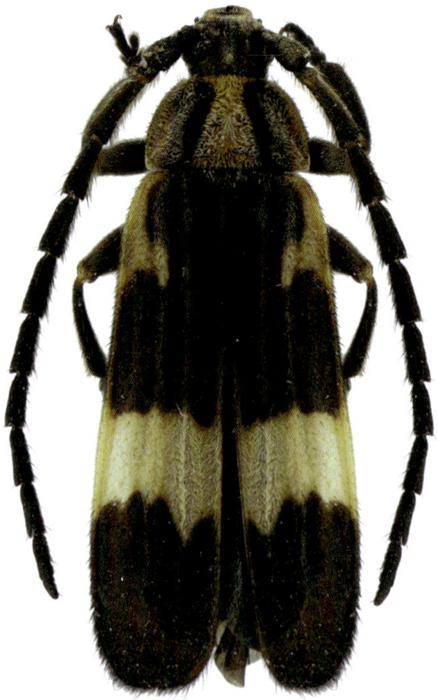

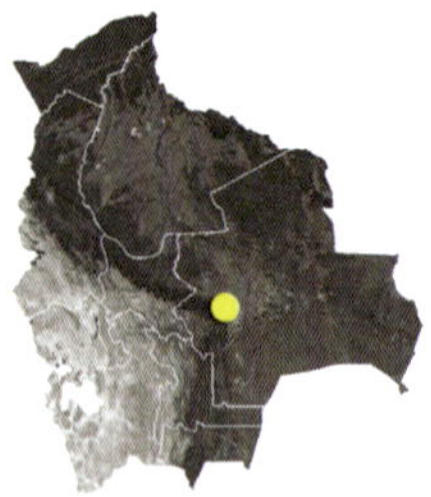

Deltosoma xerophila Di Iorio, 1995

Cerambycinae: Pteroplatini

This is one of the lycid beetle mimics, medium sized (11–15 mm) and orange and black colored. All of its black antennomeres are distinctly compressed and carinate, helping to characterize the species. It is uncommonly collected but normally diurnal in habits and collected in flight. It is widely distributed in northern South America (Ecuador, French Guiana, and northwestern Brazil) and Bolivia, where it has been recorded from one midlevel elevation locality (Refugio Los Volcanes, adjacent to Amboró National Park) in Santa Cruz Department.

Pteroplatus anchora Belon, 1903

Cerambycinae: Pteroplatini

This is a small (12–14 mm), orange and black, heavily pubescent species with costate elytral margins fringed with short, stiff hairs. This diurnal species mimics other beetles in the family Lycidae in both form and habits. It is most often encountered either perched on foliage along forest edges or simply flying along sunny trails. It is a Bolivian endemic, very uncommon, and known from only Santa Cruz Department at the Flora and Fauna Hotel near Buena Vista.

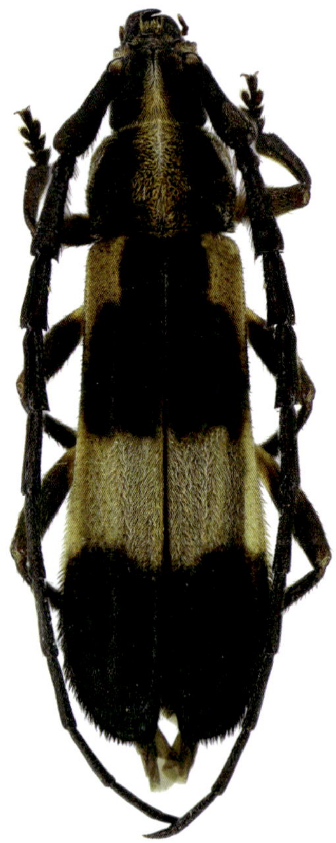

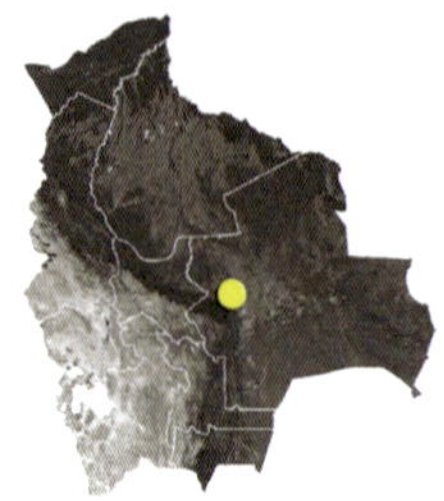

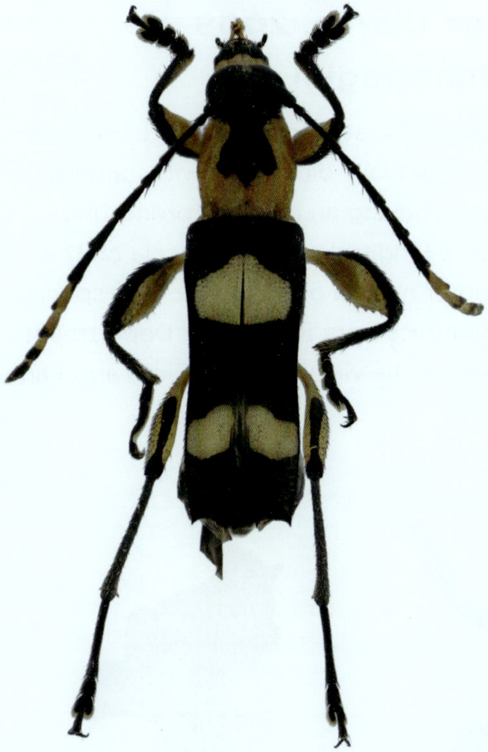

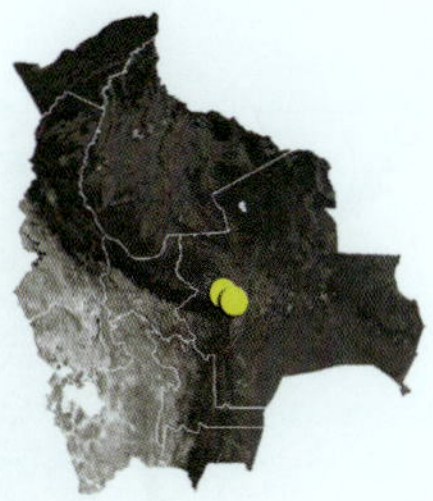

Agaone peruviensis (Fisher, 1952)

Cerambycinae: Rhinotragini

This is a small (8–10 mm), multicolored wasp mimic with a black head, orange pronotum, and two yellow-white elytral bands bordered by contrasting black areas. It is frequently collected on woody plant blossoms or coming to freshly cut tree trunks during the early rainy season to mate and oviposit. Known distribution includes the western Amazon basin, Peru, and Ecuador. In Bolivia it is found in lowland tropical forests and has been collected from the vicinity of the Flora and Fauna Hotel near Buena Vista and the Potrerillo del Guenda Reserve in Santa Cruz Department.

Anomalotragus recurvielytra Clarke, 2010

Cerambycinae: Rhinotragini

This is a small (5–7 mm), delicate, chestnut to yellow-orange wasp mimic with short recurved elytra. It feeds on flowers of woody plants, including bush *Croton*, where it can be found in large numbers and easily collected. It is known only from Bolivia. It is found from lowland tropical forests to midelevation locales of 1,000 m or more in Santa Cruz Department. It has been collected in the vicinity of the Flora and Fauna Hotel near Buena Vista, Bella Vista road near Cuevas, and north of Camiri.

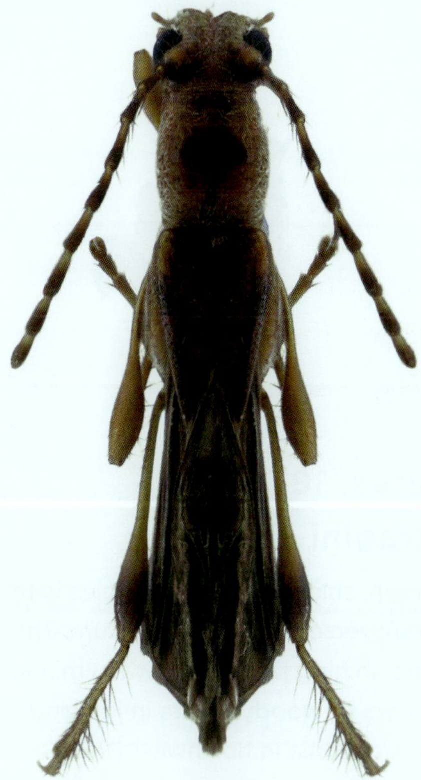

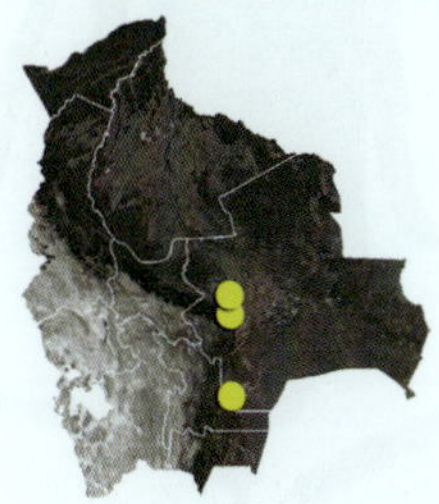

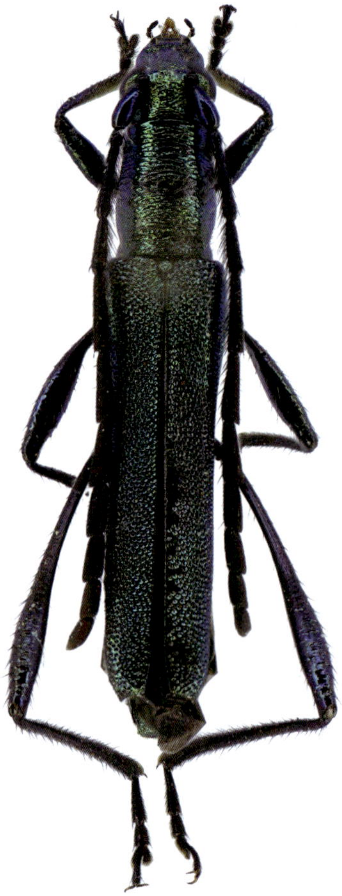

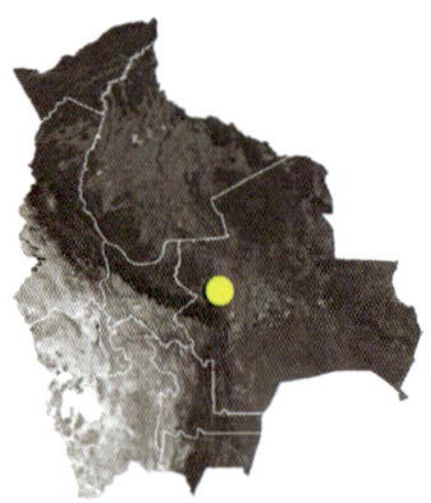

Chrysaethe aurata (Bates, 1870)

Cerambycinae: Rhinotragini

This is a small (9–11 mm), blue-green, shining metallic species with coarsely punctate elytra and a transversely rugose pronotum. The antennomeres, tibiae, and tarsi are shining black. This wasp mimic is found feeding on the small flowers of woody plants in lowland tropical forests. It is known from Amazonia in northwest Brazil and from Bolivia, where it is known from Santa Cruz Department in the vicinity of the Flora and Fauna Hotel near Buena Vista.

Ecliptoides julietae Clarke, 2009

Cerambycinae: Rhinotragini

This is a very small (5–6 mm), delicate, flower visitor of various woody plants. The species is sexually dimorphic with, among several differences, the males having an all-black pronotum. Females (shown here) have a longitudinal, black fascia on the pronotum and along the outer margin of the elytra. This species is one of many Bolivian endemics from Santa Cruz Department and has been collected around the vicinity of the Flora and Fauna Hotel and Buena Vista.

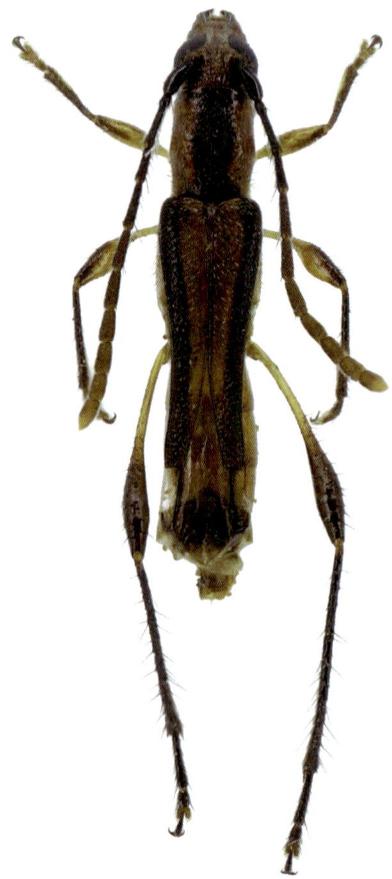

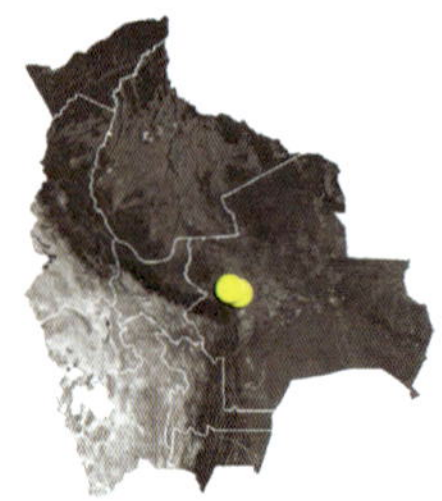

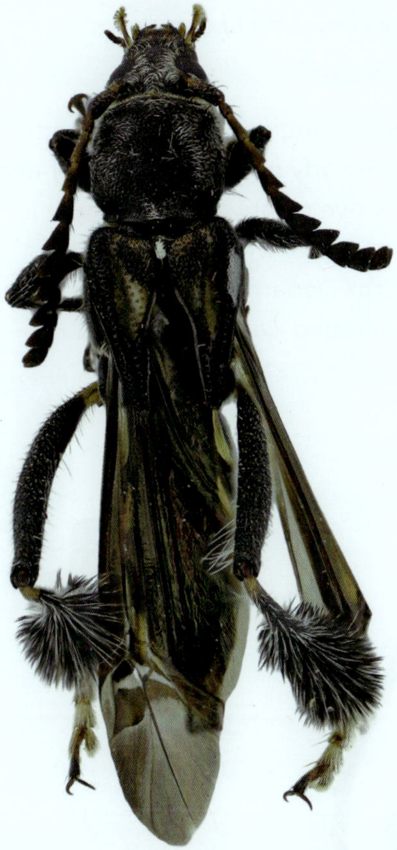

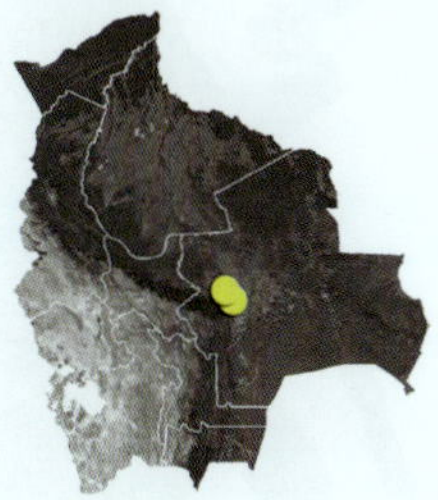

Epimelitta barbicrus (Kirby, 1818)

Cerambycinae: Rhinotragini

This is another of the small (10–12 mm), flower-feeding wasp and bee mimics. It is dark in color with short elytra and large black brushes of stiff hairs on the hind tibiae. It is widely distributed in South America from southeastern Brazil, Argentina, and Paraguay to central Bolivia. In Bolivia it is known from lowland habitats in Santa Cruz Department, including the vicinity of the Flora and Fauna Hotel near Buena Vista and the Potrerillo del Guenda Reserve.

Erythroplatys boliviensis Clarke, 2012

Cerambycinae: Rhinotragini

This is a beautiful, broad-bodied, medium-sized (14–16 mm), tricolored (reddish-orange, black, and yellow-orange) species with dark, short antennae (distal segments lighter in color and slightly serrate) and coarsely punctate elytra. As is typical for the rhinotragines, it visits flowers of woody plants. This species was recently described and is another of the Bolivian endemics from Santa Cruz Department, where it is known from the vicinity of the Flora and Fauna Hotel and the Potrerillo del Guenda Reserve.

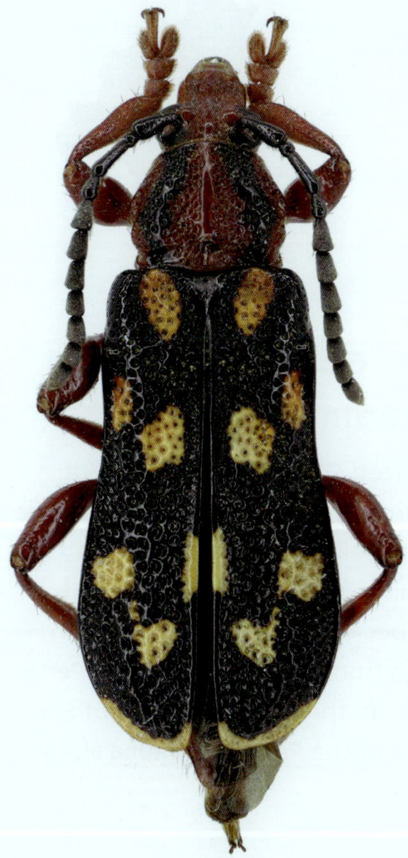

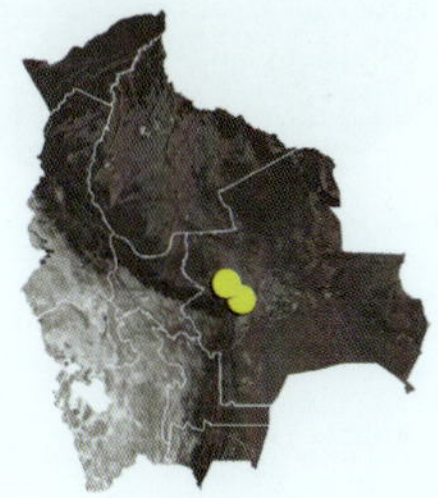

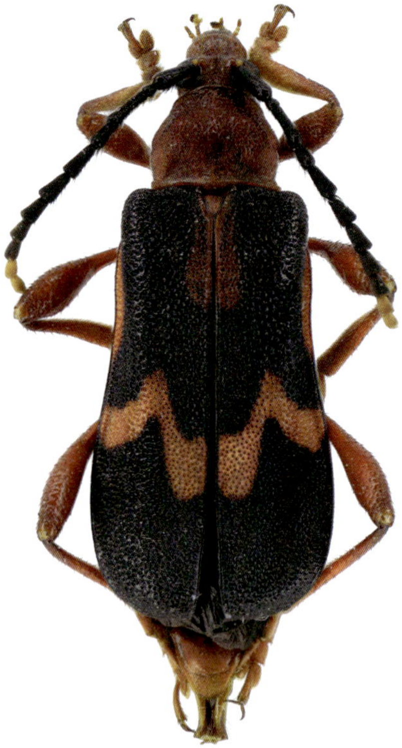

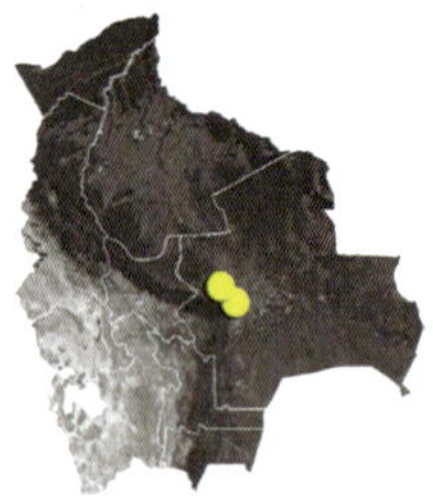

Erythroplatys simulator Gounelle, 1911

Cerambycinae: Rhinotragini

This is a broad-bodied, medium-sized (11–17 mm) species. But, unlike the previous species, *Erythroplatys boliviensis*, it is bicolored (deep red and black) and lacks the yellow-orange fasciae on the elytra, instead having deep-red markings. Like its Bolivian congener, it is a most attractive species and a pollen feeder on woody plant blossoms. It is known from east-central Brazil, Peru, and Bolivia, where it is recorded from the lowland forests of Santa Cruz Department from the vicinity of the Flora and Fauna Hotel and the Potrerillo del Guenda Reserve.

Grupiara viridis (Gounelle, 1911)

Cerambycinae: Rhinotragini

This is a small (10 mm), shining, green-bodied wasp mimic with dark legs and antennae. It is also a woody plant flower feeder and can be locally common when attractive plants are in bloom. It is known from southeastern Brazil to Bolivia, where it is found in lowland tropical forests in Santa Cruz Department at the Flora and Fauna Hotel near Buena Vista and around Camiri.

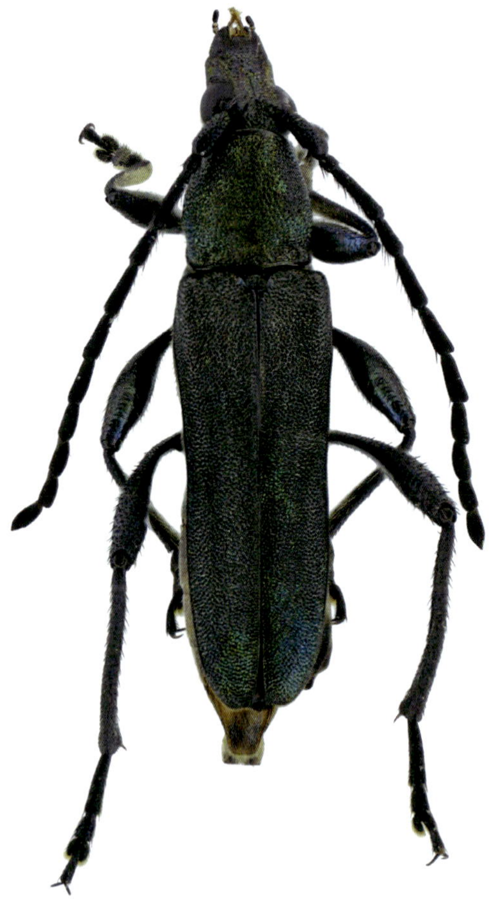

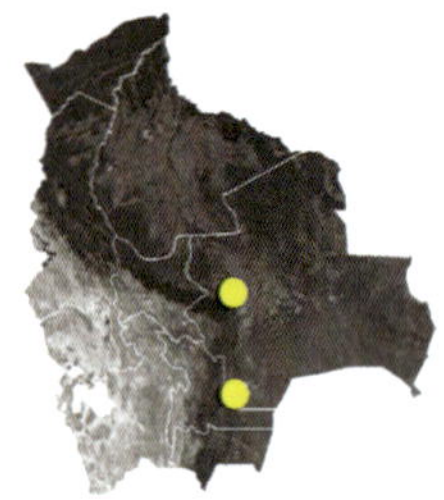

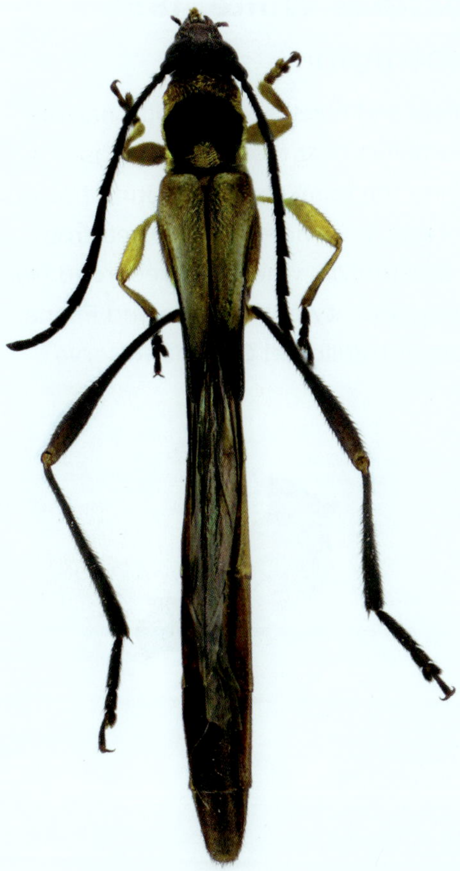

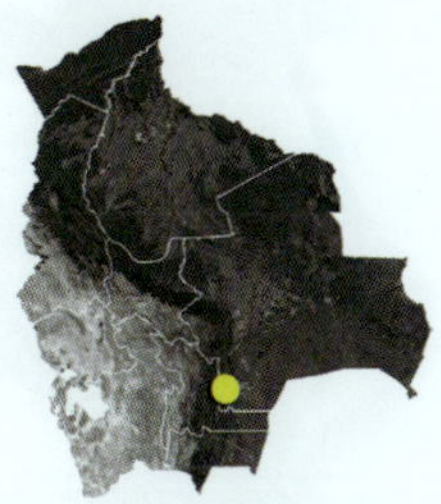

Monneus longiventris Magno, 2001
Cerambycinae: Rhinotragini

This is a very narrow, parallel-sided, pale species of moderate length (17–19 mm), with a remarkably elongate abdomen and beautiful pronotum and scutellum with long, silky, golden pubescence surrounding a black, elevated, scabrous-looking central disk. This Bolivian endemic is known only from southern Santa Cruz Department (near the border of Chuquisaca Department) from Incahuasi, east of Muyupampa.

Neoregostoma bettelai Clarke, 2010
Cerambycinae: Rhinotragini

This medium-sized (16–18 mm) species has striking orange and black coloration. The pronotum has a bold, longitudinal black fascia down the middle, and the elytra have an I-shaped macula at the basal halves and broadly black apices. This species was named in honor of Paolo Bettella, former entomology chief at the Museo Noel Kempff Mercado and founder of the Potrerillo del Guenda Nature Reserve. Fittingly, the only known specimens were collected at this property in Santa Cruz Department.

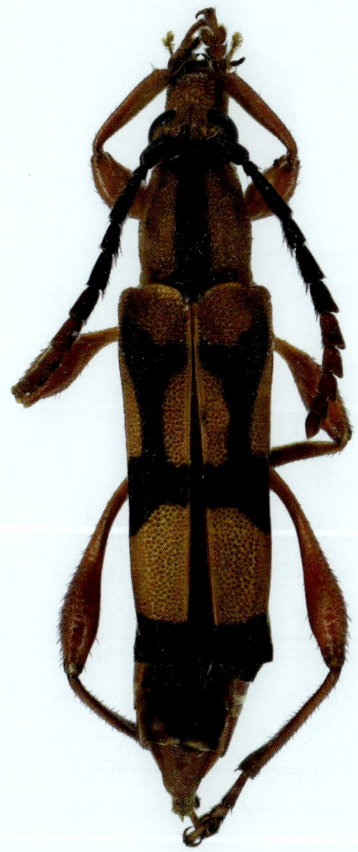

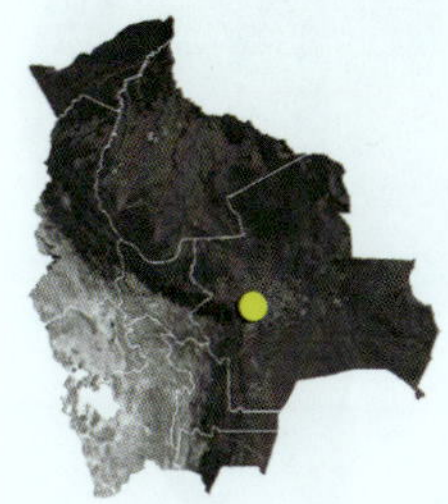

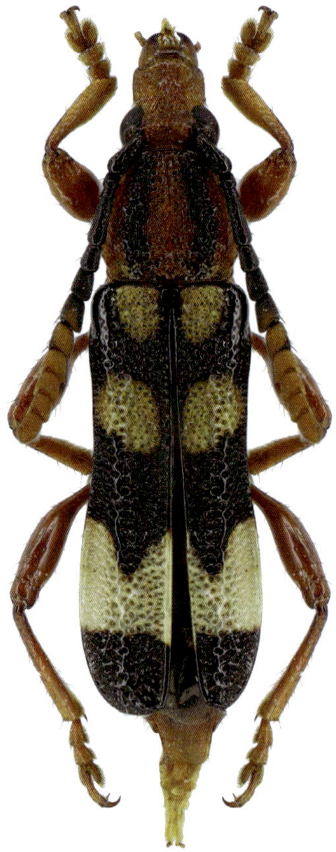

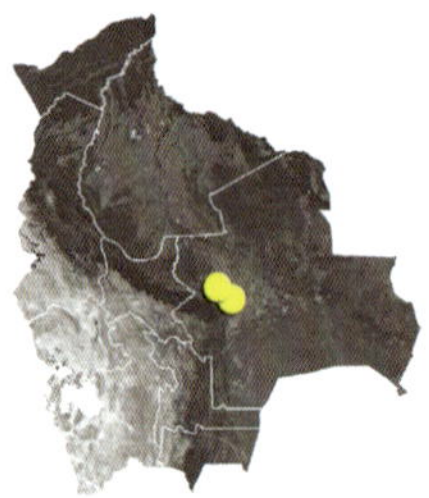

Neoregostoma giesberti Clarke, 2007

Cerambycinae: Rhinotragini

This is a small (8–10 mm), tricolored species (red, pale yellow, and black) and one of the smallest species in the genus. Unlike most cerambycids whose adult emergence is at the beginning or through the rainy season, this species appears at or near its end and feeds on the late-blooming vine *Gouania mollis* Reisseck (locally known as *Bejunco hoja lanuda*) in April and May (Clarke, 2007). It is another of the Santa Cruz Department endemics known from the vicinity of the Flora and Fauna Hotel near Buena Vista and the Potrerillo del Guenda Reserve.

Odontocera compressipes White, 1855

Cerambycinae: Rhinotragini

This is a small (9–10 mm), yellow and black species with narrow, translucent elytra that only partially cover the folded wings underneath. It is common during the rainy season, visiting flowers of woody plants in lowland forests. It is widely distributed from Ecuador, Peru, central Brazil, and Bolivia, where it is recorded from Santa Cruz Department from the vicinity of the Flora and Fauna Hotel near Buena Vista and the Potrerillo del Guenda Reserve.

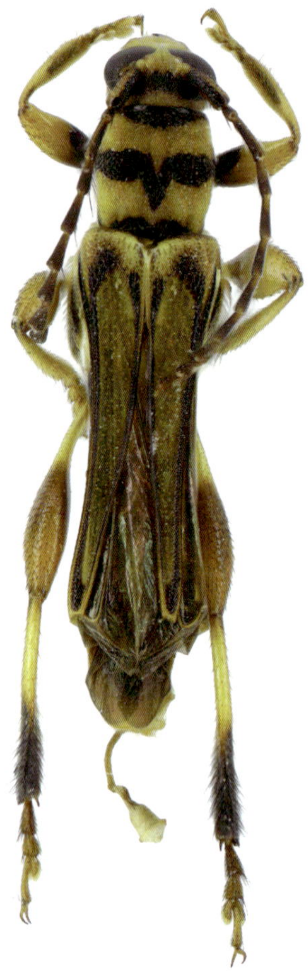

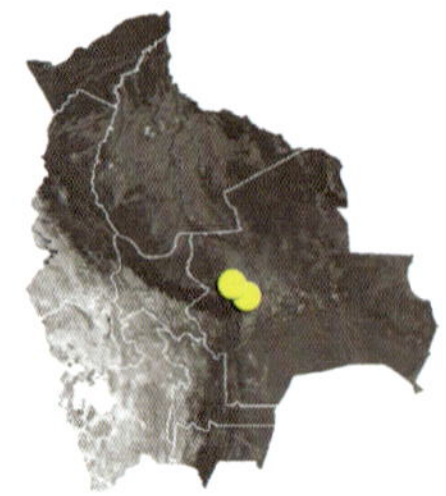

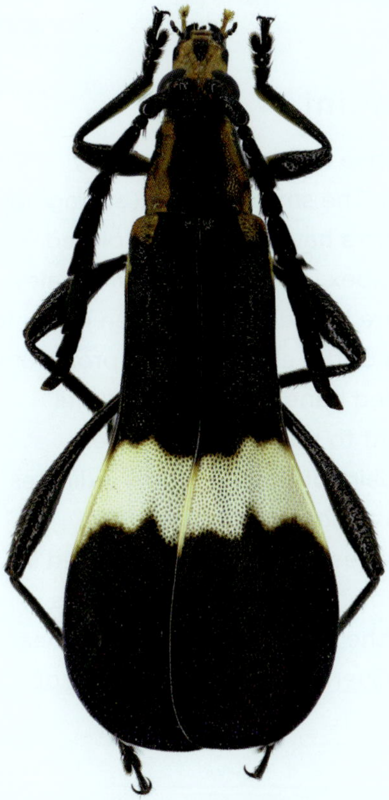

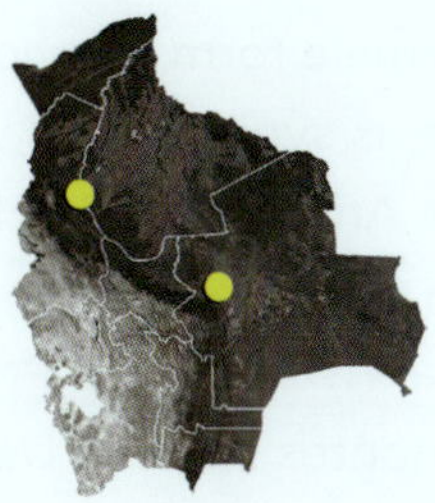

Ornistomus simulatrix Clarke, 2012

Cerambycinae: Rhinotragini

This is another of the strikingly marked, moderate-sized (16–18 mm) rhinotragines endemic to Bolivia. This black species with a broad, central, cream-colored elytral fascia has overlapping wings, which are characteristic of the genus and help immediately to identify it. It was described from Santa Cruz Department at the Flora and Fauna Hotel near Buena Vista and is also known from Beni Department (near Madidi National Park). This rare species is attracted to the flowers of *Cupania cinerea*, locally known as *sama blanca* (Clarke, 2012).

Rhinotragus antonioi Clarke, 2012

Cerambycinae: Rhinotragini

This is a small (11–13 mm), black and yellow species with two transverse, cream-colored elytral fasciae, one basally and one at the distal third. This species, endemic to Bolivia, is a flower visitor to woody plants, and is most often seen in November and December. Known only from Santa Cruz Department, it has been recorded from the Flora and Fauna Hotel near Buena Vista and adjacent areas.

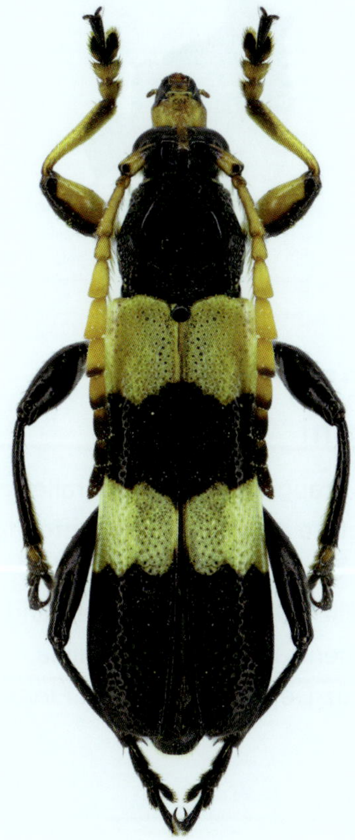

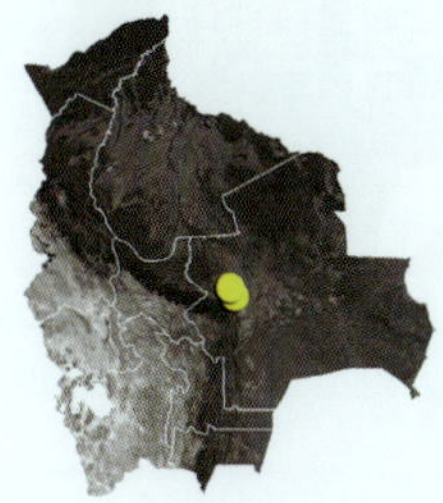

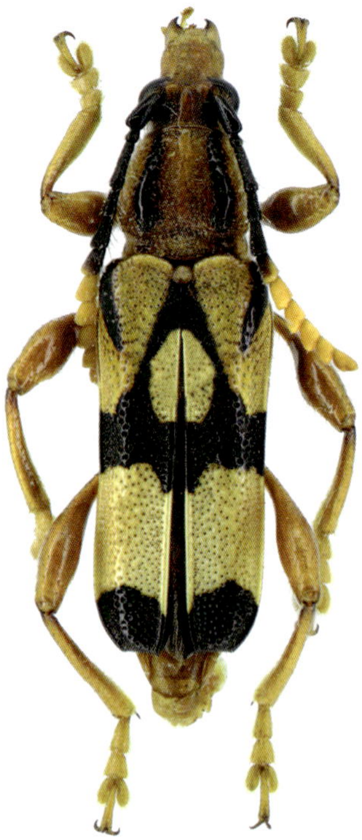

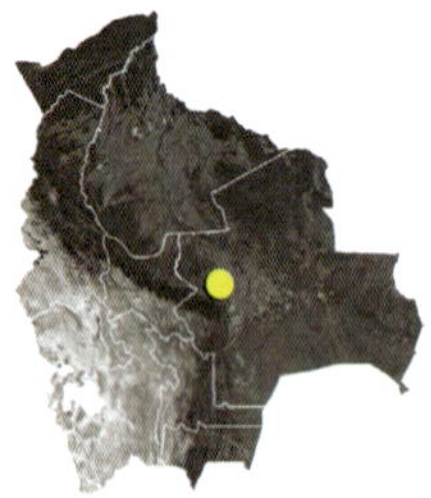

Rhinotragus lucasii Thomson, 1860
Cerambycinae: Rhinotragini

This is a moderate-sized (14–17 mm), orange and black, parallel-sided, stout species with evenly punctate elytra. Its basal antennal segments are shining black while the distal segments are dull yellow orange and somewhat serrate. It is widely distributed and recorded from French Guiana, north-central Brazil, and Bolivia, where it is known only from Santa Cruz Department at the Flora and Fauna Hotel near Buena Vista.

Tomopterus staphylinus Audinet-Serville, 1833
Cerambycinae: Rhinotragini

This is a small, common wasp and bee mimic that is sexually dimorphic in coloration and design. The smaller males (averaging 5 mm) are dark, with their short elytra having an ivory-colored, narrow band from the base to the apex. The much larger females (averaging 8 mm), pictured here, are broader and have an ivory- to rust-colored, wide band the length of the elytra and the pronotum margined in a narrow, similar-colored band. Interestingly, until recently the two sexes were thought to be different species. However the discovery of mating pairs of the two forms indicates the female form, previously known as *Tomopterus larroides* White, is synonymous with this species. It is widely distributed in South America, and in Bolivia is known from lowland localities in Santa Cruz Department, including the Flora and Fauna Hotel near Buena Vista and the Potrerillo del Guenda Reserve, and west of Villamontes in Tarija Department.

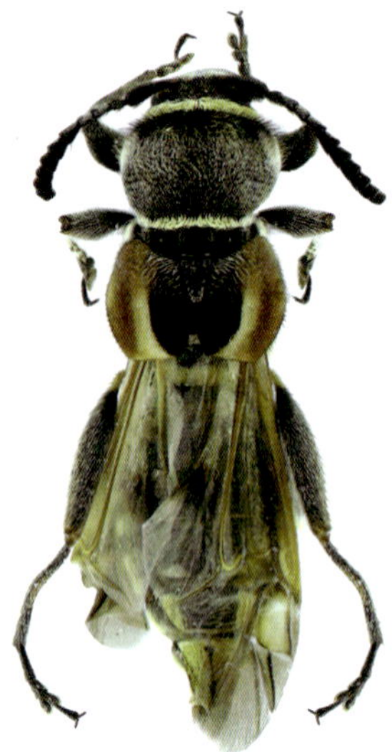

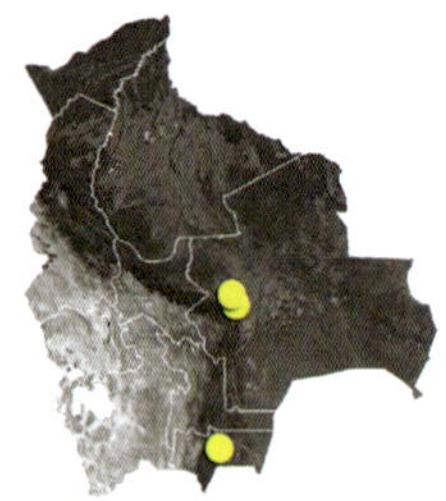

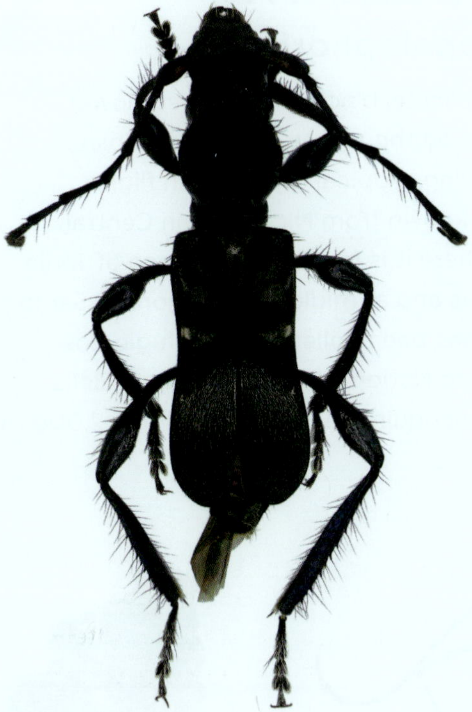

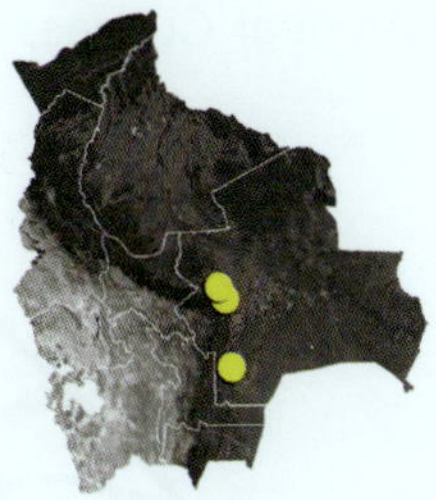

Argyrodines aurivillii (Gounelle, 1905)

Cerambycinae: Rhopalophorini

This is a small (8–10 mm), dark, flower visitor occasionally collected on bush *Croton* blossoms. Resembling a large ant, it can be mistaken for that and overlooked by the collector. It is delicately textured with differing punctures and pubescence and an ivory fascia on the dorsal surface near midelytra. The antennal scape is unusually colored, being shining black with a longitudinal blood red marking on its inner margin. It is found in a relatively small area from east central Brazil to Bolivia. In Bolivia it is known from Santa Cruz Department at the Potrerillo del Guenda Reserve, in the vicinity of the Flora and Fauna Hotel near Buena Vista, and at Cabezas between Santa Cruz and Camiri.

Brachylophora auricollis (Bruch, 1918)

Cerambycinae: Rhopalophorini

This is a small (7–9 mm), very narrow, delicate, and long-legged black wasp mimic with lateral, golden, pubescent bands on the pronotum. It is another of the many visitors to *Croton* blossoms. It is known from eastern Brazil, Argentina, and Paraguay and only recently recorded in Bolivia. It has been collected there at lower elevations in Santa Cruz Department north of Camiri, at Cabezas, and in Tarija Department west of Villamontes.

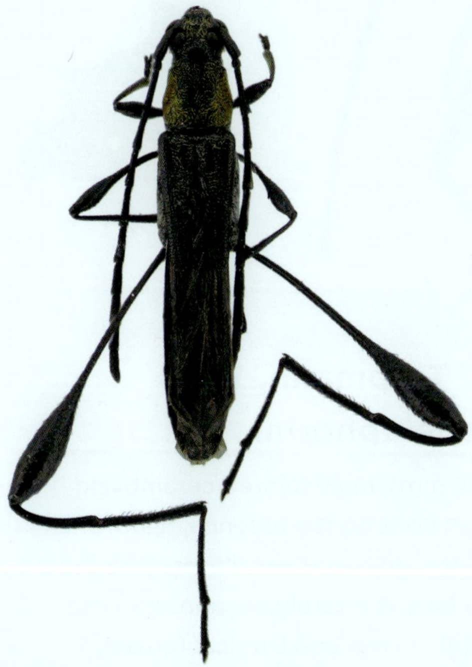

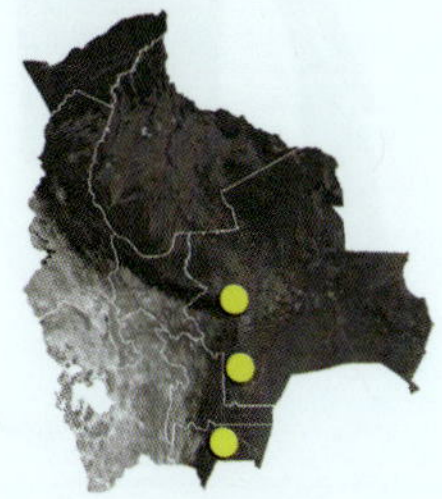

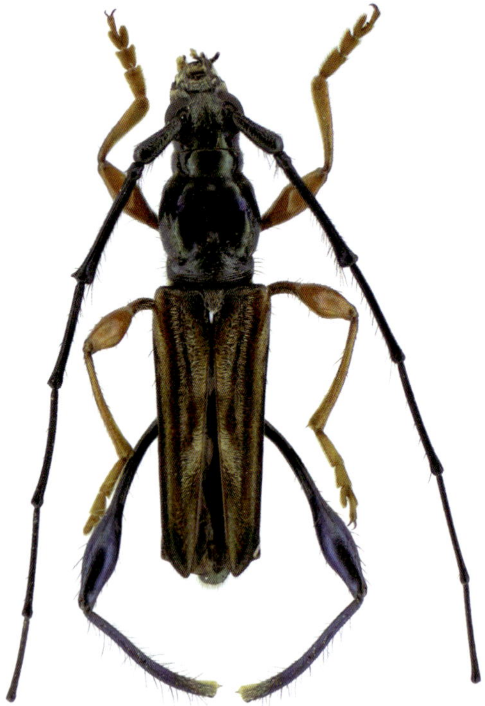

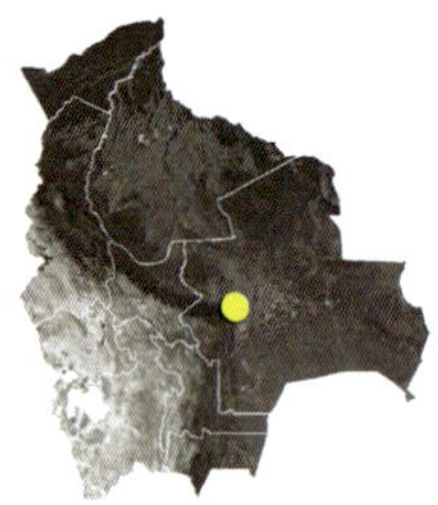

Closteropus herteli Tippmann, 1960

Cerambycinae: Rhopalophorini

This is a medium-sized (12–14 mm), multicolored cerambycid with blue-green metallic reflections on the antenna, head, and pronotum; shining brown elytra; front and middle legs shining orange; and blue-green hind legs. It is rarely encountered but sometimes is attracted to lights in lowland tropical forests. It is known from Ecuador, Peru, and Bolivia, where it has been collected in only Potrerillo del Guenda Reserve in Santa Cruz Department.

Coremia plumipes (Pallas, 1772)

Cerambycinae: Rhopalophorini

This is a small (9–10 mm), delicate, black, long-legged flower visitor with long-haired tufts on the apices of the hind tibiae. It resembles a large, slow-flying, tropical mosquito in flight. It is a wide-ranging species known from Nicaragua in Central America south to Bolivia, where it is found in the Chaco of Tarija Department near Villamontes and at midlevel elevations in Santa Cruz Department, where it has been collected at Refugio Los Volcanes, adjacent to Amboró National Park, the road to Bella Vista near Cuevas, and the Vicoquin area above Achira at 2,000 m.

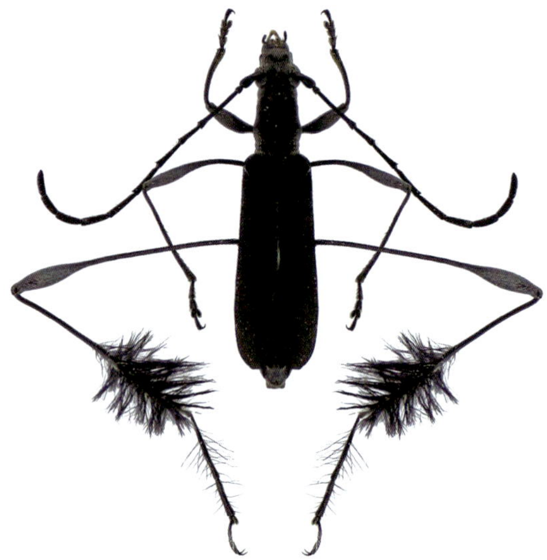

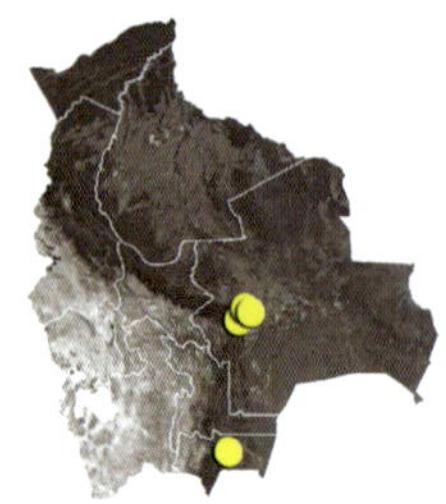

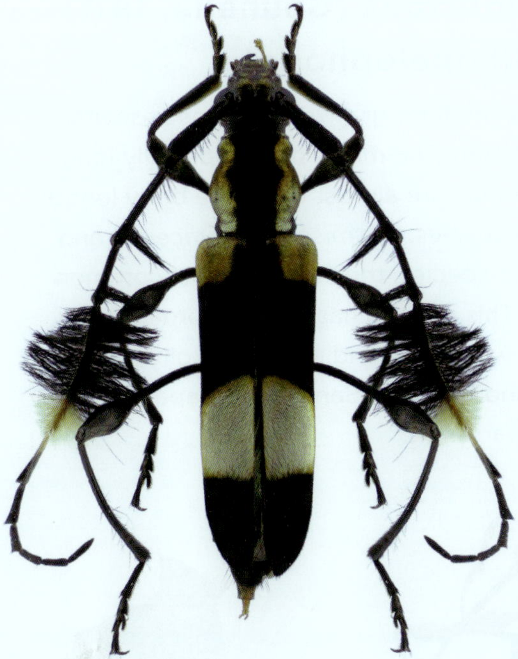

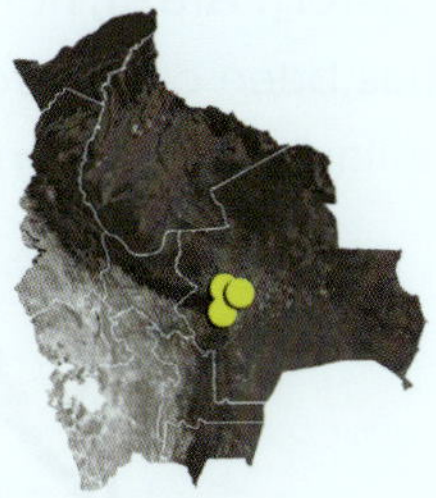

Cosmisoma ammiralis (Linnaeus, 1767)

Cerambycinae: Rhopalophorini

This is a striking black and whitish-orange midsized beetle (11–13 mm long) with an orange pronotum and pronounced pubescent tufts on the antennae—especially on segments five and six. It is most commonly encountered flying to or on freshly cut wood along forested roads and trails. Less commonly, it is found visiting flowers of woody plants. It is widely distributed in northern South America and the lowlands of Bolivia in Santa Cruz Department. It has been collected there in the vicinity of the Flora and Fauna Hotel near Buena Vista, the Potrerillo del Guenda Reserve, and Refugio Los Volcanes, adjacent to Amboró National Park.

Cosmisoma brullei (Mulsant, 1863)

Cerambycinae: Rhopalophorini

This is a narrow, black species (6–8 mm long) with an orange pronotum and small pubescent tufts on the antennae. It has been recorded from the southern Amazon Basin to Paraguay and Bolivia. In Bolivia it has been found in the Chaco of southern Santa Cruz Department around Camiri and Cabezas and in Tarija Department near Villamontes, where it is a common visitor to *Croton* blossoms.

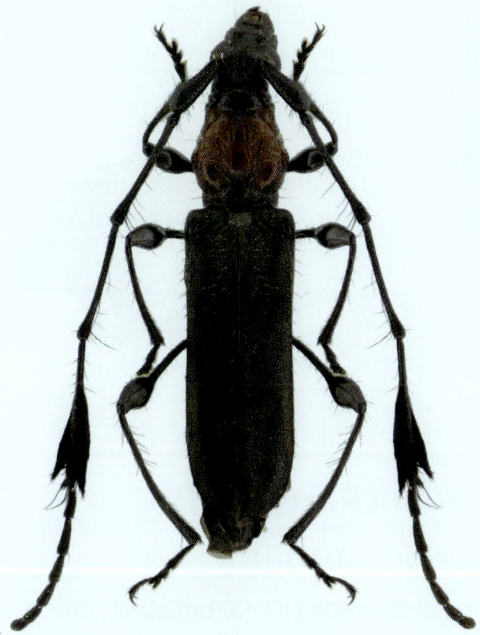

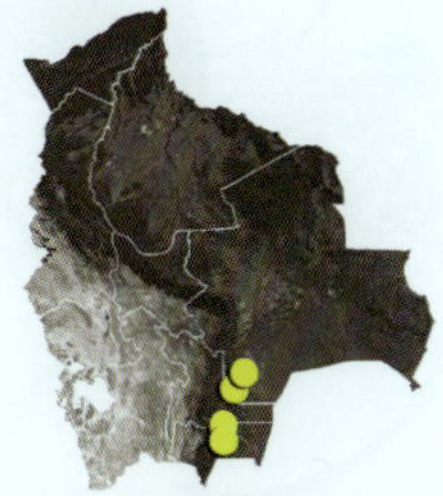

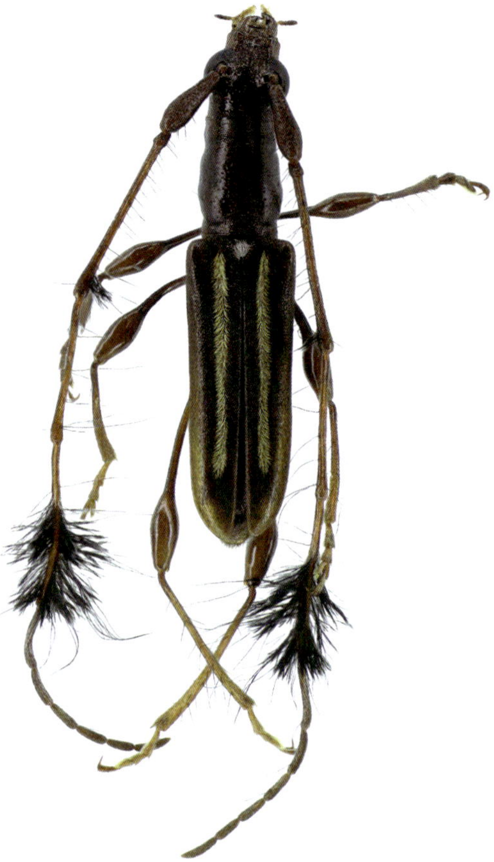

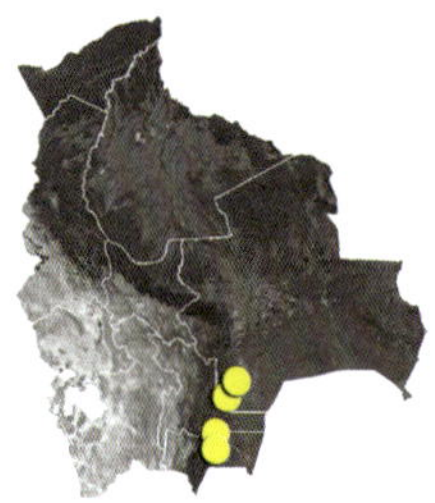

Cosmisoma ochraceum (Perty, 1832)

Cerambycinae: Rhopalophorini

This is an orange-brown, midsized beetle (10–13 mm) that, like others in the genus, has conspicuous pubescent tufts on the antennae. It also has a distinctive golden-yellow, longitudinal, pubescent fascia down the middle of each elytron. It is diurnal, like others in the tribe, and collected in similar ways. It is distributed in Brazil, Argentina, Paraguay, and Bolivia, where it has been collected only at the Potrerillo del Guenda Reserve in Santa Cruz Department.

Dirocoremia simplicipes (Gounelle, 1911)

Cerambycinae: Rhopalophorini

This is a small (6–7 mm), medium-gray to black species with dark red head and pronotum. The males have especially long hind legs, and their antennae are almost twice the body length. Along with the previous species, *Cosmisoma ochraceum,* and some of the *Chrysoprasis* species, they are the most common visitors to bush *Croton*. This species has been recorded from eastern Brazil, northern Argentina, and southern Bolivia, where it has been collected around Camiri in Santa Cruz Department and Villamontes in Tarija Department.

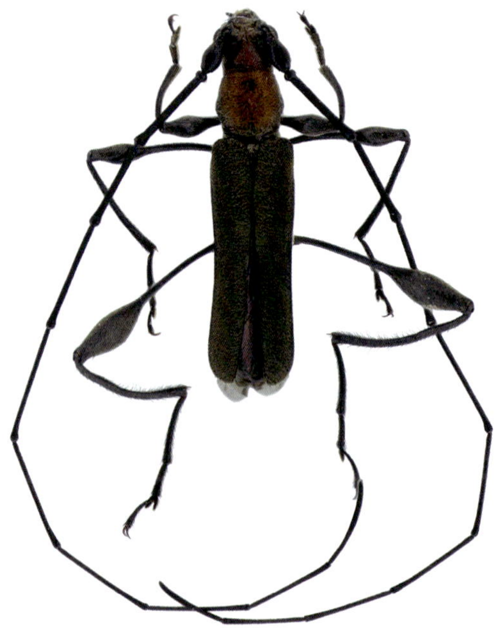

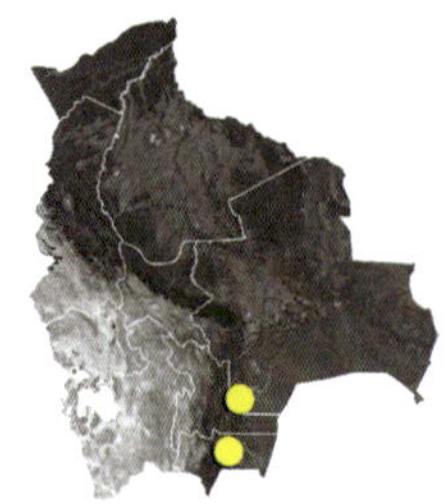

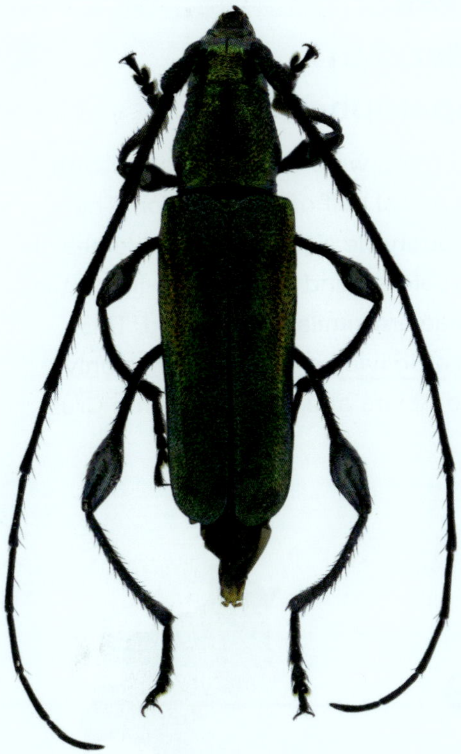

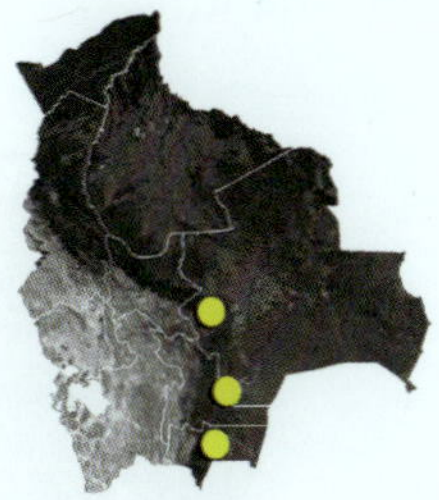

Ischionodonta iridipennis (Chevrolat, 1859)
Cerambycinae: Rhopalophorini

This is a small (8–9 mm), bright green species with darker appendages and a lateral, reddish-bronze margin along the basal half of the elytra. It is typically found feeding in the blossoms of bush *Croton* early in the rainy season and can be locally abundant. It is distributed in Argentina, Brazil, Uruguay, Venezuela, and Bolivia. In Bolivia it is known from Santa Cruz Department north of Camiri, south of Cabezas, and near Samaipata and from Tarija Department north and west of Villamontes.

Lathusia ferruginea (Bruch, 1908)
Cerambycinae: Rhopalophorini

This is another of the small (8–10 mm), common, flower visitors to woody plants. It is orange red with black femora and small pubescent tufts on the hind legs. Its flattened body shape is unusual among the flower visitors. It is a strong flyer that, when disturbed, quickly disappears by dropping slightly then speeding away. It has been recorded in South America to eastern Brazil, northern Argentina, and Paraguay and in Bolivia from southern Santa Cruz (near Cabezas and north of Camiri), Chuquisaca, and Tarija (north and west of Villamontes) Departments.

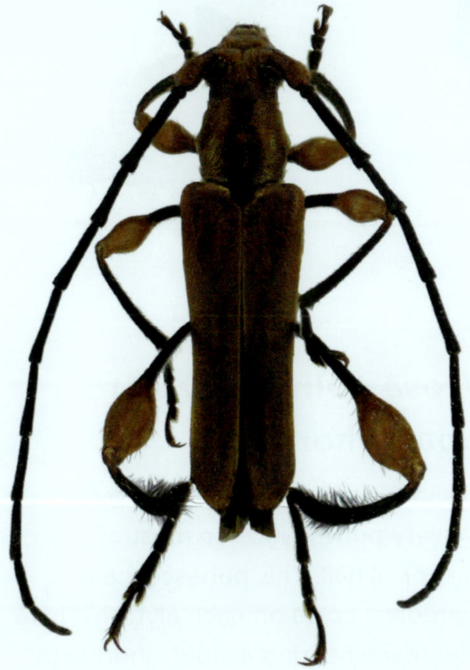

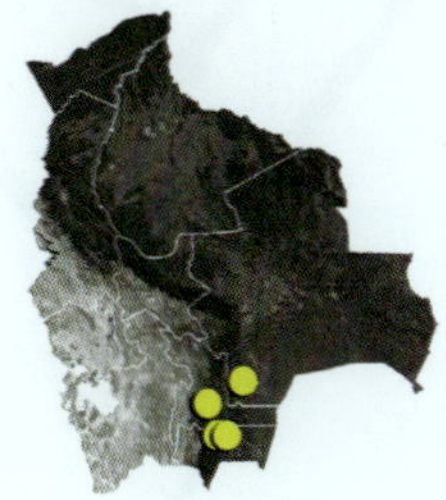

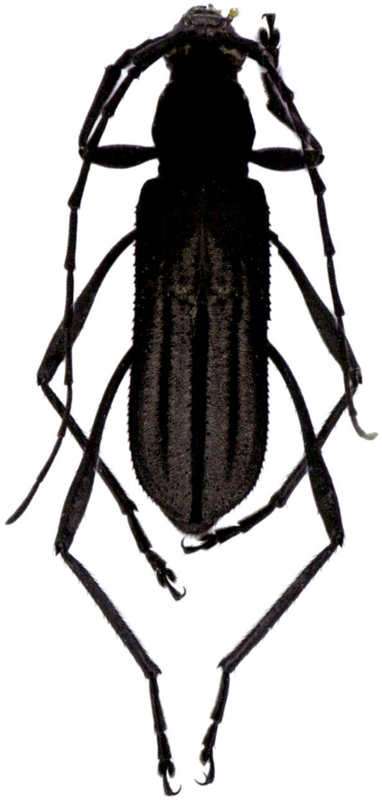

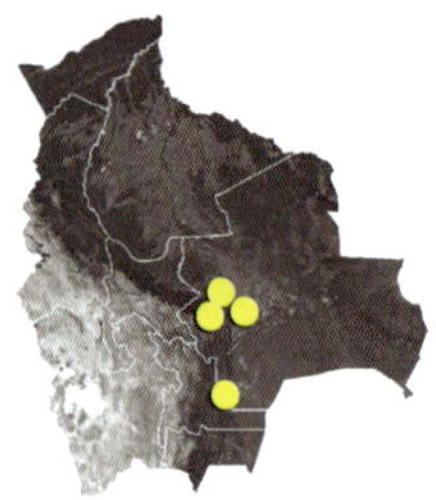

Listroptera tenebricosa (Olivier, 1790)

Cerambycinae: Rhopalophorini

This unusual-appearing, small species (8–11 mm) is completely black with somewhat grey, velvety pubescence on most of the elytra, most visible on the apical half. This pubescence is interrupted by a distinct longitudinal costa on each elytron. There are distinctive, raised tubercles, each bearing a stout, short seta along the outer edge of the elytra. This species is known from many localities in Santa Cruz Department, including Camiri, Potrerillo del Guenda Reserve, Refugio Los Volcanes, and the Flora and Fauna Hotel near Buena Vista, but is never very common.

Meringodes solangeae Wappes & Lingafelter, 2011

Cerambycinae: Rhopalophorini

This is a medium-sized (14–15 mm), wine-colored species with a golden, linear fascia on the apical half of each elytron. The antennae, from the scape through the sixth segment, are heavily margined with areas of wine-colored and black brush-like setae. The distal half of the hind tibiae are similarly margined. This strikingly beautiful species is a Bolivian endemic, known only from the Vicoquin area above Achira at 2,000 m in Santa Cruz Department.

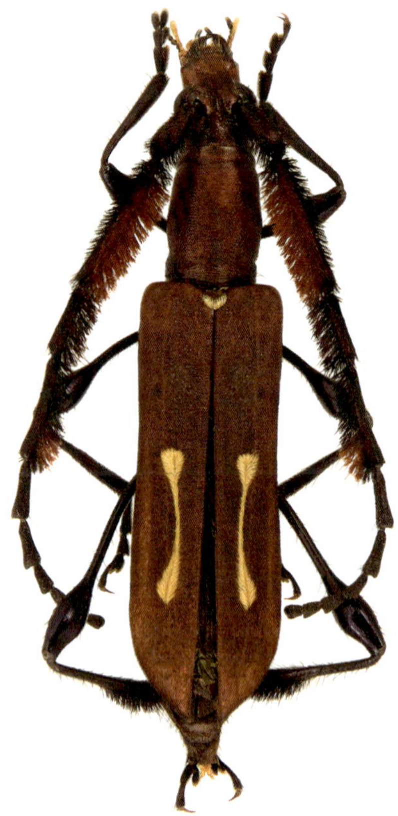

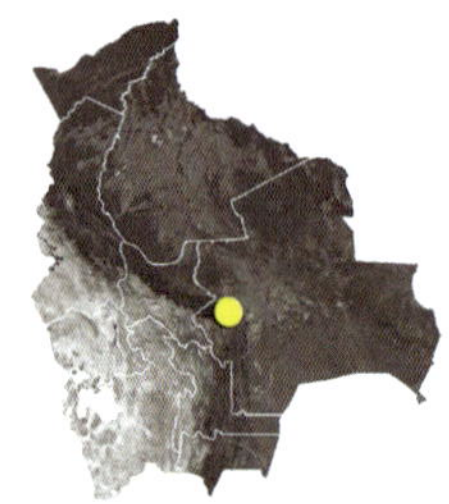

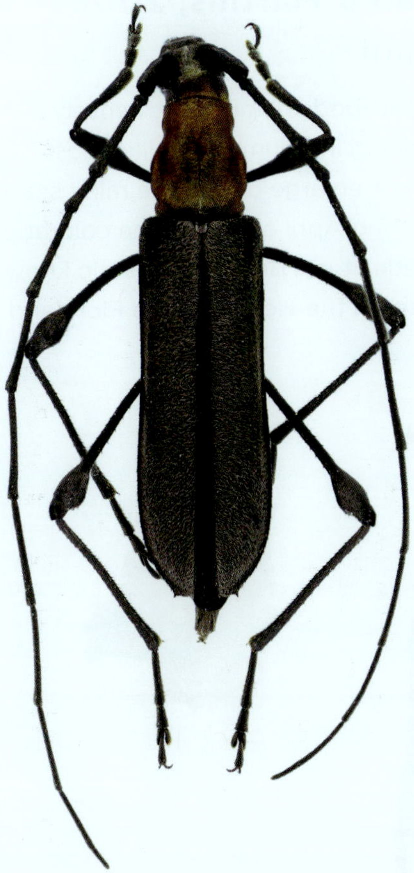

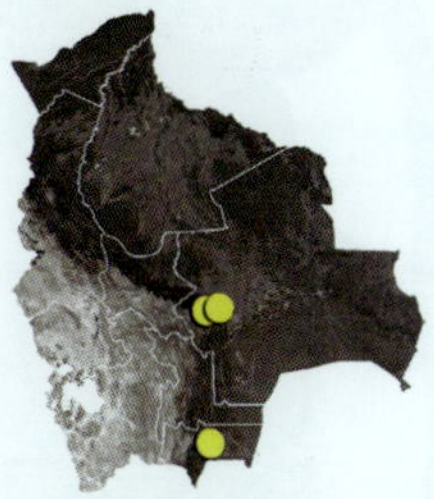

Rhopalophora collaris (Germar, 1824)

Cerambycinae: Rhopalophorini

This is a small (9–11 mm), narrow, black species with an orange pronotum and minutely gray pubescent elytra. It is a common visitor to flowering trees. Widely distributed in South America, it is known from Argentina, southeastern Brazil, Paraguay, Uruguay, and Bolivia, where it is recorded from Santa Cruz Department in Samaipata and Refugio Los Volcanes and in Tarija Department near Villamontes.

Marupiara castanea Martins & Galileo, 2006

Cerambycinae: Smodicini

This is a moderate-sized (9–16 mm, excluding the everted genitalia), flat-bodied, shining brown species. It is rather plain, with no ornamentation. The pronotum has rather densely placed, large punctures, and the antennae are shorter than the length of the body. This Bolivian endemic is attracted to lights, but is very uncommon and is known only from the Vicoquin area above Achira at 2,000 m and Refugio Los Volcanes, both adjacent to Amboró National Park, in Santa Cruz Department.

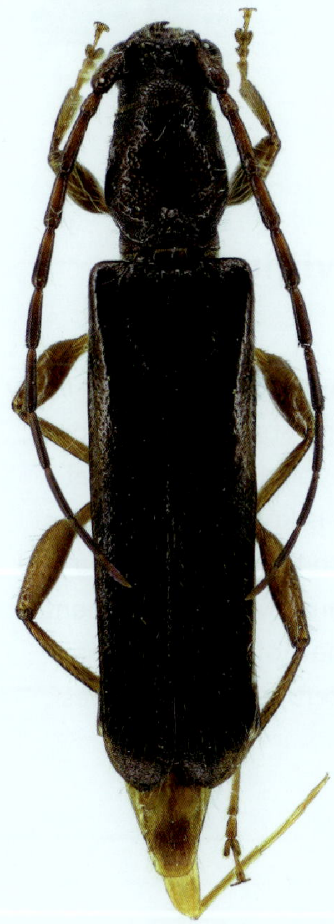

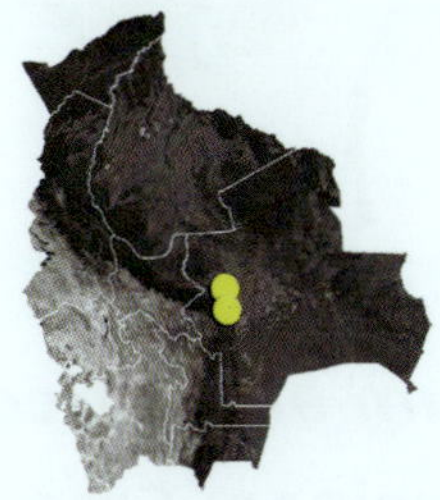

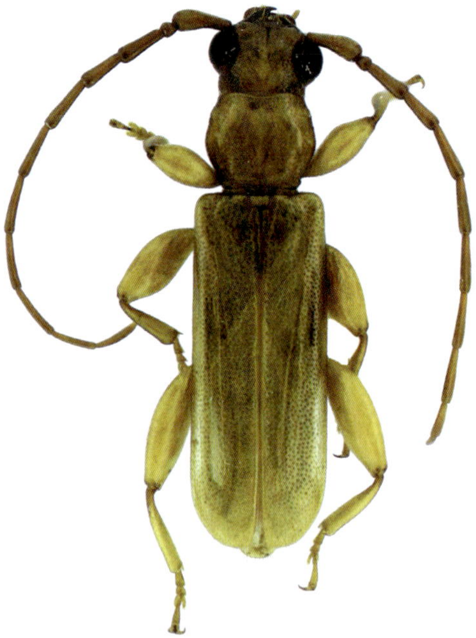

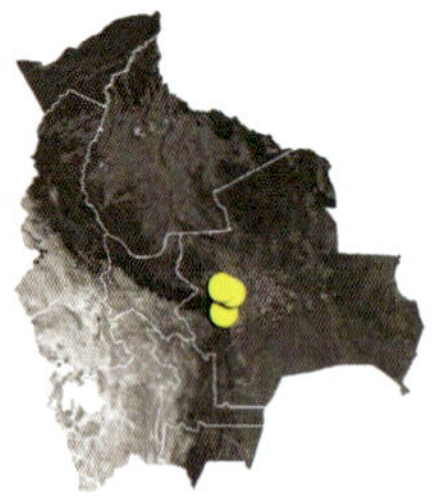

Smodicum angusticolle Aurivillius, 1919
Cerambycinae: Smodicini

This is a small species (7–8 mm) with a tan-colored, flattened body and appendages. It is one of several similarly shaped and colored species in the genus found in Bolivia. It is typically collected at lights, sometimes in numbers, and nothing is known of its habits or host. It is known from central Brazil north to French Guiana and in Bolivia only from lowland locations in Santa Cruz Department, including the vicinity of the Flora and Fauna Hotel near Buena Vista, Potrerillo del Guenda Reserve, and Refugio Los Volcanes, adjacent to Amboró National Park.

Thyrsia piranga Galileo & Martins, 2006
Cerambycinae: Thyrsiini

This is a small (7–8 mm) but broad-bodied, minutely, densely punctate species with feebly shining integument. Unlike most cerambycids, this species does not emerge during the rainy season but after it, with adults found during April and May. The coloration is very much like a lampyrid beetle. It is a Bolivian endemic known only from Santa Cruz Department in the vicinity of the Flora and Fauna Hotel near Buena Vista.

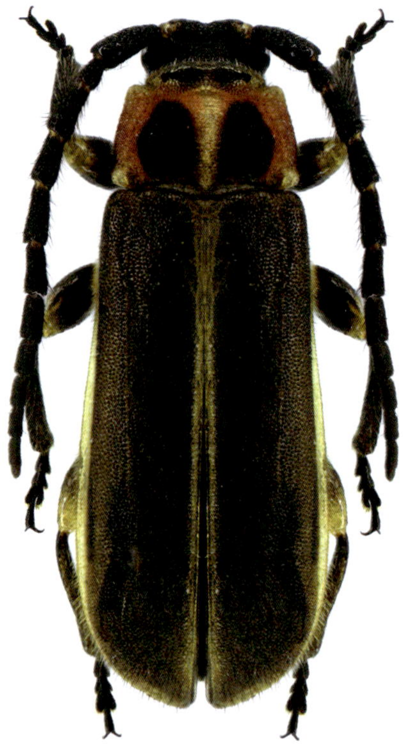

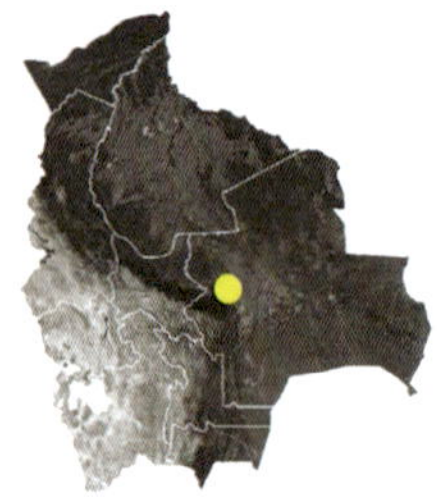

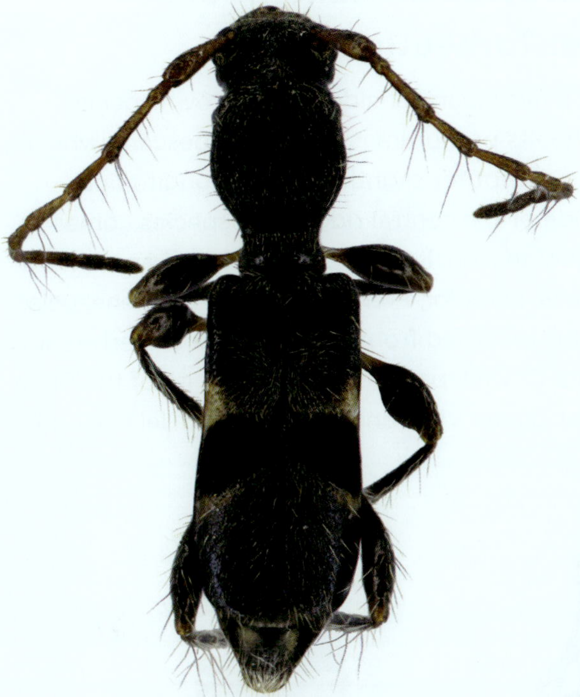

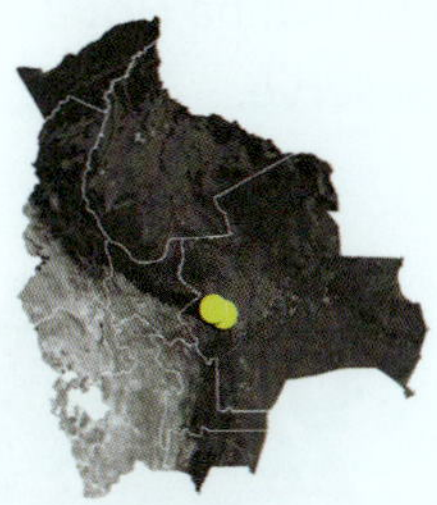

Epropetes bolivianus Galileo & Martins, 2008

Cerambycinae: Tillomorphini

This is a tiny (5–6 mm), ant-like, mostly dark species with lighter brown legs and a basally, sharply constricted pronotum. Although uncommon, it has been found on flowers of woody shrubs and by beating small branches still bearing wilted or dry leaves. It is a Bolivian endemic and known from localities in Santa Cruz Department, including the Vicoquin area above Achira, the road to Bella Vista near Cuevas, and Refugio Los Volcanes, adjacent to Amboró National Park.

Coccoderus amazonicus Bates, 1870

Cerambycinae: Torneutini

This is a moderately large (21–23 mm), tan longhorned woodborer with six pale yellow spots on the elytra and two small, black pronotal spots. The males have antennae one and one-half times their body length and are less commonly encountered than the females. This species has been collected by beating freshly dead standing trees but is more typically collected at lights near or in the forest. It is a widespread species found from Costa Rica to French Guiana and the Brazilian Amazon. In Bolivia it is known from a few lowland localities in Santa Cruz Department: the vicinity of the Flora and Fauna Hotel near Buena Vista, the Potrerillo del Guenda Reserve, and Refugio Los Volcanes, adjacent to Amboró National Park.

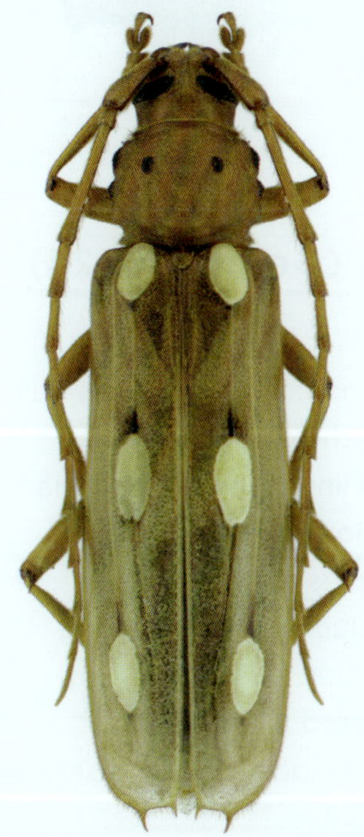

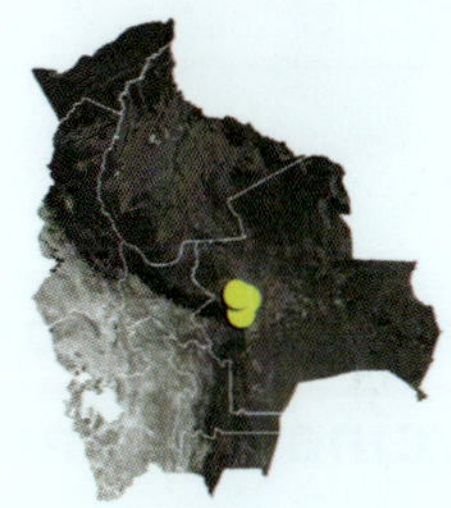

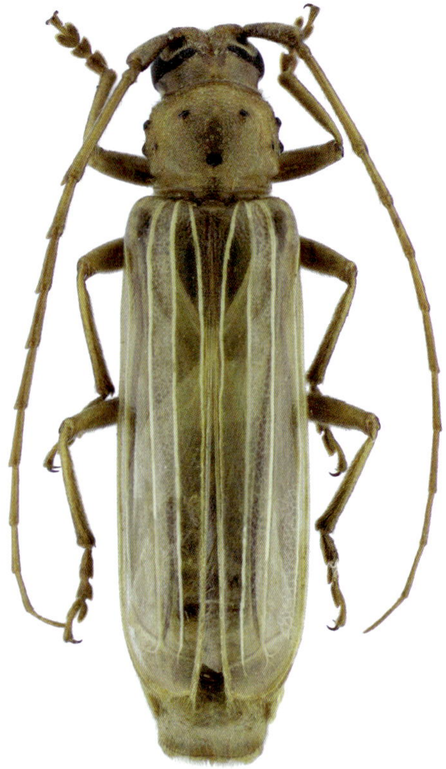

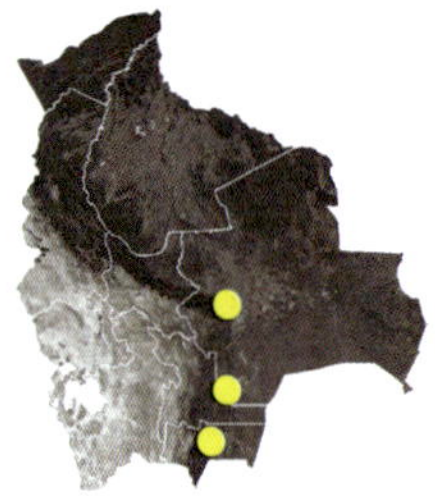

Coccoderus novempunctatus (Germar, 1824)
Cerambycinae: Torneutini

This species is similar in size (20–22 mm) and color to the previous species, *Coccoderus amazonicus,* but lacks the elytral spots, instead having pale yellow to ivory-colored lines the length of the elytra. It was only recently recorded from Bolivia, where it has been collected by beating dead standing trees and at lights. It is known from southern South America, including eastern Brazil, and is found at a few low- to midlevel elevation (below 1,500 m) localities in Bolivia. There, it has been collected in Santa Cruz Department north of Camiri and the Potrerillo del Guenda Reserve and from Tarija Department west of Villamontes.

Diploschema weyrauchi Lane, 1966
Cerambycinae: Torneutini

This is a large (34–38 mm), parallel-sided, dark brown species with light beige elytra. Its pronotum is densely pubescent, which obscures the surface except for slightly elevated longitudinal tubercles on each side of the central ridge. This species comes to lights early in the rainy season. It is widely distributed and known from Amazonian Brazil, Ecuador, French Guiana, Peru, Venezuela, and Bolivia, where it is recorded from Cochabamba Department in Villa Tunari and Santa Cruz Department in the vicinity of the Flora and Fauna Hotel near Buena Vista and the Potrerillo del Guenda Reserve.

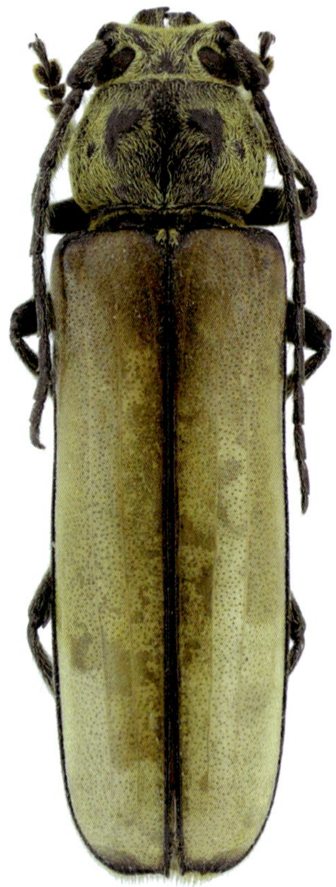

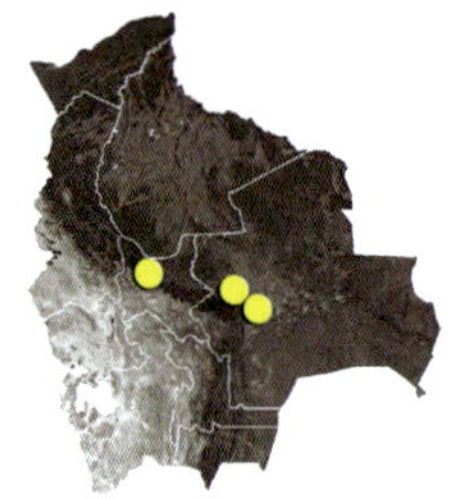

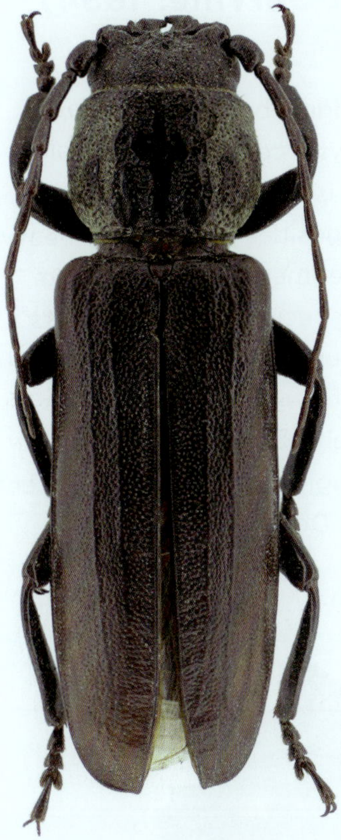

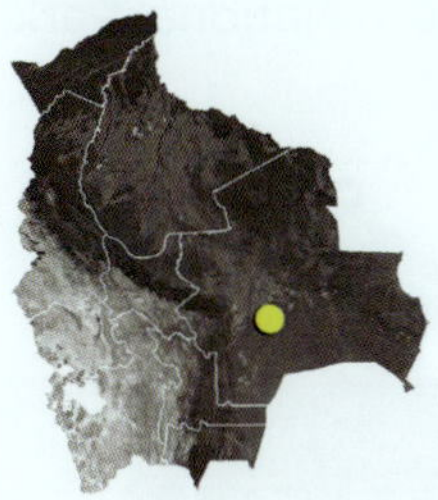

Dragomiris quadricornutus Gounelle, 1913

Cerambycinae: Torneutini

This is a large (38–42 mm), dark brown species with minutely punctate and costate elytra and short antennae (not attaining the midpoint of the elytra). The pronotum has indistinct, short, golden pubescence at the sides and otherwise ornamented with several smooth calli. This plain-appearing species is occasional at lights but very uncommon and known only from Argentina and Santa Cruz Department in Bolivia.

Gigantotrichoderes flabellicornis (Zajciw, 1965)

Cerambycinae: Torneutini

This is a large (26–32 mm), robust, parallel-sided, dark brown species with tan elytra and short pectinate antennae. It is uncommon and typically encountered only at lights. A widespread species, it is known from Costa Rica to Colombia, the Guianas, northern Brazil, and Bolivia, where it is known from the vicinity of the Flora and Fauna Hotel near Buena Vista and the Potrerillo del Guenda Reserve in Santa Cruz Department.

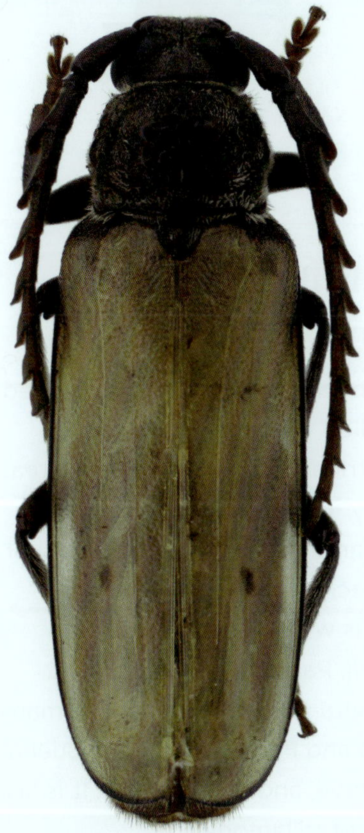

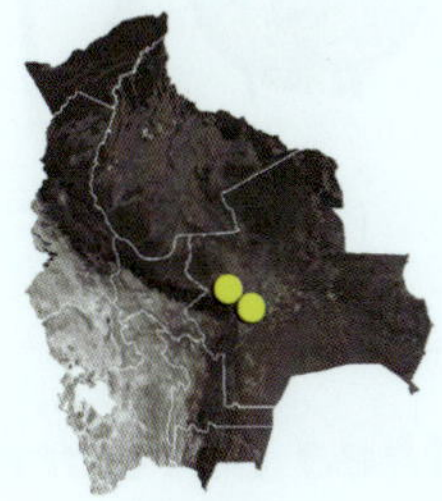

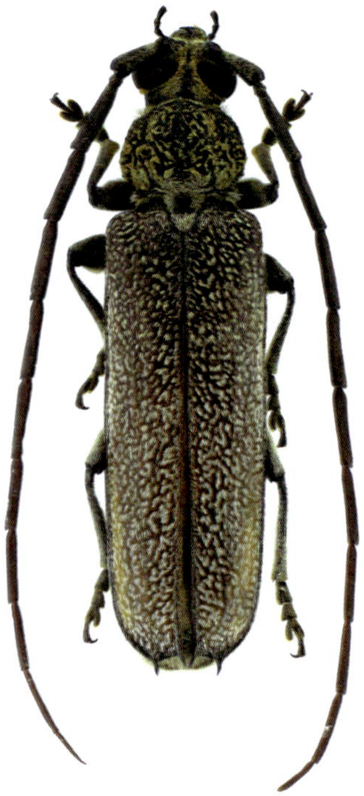

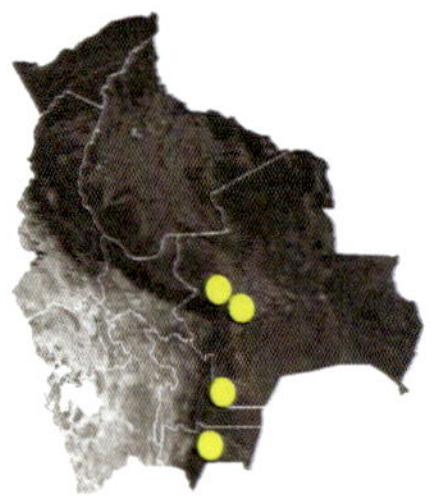

Praxithea derourei (Chabrillac, 1857)

Cerambycinae: Torneutini

This is a large (25–27 mm), brown species with deep pockmarked elytra. The elytral depressions are filled with short, whitish, recumbent setae. Its pronotum is intricately rugose with the areas between the rugosities densely hairy. This species is unusual in that specimens have 12-segmented antennae instead of the usual 11. In lowland tropical forests this species is relatively common at lights early in the rainy season. It is widely distributed and known from Argentina, eastern Brazil, Paraguay, Uruguay, and Bolivia, where it is recorded from Santa Cruz Department at many localities, including Camiri, the Flora and Fauna Hotel near Buena Vista, the Potrerillo del Guenda Reserve, and near Cuevas. It is also known from Tarija Department west of Villamontes.

Praxithea fabricii (Audinet-Serville, 1834)

Cerambycinae: Torneutini

This is a large (35–37 mm), medium reddish-brown species with densely punctate and lightly costate elytra covered in short, recumbent pubescence. The pronotum is intricately rugose with long, thick pubescence between the rugosities (particularly heavy on the disk). It is notable in having 12-segmented antennae, like its previous congener, *P. derourei*. This species is found at lights well into the rainy season, but never abundant. It is known only from southeastern Brazil and Bolivia, where it is recorded from the vicinity of the Flora and Fauna Hotel near Buena Vista, the Potrerillo del Guenda Reserve, and Refugio Los Volcanes, adjacent to Amboró National Park, in Santa Cruz Department.

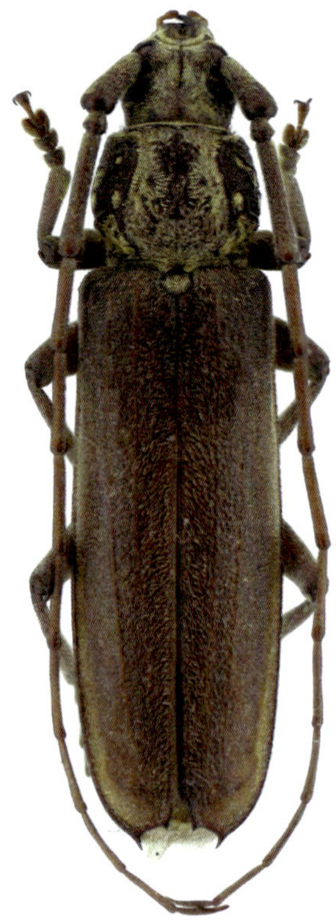

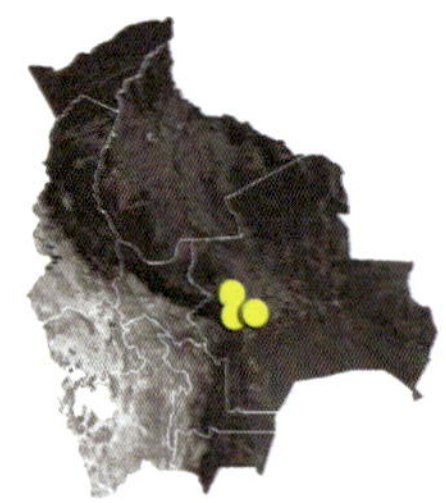

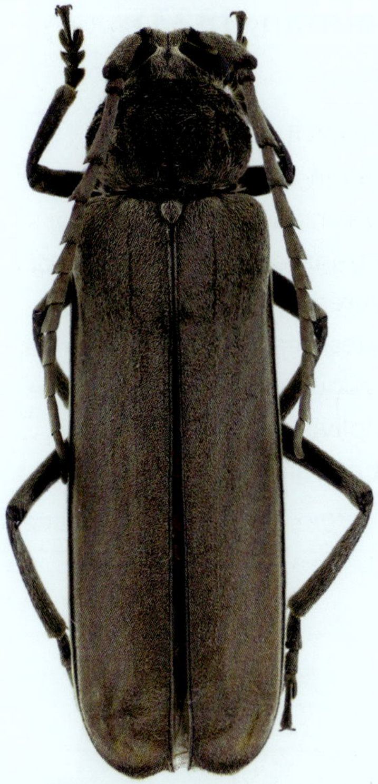

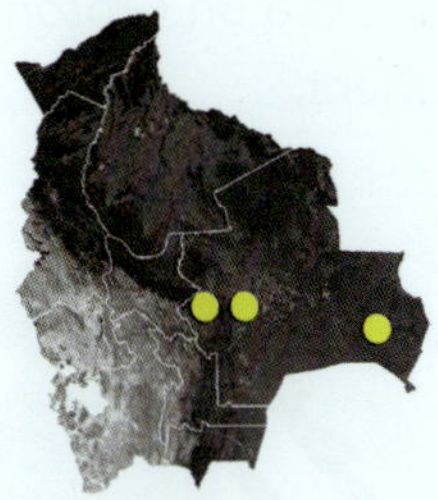

Psygmatocerus wagleri Perty, 1828

Cerambycinae: Torneutini

This is a large (33–38 mm), dusky brown species that differs from most cerambycids by having 12-segmented antennae. Additionally, the sexes have very different antennal forms; the females have slightly serrate distal segments and the males have each antennomere expanded into a long, flat blade, collectively called flabellate antenna. This species is rare in collections but does come to lights and is widely distributed, being recorded from Argentina, southeastern Brazil, Paraguay, and Bolivia. In Bolivia it has been collected at widely separate localities in Santa Cruz Department, including north of Pampagrande, the Potrerillo del Guenda Reserve, and east of Roboré.

Torneutes pallidipennis Reich, 1837

Cerambycinae: Torneutini

This is a very large, sexually dimorphic species (females range in length from 38 to 52 mm and males can approach 75 mm). The species is parallel sided, dark reddish brown, and with light tan elytra (these are somewhat translucent in the females). The antennae are very short, not attaining the basal third of the elytra in either sex. This species is rarely encountered but is attracted to lights. It is known from Argentina, Brazil, Paraguay, Uruguay, and Bolivia. In Bolivia it has been collected in Santa Cruz Department north of Pampagrande, north of Camiri, and historically near the city of Santa Cruz de la Sierra.

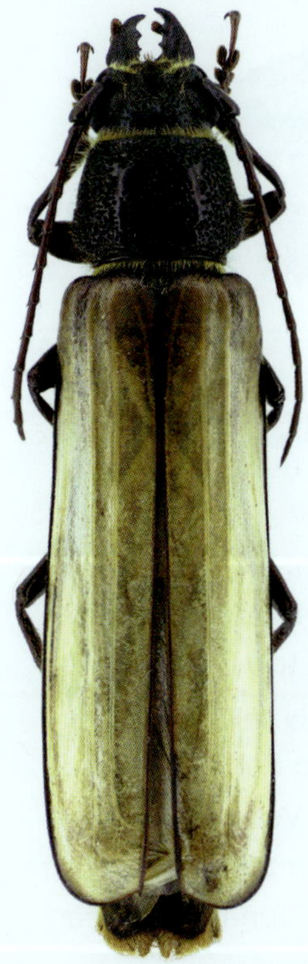

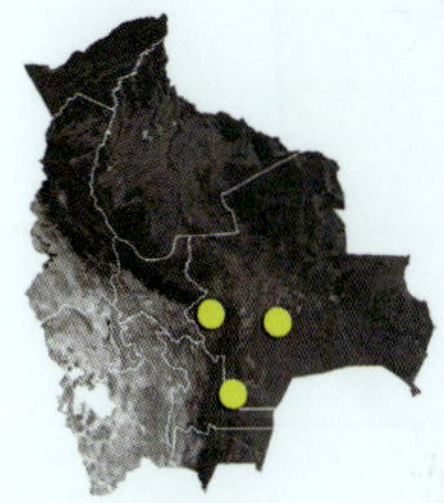

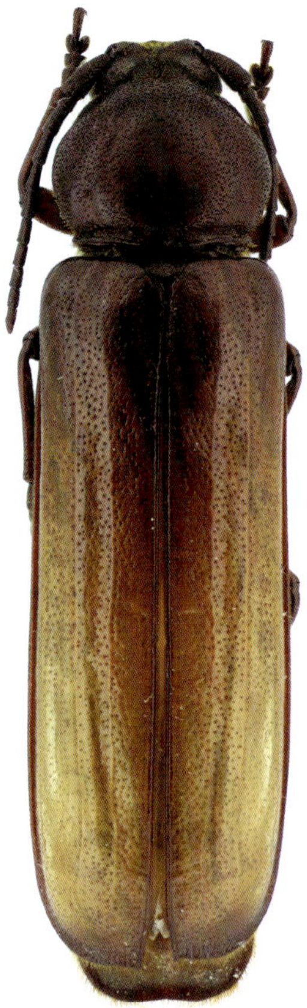

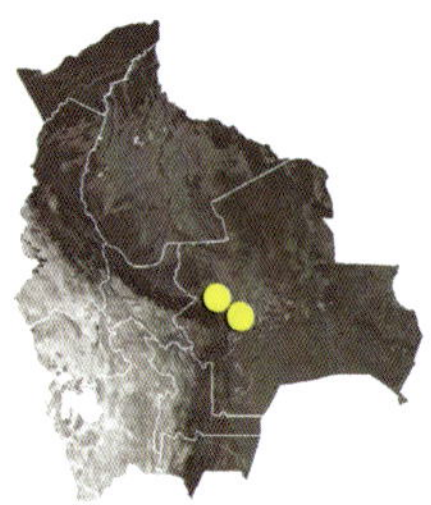

Xenambyx lansbergei (Thomson, 1865)

Cerambycinae: Torneutini

This is a large (35–40 mm), dark reddish-brown, tubular-looking species with costate and densely but finely punctate elytra. Its roundish, swollen pronotum is globular like the top of a ball. It comes to lights but is very uncommon. Widely distributed, it is known from Central America (Guatemala to Panama) and Colombia, the Guianas, central Brazil, Ecuador, Venezuela, and Bolivia. In Bolivia it is found in lowland forests in Santa Cruz Department from the vicinity of the Flora and Fauna Hotel near Buena Vista and the Potrerillo del Guenda Reserve.

Ancylocera cardinalis (Dalman, 1823)

Cerambycinae: Trachyderini

This is a very distinct species (8–11 mm) in its color and proportions. The body is boldly colored in bright orange or red, while the appendages are black. The pronotum is much longer than wide and completely covered in transverse, curved rows of punctures. The elytra are likewise covered in punctures, though not in distinct rows. The species is widespread in South America, being known from Brazil, Paraguay, Uruguay, and Argentina. We record it here for the first time from Bolivia, where a specimen was collected on *Croton* flowers west of Samaipata in Santa Cruz Department.

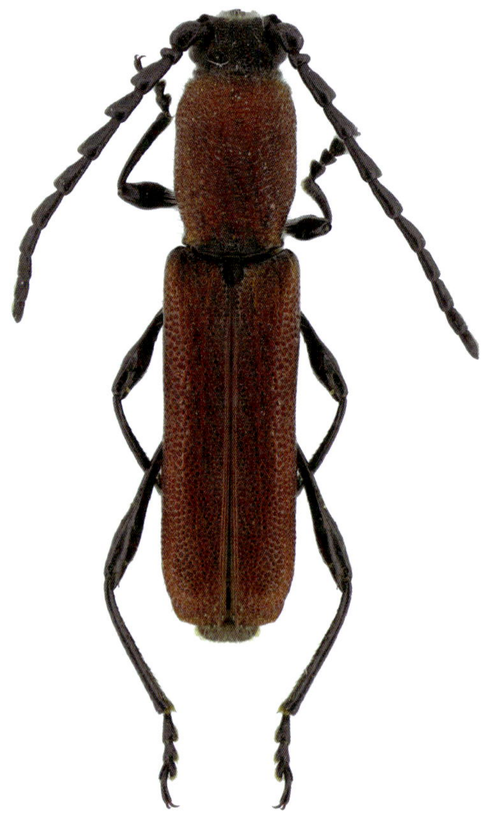

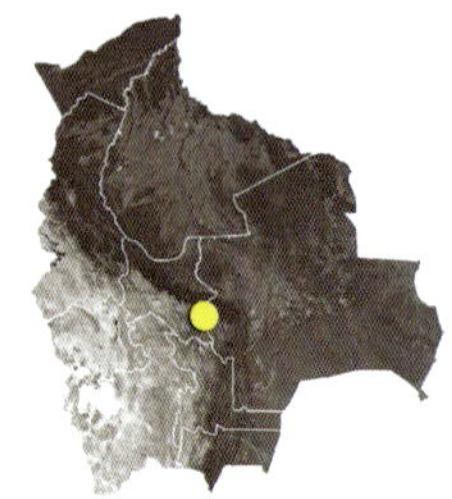

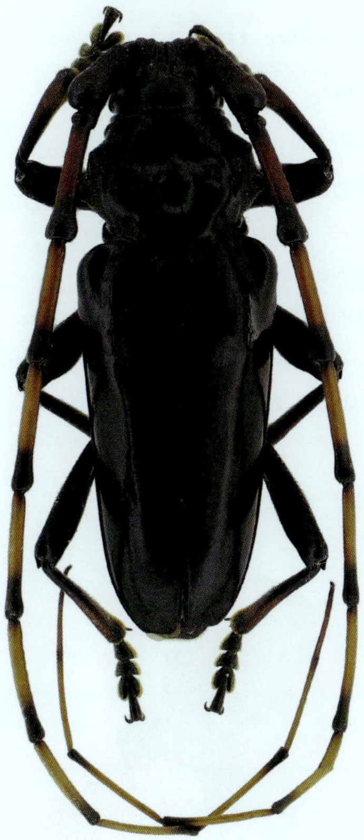

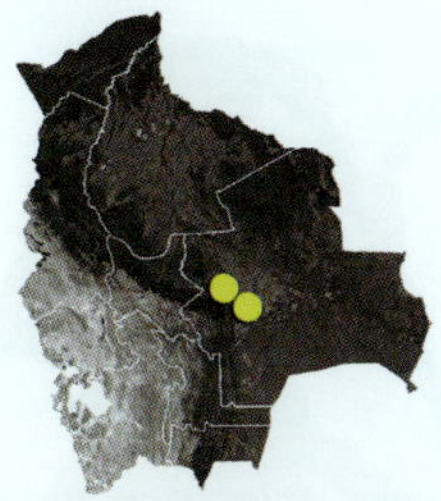

Andraegoidus rufipes (Fabricius, 1787)

Cerambycinae: Trachyderini

This is a large (23–26 mm), shining black, impunctate species with orange and black annulate antennomeres, tibiae, and tarsi. It is variable in integument color as some individuals have a small red fascia on the basal half of each elytron and differing tibial and tarsal color patterns. It is a strong flyer and can be found during the day buzzing around freshly cut trunks in agricultural *chaqueos*. It is a widespread species known from Argentina, Brazil, Paraguay, and Bolivia, where it is has been recorded from Santa Cruz Department in the vicinity of the Flora and Fauna Hotel near Buena Vista and the Potrerillo del Guenda Reserve.

Andraegoidus variegatus (Perty, 1832)

Cerambycinae: Trachyderini

This is a large (26–30 mm), shining, diurnal species and, as its name implies, is a highly variable beetle in color and dorsal pattern. Some individuals are unmarked with rust-red elytra while others are much darker and have four light tan spots on the elytra. All, however, have light tan antennae with black annulations. Males typically have very long antennae, nearly twice their body length. They can be found on hot, sunny days during the rainy season coming to partially burned logs in agricultural *chaqueos*. It is recorded from southern South America, including Brazil and Bolivia, where it is known from Beni Department near Riberalta and Santa Cruz Department in the Potrerillo del Guenda Reserve and Refugio Los Volcanes, adjacent to Amboró National Park.

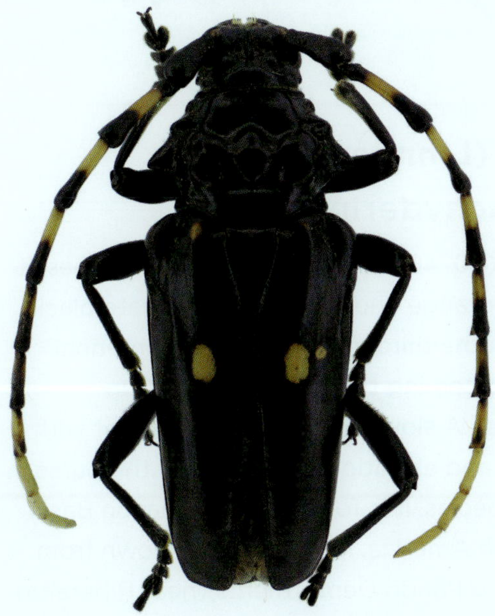

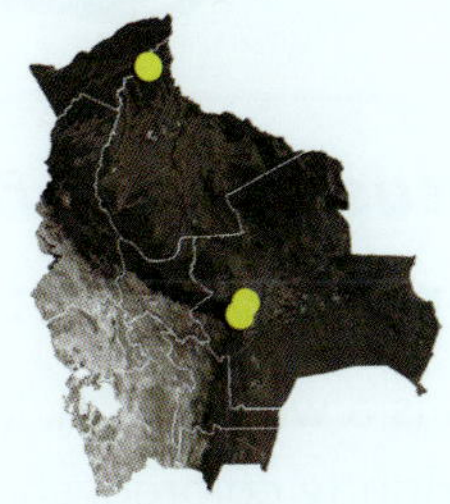

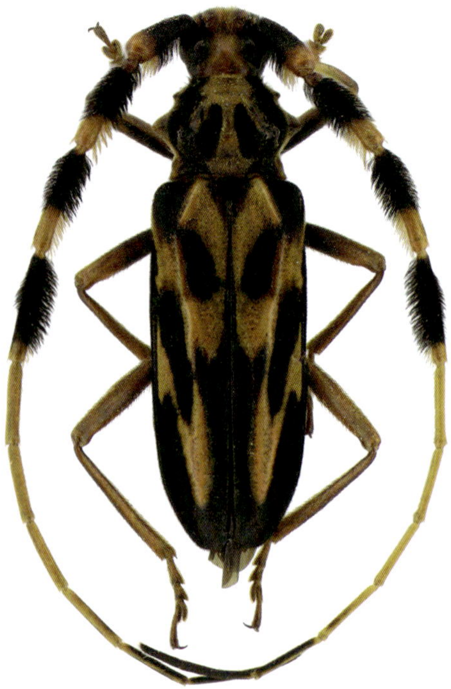

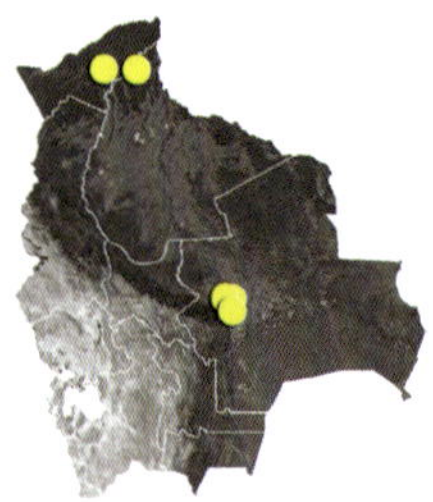

Batus barbicornis (Linnaeus, 1764)

Cerambycinae: Trachyderini

This is a well-known, large (30–40 mm), and distinctive species. It is brightly colored, shining orange and black with elongate black tufts of stiff pubescence on the third, fourth, and fifth antennal segments. It is encountered coming to or feeding on fallen fermenting fruit in the forest. A slow flyer, it is easily caught with the net until it gains speed and elevation, then eluding the pursuer. A lowland tropical forest species, it is found in the Amazon Basin and much of northern South America. In Bolivia it is known from lowland localities in Beni and Pando Departments (near Riberalta) and in Santa Cruz Department at Refugio Los Volcanes, Potrerillo del Guenda Reserve, and the vicinity of the Flora and Fauna Hotel near Buena Vista.

Callancyla croceicollis (White, 1855)

Cerambycinae: Trachyderini

This is a small (9–10 mm), parallel-sided, densely punctate, black species with an orange to reddish-orange pronotum. The dorsal surface is also sparsely covered with fine, short pubescence. It is a diurnally active species that is a flower visitor, with adult emergence after the rainy season in April and May. It has been recorded from Brazil, French Guiana, and Bolivia, where it is known from the Flora and Fauna Hotel near Buena Vista in Santa Cruz Department.

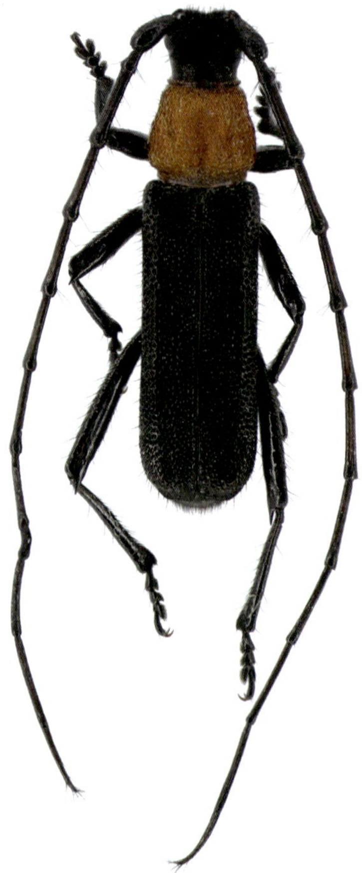

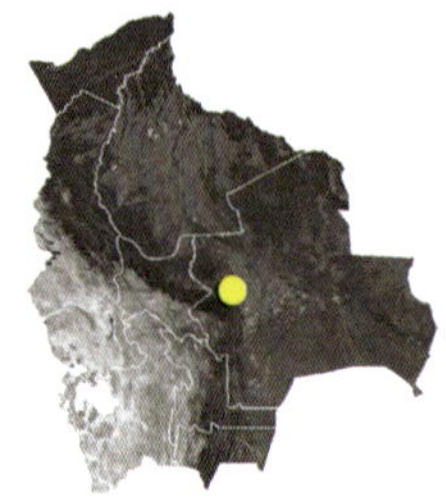

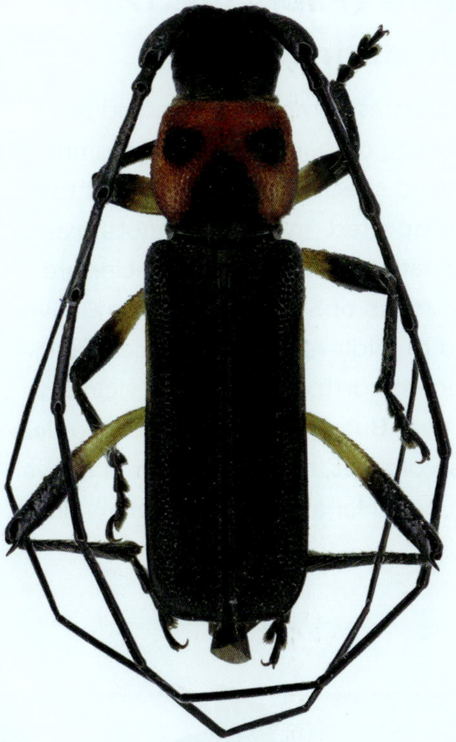

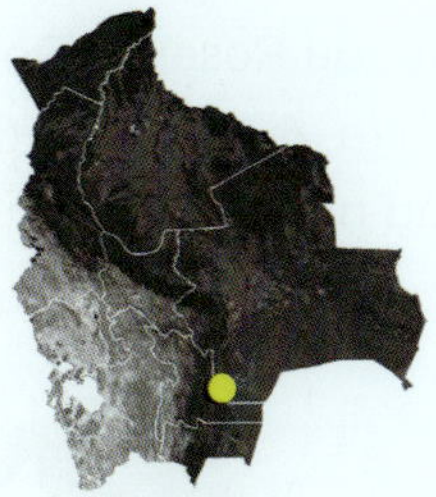

Callancyla tucumana Viana, 1971

Cerambycinae: Trachyderini

This is a small (9–10 mm), black species with an orange and black pronotum and femora basally yellow. Like its congener *C. croceicollis*, the dorsal surface is densely punctate, but in this species the punctures are much smaller and the elytra are devoid of pubescence. Adults of this species can be found in December visiting the flowers of bush *Croton*. It is known from Argentina and Bolivia, where it has been recorded from Santa Cruz Department north of Camiri.

Ceragenia bicornis (Fabricius, 1801)

Cerambycinae: Trachyderini

This is another of the midsized (18–20 mm), lightly colored, diurnal species inhabiting partially burned agricultural *chaqueos*. It can be seen in these habitats crawling on or flying to stacks of freshly cut branches or partially burned stumps. It is also attracted to fermenting baits, including very ripe bananas and *plátanos*. It is occasionally attracted to lights when in proximity to its resting site. In South America it is found in the northern countries south to Brazil and Bolivia, where it has been commonly recorded in lowland localities in Santa Cruz Department at the Potrerillo del Guenda Reserve and the vicinity of the Flora and Fauna Hotel near Buena Vista.

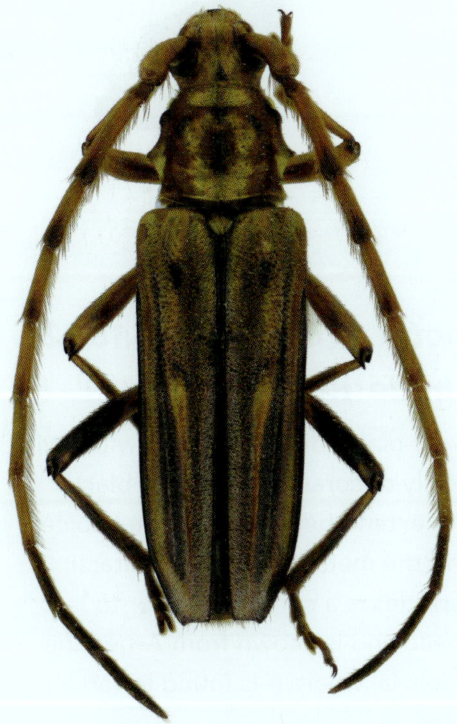

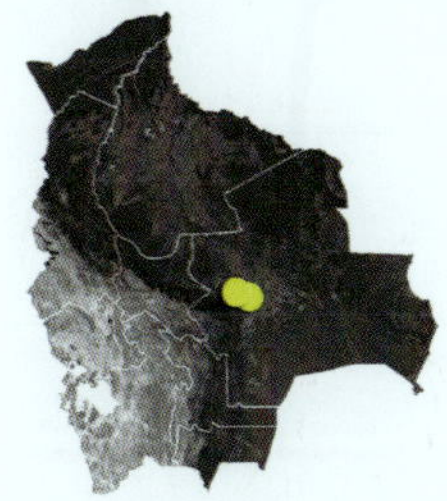

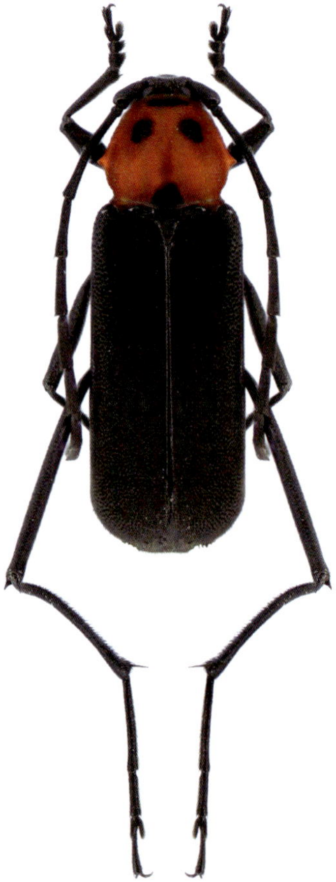

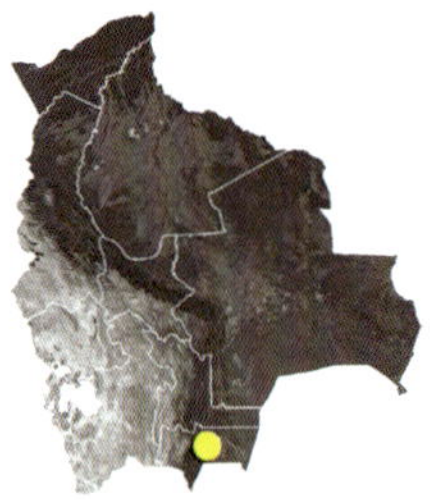

Chevrolatella tripunctata (Chevrolat, 1862)
Cerambycinae: Trachyderini

This is a small (9–11 mm), velvety black, diurnal species with a vivid orange pronotum that is typically decorated with three black, circular spots. The hind femora extend beyond the elytral apices by almost half their length, and the metatibiae and metatarsi, likewise, are quite long. This species is a common visitor to bush *Croton* flowers in the Chaco and is known from Argentina, Paraguay, central Brazil, and Bolivia, where it is found in Tarija Department west of Villamontes.

Chydarteres dimidiatus (Fabricius, 1787)
Cerambycinae: Trachyderini

This is another of the diurnal species commonly attracted to fermenting baits or found crawling on partially burned stumps in agricultural cuts. A medium-sized species (16–18 mm), light tan and black with distinctive multispotted (in black) pronotum and bicolored antennae. This nominate form and its subspecies are widely distributed throughout much of South America. In Bolivia it is known from most lowland to midlevel elevation localities in Santa Cruz Department, including north of Camiri, the vicinity of the Flora and Fauna Hotel near Buena Vista, the Potrerillo del Guenda Reserve, Refugio Los Volcanes, the Vicoquin area above Achira at 2,000 m, the region near Mairano, and the Bella Vista road near Cuevas.

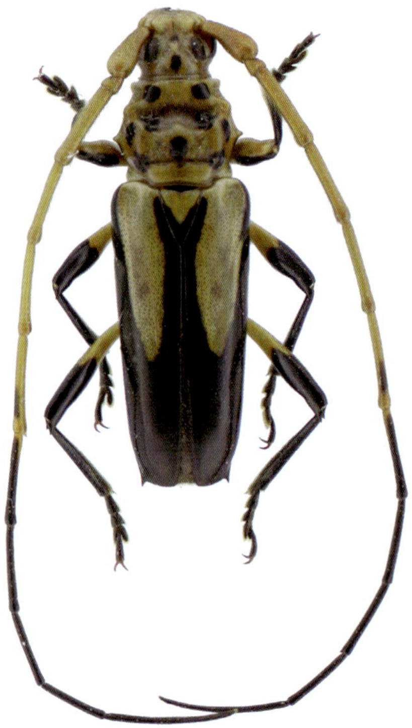

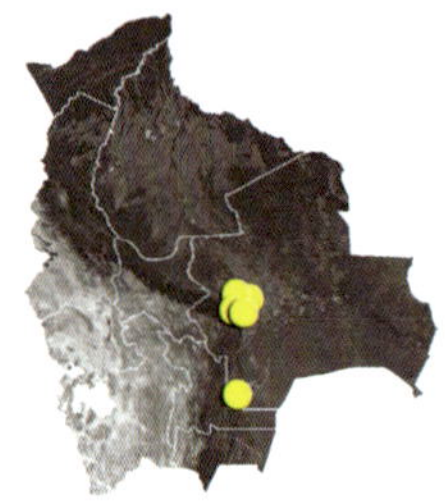

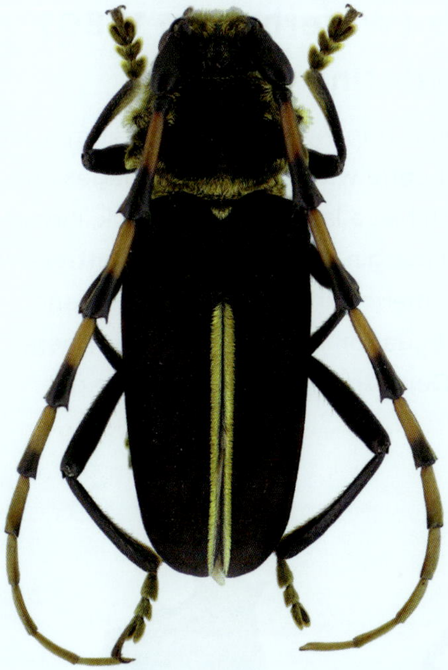

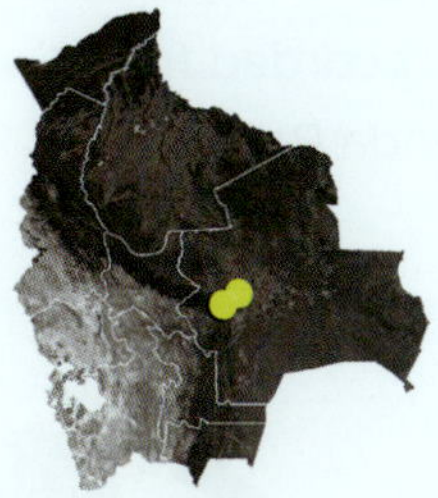

Dorcacerus barbatus (Olivier, 1790)

Cerambycinae: Trachyderini

This is a distinctive, dark reddish-brown species (20–22 mm) with annulate antennae and a yellow, heavily pubescent carpet covering its face. It can be locally common in agricultural *chaqueos*, flying to partially burned, large tree trunks to mate and oviposit. This widespread cerambycid is found throughout Central America and south to Argentina, Brazil, and Bolivia. In Bolivia it is recorded from various localities in Santa Cruz Department including Potrerillo del Guenda Reserve and Bermejo.

Eriphus haematoderus Chevrolat, 1862

Cerambycinae: Trachyderini

This is a small (10–11 mm), velvety black, diurnal species with an orange pronotum that is typically decorated with two black, circular fasciae (one apically and one basally). This species is similar to *Chevrolatella tripunctata* but is readily separable by its shorter, stouter antennomeres. It is also an inhabitant of wetter lowland tropical forests instead of the drier Chaco areas. It is a Bolivian endemic and known only from Santa Cruz Department from the vicinity of the Flora and Fauna Hotel near Buena Vista and the Potrerillo del Guenda Reserve.

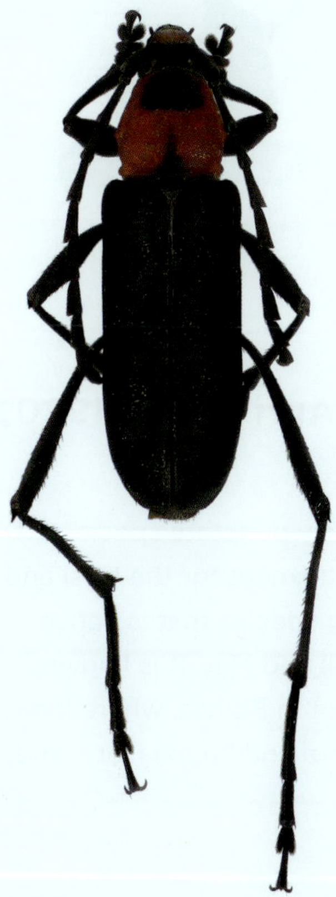

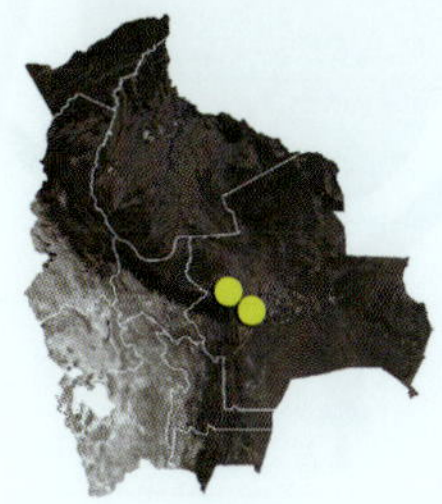

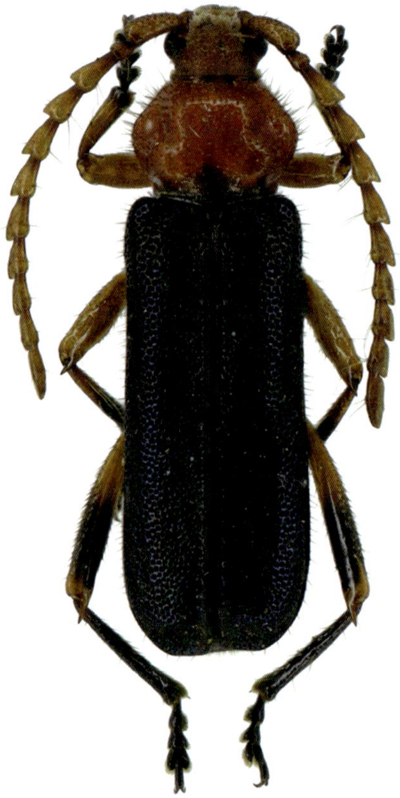

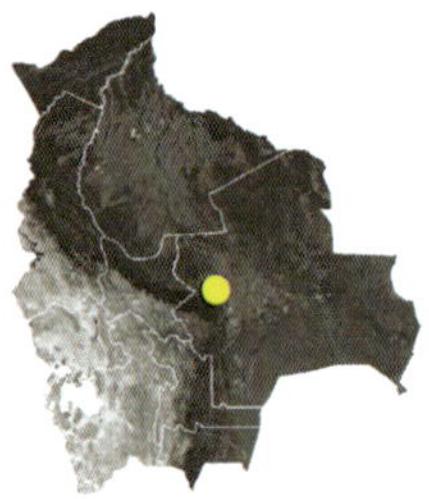

Exallancyla tuberculicollis (Aurivillius, 1920)

Cerambycinae: Trachyderini

This is a small (9–10 mm), dark blue, densely punctate species with orange head, pronotum, and appendages (except for the tarsi and hind femora). It is a diurnal, flower-visiting species that is active toward the end of the rainy season in April and May. It is known only from Peru and Santa Cruz Department in Bolivia, where it has been collected from the vicinity of the Flora and Fauna Hotel and Buena Vista.

Galissus cyaneipennis (Waterhouse, 1880)

Cerambycinae: Trachyderini

This species, formerly in the genus *Panchylissus*, is a medium-sized (19–24 mm), shining royal blue beetle with black appendages and reddish-orange pronotum, which has a large, shining black, circular macula connected to the apical margin. This colorful and attractive species is known from Central America, including Honduras and countries to the south, eastern Ecuador, Peru, and Bolivia, where it is recorded from Santa Cruz Department in the Potrerillo del Guenda Reserve.

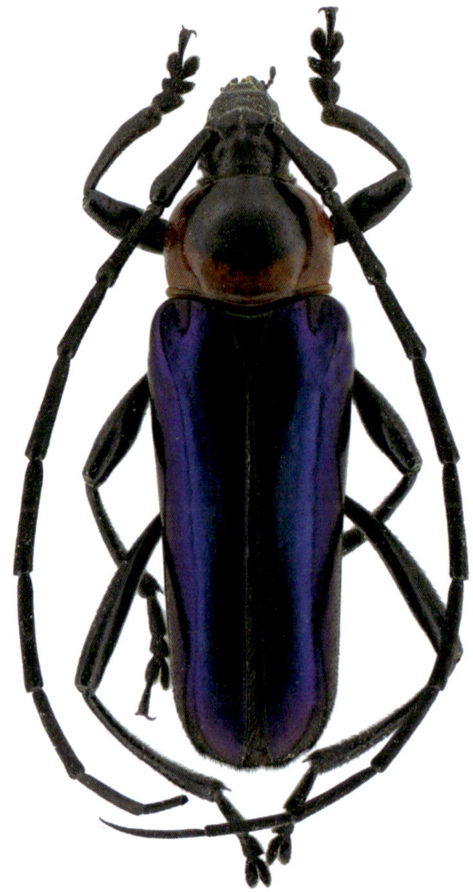

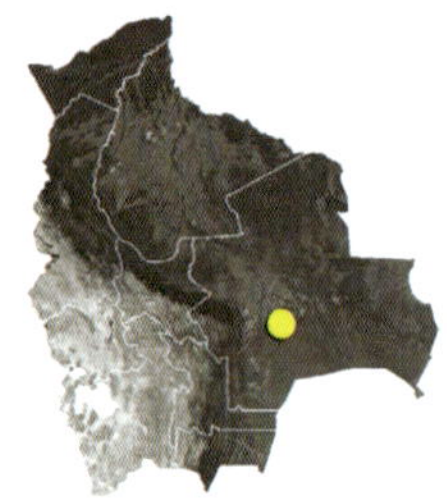

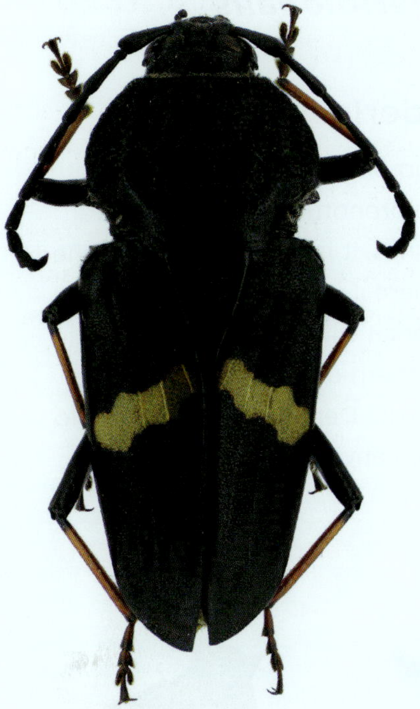

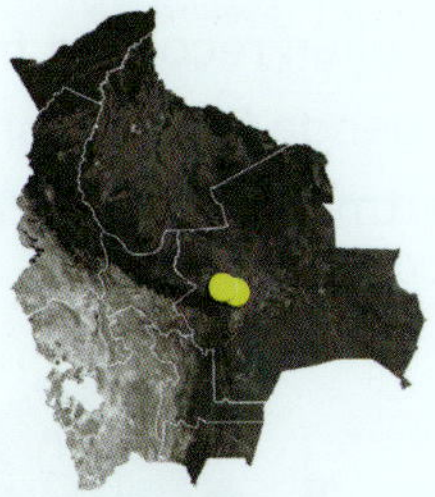

Megaderus stigma (Linnaeus, 1758)

Cerambycinae: Trachyderini

This is a uniquely shaped, large (17–28 mm), diurnal cerambycid being broadly compressed and having a heavily punctate pronotum at least as wide and about half as long as the elytra. It is dark brown with an oblique, tan elytral fascia and bright red-orange tibiae. The males are much smaller than the females and have a reduced elytral fascia. It is a common inhabitant of agricultural cuts where it prefers rather small scorched tree stubs to deposit its eggs. Widely distributed, it is found from Central America south to Argentina, Brazil, and Bolivia, where it is found in lowland forest cuts in Santa Cruz Department. It has been collected from the vicinity of the Flora and Fauna Hotel near Buena Vista and the Potrerillo del Guenda Reserve.

Micropelta rugosicollis Zajciw, 1961

Cerambycinae: Trachyderini

This is a medium-sized (15–17 mm), shining metallic green species with dark antennae (except for the bicolored scape) and legs (except for the bicolored femora). This beautiful species is a strong flyer and visits the flowers of lowland forest trees during the rainy season. It is known from western Brazil, Peru, and Bolivia, where it has been collected in Santa Cruz Department from the Potrerillo del Guenda Reserve and the Province of Velasco.

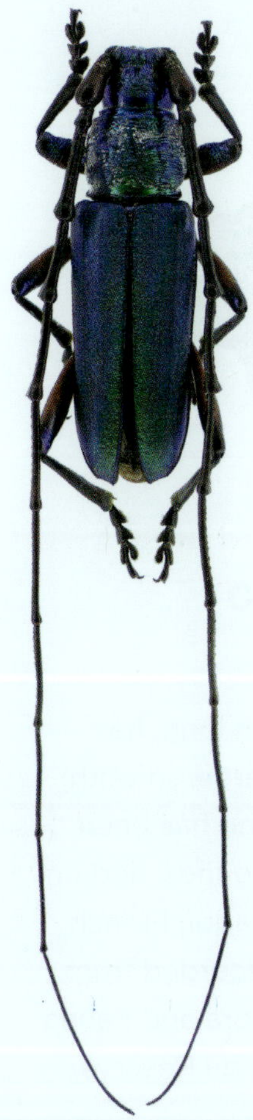

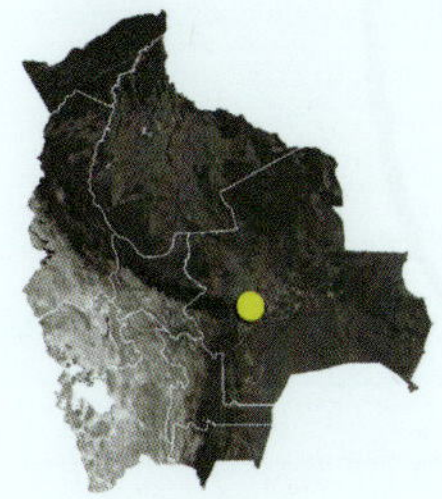

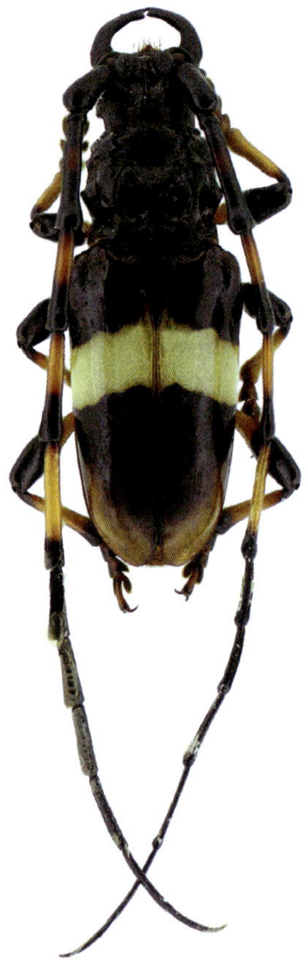

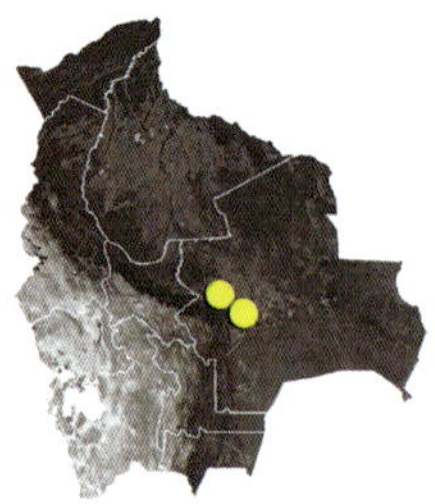

Monneellus rhodopus (Bates, 1870)

Cerambycinae: Trachyderini

This shining brown and light tan species, 10–18 mm long, has annulate antennae and, in the males, distinctive, large, smooth, circular mandibles. It is a lowland forest species and has been collected on recently cut wood, sapping at tree wounds, and on tree blossoms. It is known from central Brazil, Ecuador, French Guiana, Peru, Venezuela, and Bolivia, where it is recorded from Santa Cruz Department from the vicinity of the Flora and Fauna Hotel near Buena Vista and the Potrerillo del Guenda Reserve.

Oxymerus aculeatus meridionalis Huedepohl, 1979

Cerambycinae: Trachyderini

This is a midsized (13–15 mm), diurnal species, tan in color with black antenna, black spots on the pronotum, and dark elytral striae. Like many other trachyderines, it is attracted to scorched wood in agricultural clearings, where it can be locally common. It is found in southern South America with lowland localities in Bolivia representing the most northern distribution. In Bolivia it has been recorded from Santa Cruz Department in Potrerillo del Guenda Reserve and the Camiri environs and from Tarija Department west of Villamontes.

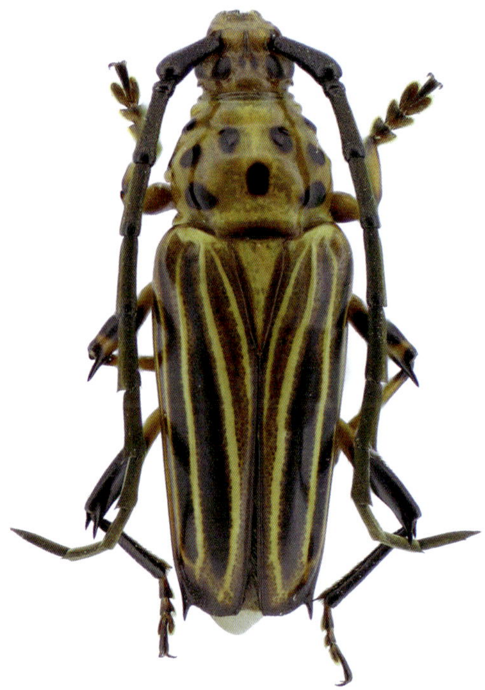

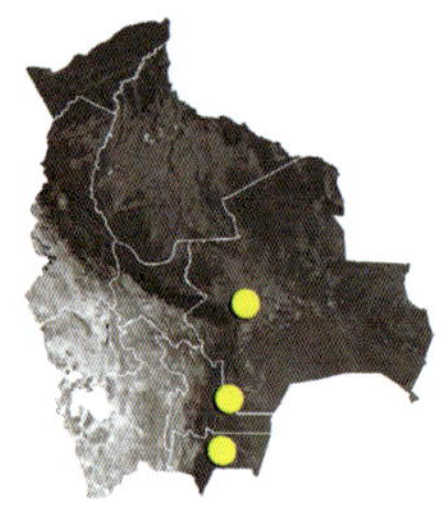

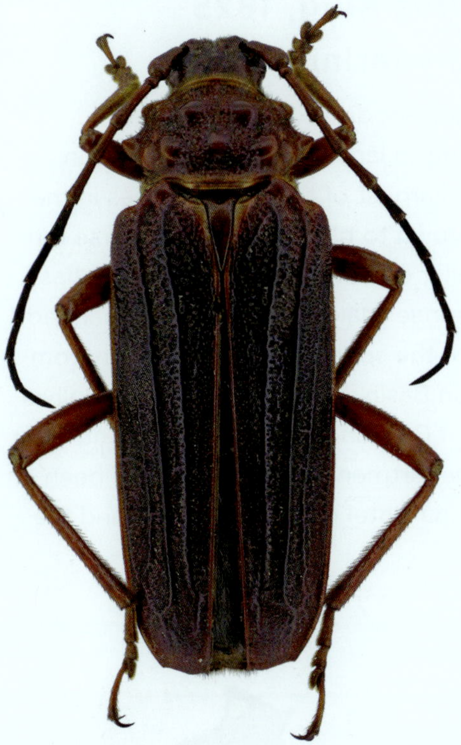

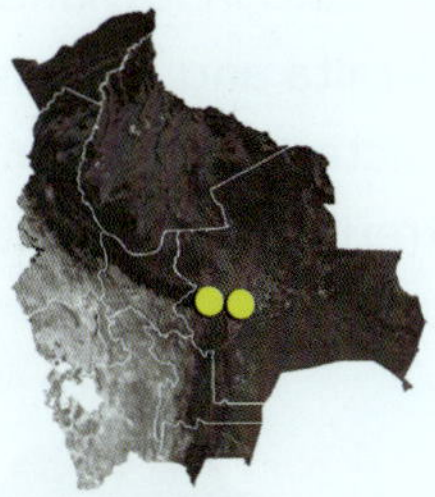

Phaedinus lanio Guérin-Méneville, 1838

Cerambycinae: Trachyderini

This remarkable species is substantial in size (35–40 mm) and in color (a vivid orange to red). The elytra have two or three raised costal ridges and are rather densely punctate between them. The pronotum is highly tuberculate with five dorsal and four lateral tubercles, in total. This species is known from French Guiana, Brazil, and just recently in Bolivia, where specimens have been collected in Santa Cruz Department at the Potrerillo del Guenda Reserve in the very early rainy season in October and at Macuñucu Camp on the north side of Amboró National Park.

Phoenidnus lissonotoides Pascoe, 1866

Cerambycinae: Trachyderini

This is a medium-sized (11–15 mm), black species with a broad, transverse, red fascia on the basal half of the elytra and femora. Although rarely encountered, it has been collected visiting flowers of forest trees in November during the rainy season. It is widely distributed and known from Panama, Colombia, central Brazil, and Bolivia, where it has been found in Santa Cruz Department at the Flora and Fauna Hotel near Buena Vista and the Potrerillo del Guenda Reserve.

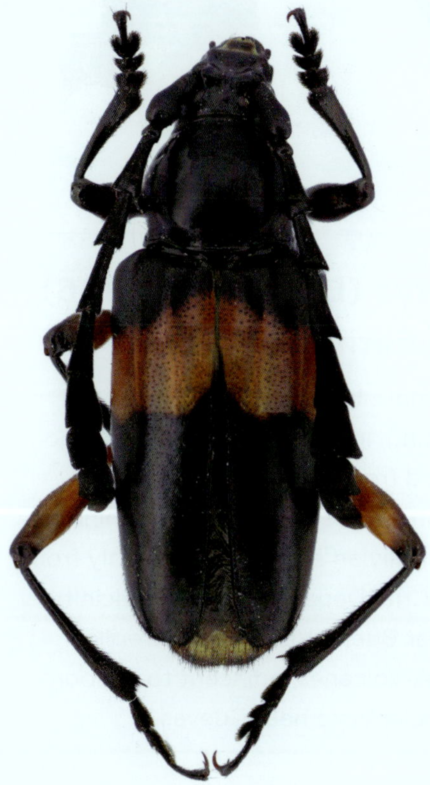

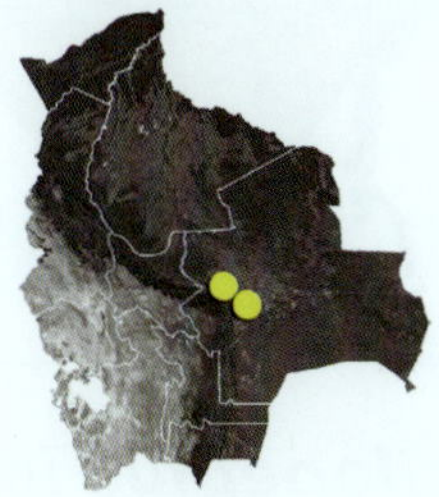

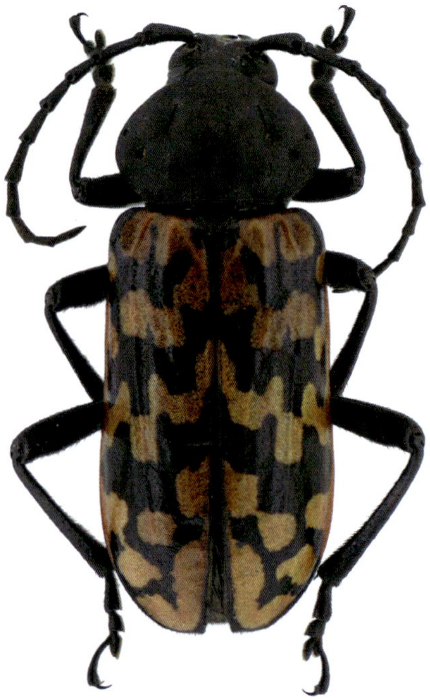

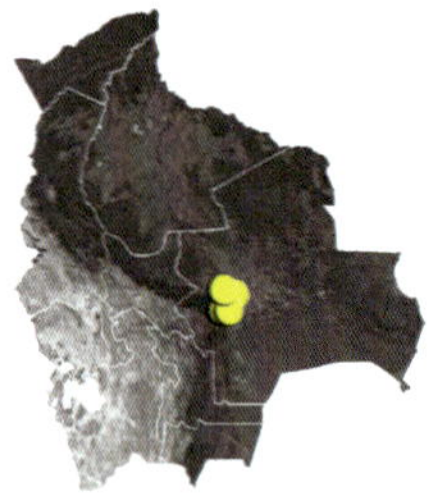

Poecilopeplus intricatus (Blanchard, 1846)

Cerambycinae: Trachyderini

This is a relatively large (22–24 mm), broad-bodied, black longhorned beetle with contrasting, intricate, orange-patterned elytra. Males can be distinguished from the females by their broader, closely punctate pronotum versus the females' shining, narrower form. It is an endemic Bolivian species known only from low and midelevations in Santa Cruz Department in the vicinity of the Flora and Fauna Hotel near Buena Vista, the Potrerillo del Guenda Reserve, Refugio Los Volcanes, adjacent to Amboró National Park, and the road to Bella Vista near Cuevas.

Trachyderes cingulatus Klug, 1825

Cerambycinae: Trachyderini

This is a large (22–29 mm), stout-bodied, diurnal species occasionally encountered flying to large logs or tree trunks in agricultural *chaqueos*. It is the largest of the three species in the genus in Bolivia and is dark brown to black with a transverse tan fascia at midelytra and a smaller one at the elytral apices. It has tan femora and annulate antennae. Males have greatly expanded scapes, which is diagnostic for the species. It is also known from Peru and the Brazilian Amazon Basin. In north and central Bolivia it is found in lowland forests of Beni and Pando Departments near Riberalta and in Santa Cruz Department, where they have been collected at the Flora and Fauna Hotel near Buena Vista and the Potrerillo del Guenda Reserve.

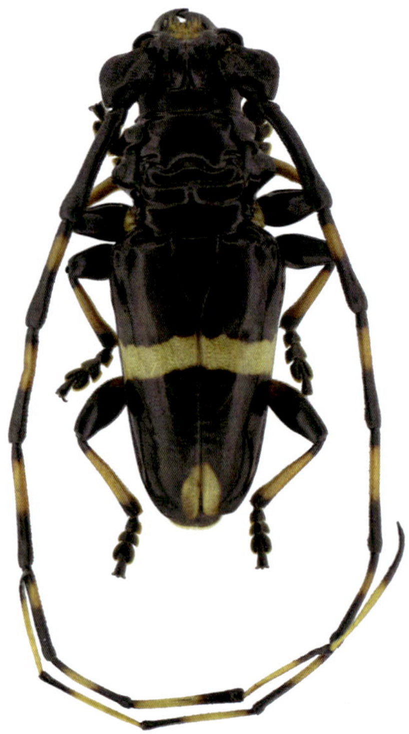

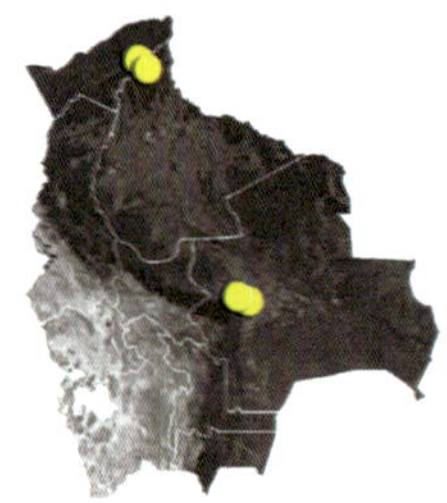

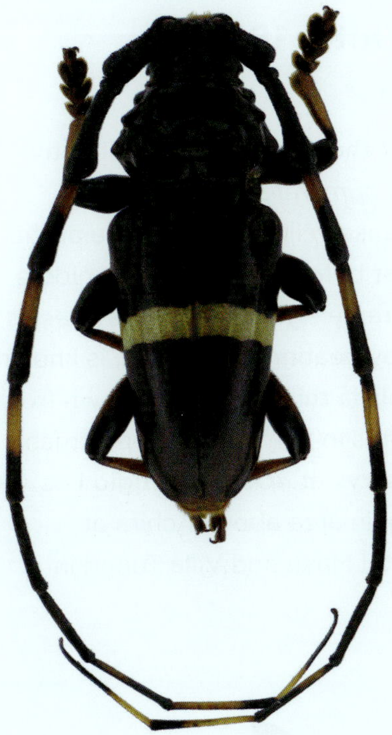

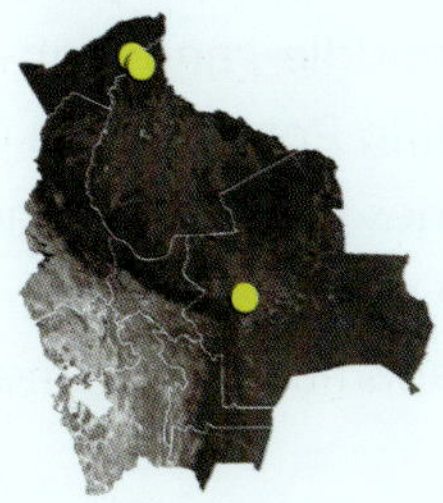

Trachyderes succinctus succinctus (Linnaeus, 1758)

Cerambycinae: Trachyderini

This is a large (22–26 mm), diurnal trachyderine closely related to *Trachyderes cingulatus* but differing in its somewhat smaller size, narrower body, lack of an apical elytral tan fascia, and dark orange tarsi, which are black in *T. cingulatus*. It is especially fond of sap flows on weakened or wounded trees, where it can be found eagerly ingesting the sugary liquid. This widely distributed cerambycid is recorded from the West Indies through South America to Argentina. In Bolivia it is the most frequently encountered species of its genus and has been found in lowland tropical forests in Beni and Pando Departments near Riberalta and in Santa Cruz Department at the Potrerillo del Guenda Reserve.

Ethemon imbasale Tippmann, 1960

Cerambycinae: Unxiini

This is a medium-sized (13–15 mm), shining orange species with dusky blue, finely pubescent elytra. It is a high-altitude montane species typically encountered along the wooded edge of agricultural *chaqueos*, either sitting on foliage or collected by beating. It is a Bolivian endemic known from La Paz Department and Santa Cruz Department in the Vicoquin area above Achira at 2,000 m and northeast of Mairana.

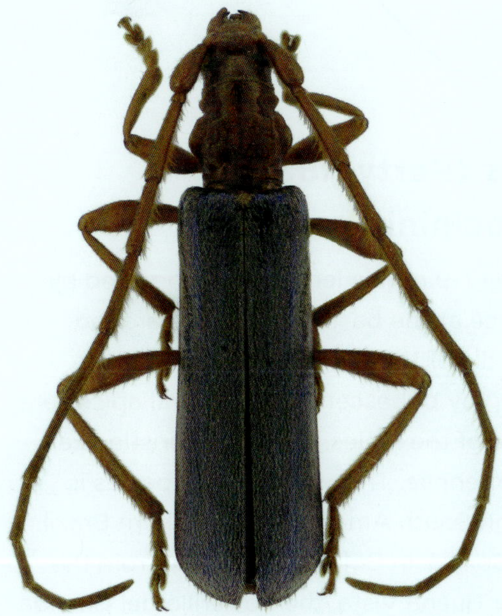

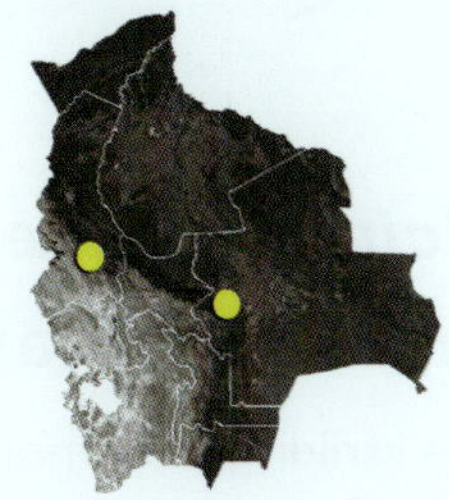

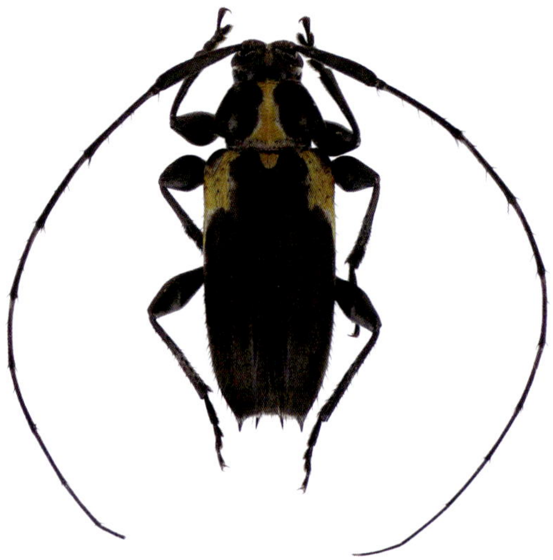

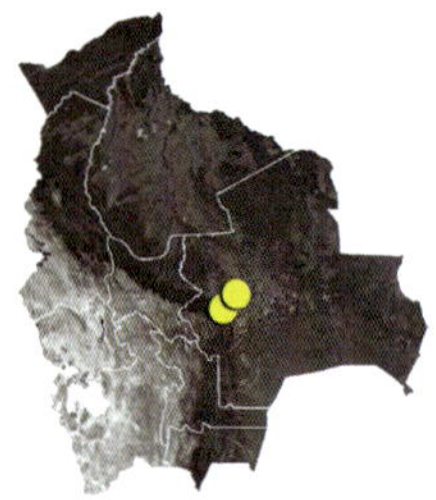

Alcidion humeralis (Perty, 1832)

Lamiinae: Acanthocinini

This striking acanthocinine (7–9 mm in length) is recognized by the vivid orange pubescence at the base of the elytra around the humeri and on the middle of the pronotum and scutellum. Otherwise, it is black with gray pubescence. The elytral apices are acute, as is the last ventrite of the males. Scattered, bristle-like setae are present on the antennae. This widespread species is known from Mexico through South America into southern Brazil, Paraguay, Argentina, and Bolivia. In Bolivia it is known from the Flora and Fauna Hotel near Buena Vista, the Potrerillo del Guenda Reserve, and the Vicoquin area above Achira at 2,000 m, all in Santa Cruz Department.

Alcidion sulphurifer (White, 1855)

Lamiinae: Acanthocinini

This is another beautiful member of the genus *Alcidion*, slightly larger than the previous species, *A. humeralis* (8–12 mm). It is uniformly dark but with spots of dense yellow pubescence at the base and apex of the pronotum, just before and after the middle of the elytra, and usually at the extreme elytral apex. This species, often encountered in the daytime by beating vegetation, is known from Brazil, Peru, and Bolivia. In Bolivia this species is known from low-, middle-, and high-elevation sites in Santa Cruz Department, including Potrerillo del Guenda Reserve at 400 m, Refugio Los Volcanes at 1,200 m, and the Vicoquin area above Achira at 2,000 m. It is also known from Cristal Mayu and Villa Tunari in Cochabamba Department.

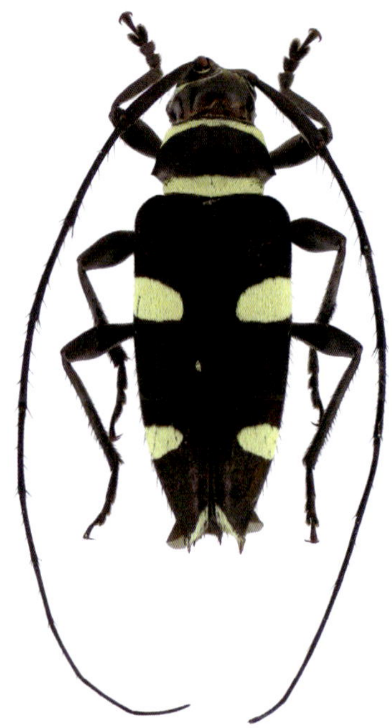

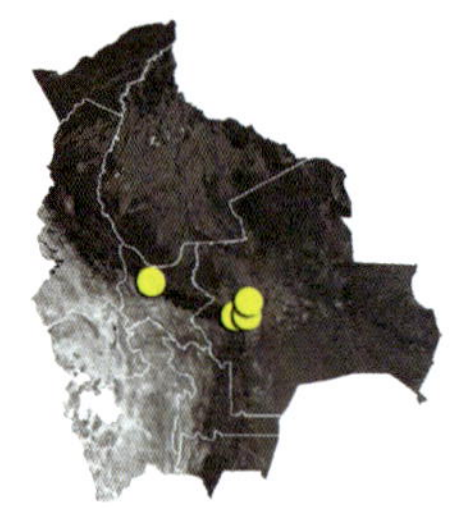

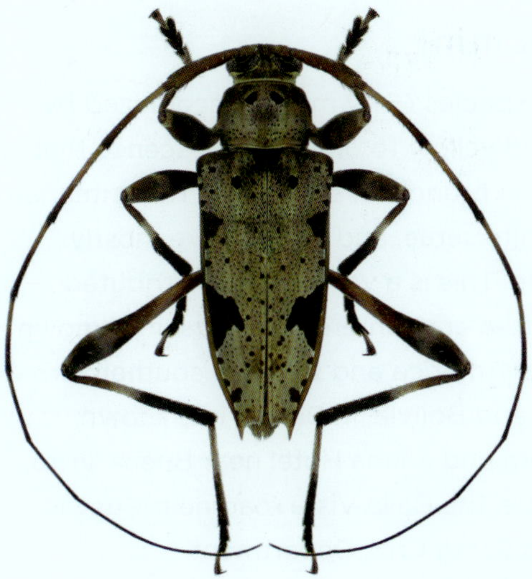

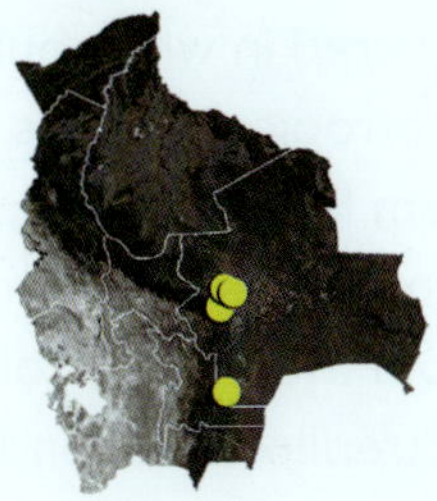

Anisopodus batesi Gilmour, 1965

Lamiinae: Acanthocinini

The large genus *Anisopodus* includes many cryptically colored species with very long hind legs. This species (11–14 mm long) is recognized by the pattern of white pubescence of the elytra and pronotum that is interrupted by various black spots: two at the anterior margin of the pronotum, near the midline, and two large ones on the lateral margins of the elytra, along with many smaller spots. This species is recorded from Ecuador, French Guiana, Brazil, Peru, and Bolivia. It is known from the grounds of the Flora and Fauna Hotel near Buena Vista, Potrerillo del Guenda, Refugio Los Volcanes, and the Camiri environs in Santa Cruz Department in Bolivia.

Anisopodus humeralis Bates, 1863

Lamiinae: Acanthocinini

This 7–10 mm long species is recognized by the triangular patch of pale pubescence on the humeri, along with the diffuse pattern of pale pubescence elsewhere on the elytra. The hind legs of this species are not as long as in other species of *Anisopodus*. This species has somewhat of an unusual distribution in that it is known from only Costa Rica in Central America and Brazil, Ecuador, and Bolivia in South America. It is known from the grounds of the Flora and Fauna Hotel near Buena Vista and Potrerillo del Guenda in Santa Cruz Department and Villa Tunari in Cochabamba Department in Bolivia.

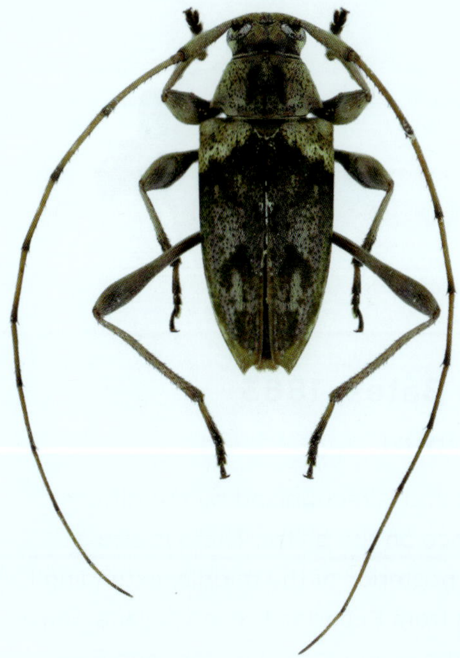

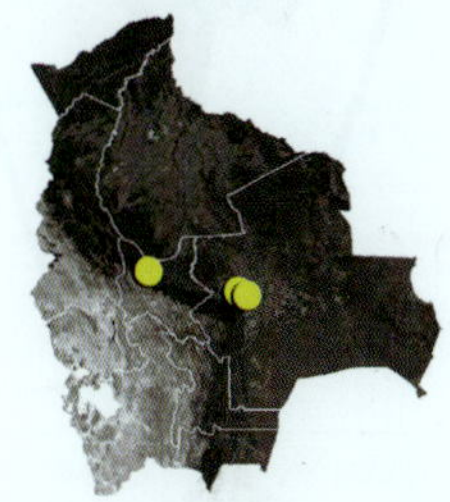

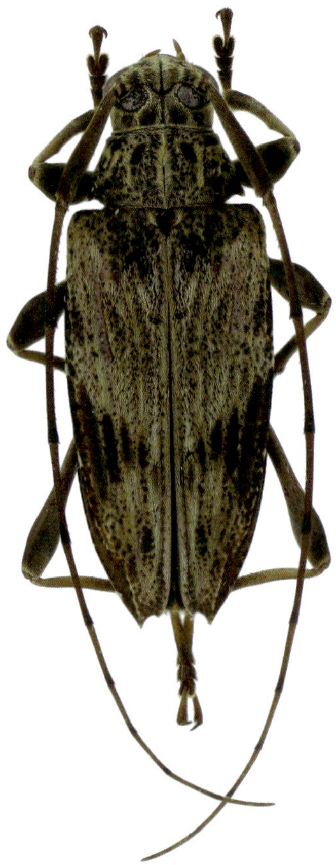

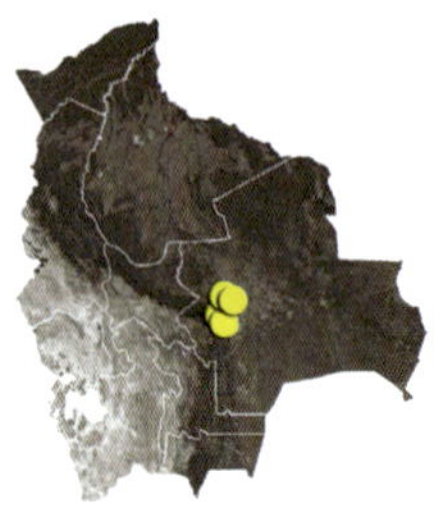

Anisopodus ligneus Bates, 1863

Lamiinae: Acanthocinini

This species is 10–12 mm long and is recognized by the diffuse, longitudinal lines of pubescence on the elytra. There is also a vague, triangular macula just posterior of the middle, extending to the lateral margin. It is known from Ecuador, French Guiana, Brazil, and Bolivia. It is known from the grounds of the Flora and Fauna Hotel near Buena Vista, Potrerillo del Guenda Reserve, Refugio Los Volcanes, adjacent to Amboró National Park, and the Bella Vista road near Cuevas in Santa Cruz Department in Bolivia.

Atrypanius conspersus (Germar, 1824)

Lamiinae: Acanthocinini

This beautiful and common species (6–12 mm) is recognized by the combination of white and yellow to orange pubescence that forms bands and spots on the pronotum and elytra. The antennae are basally annulate with white setae, and the legs are mostly covered in white pubescence. This is a very widely distributed neotropical species, most often encountered at lights. It is known from Mexico through Central America and south to southern Brazil, Paraguay, Argentina, and Bolivia. In Bolivia it is known from the grounds of the Flora and Fauna Hotel near Buena Vista, Potrerillo del Guenda Reserve, the Bella Vista road near Cuevas, and Refugio Los Volcanes in Santa Cruz Department.

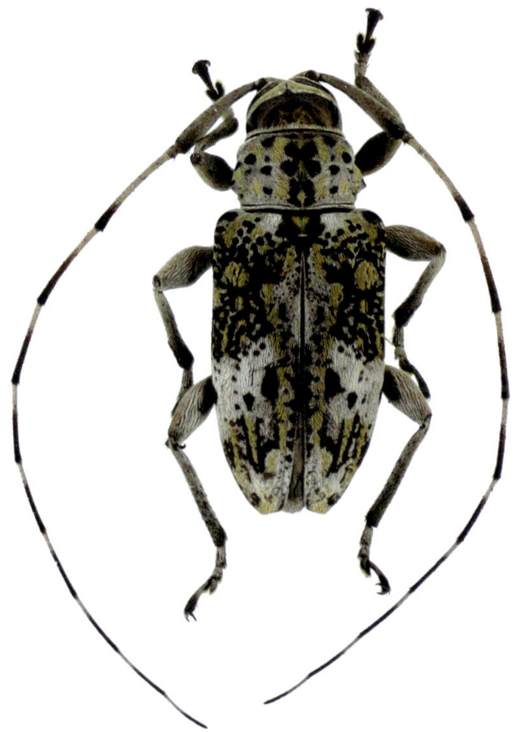

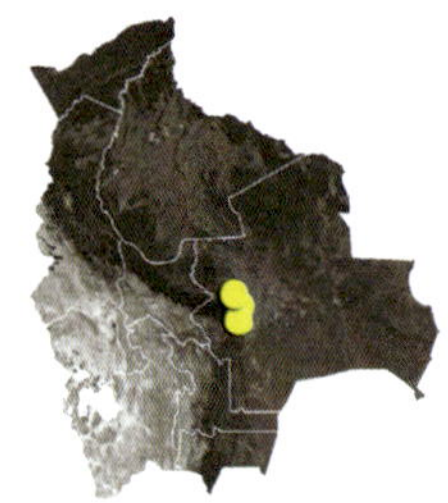

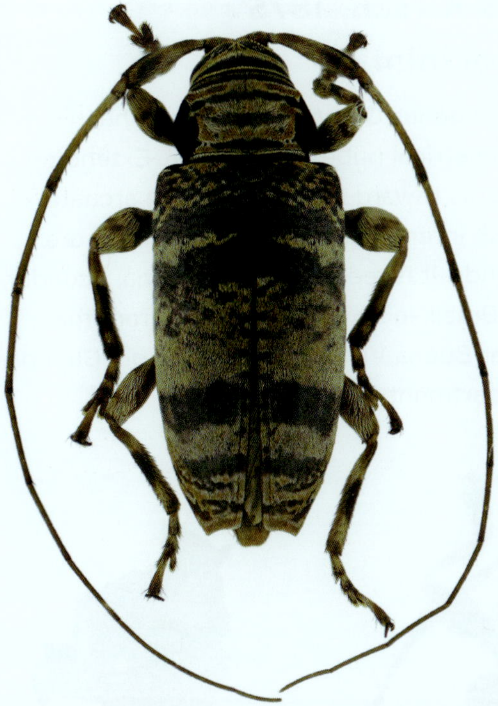

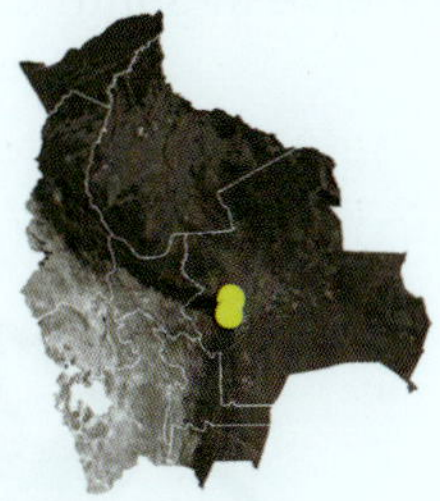

Atrypanius scitulus (Germar, 1824)

Lamiinae: Acanthocinini

This small species, 8–10 mm, is very distinct with its alternating light and dark brown, transverse bands of pubescence on the pronotum and elytra. The sides of the pronotum are black. This nocturnal species is attracted to lights but is more commonly collected by beating *chaqueos* at night. It is known from Ecuador, French Guiana, Brazil, Peru, and Bolivia. In Bolivia it has been collected at Potrerillo del Guenda Reserve and Refugio Los Volcanes in Santa Cruz Department.

Baryssinus silviae Martins & Monné, 1974

Lamiinae: Acanthocinini

This is a distinctive member of the Acanthocinini, 8–13 mm, with a pronounced V-shaped pattern of white to gray pubescence at the basal half of the elytra that extends down along the suture to the apex. Elsewhere, the elytra are covered in a mottled green and brown pubescent vestiture. The pronotum has an incomplete brown to black ring at the center and large, broad, posteriorly directed lateral tubercles. It occurs in Bolivia, Peru, and Brazil. In Bolivia it is known from the grounds of the Flora and Fauna Hotel near Buena Vista, Potrerillo del Guenda Reserve, Refugio Los Volcanes, and the Bella Vista road near Cuevas, all in Santa Cruz Department.

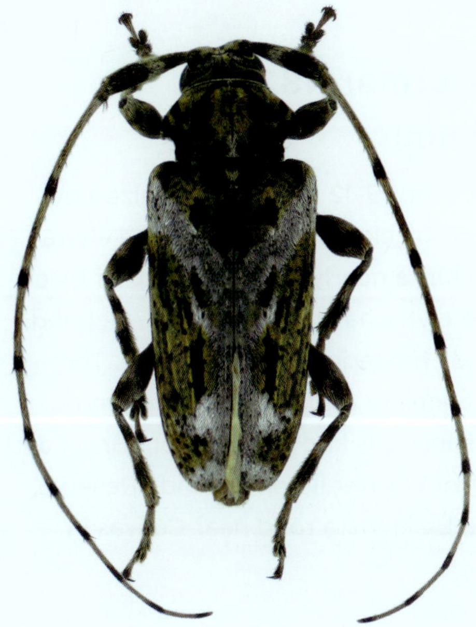

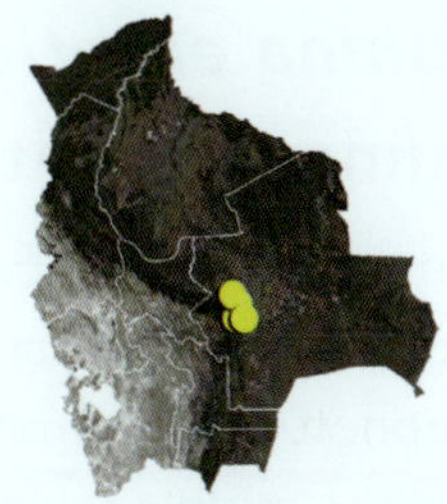

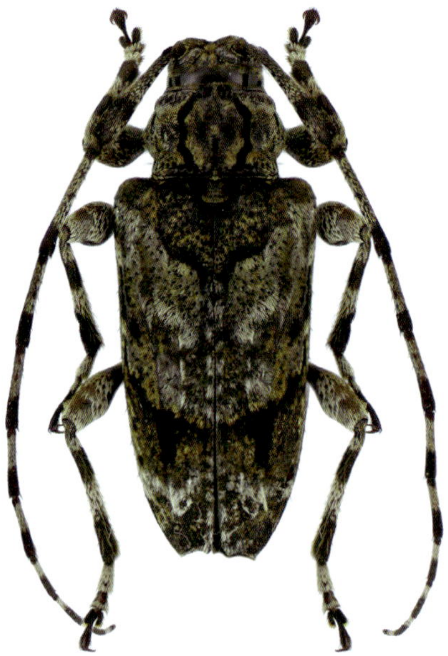

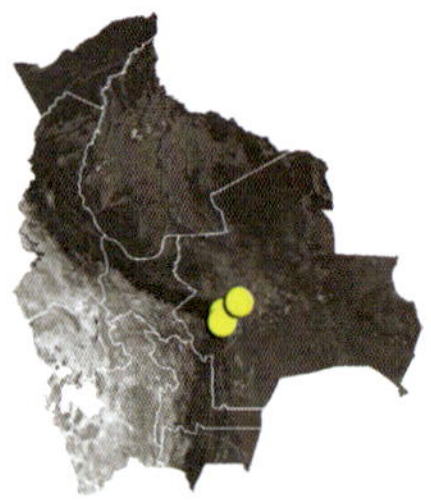

Carphina elliptica (Germar, 1824)

Lamiinae: Acanthocinini

This moderate-sized acanthocinine (9–12 mm) is recognized by the pattern of white, tawny, and black pubescence on the elytra and pronotum. The pronotum has a nearly complete ring of black around the middle region. The elytra have an arcuate black band, just behind middle, that extends posteriorly at the middle. The antennae are distinctly black annulate at the apex and extreme base of most segments. This species is known only from Brazil and Bolivia. In Bolivia it is known from Potrerillo del Guenda Reserve, Refugio Los Volcanes, and the Bella Vista road near Cuevas in Santa Cruz Department.

Carphina petulans Kirsch, 1875

Lamiinae: Acanthocinini

This striking member of the genus *Carphina* (8–10 mm) has pronounced and bold geometrical pubescent fasciae. Extending from just behind the humerus toward the suture is one arcuate white band with a yellowish interior, and at the elytral apex are two narrow, transverse bands. It is recorded from Bolivia, Ecuador, French Guiana, Peru, and Brazil. In Bolivia it is known from the Flora and Fauna Hotel near Buena Vista and Potrerillo del Guenda Reserve in Santa Cruz Department.

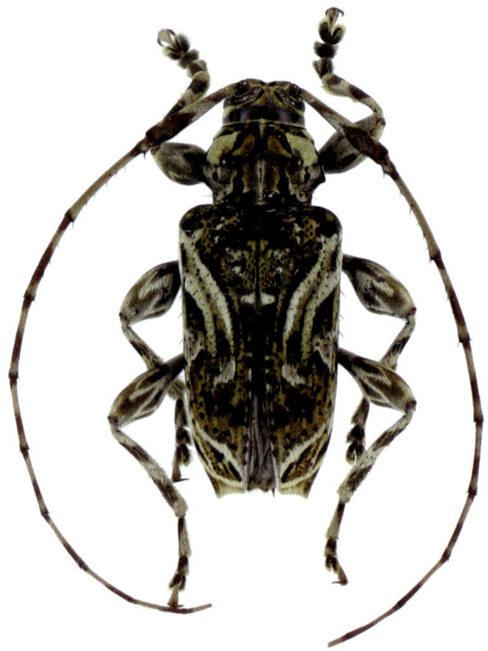

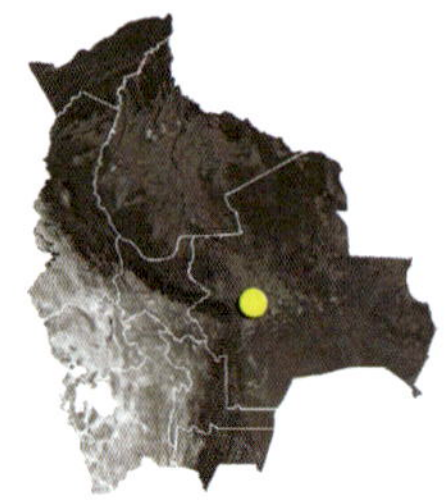

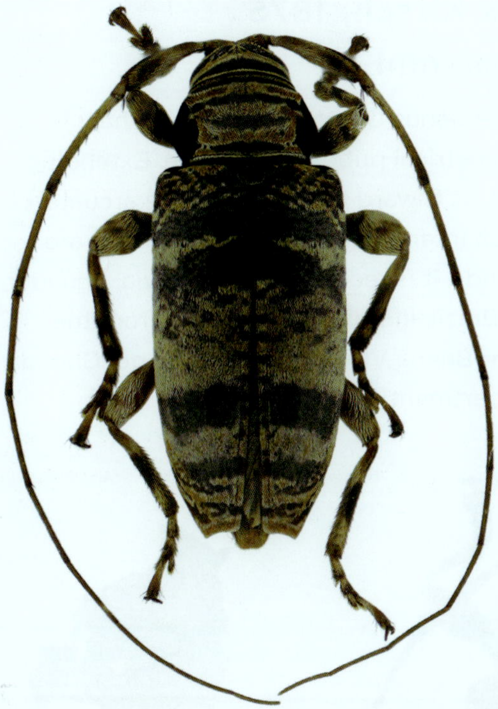

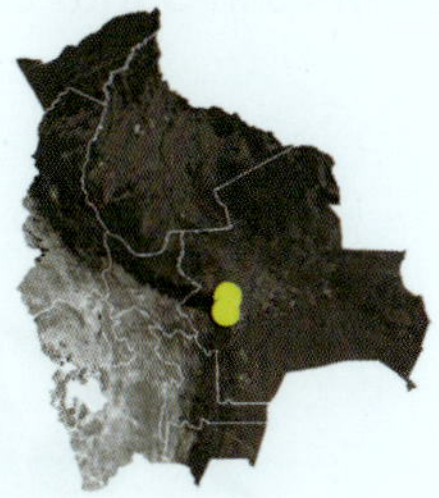

Atrypanius scitulus (Germar, 1824)

Lamiinae: Acanthocinini

This small species, 8–10 mm, is very distinct with its alternating light and dark brown, transverse bands of pubescence on the pronotum and elytra. The sides of the pronotum are black. This nocturnal species is attracted to lights but is more commonly collected by beating *chaqueos* at night. It is known from Ecuador, French Guiana, Brazil, Peru, and Bolivia. In Bolivia it has been collected at Potrerillo del Guenda Reserve and Refugio Los Volcanes in Santa Cruz Department.

Baryssinus silviae Martins & Monné, 1974

Lamiinae: Acanthocinini

This is a distinctive member of the Acanthocinini, 8–13 mm, with a pronounced V-shaped pattern of white to gray pubescence at the basal half of the elytra that extends down along the suture to the apex. Elsewhere, the elytra are covered in a mottled green and brown pubescent vestiture. The pronotum has an incomplete brown to black ring at the center and large, broad, posteriorly directed lateral tubercles. It occurs in Bolivia, Peru, and Brazil. In Bolivia it is known from the grounds of the Flora and Fauna Hotel near Buena Vista, Potrerillo del Guenda Reserve, Refugio Los Volcanes, and the Bella Vista road near Cuevas, all in Santa Cruz Department.

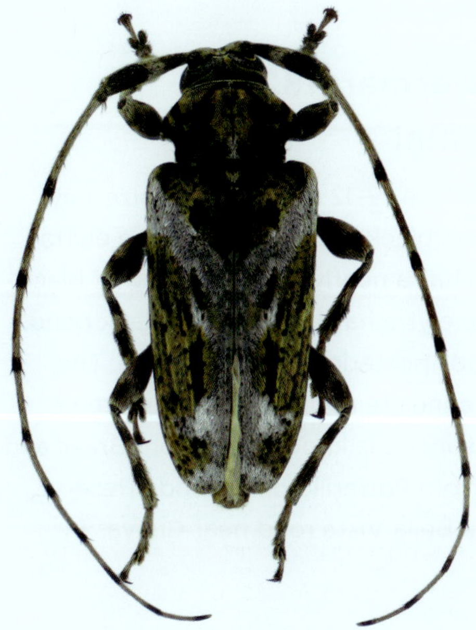

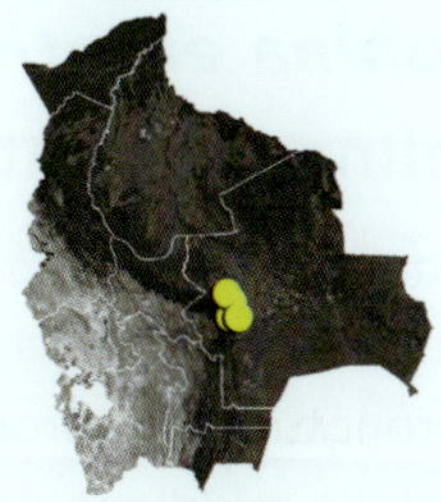

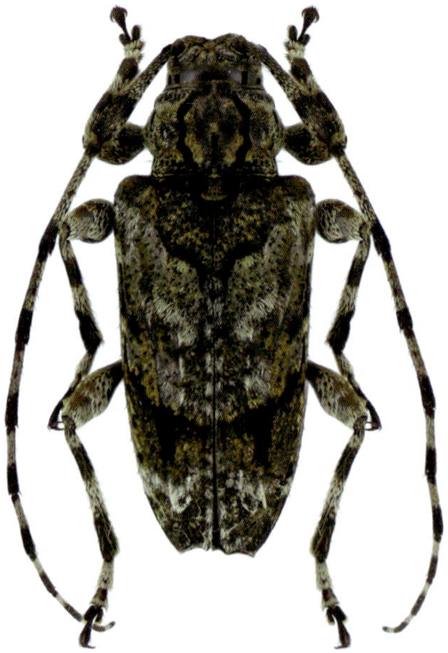

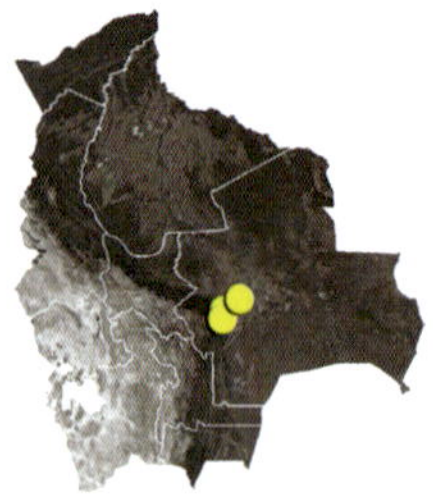

Carphina elliptica (Germar, 1824)

Lamiinae: Acanthocinini

This moderate-sized acanthocinine (9–12 mm) is recognized by the pattern of white, tawny, and black pubescence on the elytra and pronotum. The pronotum has a nearly complete ring of black around the middle region. The elytra have an arcuate black band, just behind middle, that extends posteriorly at the middle. The antennae are distinctly black annulate at the apex and extreme base of most segments. This species is known only from Brazil and Bolivia. In Bolivia it is known from Potrerillo del Guenda Reserve, Refugio Los Volcanes, and the Bella Vista road near Cuevas in Santa Cruz Department.

Carphina petulans Kirsch, 1875

Lamiinae: Acanthocinini

This striking member of the genus *Carphina* (8–10 mm) has pronounced and bold geometrical pubescent fasciae. Extending from just behind the humerus toward the suture is one arcuate white band with a yellowish interior, and at the elytral apex are two narrow, transverse bands. It is recorded from Bolivia, Ecuador, French Guiana, Peru, and Brazil. In Bolivia it is known from the Flora and Fauna Hotel near Buena Vista and Potrerillo del Guenda Reserve in Santa Cruz Department.

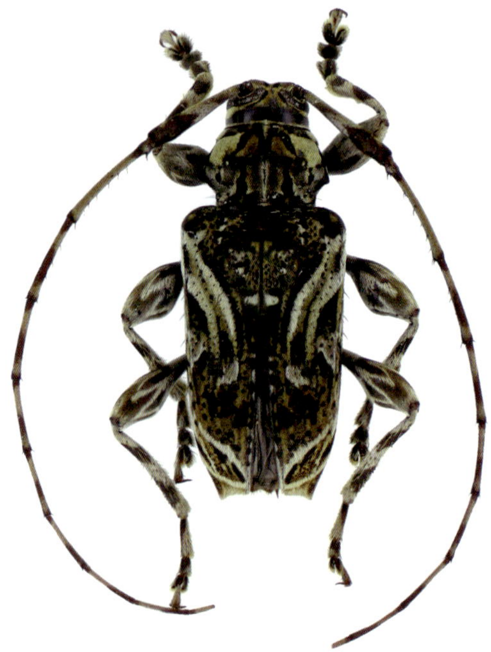

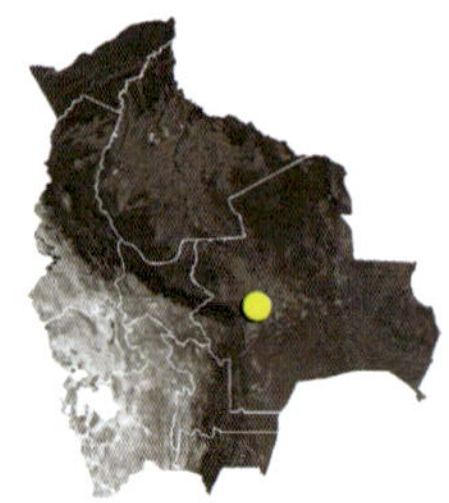

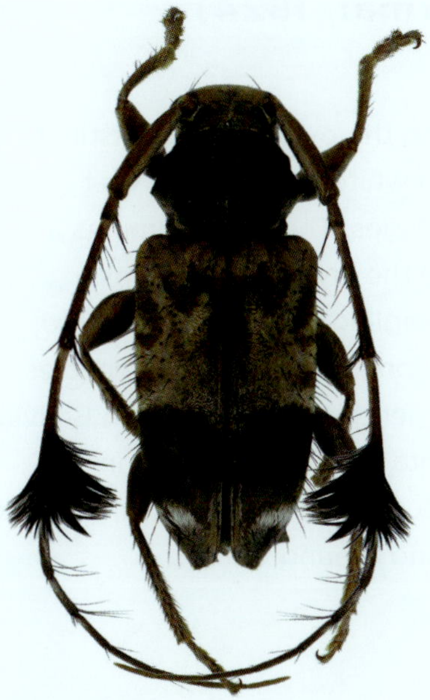

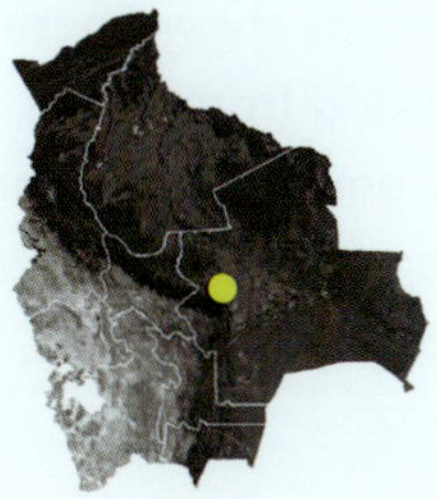

Cosmotoma adjuncta (Thomson, 1860)

Lamiinae: Acanthocinini

This is an especially attractive, small (5–6 mm) acanthocinine with a dark, minutely punctate pronotum and tan elytral integument with a pair of raised linear basal crests that are decorated dorsally with a row of stiff, long, black setae. These are followed by a network of narrow, silvery pubescent fasciae that give way to an area of darker integument, mostly covered in recumbent pubescence, and finally to a whitish, rounded pubescent macula on each elytron before the apices. In addition, the fourth antennal segment has a large, black setal brush covering its internal edge on the distal third. Quite a gaudy package for such a small beetle. This species is widespread and known from Costa Rica and most of northern South America including Bolivia. It is recorded from Santa Cruz Department in Bolivia on the grounds of the Flora and Fauna Hotel near Buena Vista.

Eucharitolus dorcadioides (White, 1855)

Lamiinae: Acanthocinini

This elegantly patterned small (5–8 mm) acanthocinine is recognized by the pattern of white pubescence and dark maculae on the elytra. The antennae (which are twice as long as the body in males, as shown) are mostly yellowish, as are the legs except for the darkened middle and hind tibiae. The elytral apices are obliquely truncate. This species is known from Brazil, French Guiana, Peru, and Bolivia. In Bolivia it has been collected at the Flora and Fauna Hotel near Buena Vista and the Potrerillo del Guenda Reserve in Santa Cruz Department.

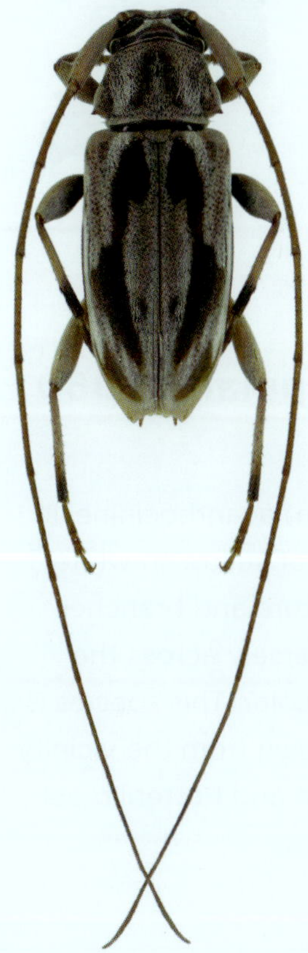

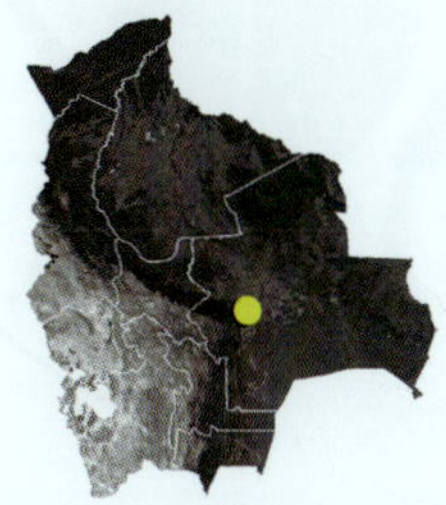

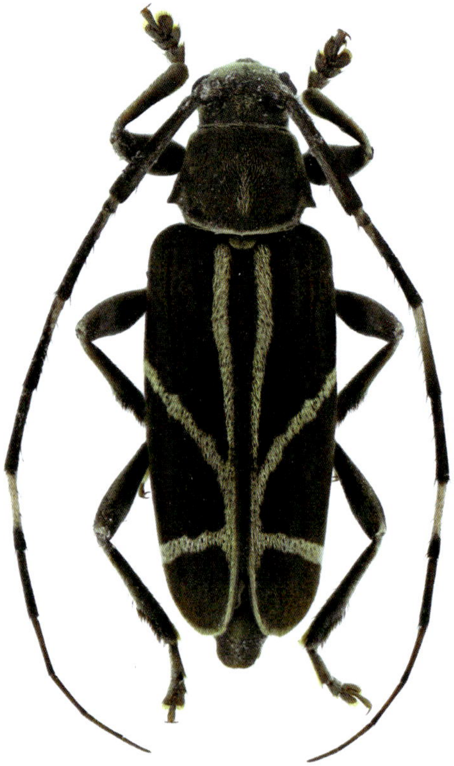

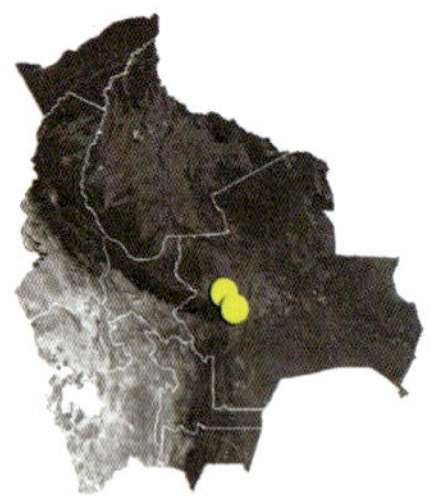

Eucharitolus geometricus (Tippmann, 1960)

Lamiinae: Acanthocinini

This is a very boldly patterned, small (5–8 mm) acanthocinine. It is identified by the geometrical arrangement of yellow to white pubescence that extends along the elytral suture and branches off across the middle of the elytra and transversely across the apical fourth. Otherwise, it is mostly black in color. This species is distributed in Peru and Bolivia, where it is known from the vicinity of the Flora and Fauna Hotel near Buena Vista and Potrerillo del Guenda Reserve in Santa Cruz Department.

Eutrypanus dorsalis (Germar, 1824)

Lamiinae: Acanthocinini

This species (9–14 mm long, excluding the ovipositor) is distinct with its pattern of yellow, brown, and white pubescence that covers most of the body and appendages. The pronotum has a distinct dark fascia at the center, and the elytra have a similar black fascia just behind the middle, angling up along the suture. Females, as shown here, have a very long ovipositor. The species is recorded from Brazil, Paraguay, Argentina, and Bolivia. In Bolivia it has been collected throughout Santa Cruz Department in Pampagrande, Buena Vista, Potrerillo del Guenda Reserve, Refugio Los Volcanes, Bermejo, Samaipata, and around Camiri.

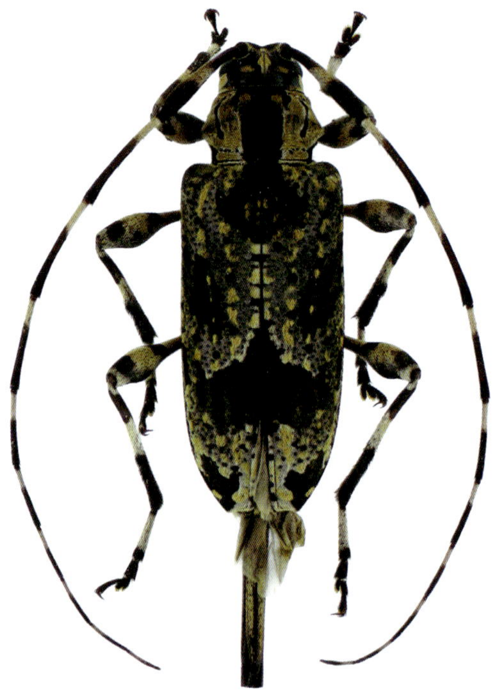

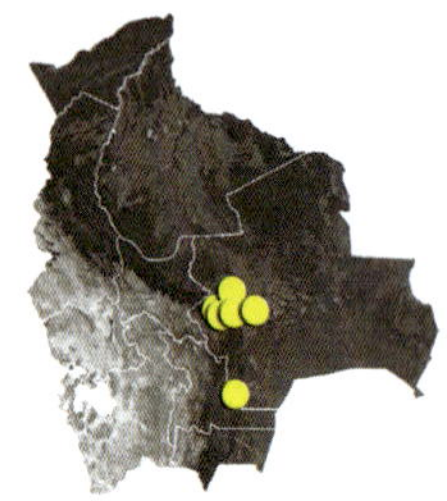

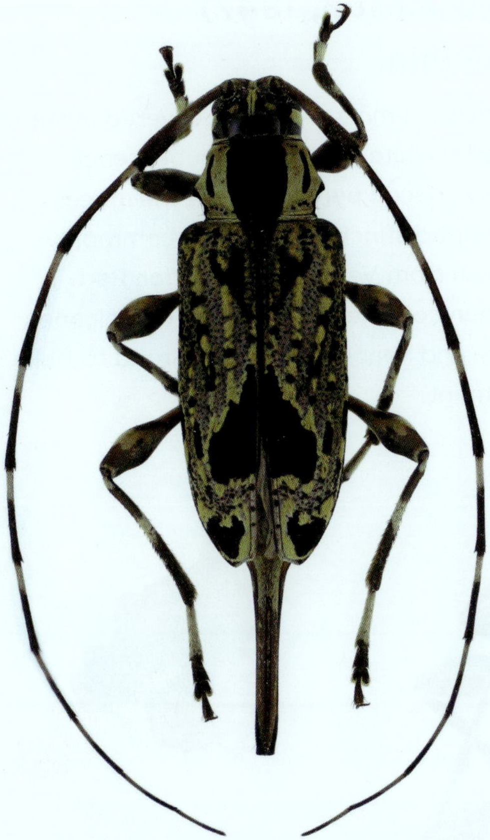

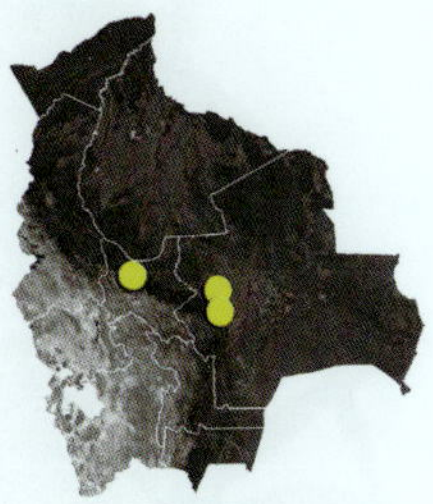

Eutrypanus triangulifer Erichson, 1847

Lamiinae: Acanthocinini

This species (8–12 mm long, excluding the ovipositor) is, like its preceding congener, *E. dorsalis*, covered in beautiful yellow, white, and dark brown pubescence. The pronotum has a very broad, dark fascia at the center, margined with yellow pubescence. Females have a very pronounced ovipositor. This species is known from Venezuela, Ecuador, Peru, Brazil, and Bolivia, where it has been collected in Cochabamba Department from Cristal Mayu and Santa Cruz Department from Buena Vista and Refugio Los Volcanes.

Granastyochus elegantissimus (Tippmann, 1953)

Lamiinae: Acanthocinini

This is a very distinctive acanthocinine, over 20 mm long, with green, white, and brown pubescent patterns on the pronotum and elytra. These form three roughly transverse bands on the elytra, expanding posteriorly once reaching the suture. Females have an elongate ovipositor like species in *Eutrypanus*. This species is known from Peru but rare in Bolivia, where one specimen was recorded in Wappes et al. (2006) but without specific locality data.

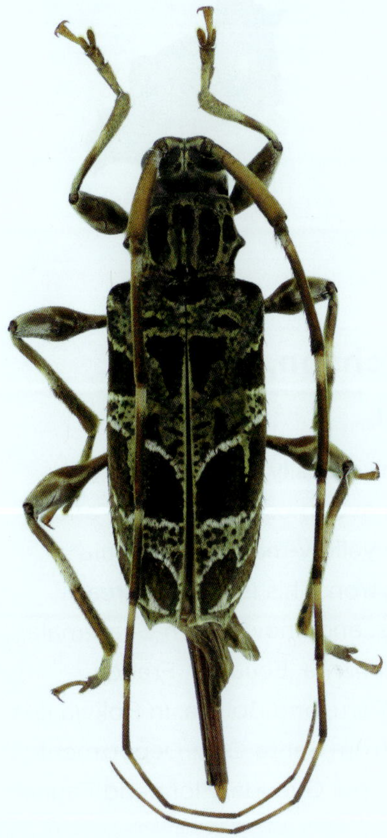

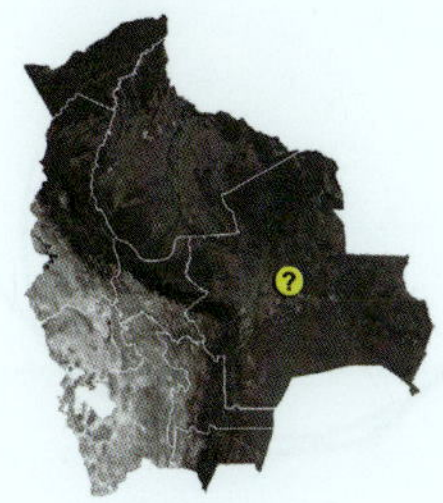

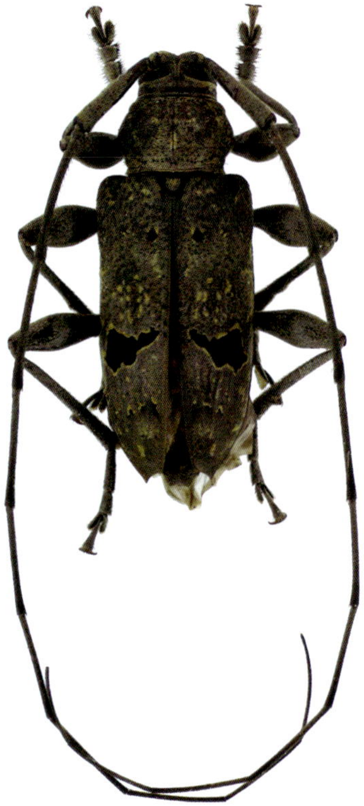

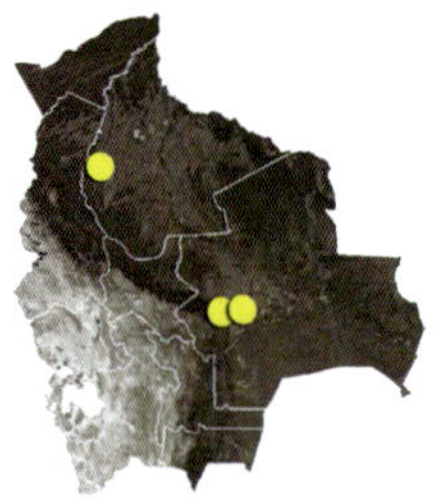

Hylettus coenobita (Erichson, 1847)

Lamiinae: Acanthocinini

This large acanthocinine (13–18 mm) is easily recognized by its uniform gray-green pubescence over most of the body and the single bold, black and narrowly yellow-margined macula just posterior of middle on each elytron. This is a widespread species, often encountered at lights, and known from Guatemala, Honduras, Nicaragua, Costa Rica, Panama, Ecuador, French Guiana, Guyana, Venezuela, Brazil, Peru, and Bolivia. In Bolivia it is known from Beni Department and from Santa Cruz Department in the specific localities of Potrerillo del Guenda, Flora and Fauna Hotel, and Refugio Los Volcanes.

Hylettus eremita (Erichson, 1847)

Lamiinae: Acanthocinini

This medium-sized species (10–14 mm) is densely covered in subtle patterns of green-brown, ashy white, and yellow pubescence. The elytra have large, widely spaced punctures and the apices are obliquely truncate, without distinct spines. Most commonly collected at lights, it is known from Venezuela, Ecuador, Peru, and Bolivia, where it has been collected at Refugio Los Volcanes, Vicoquin area above Achira, and near Cuevas on the road to Bella Vista in Santa Cruz Department.

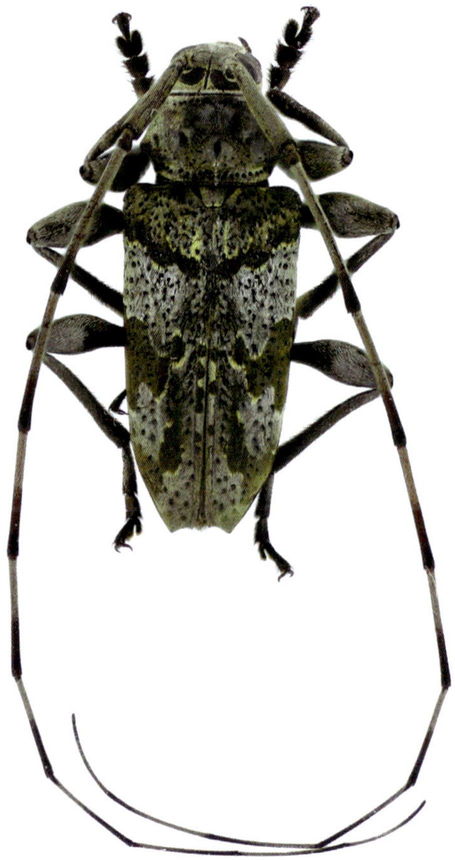

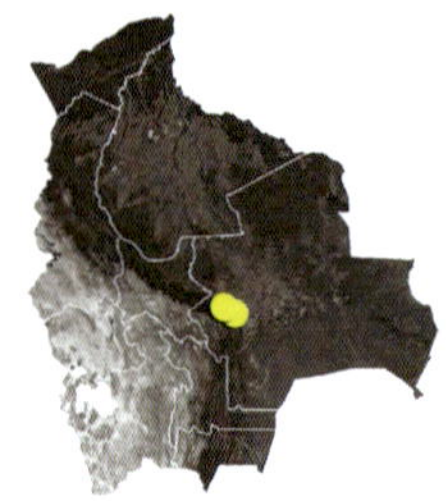

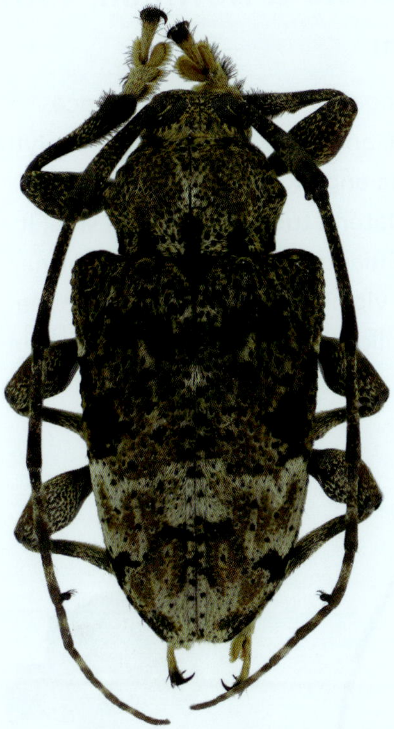

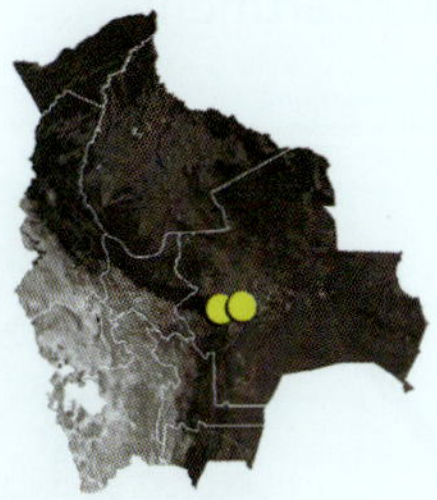

Lagocheirus plantaris Erichson, 1847

Lamiinae: Acanthocinini

This is one of the broader and more robust species of Acanthocinini (14–17 mm long; 7–8 mm wide). Superficially, it resembles species of the genus *Psapharochrus* in the related tribe Acanthoderini. It has dense pubescence forming distinct mottling on the legs, antennae, and pronotum. The elytra have a combination of white, tan, and black pubescence forming a cryptic pattern. It is known from Colombia, Ecuador, Peru, and Brazil but is rarely collected in Bolivia. It has been collected in the Potrerillo del Guenda Reserve, the Flora and Fauna Hotel, and Refugio Los Volcanes in Santa Cruz Department.

Lasiolepturges zikani Melzer, 1928

Lamiinae: Acanthocinini

This is a very delicate, small species (5–8 mm) with a distinctive, posteriorly placed, black macula on each elytron. The antennae and tibiae especially are bristling with sparse but conspicuous, long, black setae. It is known from Brazil, Peru, French Guiana, and Bolivia, where it can be common at lights at low- and midelevation sites in Santa Cruz Department. It has been collected in the vicinity of the Flora and Fauna Hotel near Buena Vista, the Potrerillo del Guenda Reserve, and Refugio Los Volcanes, adjacent to Amboró National Park.

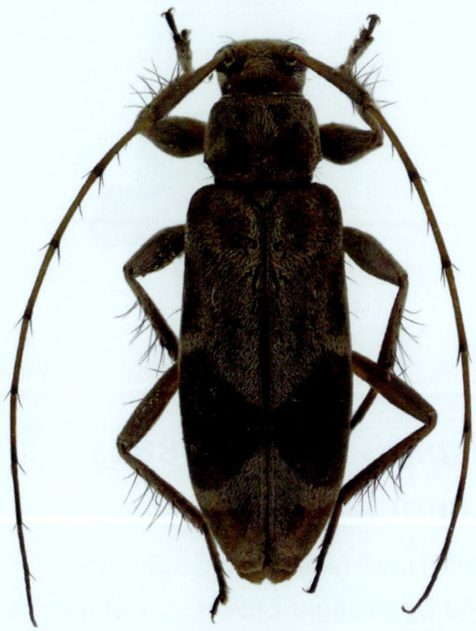

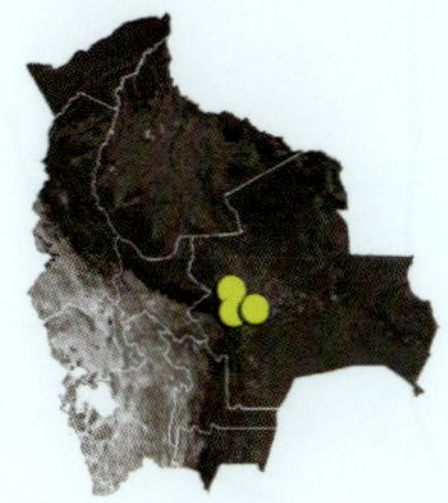

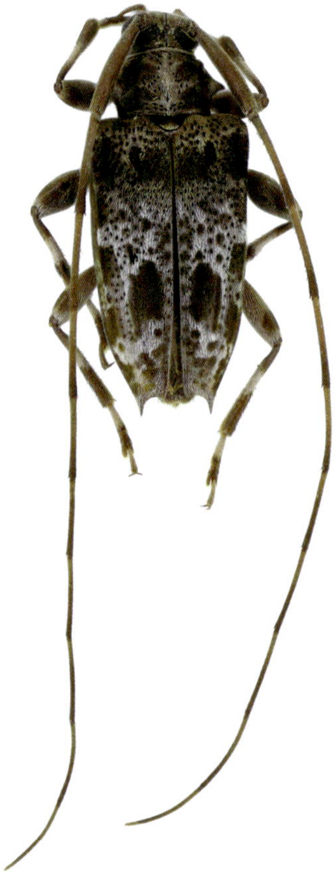

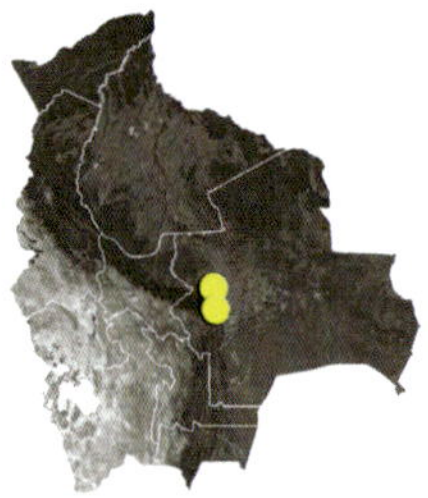

Leptocometes acutispinis (Bates, 1863)

Lamiinae: Acanthocinini

This small species (7–9 mm) was, until recently, placed in the genus *Tithonus*. It is recognized by the light brown and white pubescence that forms a nice pattern on the elytra. The outer elytral apex is strongly spined, as are the sides of the pronotum. The base of each elytron has a small, elevated crest. It is known from Mexico, Panama, Ecuador, French Guiana, Brazil, and Bolivia, where it has been collected at Refugio Los Volcanes and the Flora and Fauna Hotel near Buena Vista in Santa Cruz Department.

Leptostylus armatus Monné & Hoffman, 1981

Lamiinae: Acanthocinini

This species (9–11 mm) is densely coated in white pubescence, especially on the pronotum, elytra, and legs. The apex of the elytra has a vaguely defined darker fascia angling anteriorly along the suture. The pronotum has a small lateral tubercle on each side. It is distributed in Ecuador, French Guiana, Brazil, Peru, and Bolivia, where it has been collected in the vicinity of the Flora and Fauna Hotel near Buena Vista, the Potrerillo del Guenda Reserve, and Refugio Los Volcanes, adjacent to Amboró National Park, in Santa Cruz Department.

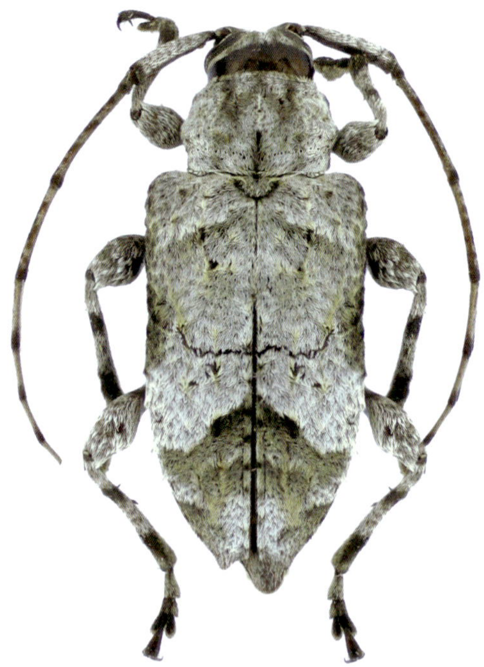

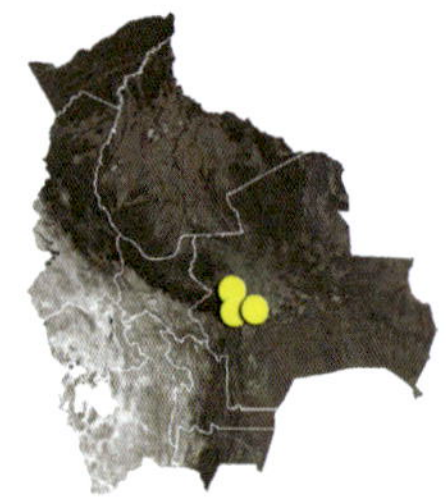

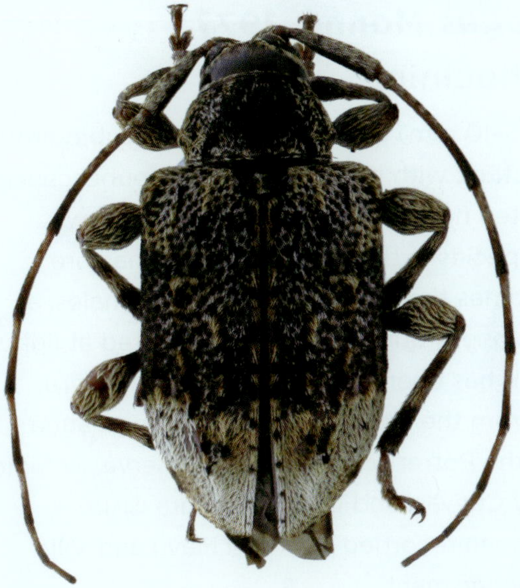

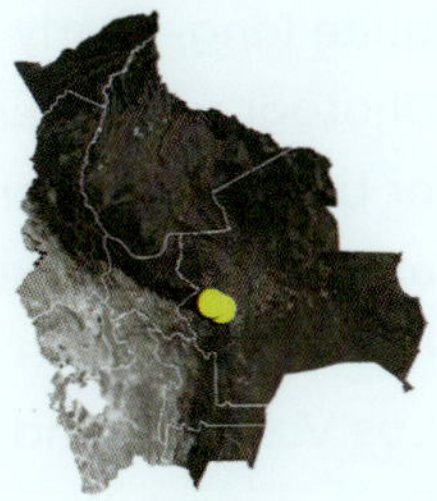

Leptostylus gnomus Monné & Hoffman, 1981

Lamiinae: Acanthocinini

This is one of the smallest members of the genus with specimens averaging about 4–5 mm long. The body is somewhat convex, tapering anteriorly and posteriorly. The elytra have white pubescent apices but, along with the pronotum and head, are darker elsewhere. This species has been recorded from Brazil, Paraguay, Argentina, and Bolivia. In Bolivia it has been collected at Refugio Los Volcanes, the Vicoquin area above Achira, and near Samaipata in Santa Cruz Department.

Lepturgantes prolatus Monné & Monné, 2008

Lamiinae: Acanthocinini

This beautiful, delicate lamiine (5–7 mm) is distinctive with its white and pale green pubescence, black outer margins of the elytra, and extremely long antennae that, in males, are almost three times the length of the body. This species is known from Brazil and Bolivia. In Bolivia it is known from the vicinity of the Flora and Fauna Hotel near Buena Vista, the Potrerillo del Guenda Reserve, and Refugio Los Volcanes in Santa Cruz Department.

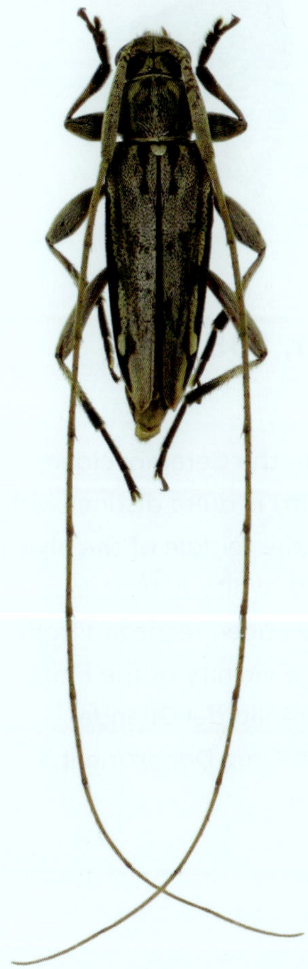

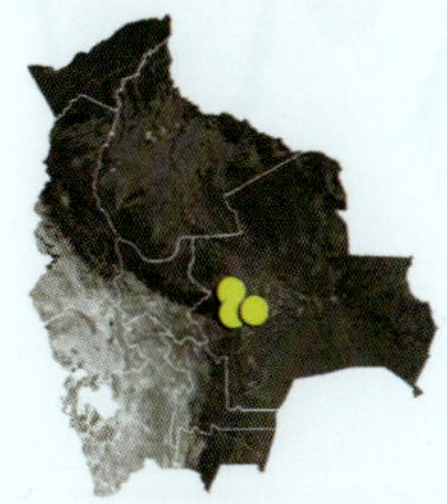

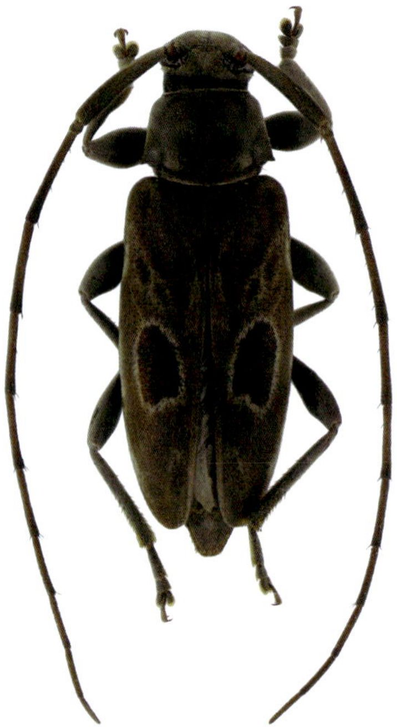

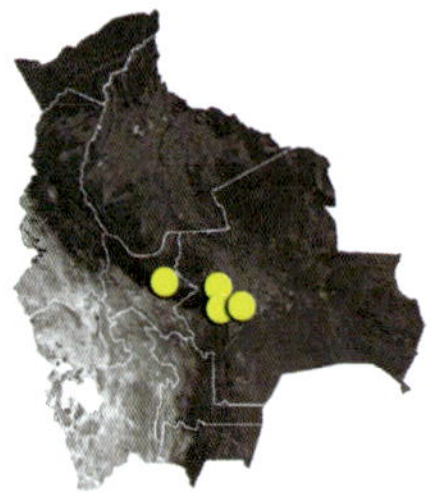

Lepturges curvilinea Gilmour, 1959

Lamiinae: Acanthocinini

The genus *Lepturges* is one of the largest in the Cerambycidae, with hundreds of species. This one (6–9 mm) is quite distinct with its two eye-spots of darker pubescence at the middle of the elytra. These eye-spots are narrowly margined with white pubescence, making them more striking. This species has been recorded from Peru and Bolivia, where it is known from the vicinity of the Flora and Fauna Hotel near Buena Vista, the Potrerillo del Guenda Reserve, and Refugio Los Volcanes in Santa Cruz Department.

Lepturges epagogus Monné, 1977

Lamiinae: Acanthocinini

This is another species (5–10 mm) of the very large and ubiquitous genus *Lepturges*. It is distinct with its pattern of ashy pubescence interrupted by a reticulated network of darker, reddish-brown integument. Like other species in the genus, the antennae are quite long, nearly three times the length of the body in males, as photographed here. This one is quite commonly collected at lights or by beating branches. It has been recorded from Brazil, Peru, and Bolivia. It is known from the vicinity of the Flora and Fauna Hotel near Buena Vista, the Potrerillo del Guenda Reserve, Refugio Los Volcanes and nearby Cuevas, and Camiri in Santa Cruz Department. It has also been recorded in Cristal Mayu and Villa Tunari in Cochabamba Department.

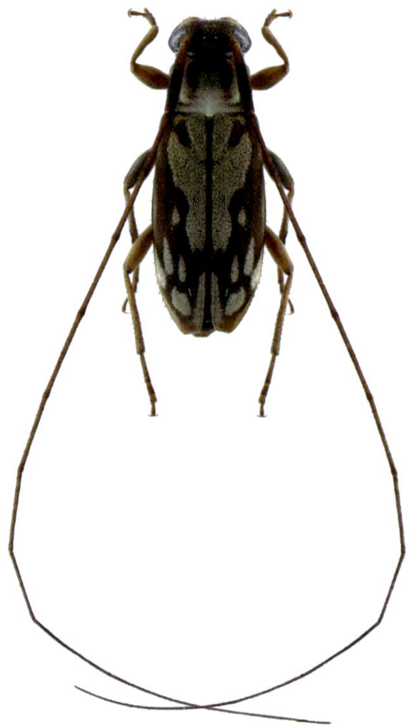

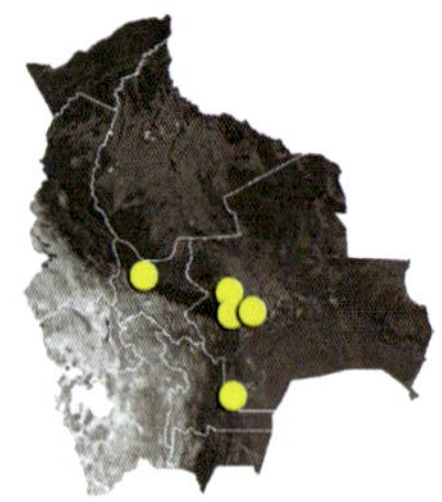

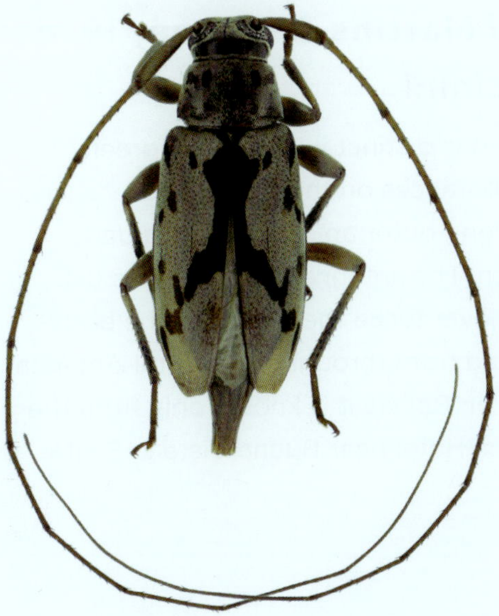

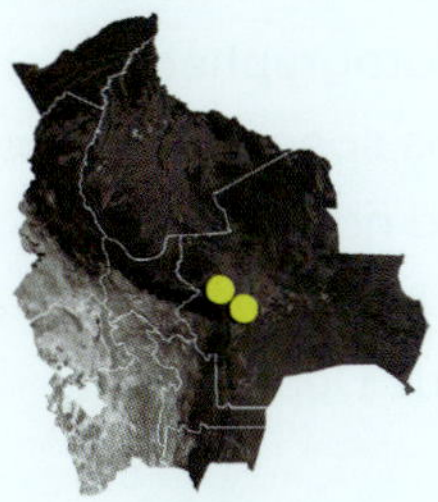

Lepturges eurynota Tippmann, 1960

Lamiinae: Acanthocinini

This species of *Lepturges* (7–9 mm long) is distinct with its dark sutural macula along the basal two-thirds of the elytra and the small, round maculae on the pronotum. The antennae are pale yellow with small, spine-like setae scattered along most segments. This species is attracted to lights and can be locally abundant. It is known from Brazil, Peru, and Bolivia, where it has been collected from the vicinity of the Flora and Fauna Hotel near Buena Vista and the Potrerillo del Guenda Reserve.

Lepturges multilineatus Melzer, 1928

Lamiinae: Acanthocinini

This 5–6 mm long *Lepturges* has distinct, longitudinal rows of pale pubescence on the elytra—four such rows on each elytron. Otherwise, the integument is dark reddish brown. This species has been recorded from Brazil and Bolivia, where it is has been collected at lights and by beating along forest edges from Potrerillo del Guenda and west of Samaipata in Santa Cruz Department.

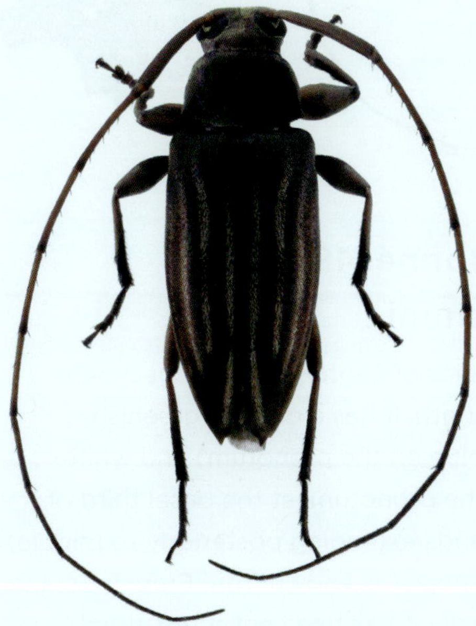

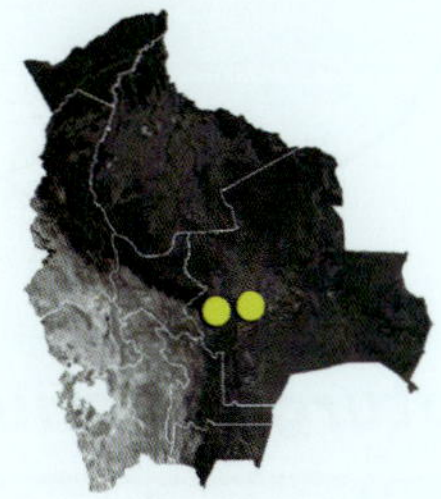

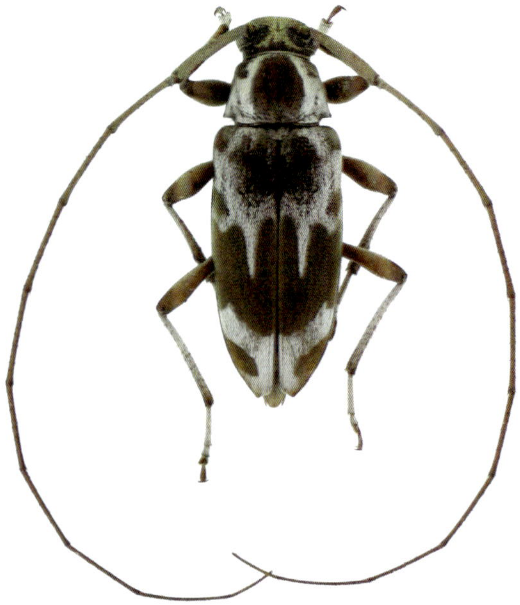

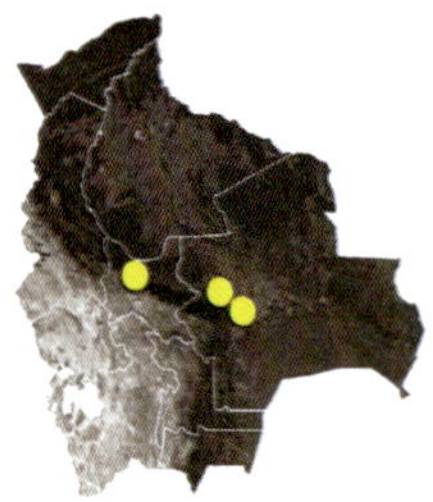

Lepturges zonula Monné, 1976
Lamiinae: Acanthocinini

This is one of the largest species of *Lepturges* with specimens ranging from 9 to 11 mm in length. It has a distinct greenish hue (sometimes even slight blue on the pronotum) and white pubescence on the sides of the pronotum, at the basal third of the elytra (with elongated bands extending posteriorly to middle), and at the elytral apex. This species is known from French Guiana, Brazil, and Bolivia, where usually it has been collected at lights in Villa Tunari (Cochabamba Department) and the Flora and Fauna Hotel and Potrerillo del Guenda (Santa Cruz Department).

Lithargyrus melzeri Martins & Monné, 1974
Lamiinae: Acanthocinini

This acanthocinine (10–12 mm) is distinct with its pale green pubescence and basal protuberances on the elytra. It is further recognized by the strong outer apical spine and dark, lateral macula on each elytron. The antennae of the males (as photographed here) are well over twice the length of the body. This species has been reported from throughout Central America and northern South America. In Bolivia it is known only from the vicinity of the Flora and Fauna Hotel near Buena Vista in Santa Cruz Department.

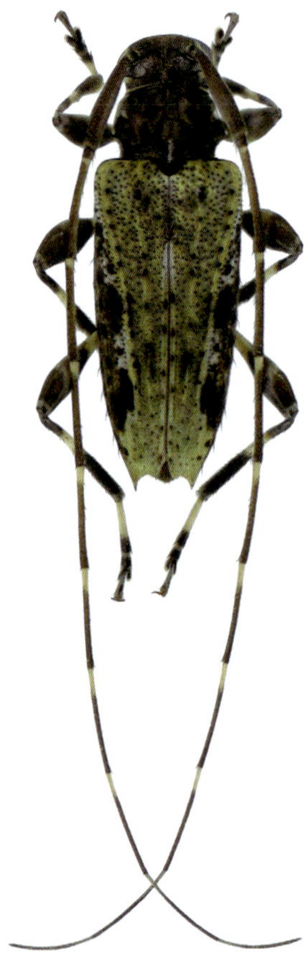

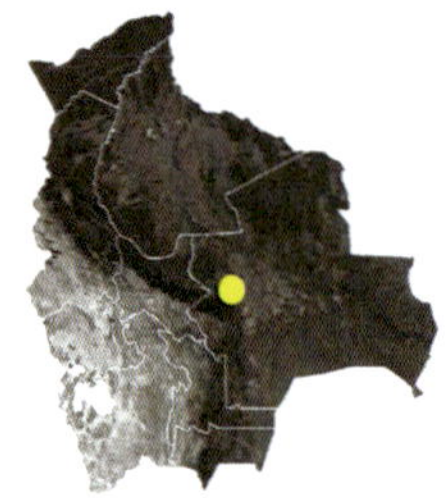

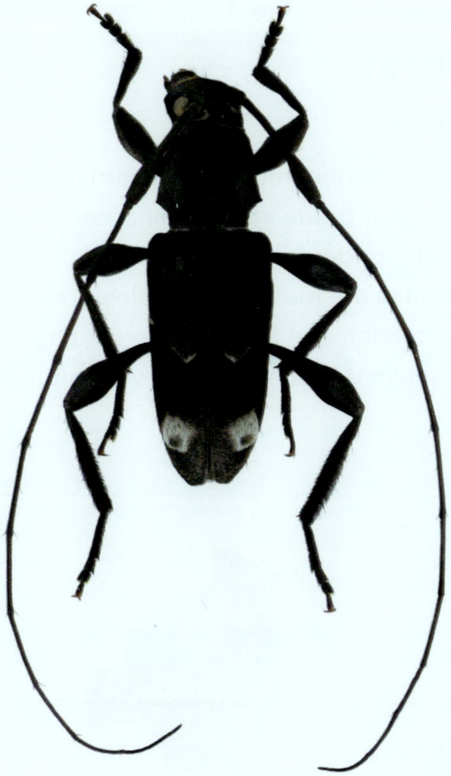

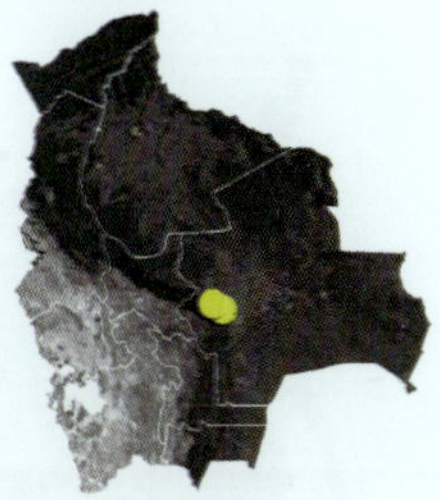

Microplia nigra Monné, 1976

Lamiinae: Acanthocinini

This is one of the most striking and unusual members of the tribe. It is 7–9 mm long and all black except for two small patches of white pubescence at the middle and apex of each elytron. The legs and scape are clavate, and the pronotum has a small lateral spine on each side. Uncommonly collected, specimens are usually taken by beating cut forested margins or small, dead standing trees. It has been recorded from Brazil, Argentina, and Bolivia, where it is known from Santa Cruz Department at Refugio Los Volcanes and the Vicoquin area above Achira.

Nealcidion cereicola (Fisher, 1936)

Lamiinae: Acanthocinini

This robust Lamiinae (10–12 mm) is characterized by its dark, heavily punctate elytral base bearing an elevated crest on each elytron. Both the pronotum, preceding this area, and the elytra following it, contrast with the base by being much lighter brown or tan in color. The apical third of the elytra are again darkened but much less so than the base. This species is known from Argentina and Bolivia, where it has been collected at Refugio Los Volcanes and Camiri in Santa Cruz Department.

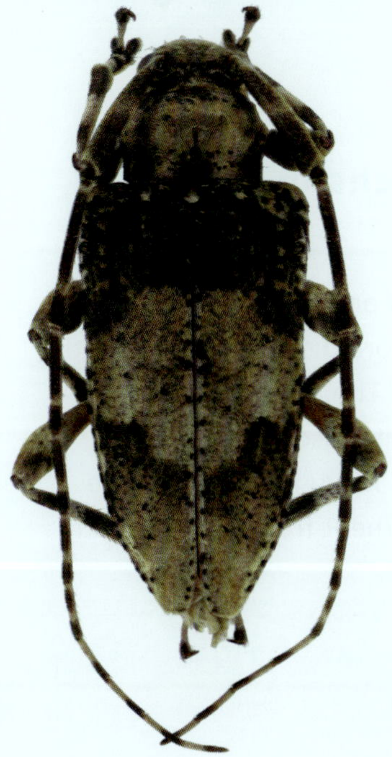

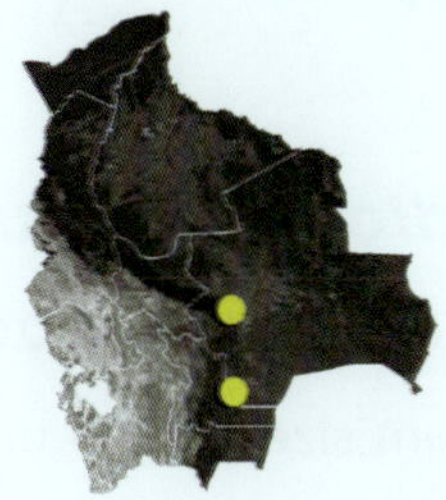

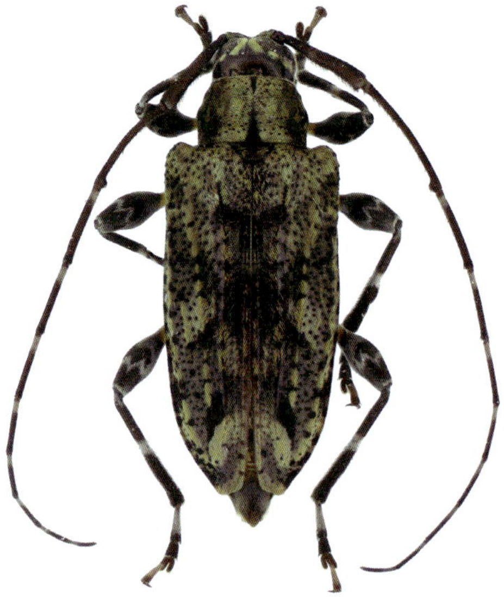

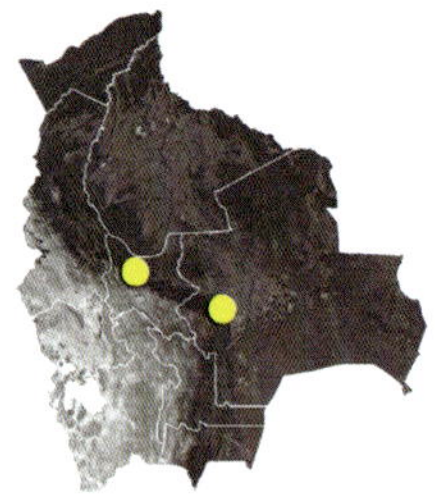

Nealcidion emeritum (Erichson, 1847)

Lamiinae: Acanthocinini

This medium-sized (10–14 mm), yellow and brown pubescent species has strongly costate elytra (more distinct basally) with coarse punctures between the costae. The clavate femora are mostly glabrous but have a zigzag white fascia on the inner club. In Bolivia it is known from Cristal Mayu in Cochabamba Department and in Santa Cruz Department from the Flora and Fauna Hotel and Refugio Los Volcanes.

Nealcidion hylaeanum (Monné & Martins, 1976)

Lamiinae: Acanthocinini

This species (9–12 mm) is beautifully decorated with alternating light and dark pubescent bands that grade into one another, making it very cryptic but striking under magnification and illumination. The legs and antennae are mostly pale yellow. This species is known from Brazil, Ecuador, and Bolivia, where it has been collected only at the Potrerillo del Guenda Reserve (Santa Cruz Department).

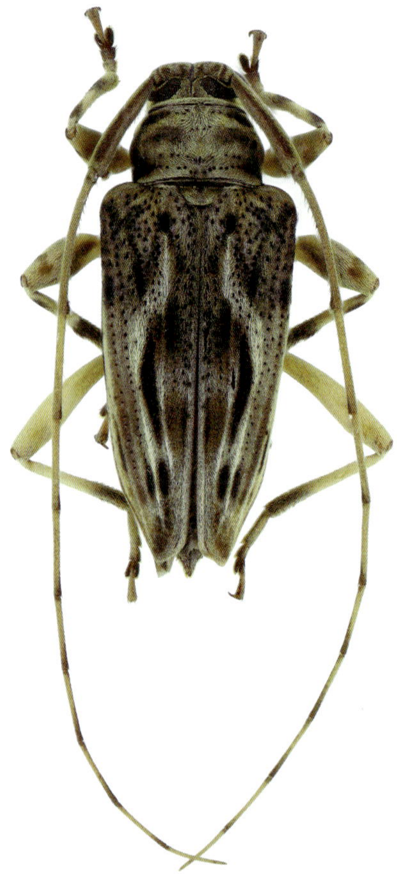

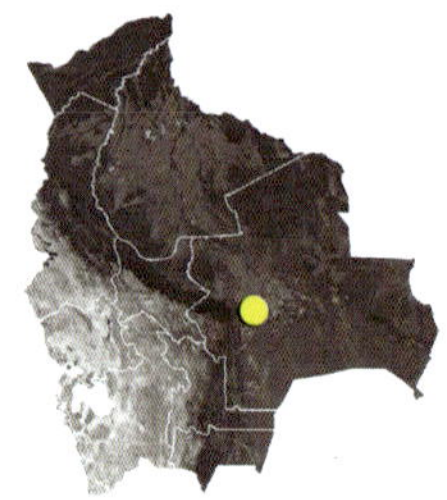

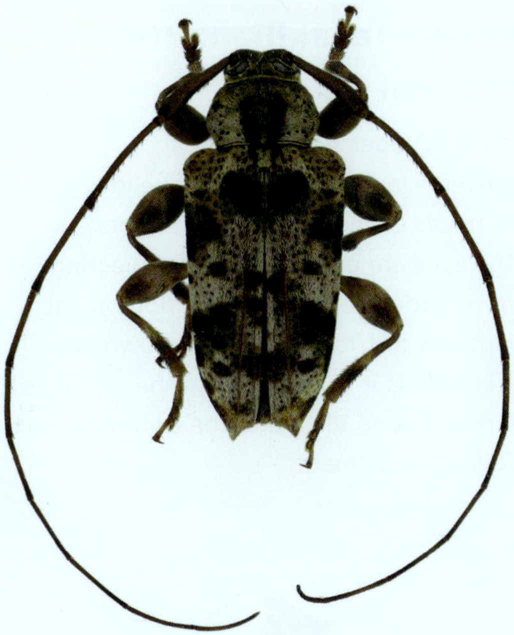

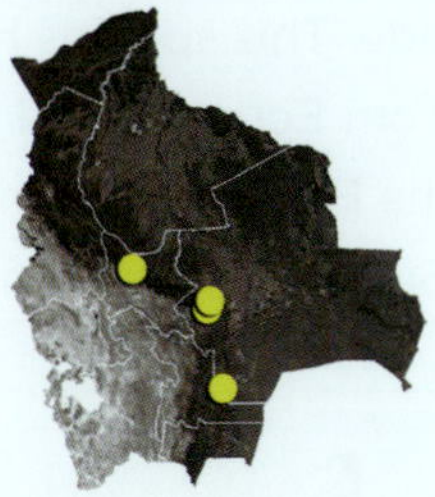

Nealcidion quinquemaculatum (Tippmann, 1960)

Lamiinae: Acanthocinini

This small, broad lamiine (7–9 mm) has dark and light pubescent maculae over most of the elytral and pronotal surface. The elytral suture is bounded on either side with a narrow, raised, glabrous crest. The femora are strongly clavate and the elytral apex is moderately acute laterally. This species is known only from Bolivia, where it has been collected in Yungas de Totora (Sehuencas) and Villa Tunari (Cochabamba Department), and from several sites in Santa Cruz Department, including the Amboró road above Achira, near Samaipata, and north of Camiri.

Neoeutrypanus generosus (Monné & Martins, 1976)

Lamiinae: Acanthocinini

This species (11–13 mm, excluding ovipositor) has the markings and morphology very typical of the tribe Colobotheini. The elytra and pronotum have alternating bands and maculae of black, yellow, and bluish-gray color. The antennae are white annulate at the bases of most segments. Females have a long ovipositor modification of the abdomen. This is a Bolivian endemic known only from Refugio Los Volcanes in Santa Cruz Department.

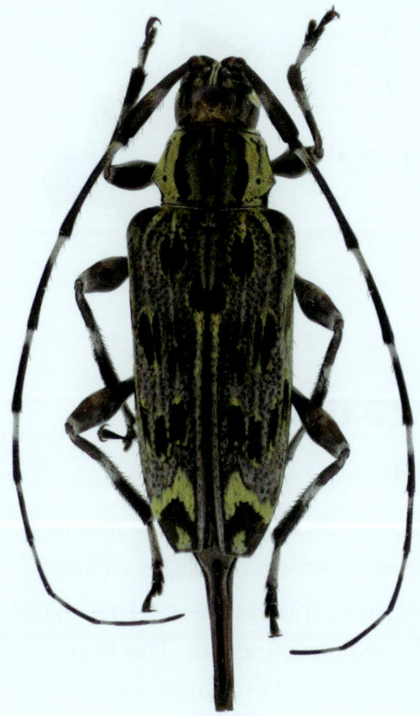

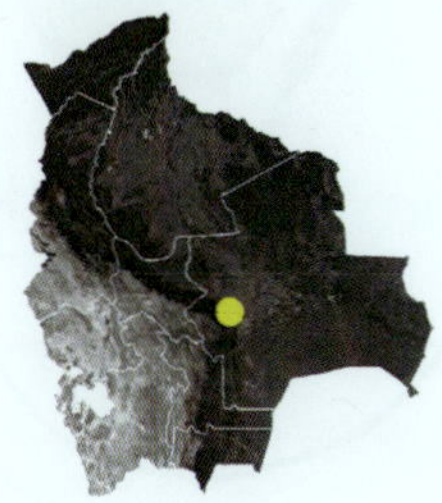

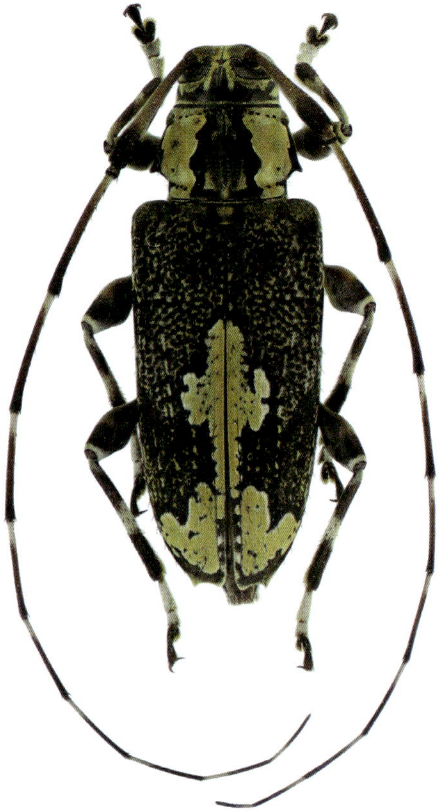

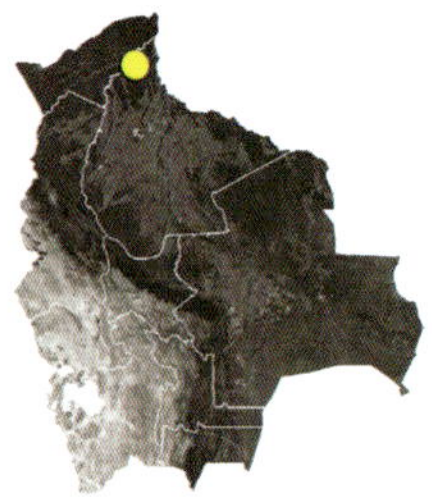

Neoeutrypanus nobilis (Bates, 1864)

Lamiinae: Acanthocinini

This is a moderate-sized (11–14 mm) species with bold maculae of dense yellow to ochraceous pubescence on the sides of the pronotum and along the suture between the elytra and the elytral apex. The basal half of the elytra is heavily punctate. The first two tarsal segments on each leg are covered in dense, white pubescence. This species ranges from Ecuador to Brazil, French Guiana, and Peru and was recently recorded in northern Bolivia near Riberalta in Beni Department, where one specimen was collected by beating branches of a tree fall.

Neseuterpia couturieri Tavakilian, 2001

Lamiinae: Acanthocinini

This small species (7–9 mm long) does not look like an acanthocinine, or possibly even a lamiine. It is mostly yellow orange in color, with black apices of the antennal segments and femoral-tibial junctures. The eyes are prominent and projecting laterally. The pronotum has a moderate lateral projection on each side. This species has a disparate distribution and is known only from Ecuador and Bolivia. In Bolivia it is known from the vicinity of the Flora and Fauna Hotel near Buena Vista and the Potrerillo del Guenda Reserve in Santa Cruz Department.

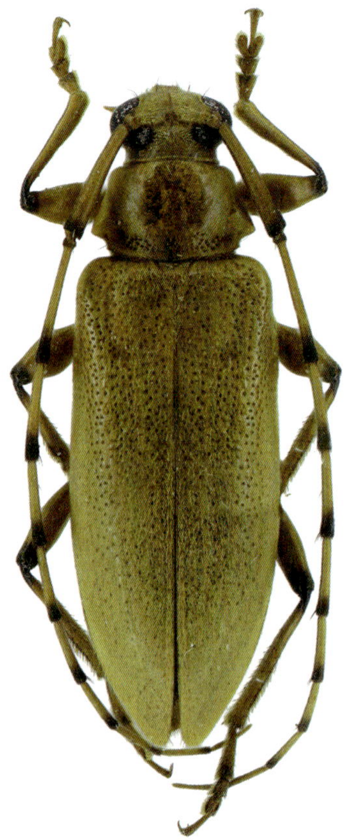

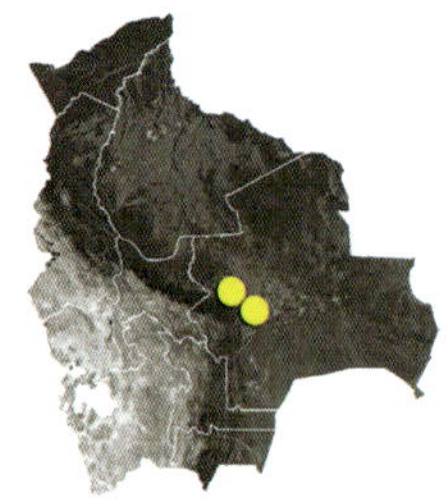

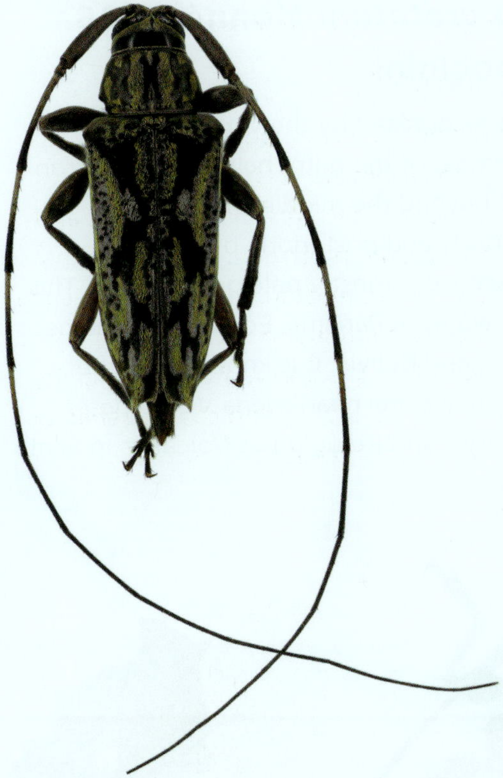

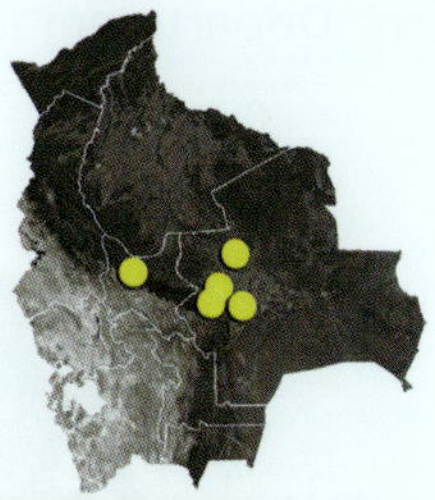

Nyssocarinus humeralis Monné, 1985

Lamiinae: Acanthocinini

This small (6–10 mm) species has dark brown integument with alternating blue-gray and yellow pubescent maculae on the pronotum and elytra. Its elytral apices are spined as is the last tergite in males. The third through sixth antennomeres are basally pale, but the remaining segments are nearly black. This species is known from Ecuador, Peru, Brazil, and Bolivia. In Bolivia it is known from Cristal Mayu and El Sacta in Cochabamba Department and the vicinity of the Flora and Fauna Hotel near Buena Vista, the Potrerillo del Guenda Reserve, and near Achira in Santa Cruz Department.

Nyssodrysternum cingillum Monné, 2009

Lamiinae: Acanthocinini

The seven narrow, dorsal white to pale yellow stripes of pubescence on the pronotum help to distinguish this small (8–9 mm), recently described species from others. The elytra are partially covered in small to large spots of similarly colored pubescence. Otherwise, the integument color is brown. This species is known only from French Guiana and Bolivia, where it has been collected in the vicinity of the Flora and Fauna Hotel near Buena Vista and the Potrerillo del Guenda Reserve in Santa Cruz Department.

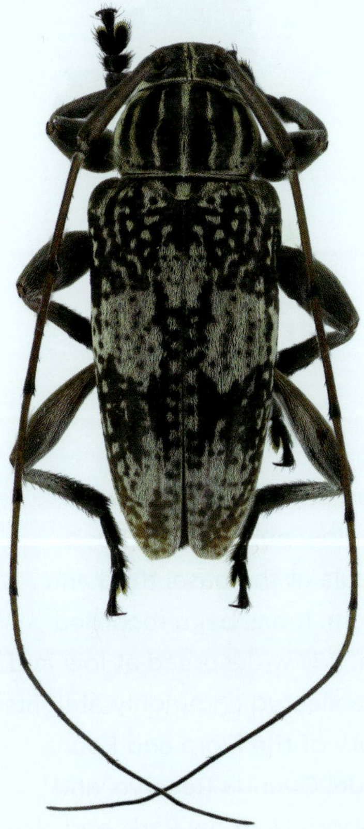

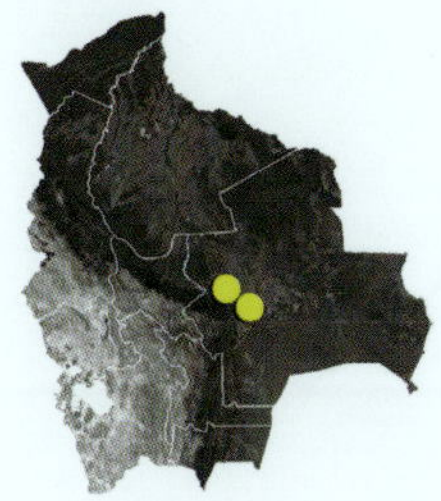

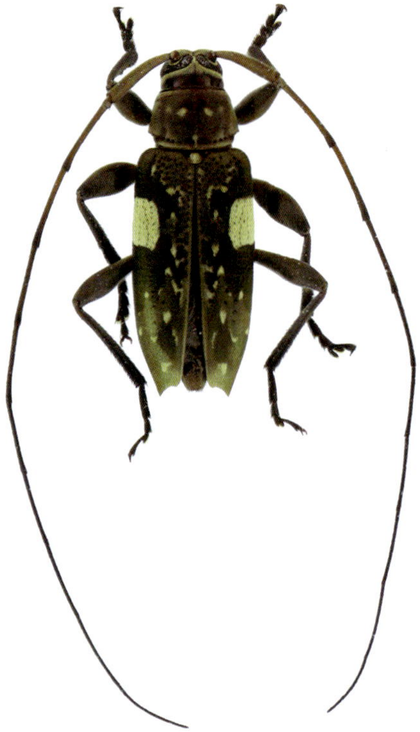

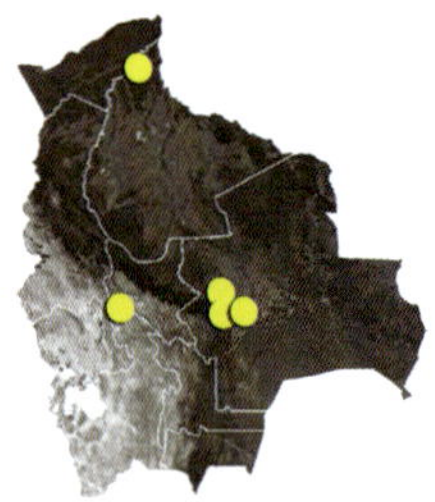

Nyssodrysternum conspicillare (Erichson, 1847)

Lamiinae: Acanthocinini

This small species (6–11 mm) has brown integument densely covered in brown to gray to whitish-yellow pubescence. Each elytron has a large whitish-yellow macula at the basal third and two to four small spots on the pronotum. It has been recorded from Peru, Ecuador, Brazil, and Bolivia. It is widespread at low and midelevations in Bolivia, having been collected commonly at lights in Santa Cruz Department in the vicinity of the Flora and Fauna Hotel near Buena Vista, the Potrerillo del Guenda Reserve, and Refugio Los Volcanes, adjacent to Amboró National Park, and also in Beni (Riberalta) and Cochabamba Departments.

Nyssodrysternum cretatum Monné, 1985

Lamiinae: Acanthocinini

This species (8–10 mm) is recognized by the very broad, white macula that extends over most of the outer half of each elytron from near the humerus to beyond the middle. This patch is irregularly margined anteriorly and posteriorly by a black macula. Otherwise, pubescence is mostly pale green or tan. This widespread species is known from Panama, Ecuador, Suriname, French Guiana, Brazil, Peru, and Bolivia. It is known from the vicinity of the Flora and Fauna Hotel near Buena Vista, the Potrerillo del Guenda Reserve, and Refugio Los Volcanes in Santa Cruz Department in Bolivia.

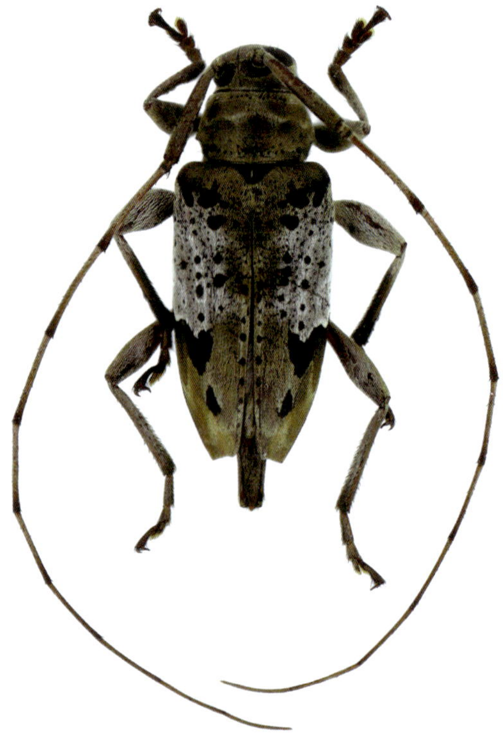

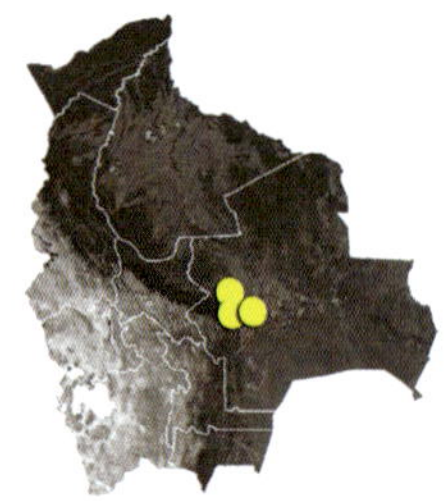

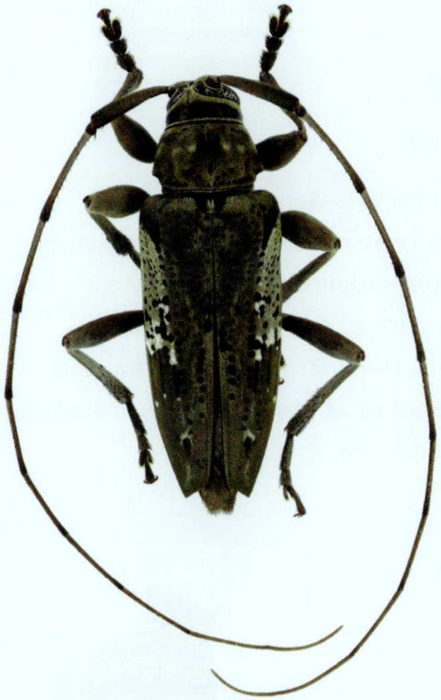

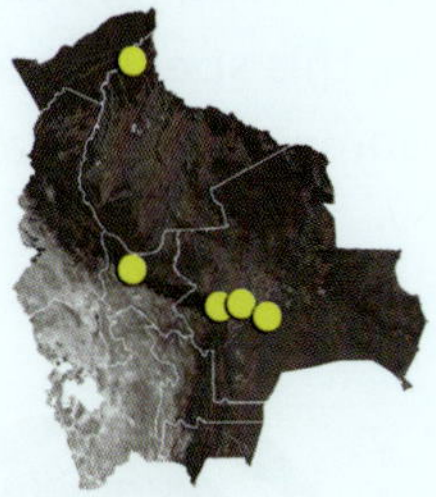

Nyssodrysternum serpentinum (Erichson, 1847)

Lamiinae: Acanthocinini

This small (8–11 mm), reddish-brown species is covered with pale green, recumbent pubescence, with the basal half of the elytra each having an irregularly shaped, diffuse, white macula toward the side. The elytral disk bears punctures of varying size, but none are setigerous. There is a small, dark brown spot anteriorly margined with white pubescence on the apical fifth of the elytra that helps characterize the species. This is among the most widespread neotropical species, having been recorded from almost all countries from Mexico to Bolivia, where specimens are most often encountered at lights. In Bolivia it has been recorded at most sites sampled in Santa Cruz Department as well as near Villa Tunari in Cochabamba Department and in Beni Department.

Nyssodrysternum spilotum Monné, 1975

Lamiinae: Acanthocinini

This dark brown, small species (7–9 mm) is densely covered in silvery gray and brown pubescence. The elytra are densely punctured basally, but less so toward the apices, and have a brown, transverse fascia just past the middle that extends from the lateral margin to just short of the suture. The elytral apices are acutely projecting but not spined. This species is known only from Bolivia and Brazil. In Bolivia it is known from Refugio Los Volcanes, Camiri, and Achira in Santa Cruz Department.

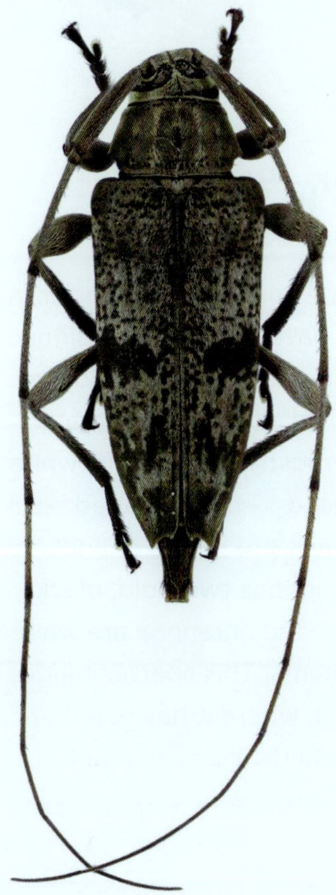

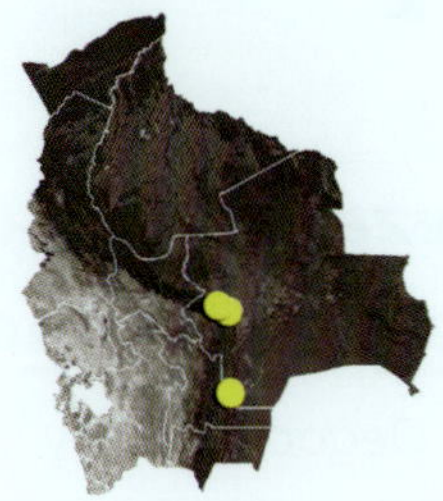

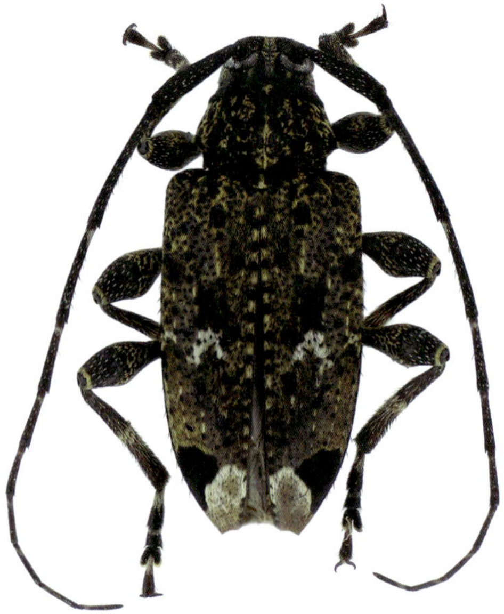

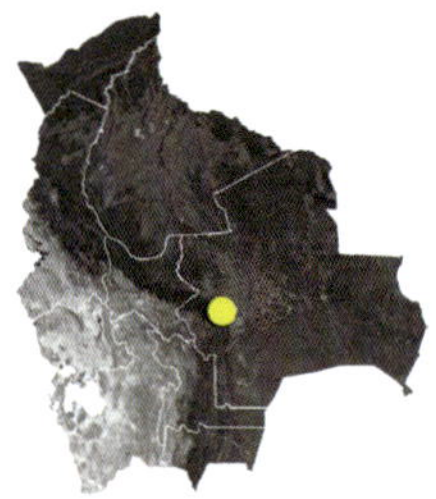

Oedopeza apicale (Gilmour, 1962)

Lamiinae: Acanthocinini

The genus *Oedopeza* likely includes some undescribed species in Bolivia, and this may represent one. We have tentatively assigned it to *O. apicale*, but there are some differences with the holotype. This species (10–12 mm) has boldly marked elytral apices that each have a black macula adjacent to a mostly white or off-white macula. The elytra each have a pronounced V- or W-shaped black macula just behind middle; otherwise, there are bands of light and medium brown pubescence. The pronotum has two bold, black maculae just behind the upper eye lobes. The antennae are white annulate at the basal third of most segments. This species ranges from northern South America into Bolivia, where it has been collected at Refugio Los Volcanes in Santa Cruz Department.

Oedopeza maculatissima Monné & Martins, 1976

Lamiinae: Acanthocinini

This reddish-brown species (9–13 mm) is densely pubescent and contrastingly marked with its brown pubescence irregularly interrupted by small black, white, and yellow maculae. The antennae are narrowly white annulate at the base of each segment. Ranging from Venezuela, Peru, Brazil, Argentina, and Bolivia, this species has been collected at almost all low-elevation sites sampled in Santa Cruz Department.

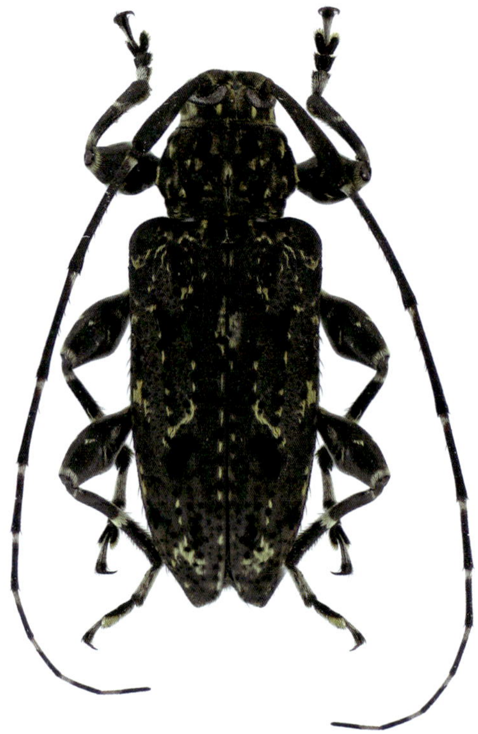

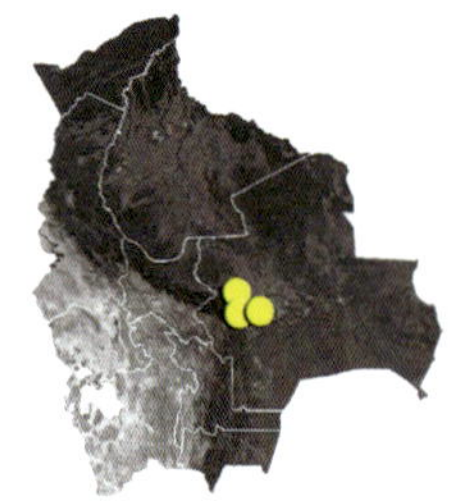

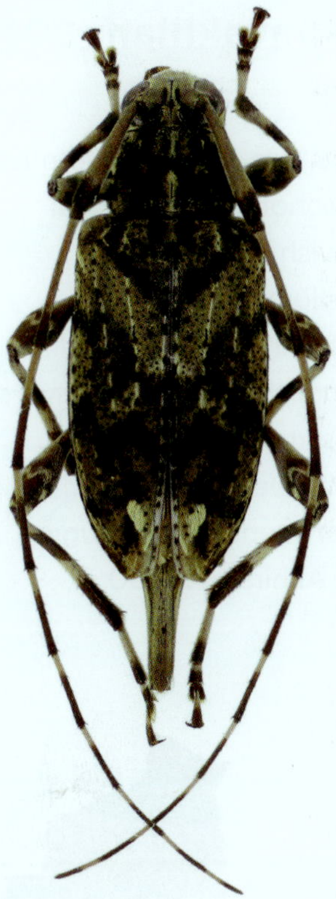

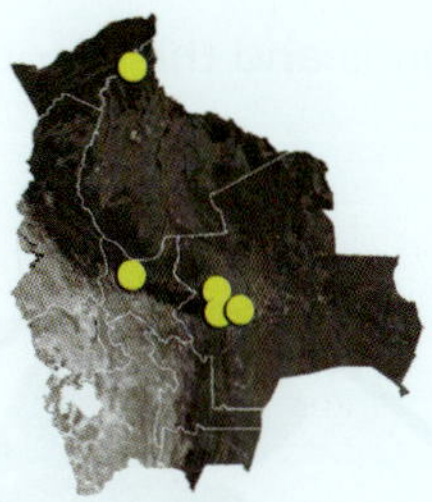

Oedopeza ocellator (Fabricius, 1801)

Lamiinae: Acanthocinini

This moderate-sized species (10–12 mm) is recognized by its apical sutural white macula, which is preceded by a black W- or chevron-shaped fascia on each elytron. This is another very widespread neotropical species that has been recorded from Mexico, through Central America, and into most countries of South America. It is also known from Cuba and Barbados in the West Indies. In Bolivia it has been recorded from the vicinity of the Flora and Fauna Hotel near Buena Vista, the Potrerillo del Guenda Reserve, and Refugio Los Volcanes (Santa Cruz Department), the Riberalta vicinity (Beni Department), and Cristal Mayu (Cochabamba Department).

Onalcidion fibrosum Monné & Martins, 1976

Lamiinae: Acanthocinini

This dark-reddish, small species (6–8 mm) is intricately marked with black punctures and pubescent maculae. Additionally, each elytron is centrally elevated along the basal margin, and this area bears a thick patch of upright, bristle-like, black setae on its apex. This species is known from Costa Rica, Panama, Venezuela, and Bolivia, where it has been collected at the Flora and Fauna Hotel near Buena Vista, the Potrerillo del Guenda Reserve, Refugio Los Volcanes, and Amboró road above Achira in Santa Cruz Department.

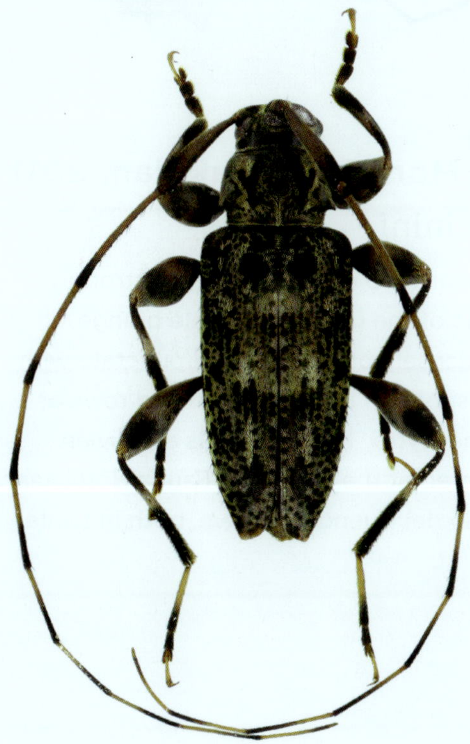

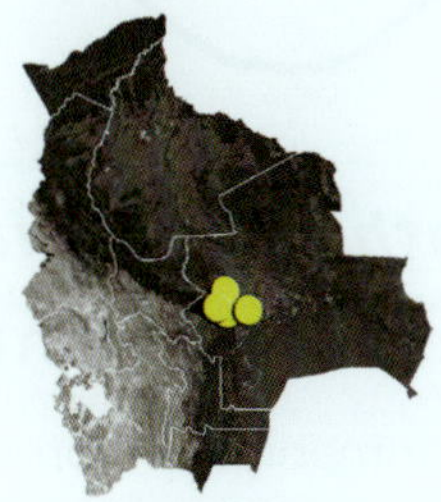

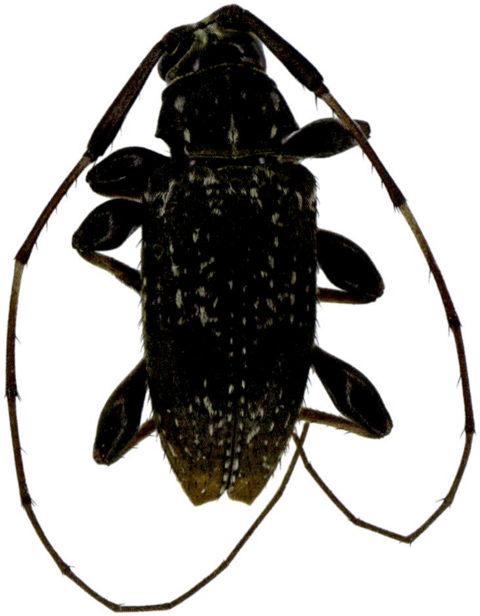

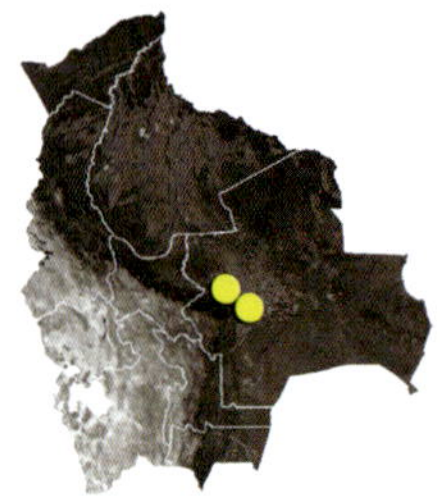

Oxathres boliviana Monné & Tavakilian, 2011

Lamiinae: Acanthocinini

This primarily dark brown, small (7–9 mm) species has strongly clavate femora (characteristic of the genus) and pale orange tibiae. The dorsal surface is densely pubescent and punctate with nine small, white pubescent spots on the pronotum and rows of similar spots the length of the elytra. This species is a Bolivian endemic and has been collected at the Flora and Fauna Hotel near Buena Vista and the Potrerillo del Guenda Reserve, both in Santa Cruz Department.

Oxathres maculosa Monné & Tavakilian, 2011

Lamiinae: Acanthocinini

More colorful than its congener *O. boliviana*, this species (7–9 mm) also has slightly less clavate femora. The pronotum has several closely spaced longitudinal bands of yellowish pubescence, separated by brown integument. The scutellum is dark but surrounded on either side and posteriorly by small maculae of yellow pubescence. The elytra have three poorly defined transverse bands of white-yellow setae. This species is known only from Bolivia, where it has been collected in La Paz Department at Inquisivi (Plazuela) and in Santa Cruz Department from Refugio Los Volcanes and the Vicoquin area above Achira at 2,000 m.

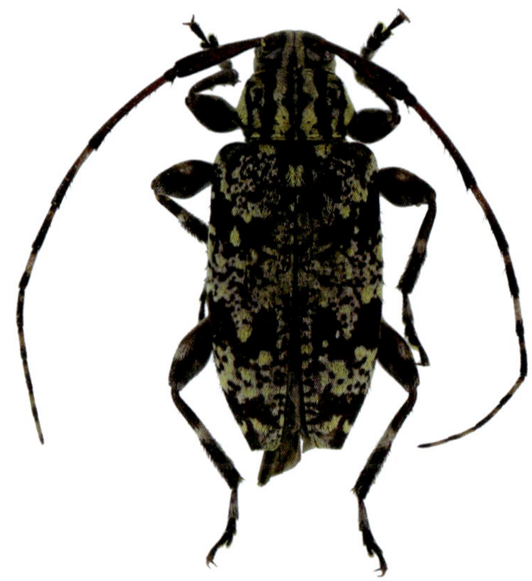

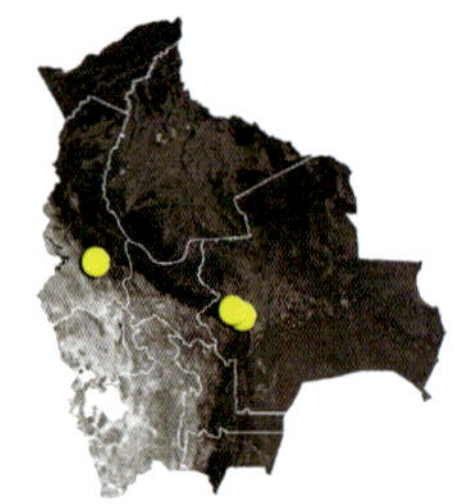

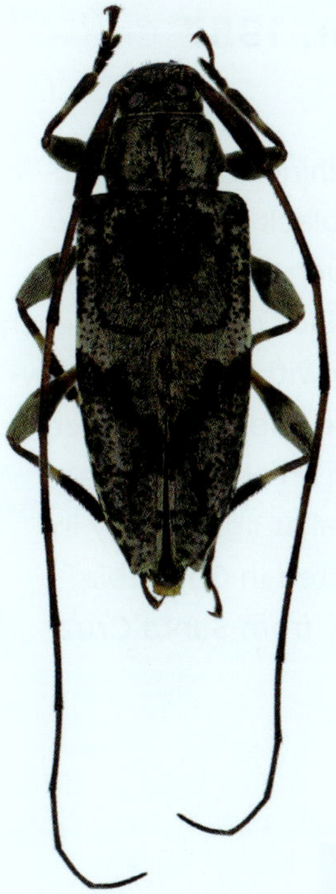

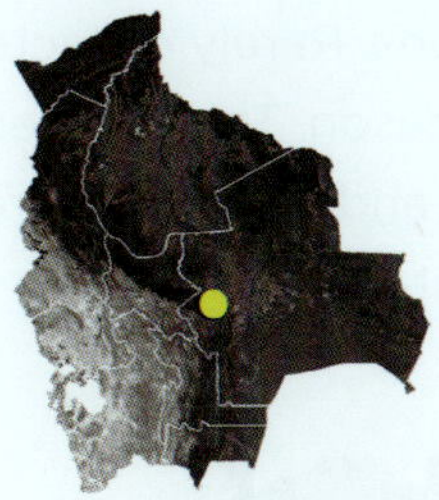

Ozineus achirae Monné & Monné, 2012

Lamiinae: Acanthocinini

This species is in another of the genera whose species have basal tubercles or crests bearing bristles on the base of each elytron (others found in Bolivia include *Baryssinus*, *Cosmotoma*, *Leptocometes*, *Onalcidion*, and *Trichillurges*). *Ozineus achirae* is a small (6–9 mm) species, mostly dark brown on the head, pronotum, and elytra but lighter on the appendages. The antennal scape is reddish brown, and the bases of most of the remaining antennomeres are yellow with the apices black. The elytra have indistinct patches of gray pubescence forming a transverse band in front of the middle and along the sutural apex. This species is endemic to Bolivia, where it has been collected only at the Vicoquin area above Achira near 1,900 m in Santa Cruz Department.

Ozineus argus Monné & Monné, 2012

Lamiinae: Acanthocinini

The antennae of this small (7–11 mm) species are similar to its congener *Ozineus achirae*, in that the scape is reddish brown and the remaining segments are pale basally and dark apically. The elytra are brown but covered with densely pubescent maculae of gray and brown. This species, like its congener *O. achirae*, is endemic to Bolivia, where it has been collected only at the Vicoquin area above Achira from 1,700 to 2,000 m in Santa Cruz Department by beating along the edge of agricultural cuts at forest edges.

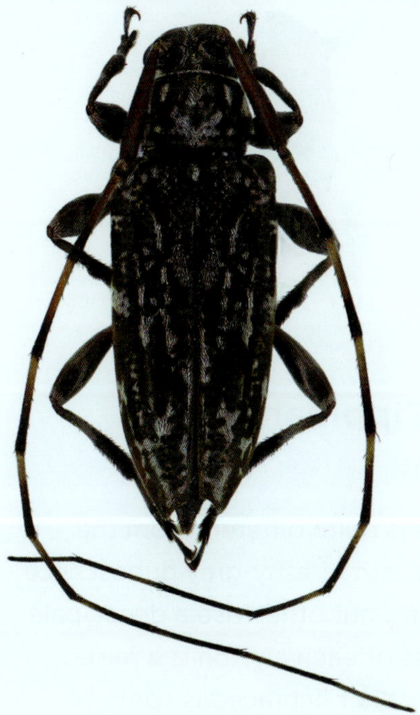

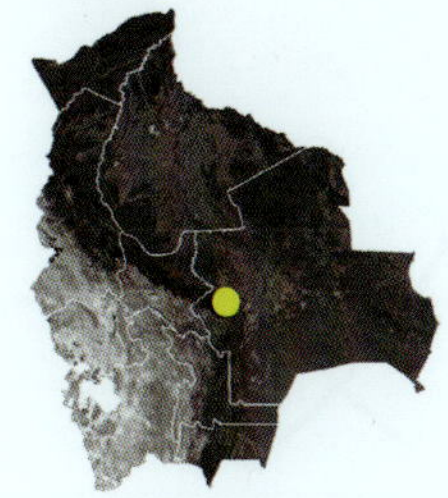

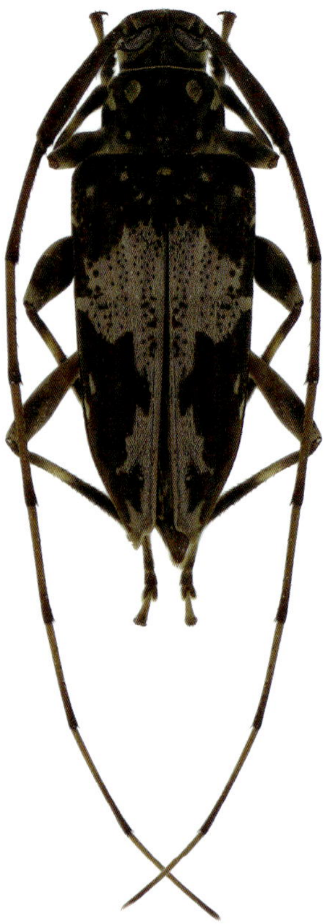

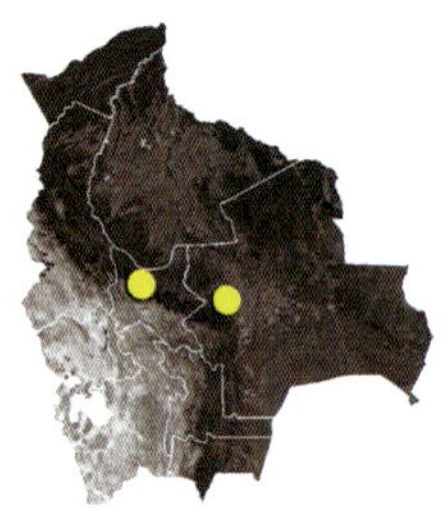

Ozineus nyssodroides Tippmann, 1960

Lamiinae: Acanthocinini

This 9–11 mm long *Ozineus* species is quite different from the others in that it has a very bold pattern of ashy-gray pubescence at the center and suture of the elytra but otherwise a dense pale green to tan coloration. At the base of each elytron is a raised tubercle. The pronotum has four distinct ochraceous spots of pubescence along the anterior margin. Known only from Ecuador and Bolivia, it has been collected near Buena Vista in Santa Cruz Department and in the Chaparé region of Cochabamba Department.

Paralcidion bilineatum Gilmour, 1957

Lamiinae: Acanthocinini

This is a relatively large (12–16 mm) acanthocinine with black integument and minutely pubescent gray to black pubescence. Its elytra are densely punctate basally but much less so on the apical half. It is boldly marked with parallel, black maculae on each side of the pronotal midline and on the elytra with a small, yellow-orange pubescent spot centrally on the suture and much larger, similar pubescent macula along the suture (in an arrow shape) near the apex. Rarely collected, it is occasional at lights early in the rainy season. This species has been collected in Colombia, Venezuela, and Bolivia, where it is known only from Santa Cruz Department.

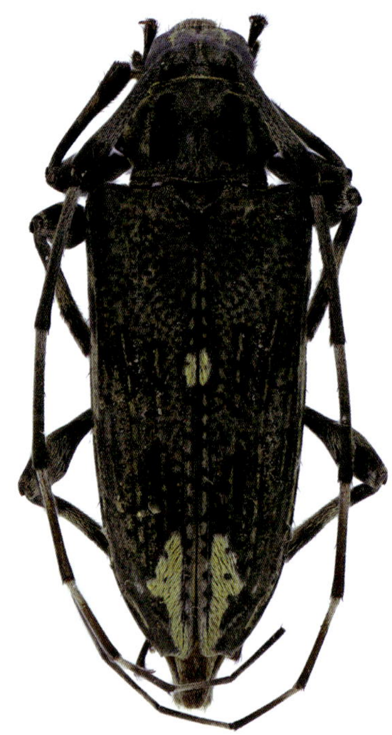

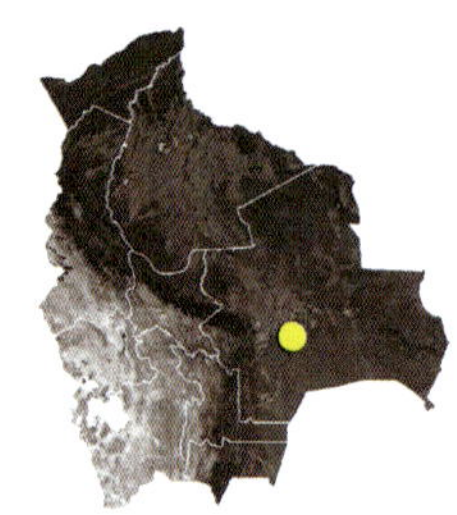

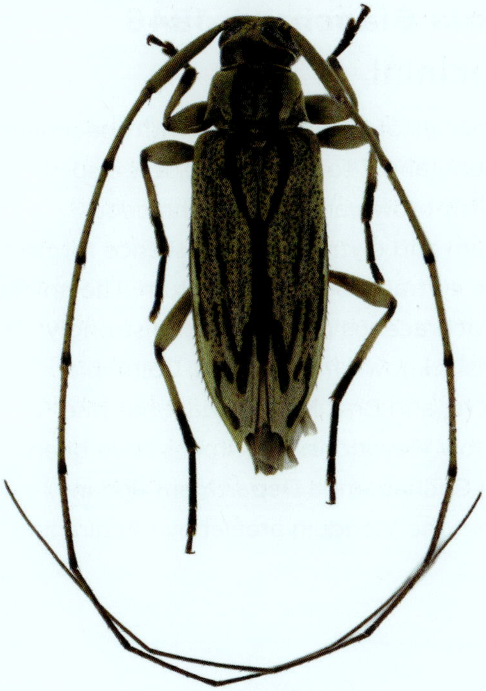

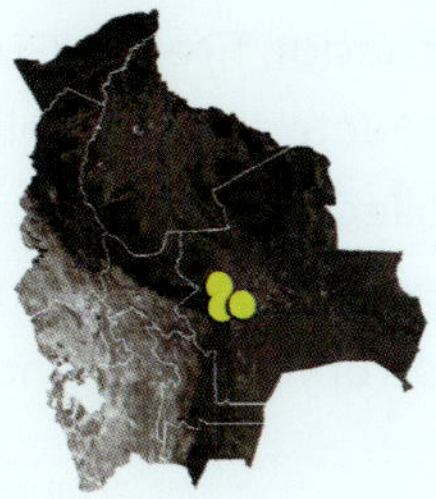

Pattalinus lineatus Monné & Martins, 1976

Lamiinae: Acanthocinini

This is a beautiful, small species (6–10 mm), with its elytral surface completely covered with pale gray pubescence with a greenish hue. Exceptions are the apical third to half of the tibiae, all the tarsi, and the extreme apex of each antennomere, which are dark brown to black. Also, there is a distinct pattern of brown lines along the suture, around the scutellum, and elsewhere on the elytra. Distinct brown punctures are scattered throughout the basal four-fifths of the elytra but are absent from the apex. This species is known only from Bolivia and Peru. In Bolivia it is recorded from the vicinity of the Flora and Fauna Hotel near Buena Vista, the Potrerillo del Guenda Reserve, and Refugio Los Volcanes, adjacent to Amboró National Park, in Santa Cruz Department.

Sporetus seminalis Bates, 1864

Lamiinae: Acanthocinini

This is a small (7–9 mm), attractive species with dark-reddish integument and a multitude of small, cream-colored pubescent maculae unevenly distributed on the elytra and pronotal surface. There are small, whiter pubescent spots on the elytral apices, and most segments of the antennae are white annulate basally. The dark tarsi and femora are beautifully marked with patches of bright white pubescence. This species is widespread in northern South America and recorded in Bolivia, where it is known from the Vicoquin area above Achira at 2,000 m, the vicinity of the Flora and Fauna Hotel near Buena Vista, and Refugio Los Volcanes in Santa Cruz Department.

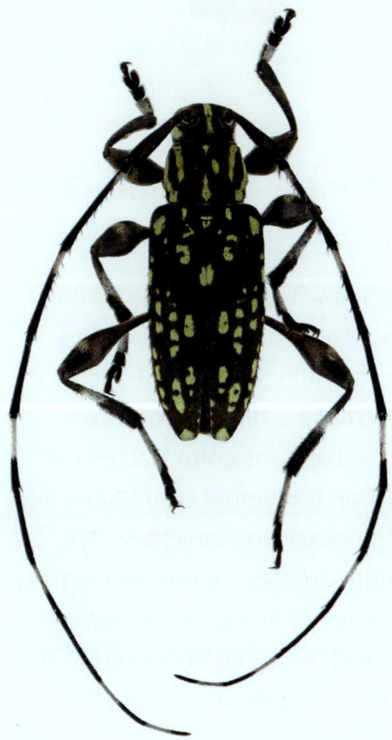

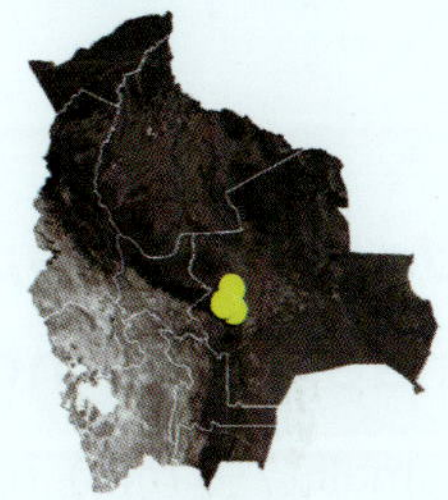

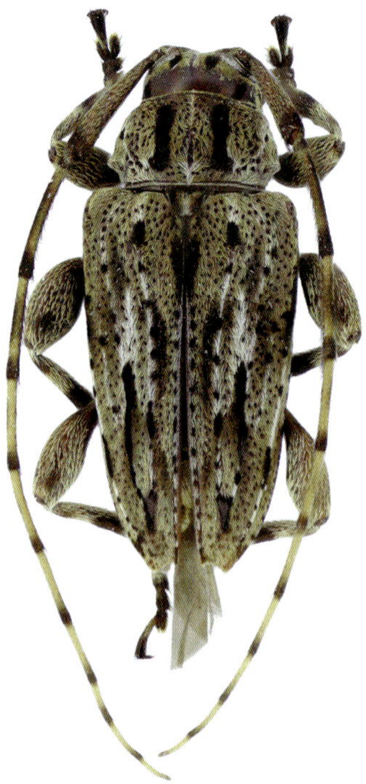

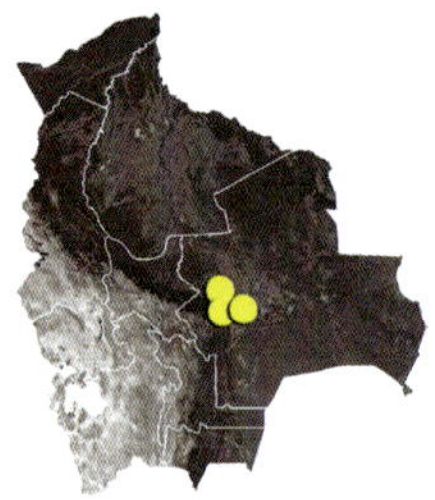

Tropidozineus wappesi Monné & Monné, 2012

Lamiinae: Acanthocinini

This is a small (7–8 mm), robust species with the surface of the elytra and pronotum closely punctate throughout and with its brown integument obscured by dense, multicolored pubescence. The pronotum is decorated with two black stripes along the midline that parallel each other. The pubescent color pattern of the elytra is extremely variable between individual specimens and defies any attempt at a meaningful summary description. The species is most often taken by beating small branches on partially burnt or cut trees but also readily comes to lights. It is known only from Bolivia and Brazil. In Bolivia it has been collected at the Potrerillo del Guenda Reserve, Bermejo, and the Flora and Fauna Hotel near Buena Vista in Santa Cruz Department.

Trypanidius andicola Blanchard, 1846

Lamiinae: Acanthocinini

This is one of the largest species of Acanthocinini, with specimens ranging from 14 to 17 mm long and 7 to 8 mm wide. It is also boldly patterned with reddish-brown and white-ochraceous pubescence on the pronotum and elytra. The pubescence forms concentric rings around the elytral base and pronotum. The apices of the elytra are uniformly ochraceous pubescent and strongly spined laterally. This species is known from Bolivia, Colombia, Ecuador, French Guiana, Peru, and Brazil. In Bolivia it has been collected in both high and low elevations. Specimens have been collected at Cristal Mayu in Cochabamba Department and in Santa Cruz Department from the Vicoquin area above Achira and Potrerillo del Guenda.

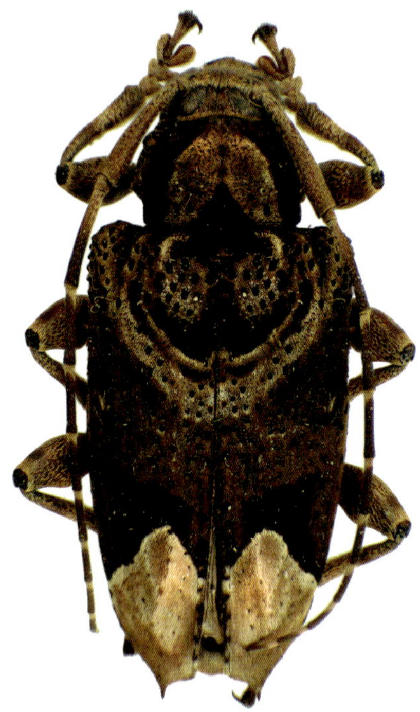

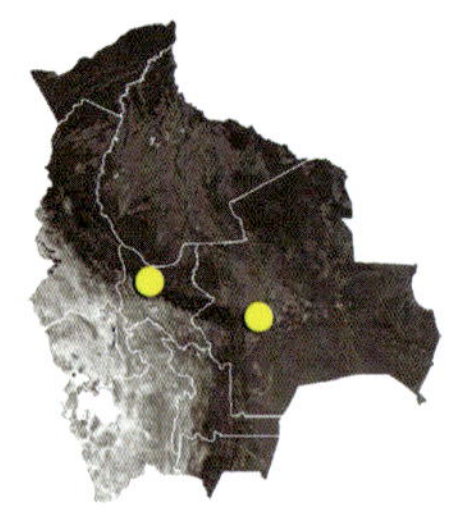

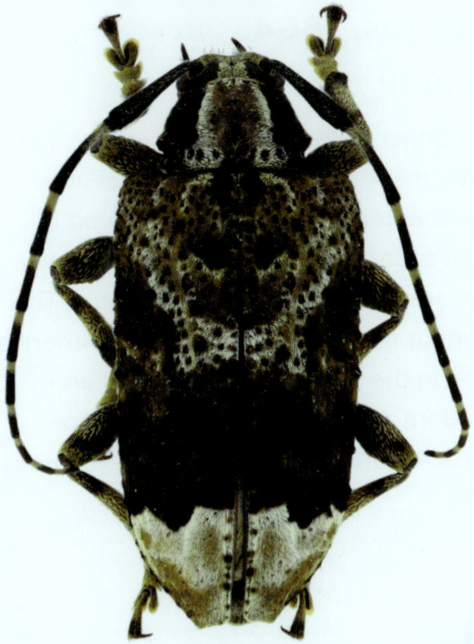

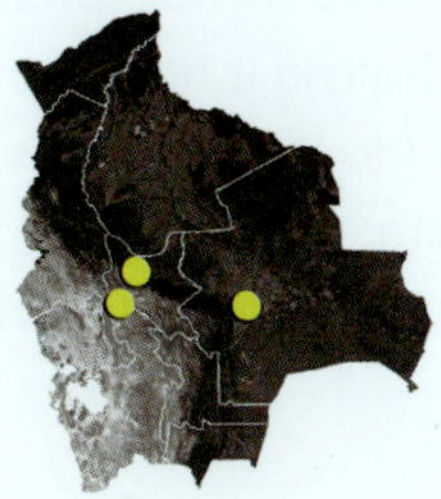

Trypanidius apicalis Aurivillius, 1921

Lamiinae: Acanthocinini

Although this robust species has similar size (16–18 mm) and an elytral apex covered in pale pubescence like its congener *T. andicola*, the apices are unarmed and truncate versus strongly spined, as in that species. In addition to its elytral punctation, this species also has rows of rather evenly spaced, black, tuberculate setal tufts on the central portion of the disk. Its pronotum is attractively marked with a light brown, central midline bordered by bright white pubescence, which in turn is sharply margined by short, coal-black pubescence. Known only from Bolivia and Colombia, this species has been collected in Villa Tunari and Cochabamba (Cochabamba Department) and Achira (Santa Cruz Department) in Bolivia.

Urgleptes ovalis (Bates, 1866)

Lamiinae: Acanthocinini

Most species in the large genus *Urgleptes* (almost 80 species) are small (4–6 mm) and drab, with many difficult to characterize. This one is among the more distinctive species, with narrow, longitudinal rows of pale pubescence on the elytra that are separated by mostly glabrous stripes of dark or reddish-brown integument. In addition, the femora are strongly convex. It is recorded from Bolivia, French Guiana, and Brazil. In Bolivia it is known from north of Camiri and the Potrerillo del Guenda Reserve in Santa Cruz Department.

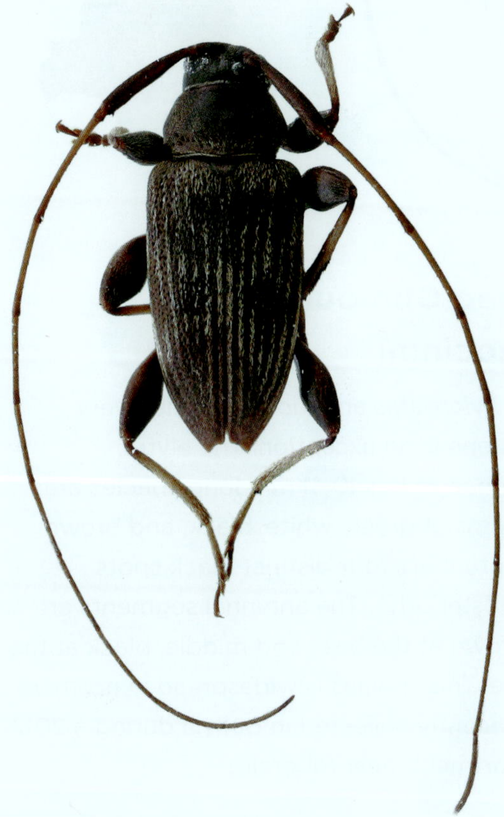

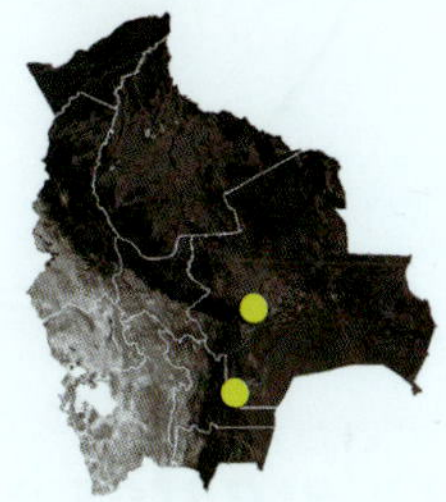

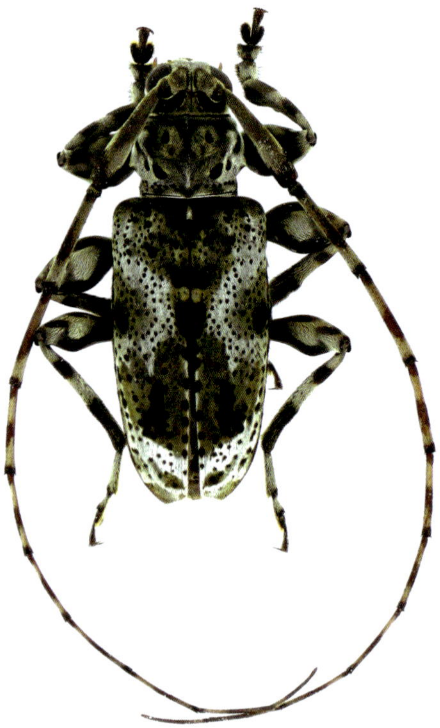

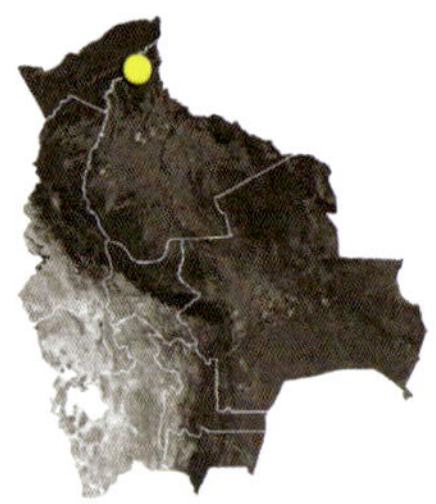

Xylergates elaineae Gilmour, 1962

Lamiinae: Acanthocinini

The species of the genus *Xylergates* are known for their very elegant patterns, and this one is no exception. The elytra, pronotum, legs, and antennae of this 10–11 mm long species are coated with delicate patterns of green, white, black, and brown pubescence. On the pronotum are four distinct black spots (two anteriorly and two posteriorly). The antennal segments are distinctly mottled light brown at the base and middle, black at the apex, and white elsewhere. This species is widespread in northern South America but only recently collected in Bolivia during a 2012 expedition to Pando Department, near Riberalta.

Xylergates lacteus Bates, 1864

Lamiinae: Acanthocinini

This species resembles its congener *X. elaineae* but is larger (14–17 mm) and more distinctly marked with white pubescence on the elytral disk. The legs and antennae are mostly covered in white or gray pubescence. The pronotum, like the previous species, has four black spots, but they are much reduced in *X. lacteus*. This species is widespread throughout north and central South America. In Bolivia it has been collected in the Riberalta area in Beni and Pando Departments and the Flora and Fauna Hotel near Buena Vista, the Potrerillo del Guenda Reserve, and Refugio Los Volcanes in Santa Cruz Department.

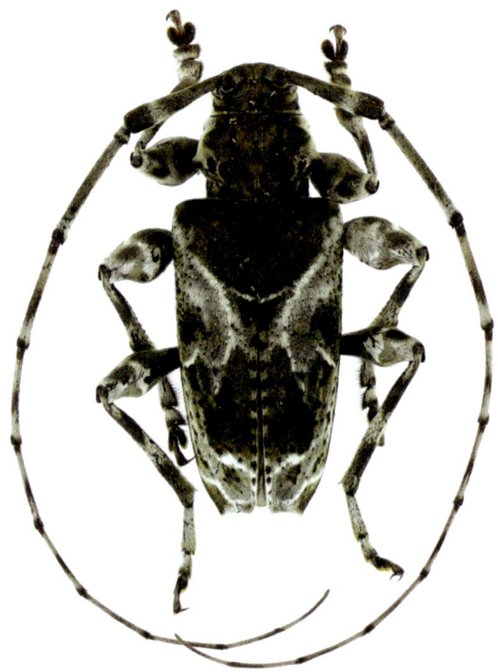

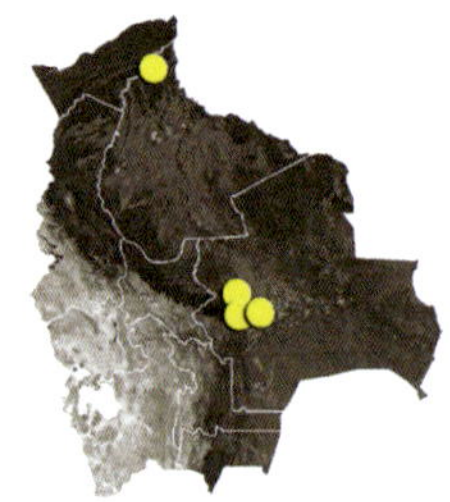

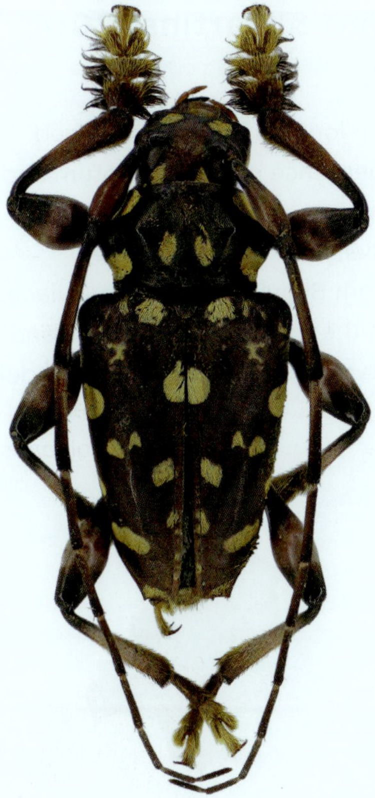

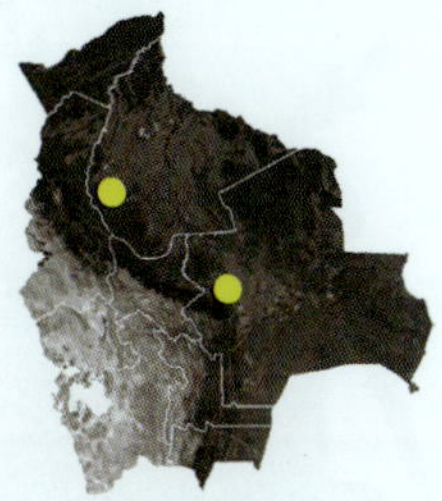

Acanthoderes thammi Bates, 1880
Lamiinae: Acanthoderini

The large genus *Acanthoderes* is represented by almost 50 species in North, Central, and South America. Although many of its species are characterized by somber colors, this one, 13–18 mm, is very distinct with its yellow spots of dense hairs scattered over the elytra, pronotum, and head. The front tarsi are also boldly marked with dense yellow and black setae. This species is known from Bolivia and Peru. In Bolivia it has been collected in Beni Department and Santa Cruz Department from the Flora and Fauna Hotel grounds.

Aegoschema moniliferum (White, 1855)
Lamiinae: Acanthoderini

This large acanthoderine (16–18 mm long) has most of the elytra, pronotum, and legs covered in dense, white pubescence. There are several incomplete rows of large punctures extending from the elytral base to beyond the middle. The pronotum has large, rounded, lateral tubercles. It is known from French Guiana, Brazil, Peru, and Bolivia, where it has been recorded from the Flora and Fauna Hotel near Buena Vista and Potrerillo del Guenda in Santa Cruz Department.

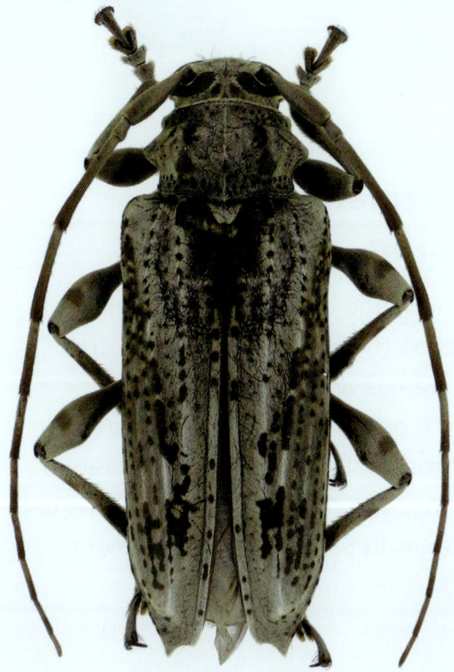

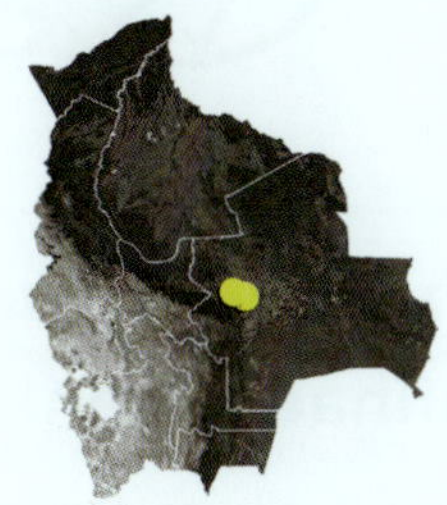

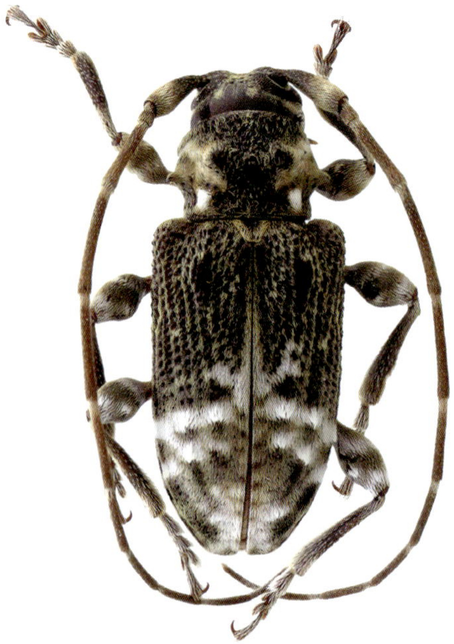

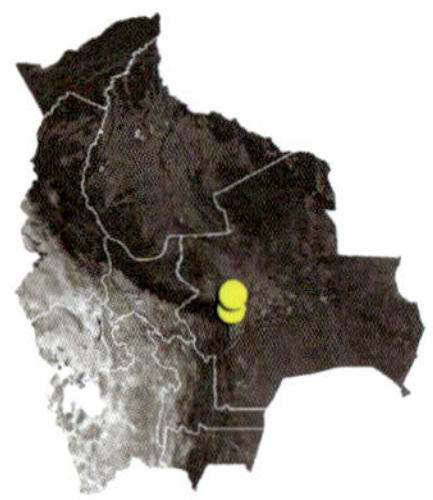

Alphus tuberosus (Germar, 1824)

Lamiinae: Acanthoderini

This cryptically colored acanthoderine (8–12 mm) is recognized by the rows of deep punctures on the basal half of the elytra while the apical half of the elytra has an irregular pattern of interconnected patches of white pubescence. It is widespread in South America and recorded from Bolivia, Brazil, Peru, Paraguay, and Argentina. In Bolivia this species is known from Potrerillo del Guenda and Refugio Los Volcanes in Santa Cruz Department.

Ateralphus lacteus Galileo & Martins, 2006

Lamiinae: Acanthoderini

This robust acanthoderine (15–17 mm long) is easily recognized by the grayish-green and white pubescence on the pronotum and elytra. There is a large, rectangular patch of white pubescence in the middle of the elytra and smaller, less defined patches posteriorly. This species is known from Bolivia and French Guiana. In Bolivia it has been collected from the grounds of the Flora and Fauna Hotel near Buena Vista (the type locality) and Potrerillo del Guenda in Santa Cruz Department.

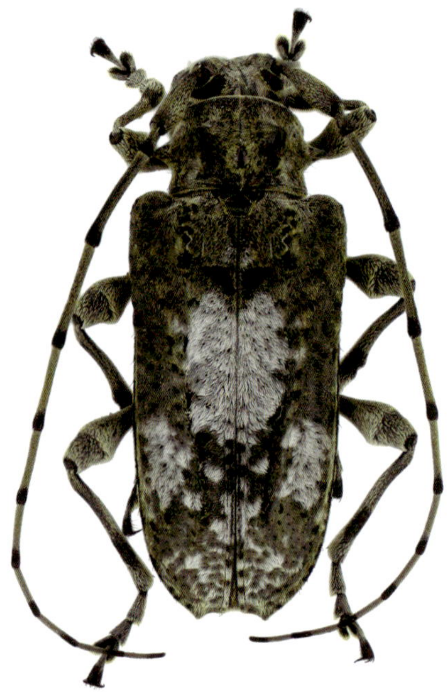

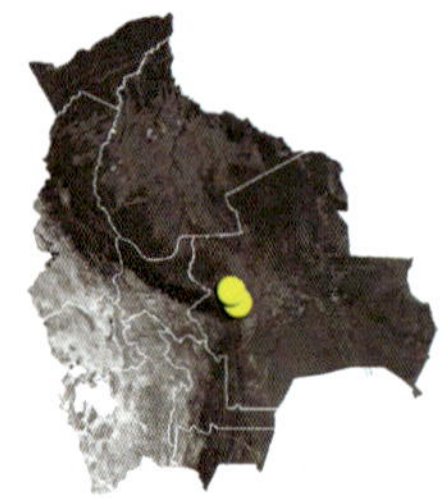

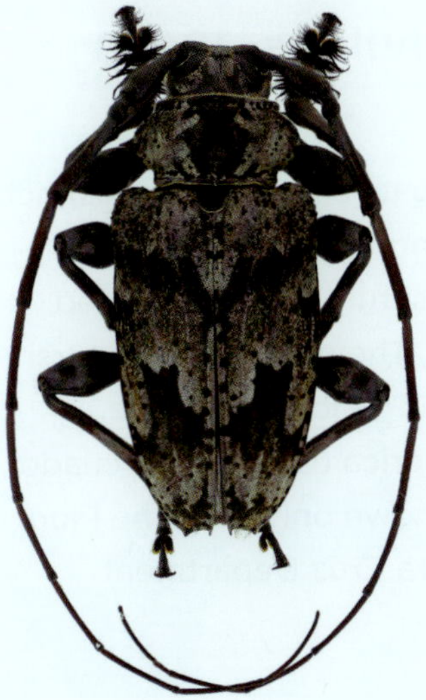

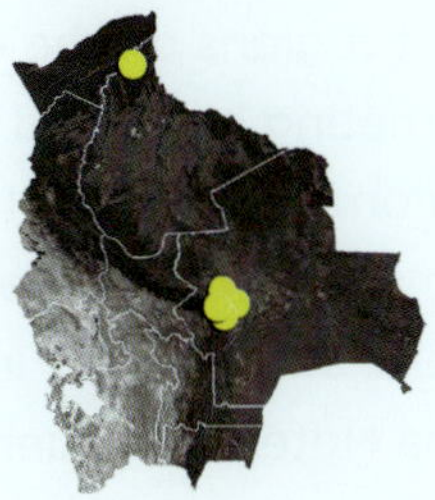

Dryoctenes scrupulosus (Germar, 1824)

Lamiinae: Acanthoderini

This broad, robust species is one of the largest acanthoderines at more than 25 mm in length. Its reddish-brown integument is clothed with light gray pubescence, forming an irregular pattern of maculae and transverse fasciae on the elytra. There are small, shining tubercles on the elytral base that transition to pubescent spots distally. The pronotum is laterally tuberculate and spined. Females of this species have expanded genae that support unusually robust mandibles. Males, as shown here, have expanded front tarsal segments that are fringed with long, silky hairs. This is a fairly common species often encountered crawling on freshly downed tree trunks or occasionally at lights. It is recorded from Costa Rica to Bolivia, where it is widely distributed in lowland forests in the vicinity of the Flora and Fauna Hotel near Buena Vista, the Potrerillo del Guenda Reserve, Refugio Los Volcanes, adjacent to Amboró National Park, and the road to Bella Vista near Cuevas in Santa Cruz Department. It has also been collected in Beni Department near Riberalta in the extreme north of the country.

Dufauxia thomasi Martins & Galileo, 2007

Lamiinae: Acanthoderini

This is a rarely encountered species of moderate size (10–16 mm), and is a Bolivian endemic that is quickly recognized by its dark elytra and white, pubescent, W-shaped fascia at the apical third. Both sexes have long antennae, with those of the males four to five times the body length. It is known from only Santa Cruz Department at Refugio Los Volcanes, adjacent to Amboró National Park, and the type locality at the Flora and Fauna Hotel near Buena Vista.

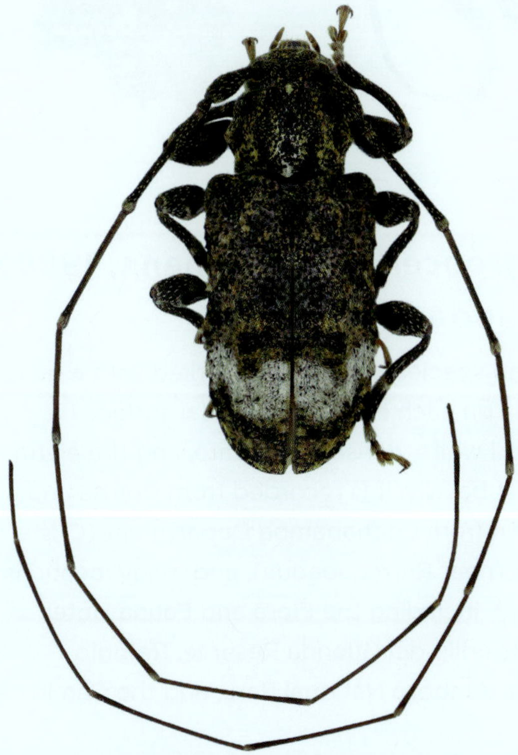

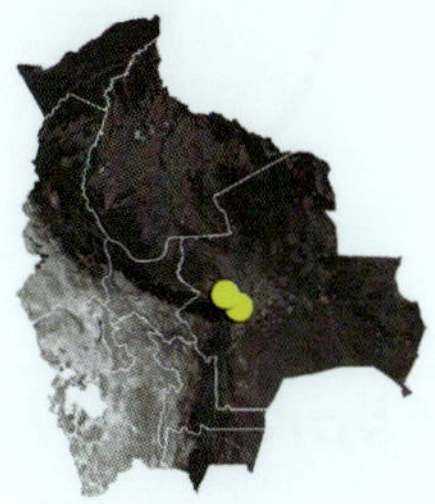

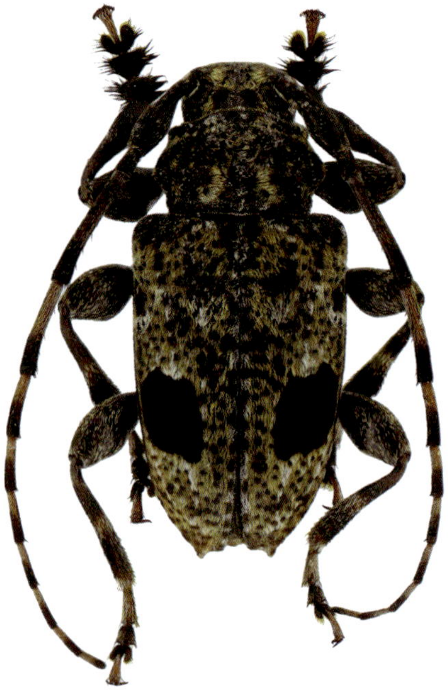

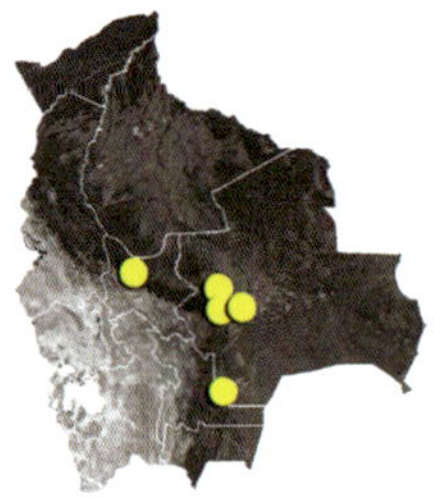

Eupromerella nigroocellata (Tippmann, 1960)

Lamiinae: Acanthoderini

This small (6–9 mm), drab species is distinctly marked with a bold, postmedial, black macula on each elytron. Its dorsal surface is also mottled in brown and white, densely punctate, and the elytra have a low, elongate crest basally. It is recorded from Bolivia and Peru. In Bolivia it is known from Cochabamba Department (Cristal Mayu), Beni Department (near Rurrenabaque), and many locations in Santa Cruz Department, including the Flora and Fauna Hotel near Buena Vista, the Potrerillo del Guenda Reserve, Refugio Los Volcanes, adjacent to Amboró National Park, and the Camiri environs.

Eupromerella orbifera (Aurivillius, 1908)

Lamiinae: Acanthoderini

This species (7–9 mm), similar in shape to its congeners, is distinct, with a concentric pattern of black, tan, and white glabrous and pubescent regions on the elytra and the suture mostly adorned with alternating black and white patches. The pubescent protarsi are black on the two basal tarsomeres and yellow elsewhere. Although widely distributed in South America (Colombia, Ecuador, Brazil, Peru, and Bolivia), in Bolivia it is known only from the Flora and Fauna Hotel near Buena Vista in Santa Cruz Department.

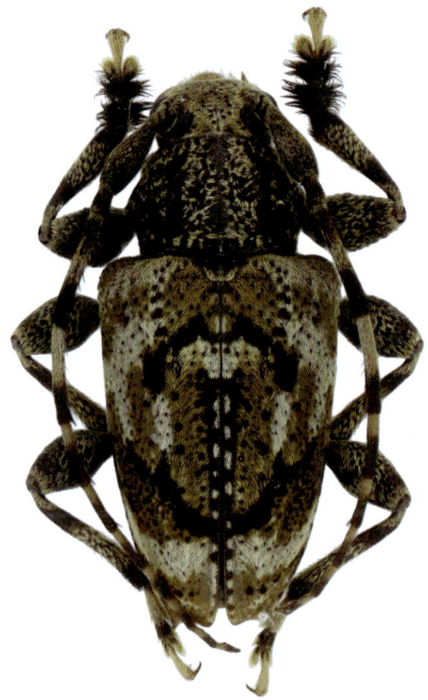

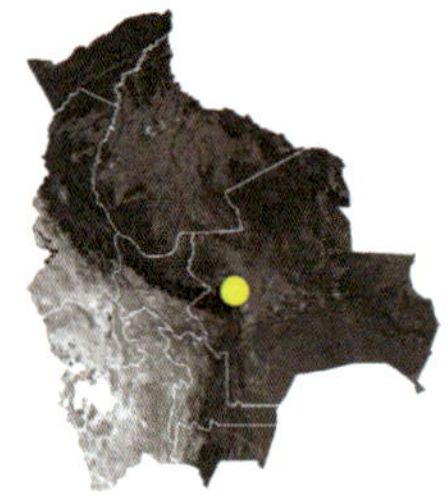

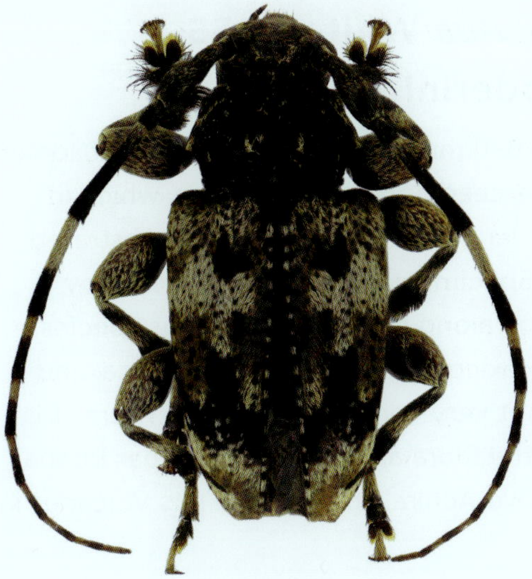

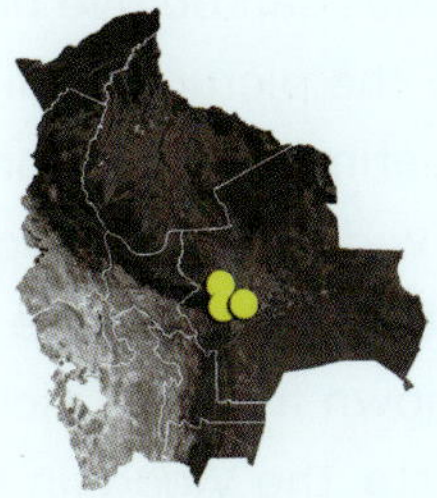

Eupromerella propinqua (Melzer, 1931)

Lamiinae: Acanthoderini

This and the following species, *Psapharochrus pseudopropinquus*, are the smallest (4–8 mm) members of *Eupromerella* recorded in Bolivia. It can be characterized by its densely punctate dorsal surface, much darker pronotum (versus the elytra), dark, inverted V-shaped macula at the apical fourth of each elytron, and transverse band of lighter whitish-gray pubescence across the basal third. This species occurs in Brazil, Argentina, Paraguay, and Bolivia, where it has been collected on the road to Bella Vista, Potrerillo del Guenda Reserve, and Refugio Los Volcanes in Santa Cruz Department.

Psapharochrus pseudopropinquus (Fuchs, 1958)

Lamiinae: Acanthoderini

Although placed in the genus *Psapharochrus*, this small (4–6 mm) species rather should be placed in *Eupromerella*. It is very similar to the previous species, *E. propinqua*, but is distinguished by a black, circular fascia that has the top half on the pronotum and the bottom half on the base of the elytra. The elytra are also distinctively marked by a narrow, black, transverse band crossing it postmedially. This is a common species routinely collected at lights. It is recorded from Bolivia and Argentina. In Bolivia it has been collected from many locations in Santa Cruz Department, including the Flora and Fauna Hotel near Buena Vista, the Potrerillo del Guenda Reserve, and Refugio Los Volcanes, adjacent to Amboró National Park. It has also been collected near Villamontes in Tarija Department.

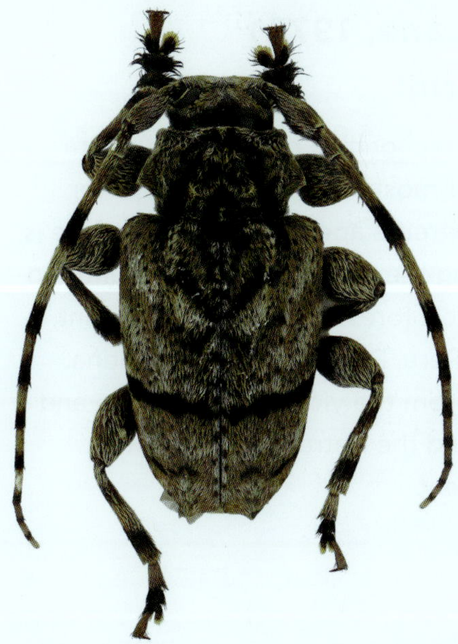

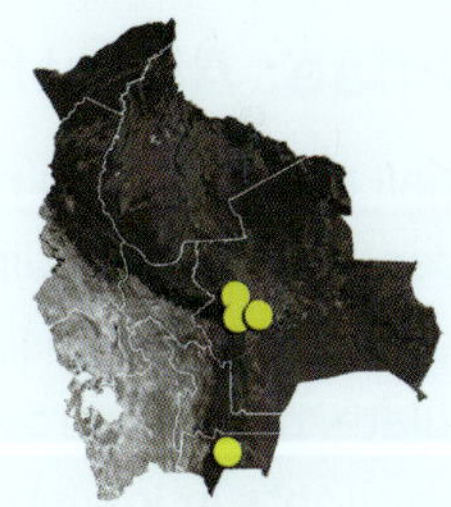

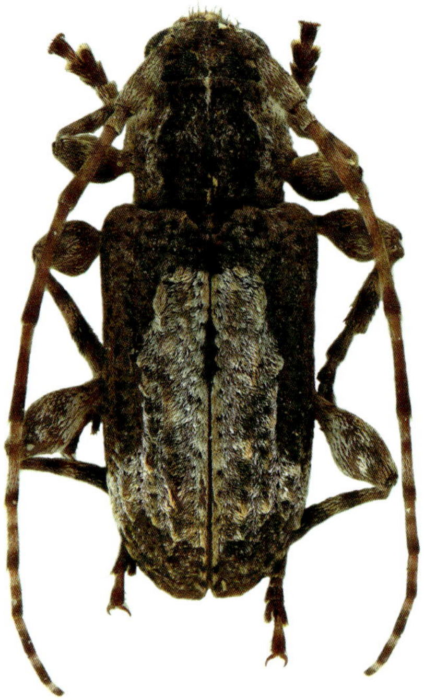

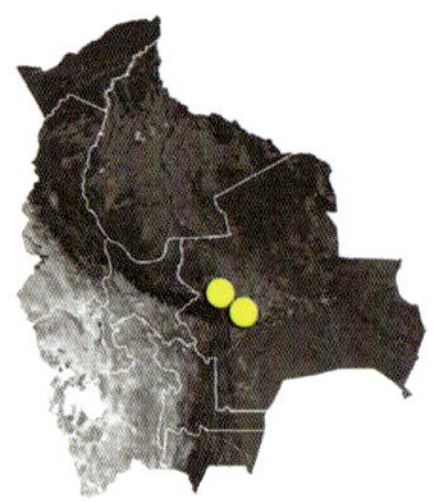

Exalphus gounellei (Lane, 1973)

Lamiinae: Acanthoderini

This small species (7–10 mm) is recognized by the large, white pubescent patch extending over most of the elytra except for the base, lateral margins, and extreme apex. The elytral surface is tuberculate, mostly in rows along the costae. Rarely encountered except at lights in lowland tropical forests and never abundant, it is known from Bolivia, Brazil, Peru, Paraguay, and Argentina. In Bolivia it has been recorded from the vicinity of the Flora and Fauna Hotel near Buena Vista and the Potrerillo del Guenda Reserve in Santa Cruz Department.

Macronemus asperulus White, 1855

Lamiinae: Acanthoderini

This cryptic, small species (6–11 mm), when resting on a leaf, looks like a bird dropping. It has exceedingly long antennae, which in males can be five times the length of the body, as demonstrated by the pictured individual. Specimens are usually collected by beating branches and foliage along forest margins. This species is widely distributed from Mexico, through Central America, and into South America but is not very common in Bolivia. There, it is known from the vicinity of the Flora and Fauna Hotel near Buena Vista, the Vicoquin area above Achira, and Refugio Los Volcanes in Santa Cruz Department.

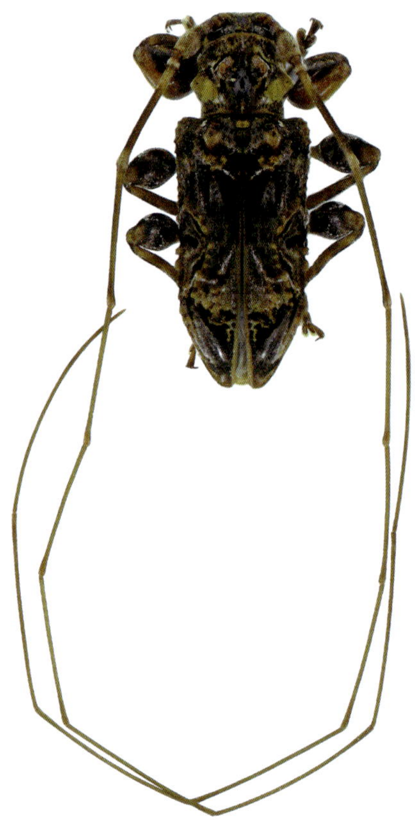

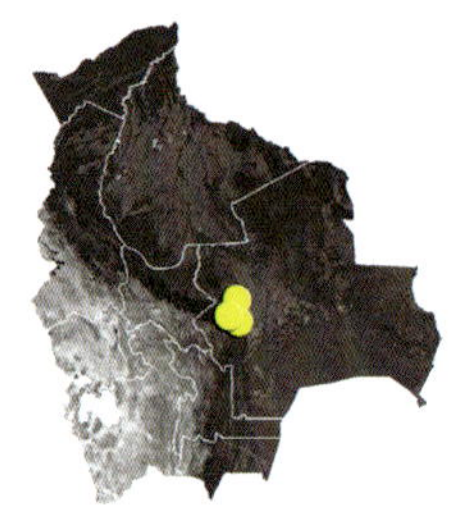

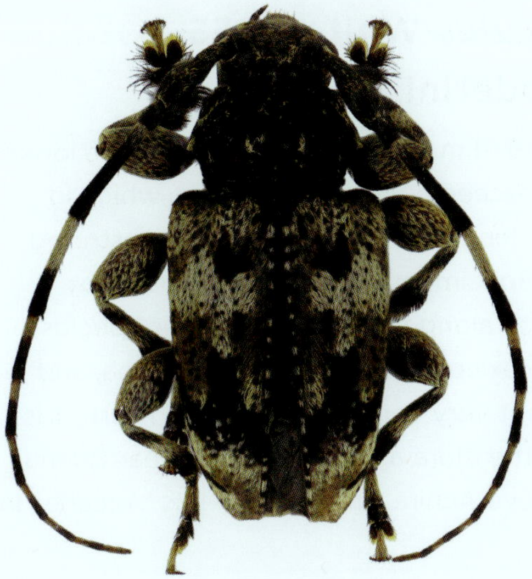

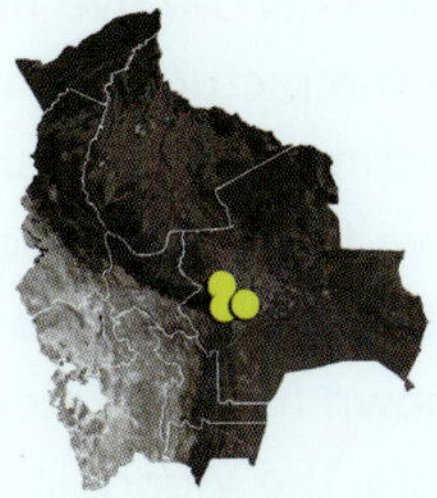

Eupromerella propinqua (Melzer, 1931)

Lamiinae: Acanthoderini

This and the following species, *Psapharochrus pseudopropinquus*, are the smallest (4–8 mm) members of *Eupromerella* recorded in Bolivia. It can be characterized by its densely punctate dorsal surface, much darker pronotum (versus the elytra), dark, inverted V-shaped macula at the apical fourth of each elytron, and transverse band of lighter whitish-gray pubescence across the basal third. This species occurs in Brazil, Argentina, Paraguay, and Bolivia, where it has been collected on the road to Bella Vista, Potrerillo del Guenda Reserve, and Refugio Los Volcanes in Santa Cruz Department.

Psapharochrus pseudopropinquus (Fuchs, 1958)

Lamiinae: Acanthoderini

Although placed in the genus *Psapharochrus*, this small (4–6 mm) species rather should be placed in *Eupromerella*. It is very similar to the previous species, *E. propinqua*, but is distinguished by a black, circular fascia that has the top half on the pronotum and the bottom half on the base of the elytra. The elytra are also distinctively marked by a narrow, black, transverse band crossing it postmedially. This is a common species routinely collected at lights. It is recorded from Bolivia and Argentina. In Bolivia it has been collected from many locations in Santa Cruz Department, including the Flora and Fauna Hotel near Buena Vista, the Potrerillo del Guenda Reserve, and Refugio Los Volcanes, adjacent to Amboró National Park. It has also been collected near Villamontes in Tarija Department.

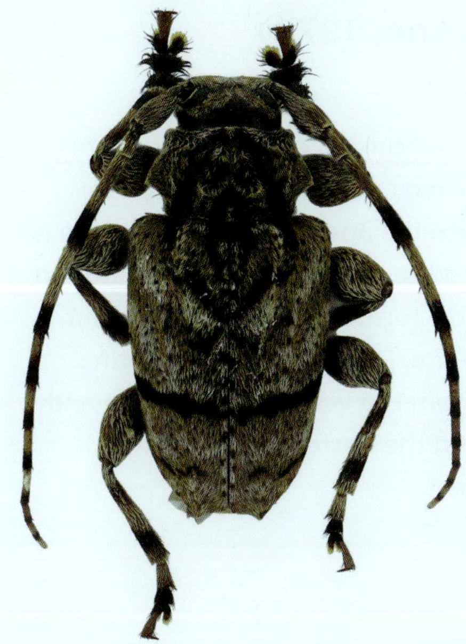

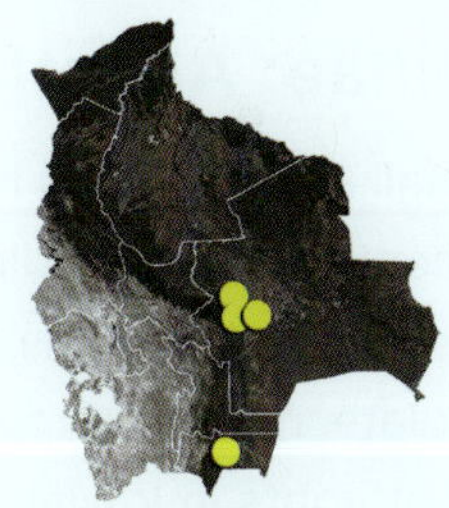

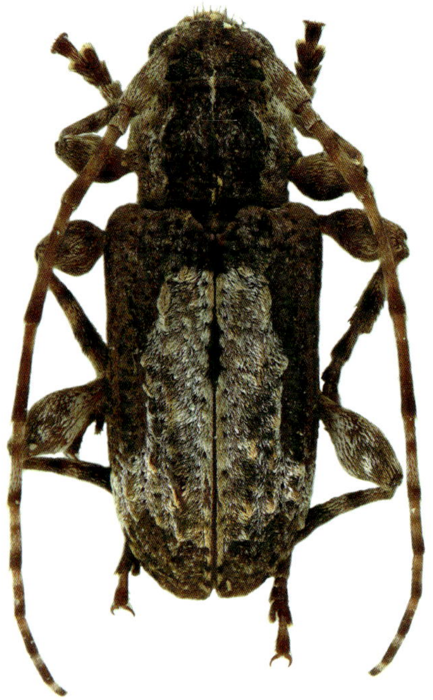

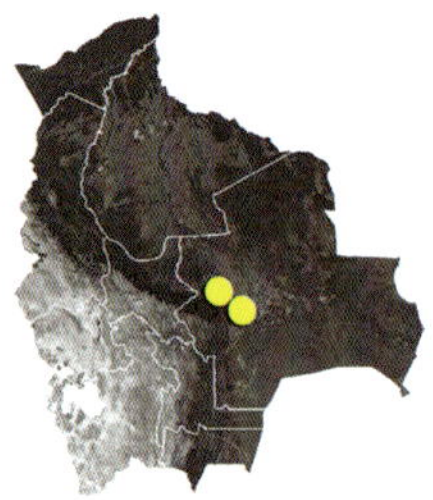

Exalphus gounellei (Lane, 1973)

Lamiinae: Acanthoderini

This small species (7–10 mm) is recognized by the large, white pubescent patch extending over most of the elytra except for the base, lateral margins, and extreme apex. The elytral surface is tuberculate, mostly in rows along the costae. Rarely encountered except at lights in lowland tropical forests and never abundant, it is known from Bolivia, Brazil, Peru, Paraguay, and Argentina. In Bolivia it has been recorded from the vicinity of the Flora and Fauna Hotel near Buena Vista and the Potrerillo del Guenda Reserve in Santa Cruz Department.

Macronemus asperulus White, 1855

Lamiinae: Acanthoderini

This cryptic, small species (6–11 mm), when resting on a leaf, looks like a bird dropping. It has exceedingly long antennae, which in males can be five times the length of the body, as demonstrated by the pictured individual. Specimens are usually collected by beating branches and foliage along forest margins. This species is widely distributed from Mexico, through Central America, and into South America but is not very common in Bolivia. There, it is known from the vicinity of the Flora and Fauna Hotel near Buena Vista, the Vicoquin area above Achira, and Refugio Los Volcanes in Santa Cruz Department.

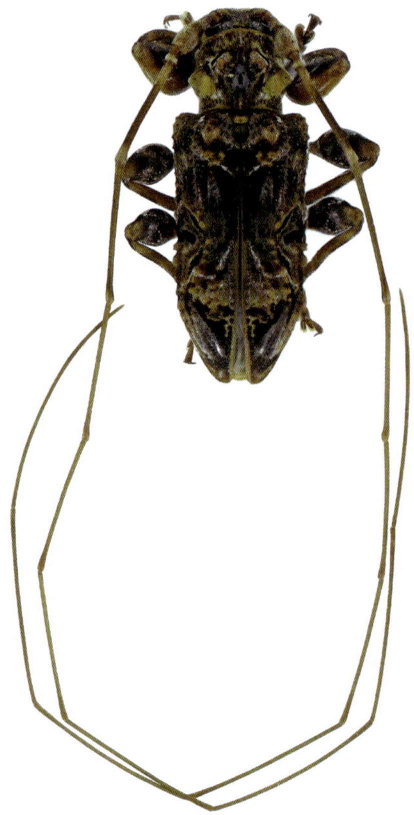

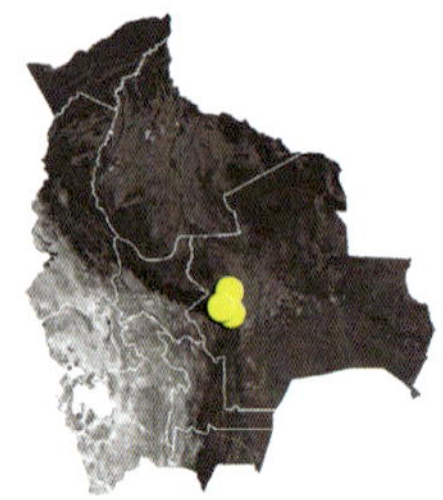

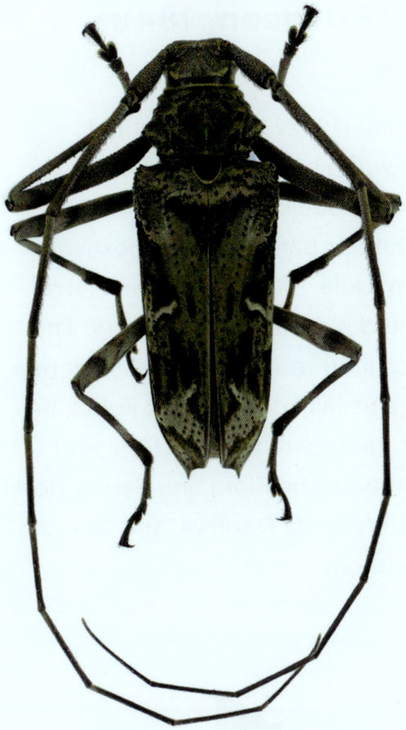

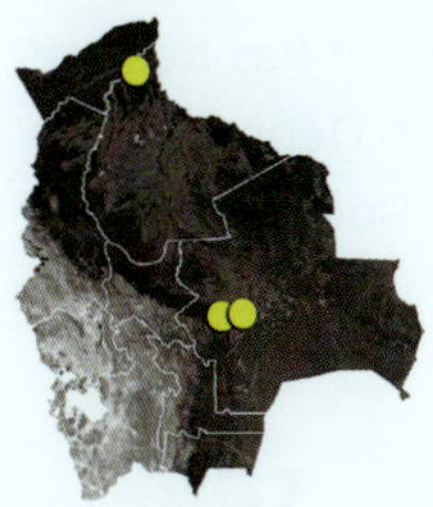

Macropophora accentifer (Olivier, 1795)

Lamiinae: Acanthoderini

At 19–40 mm long, this species and its congeners are the largest members of Acanthoderini in Bolivia. Males, like the one shown here, have particularly long front legs. The integument of the pronotum and elytra is mostly covered with pale grayish-green pubescence with scattered, diffuse, dark and pale maculae. Specimens are usually collected at lights or occasionally on fallen tree trunks at night. It occurs in Brazil, Paraguay, Argentina, Uruguay, and Bolivia, where specimens have been collected at the Flora and Fauna Hotel near Buena Vista, the Potrerillo del Guenda Reserve, and Bermejo in Santa Cruz Department and from Beni Department.

Macropophora lacordairei Lepesme, 1946

Lamiinae: Acanthoderini

This species is a little smaller than the previous one, *Macropophora accentifer*, with most specimens between 24 and 30 mm. Like the previous species, males have very long front legs. The specimen shown here is a female. The elytra have pale grayish-green pubescence throughout and a distinct pattern with a triangular, black macula at the base and a small, transverse, yellow macula at the middle of each elytron. It has been recorded from French Guiana, Brazil, Colombia, Ecuador, Peru, and Bolivia. There, it has been collected in Cochabamba Department near Guarayos and in Santa Cruz Department from Potrerillo del Guenda, the Flora and Fauna Hotel near Buena Vista, and Refugio Los Volcanes.

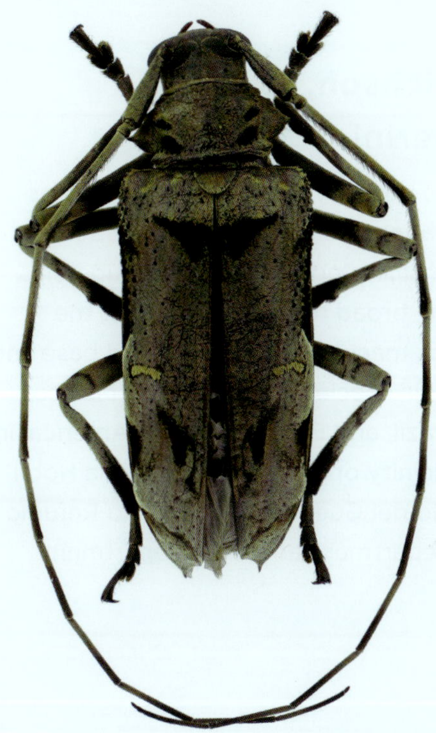

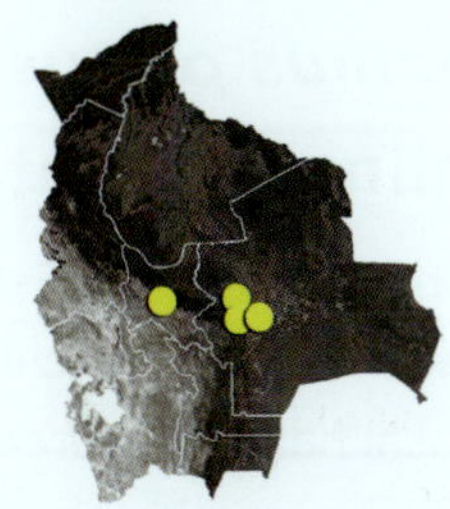

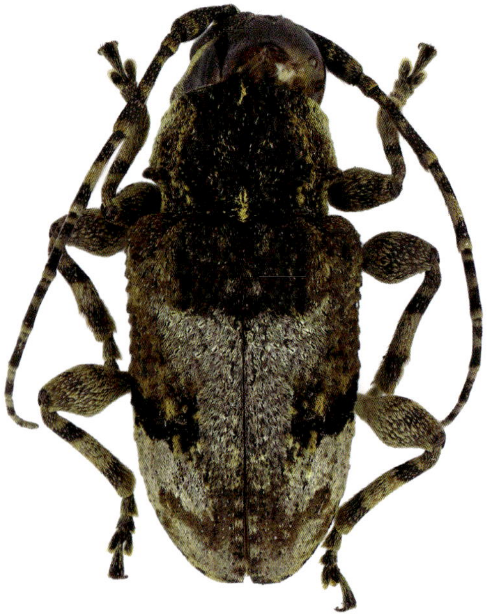

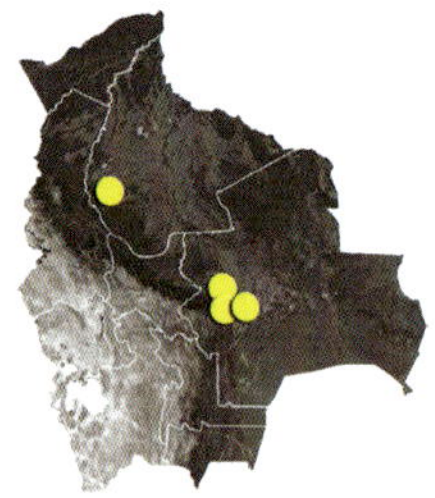

Myoxinus pictus (Erichson, 1847)
Lamiinae: Acanthoderini

This species is small (5–9 mm) but very elegantly patterned, as seen with microscopic illumination. Most of the elytra, especially the basal half, is covered in small, rounded, pubescent tubercles. The pubescence forms a white, broad hourglass shape at the middle of the elytra, with black maculae at the sides and base and over most of the pronotum. This species is known from Mexico, Central America, and Peru, Brazil, and Bolivia in South America. In Bolivia it is known from the vicinity of the Flora and Fauna Hotel near Buena Vista, the Potrerillo del Guenda Reserve, and Refugio Los Volcanes in Santa Cruz Department and Beni Department.

Myoxomorpha funesta (Erichson, 1848)
Lamiinae: Acanthoderini

This moderately large acanthoderine (14–22 mm) is distinct with its contrasting black and off-white pubescence. It is easily recognized by the somewhat skull-like pattern of white pubescence on the pronotum. White or pale yellow pubescence also forms maculae on the head, middle, and apex of the elytra. These white maculae are interrupted by round, black spots. The legs and antennae are black. It occurs throughout South America and Panama, where specimens are usually collected at lights. In Bolivia it has been recorded from the Riberalta area (Beni and Pando Departments) and the vicinity of the Flora and Fauna Hotel near Buena Vista, the Potrerillo del Guenda Reserve, and Refugio Los Volcanes in Santa Cruz Department.

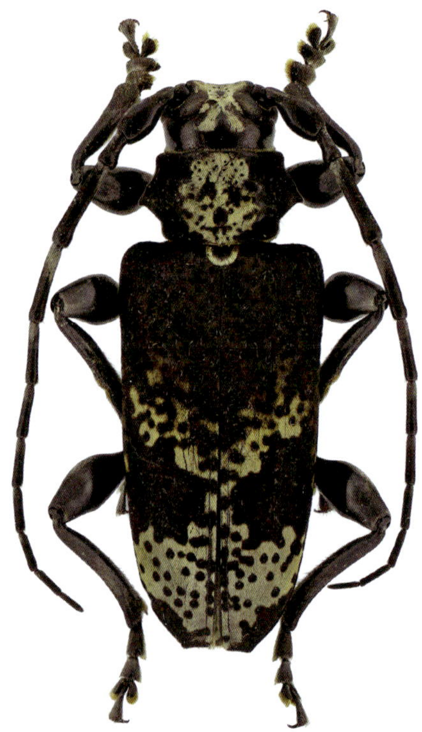

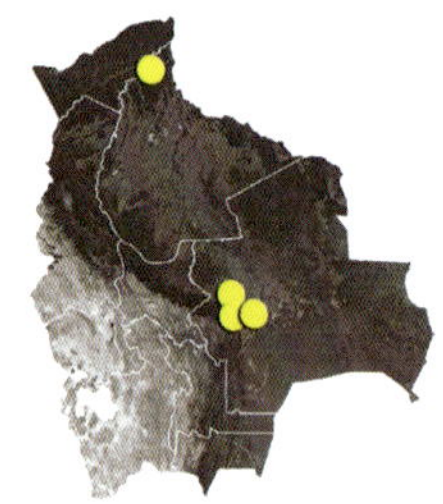

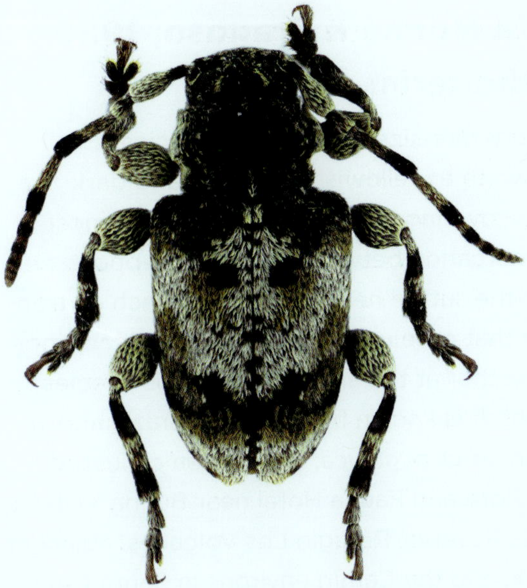

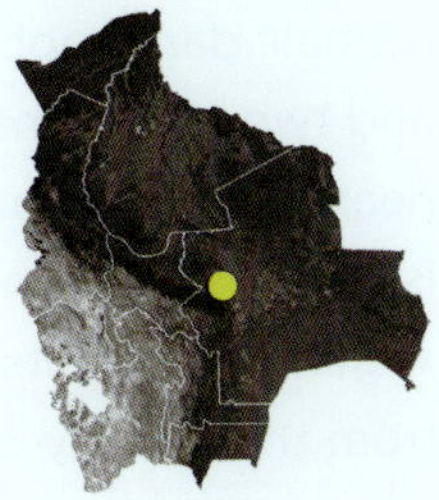

Octotapnia ceiaca Galileo & Martins, 1998

Lamiinae: Acanthoderini

The shape and pubescence of this unique little (4–7 mm) acanthoderine is somewhat similar to *Myoxinus pictus*, but its unusual antennal formula of only 8 antennomeres (typically 11 in Cerambycidae) immediately sets it apart. The basal four segments are rather typical of most Cerambycidae, but the remaining four are very reduced in length. This species is rarely found but can be attracted to lights where it is easily overlooked. It is known only from Brazil and Bolivia, where it has been collected from the vicinity of the Flora and Fauna Hotel near Buena Vista in Santa Cruz Department.

Oreodera boliviana Tippmann, 1960

Lamiinae: Acanthoderini

This moderate-sized (12–18 mm) beetle is one of the least common species in this large genus of almost 120 species. Much of its reddish-brown elytral integument is covered by a pale white, irregularly shaped fascia. The elytral surface is also moderately punctate. The pronotum of this species is laterally tuberculate, and the disk has a slightly elevated, shining tubercle basally with two larger tubercles on each side of the middle placed near the apical margin. This species is a Bolivian endemic and is known only from the Chaparé region of Cochabamba Department and the Potrerillo del Guenda Reserve in Santa Cruz Department.

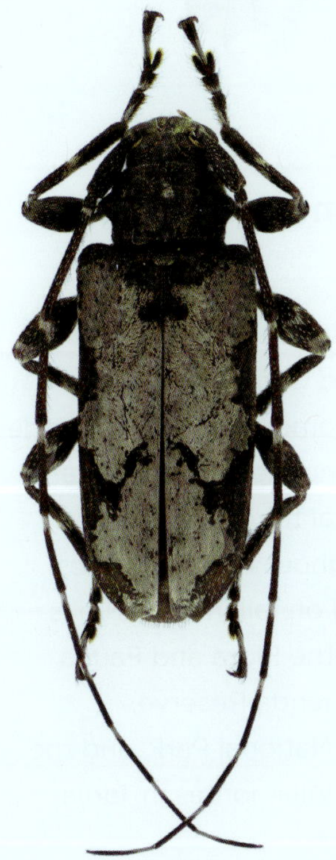

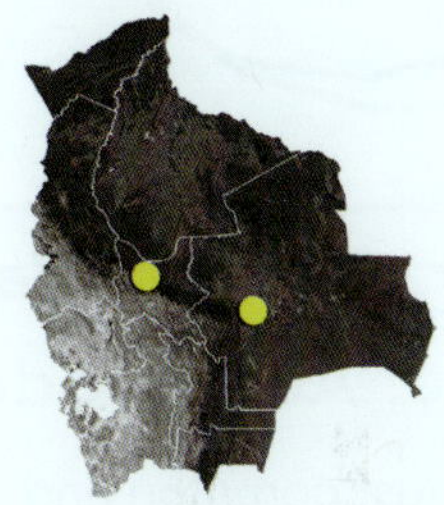

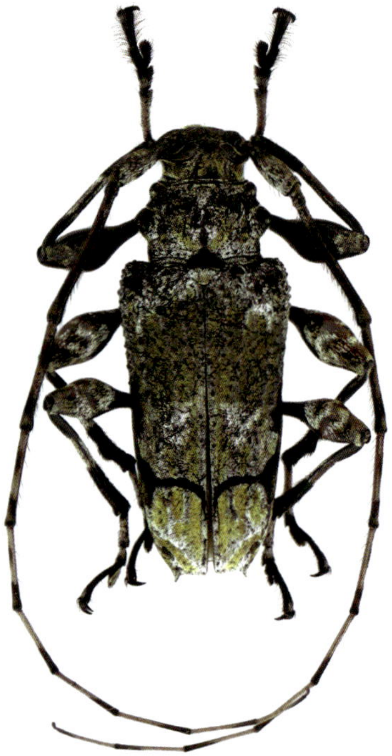

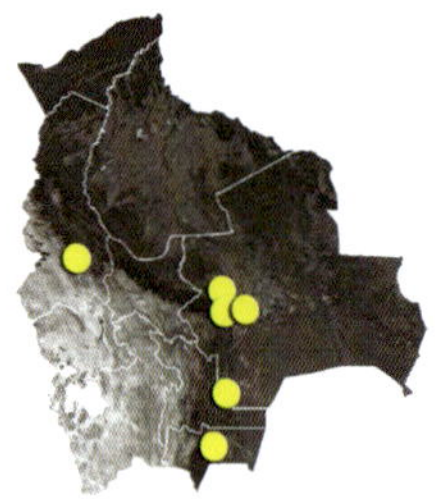

Oreodera glauca glauca (Linnaeus, 1758)

Lamiinae: Acanthoderini

This is the most common and widespread species in the genus and one of the most widespread New World Cerambycidae. Specimens are highly variable in size, and some can be rather large (12–26 mm). It is distinct, with a heavily granulate elytral base and narrow, transverse black macula at the apical third of the elytra. The species has been recorded from throughout the neotropics. In Bolivia it has been collected at lights and on fallen tree trunks and has been recorded from the vicinity of the Flora and Fauna Hotel near Buena Vista, the Potrerillo del Guenda Reserve, Refugio Los Volcanes, adjacent to Amboró National Park, and the Camiri environs in Santa Cruz Department, Villamontes in Tarija Department, and La Paz Department.

Oreodera occulta Monné & Fragoso, 1988

Lamiinae: Acanthoderini

This drab-appearing, moderate-sized (11–19 mm) species is dorsally covered in off-white to yellowish-tan, gray, and dark brown pubescence. This covering is infuscate and for the most part without a clear demarcation between colors. Exceptions are a small, dark spot along the suture near the base on each elytron and two lateral maculae that ease up onto the dorsal surface, one near the middle and the other at the apical fourth. This species is locally common at light. It is known from Bolivia, Brazil, and Argentina. Like its congener *O. g. glauca*, it has been collected from the vicinity of the Flora and Fauna Hotel near Buena Vista, the Potrerillo del Guenda Reserve, Refugio Los Volcanes, adjacent to Amboró National Park, and the Camiri environs in Santa Cruz Department and Villamontes in Tarija Department.

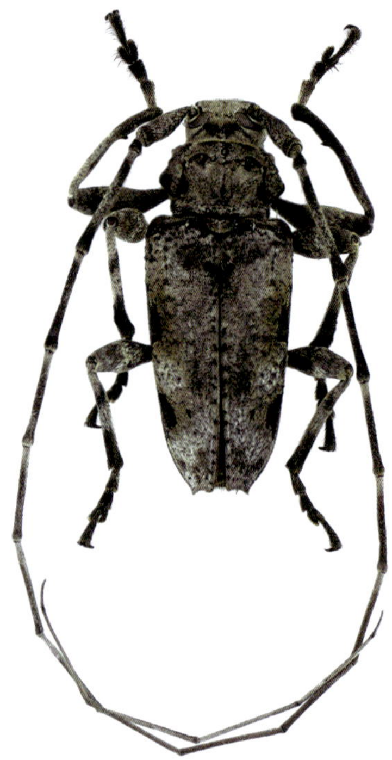

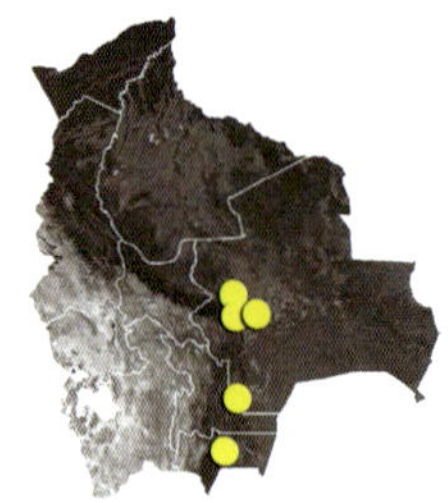

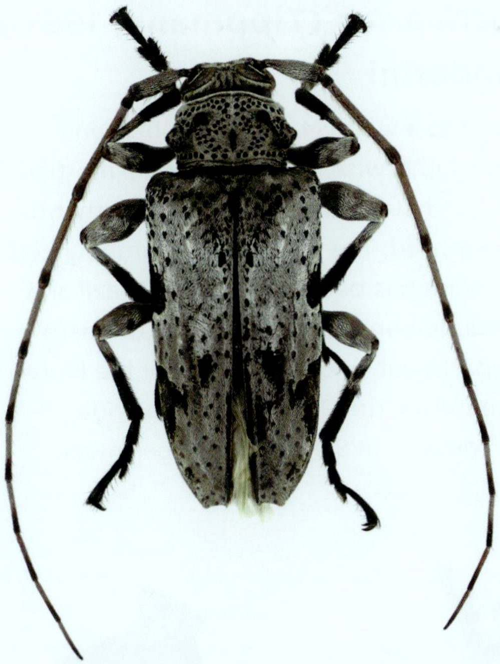

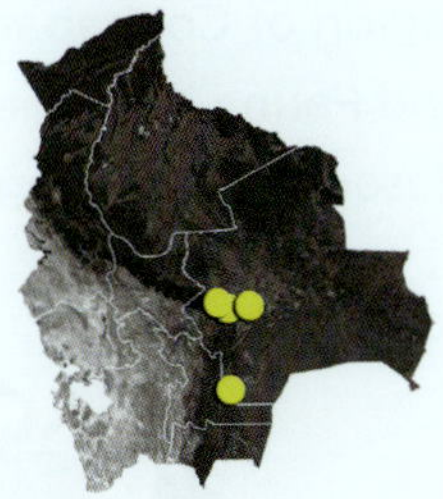

Oreodera quinquetuberculata (Drapiez, 1820)

Lamiinae: Acanthoderini

The bright white pubescence of the pronotum and elytra, interrupted by widely spaced, small, black spots, is characteristic of this *Oreodera* species. At 10–15 mm, it is moderate in size for the genus. It is recorded from Brazil, Paraguay, Argentina, and Bolivia. In Bolivia it has been collected at the Flora and Fauna Hotel near Buena Vista, the Potrerillo del Guenda Reserve, Refugio Los Volcanes, adjacent to Amboró National Park, the Amboró road above Achira, and the Camiri environs in Santa Cruz Department.

Psapharochrus atrosignatus (Melzer, 1932)

Lamiinae: Acanthoderini

Among the most striking members of the large genus *Psapharochrus*, this 9–13 mm long species is adorned with alternating, wavy, white and black fasciae on the elytra. The black fasciae are mostly glabrous, while the white fasciae are formed by very dense pubescence. The legs are mostly white pubescent but interrupted by black bands on the tibiae and femora. It is known only from Brazil and Bolivia. In Bolivia this species is known only from Potrerillo del Guenda and Refugio Los Volcanes, both in Santa Cruz Department.

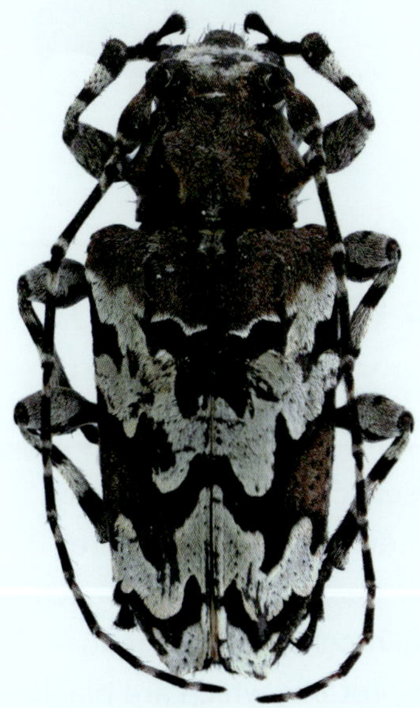

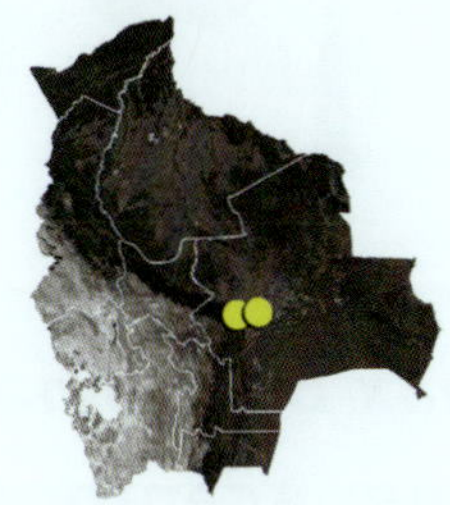

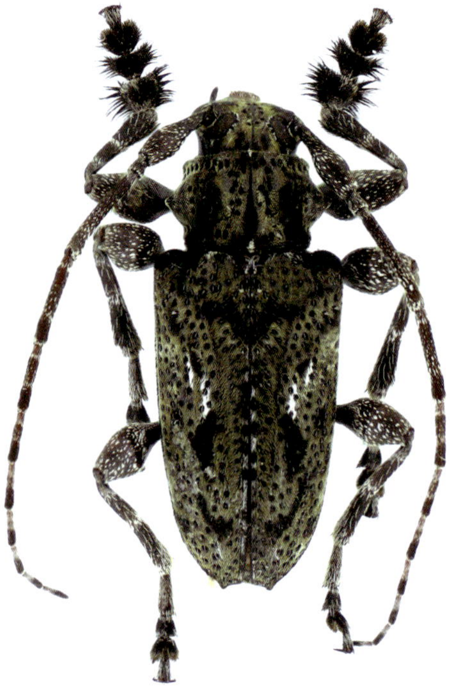

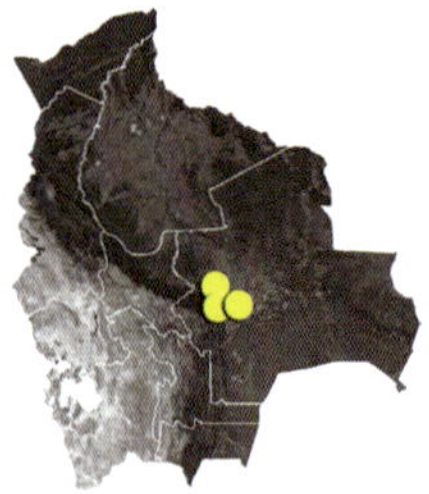

Plistonax bialbomaculatus (Zajciw, 1964)

Lamiinae: Acanthoderini

This dark brown, medium-sized (10–16 mm) species is heavily punctate on its dorsal surface and densely covered in primarily golden-brown pubescence. Its most noteworthy markings are the two small, white pubescent maculae near the center of each elytron that give it its name. It was, until recently, in the genus *Psapharochrus*. It has very pronounced protarsi, which are black and densely pubescent. The antennae and femora are speckled with very small patches of white setae. This species is known from Colombia, Peru, French Guiana, and Bolivia. In Bolivia it is known from the vicinity of the Flora and Fauna Hotel near Buena Vista, the Potrerillo del Guenda Reserve, and Refugio Los Volcanes in Santa Cruz Department.

Psapharochrus corticarius (Tippmann, 1960)

Lamiinae: Acanthoderini

This 10–14 mm long species has a white M-shaped pubescent macula on the pronotal disk. Otherwise, it is coated by a mottled golden-orange pubescence, although not too dense such that the black integument often is exposed. The elytra have a pronounced spine at the apices. This species has been collected in Brazil and Bolivia. Bolivia localities include Beni Department, the Chaparé region of Cochabamba Department, and the vicinity of the Flora and Fauna Hotel near Buena Vista, the Potrerillo del Guenda Reserve, and Refugio Los Volcanes in Santa Cruz Department.

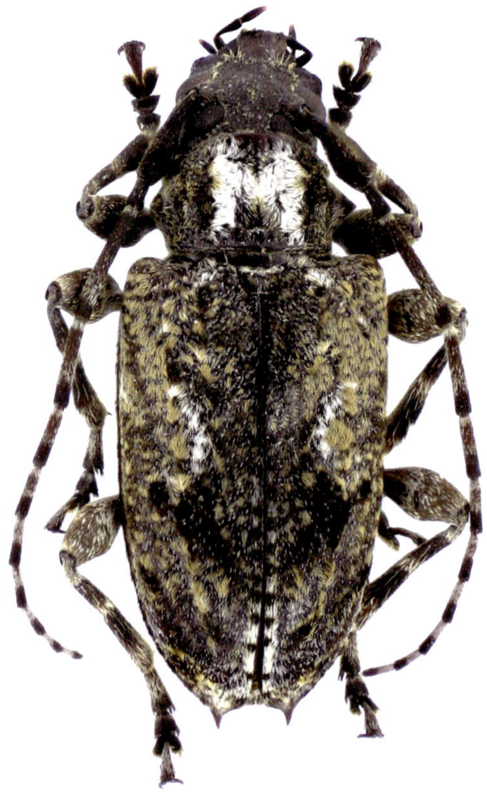

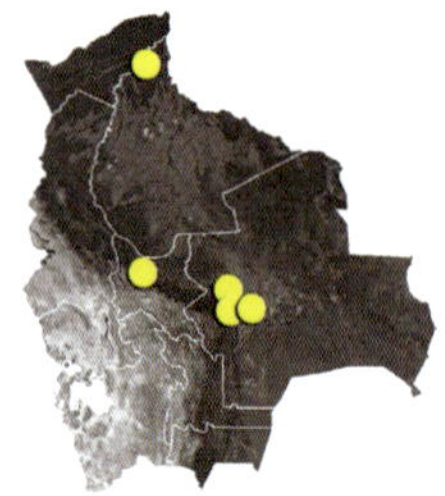

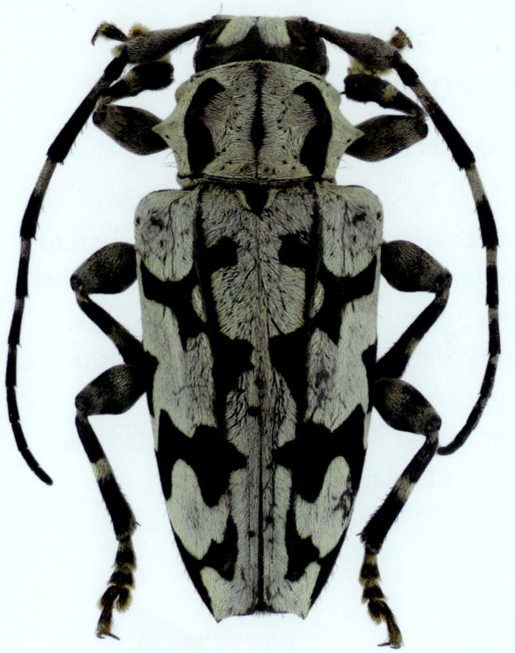

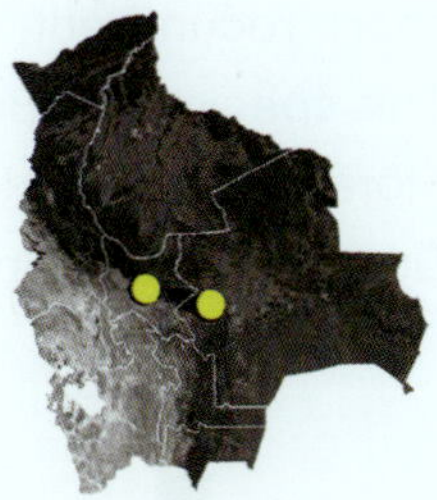

Psapharochrus homonymus (Blackwelder, 1946)

Lamiinae: Acanthoderini

This 11–14 mm long species is one of the most boldly patterned lamiines. It has dense white pubescence throughout, but this is broken up by large, reticulating, black maculae. The antennae are black except for a white scape and white annulate bases of most segments. It is distributed in Brazil, Peru, and Bolivia, where it has been collected at the Vicoquin area above Achira at 2,000 m in Santa Cruz Department. It is also recorded from Cochabamba Department.

Psapharochrus laetificus (Bates, 1880)

Lamiinae: Acanthoderini

This large (15–19 mm), black, shining species has distinct carinae running the length of the elytra on each side of the suture. The elytra and pronotum each have six to seven large ochraceous maculae, most spaced on the outer halves, each narrowly bordered by white setae. The pronotal disk has a pronounced middle ridge flanked by two less elevated ridges. This species is rarely encountered but occasional at lights. It is known from Ecuador, Peru, Brazil, and Bolivia, where it has been collected from the vicinity of the Flora and Fauna Hotel near Buena Vista in Santa Cruz Department.

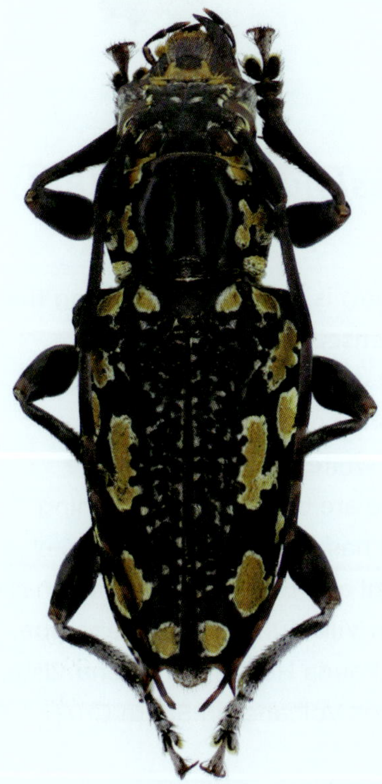

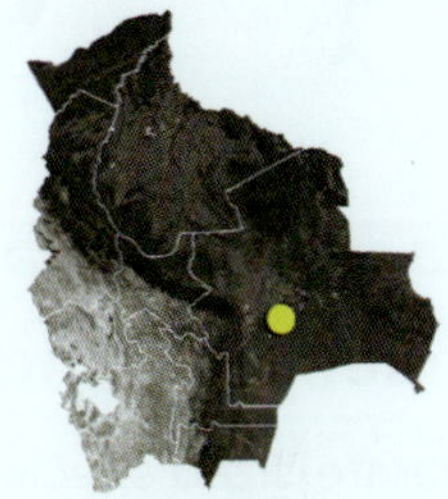

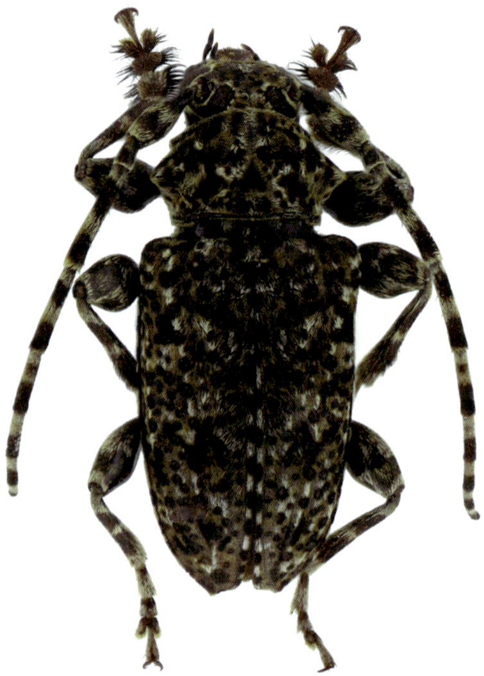

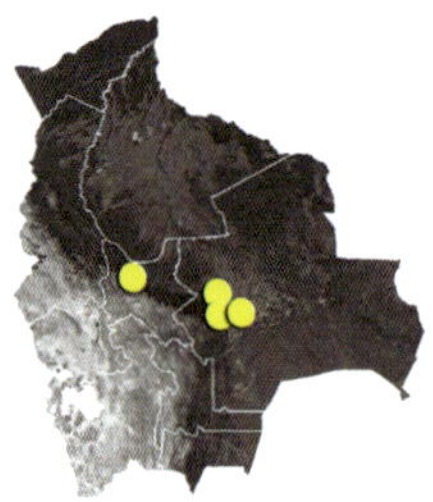

Psapharochrus maculatissimus (Bates, 1861)

Lamiinae: Acanthoderini

This moderate-sized species (9–12 mm) is very distinct among the large genus *Psapharochrus* by the dense, mottled pubescence of three colors: tan, dark brown, and white. The mottling is so complete that no large maculae of one color are present. The third and fourth antennomeres have white pubescence at the base, middle, and apex but otherwise are brown. The remaining antennomeres are white at the basal half and brown at the apex. This species is known only from Brazil and Bolivia. In Bolivia it has been collected, mostly at lights, from Villa Tunari in Cochabamba Department and from the Flora and Fauna Hotel near Buena Vista, Potrerillo del Guenda, and Refugio Los Volcanes in Santa Cruz Department.

Punctozotroctes bolivianus Martins & Galileo, 2007

Lamiinae: Acanthoderini

This small (9–11 mm), dark brown species has, as the genus name implies, the dorsum covered in nearly confluent punctures. Even the central ridge and lateral tubercles of the pronotum are punctate. In addition to the punctures the elytra have sparse, short, recumbent, gold and pale white pubescence on its surface. This species is a Bolivian endemic known from the vicinity of the Flora and Fauna Hotel near Buena Vista and the Potrerillo del Guenda Reserve in Santa Cruz Department in Bolivia.

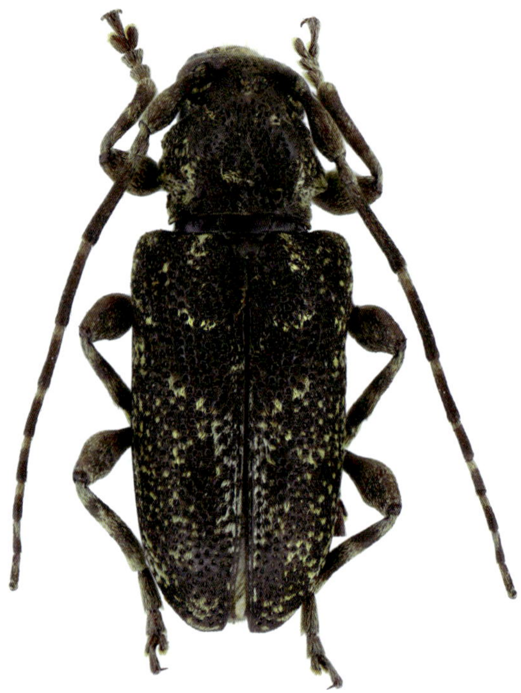

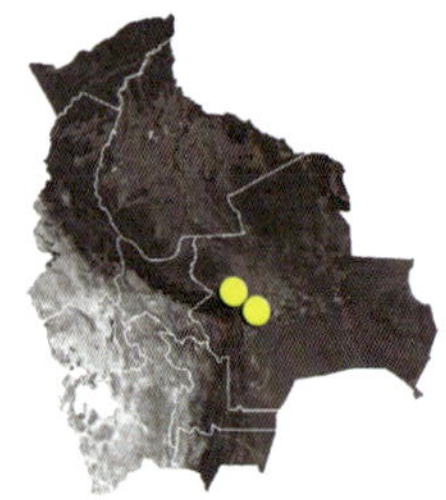

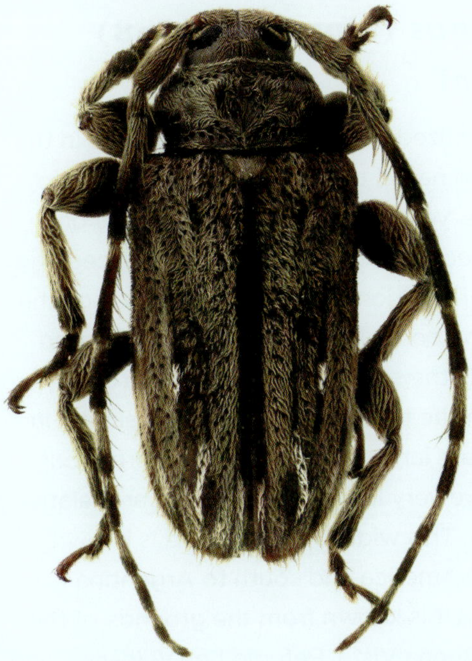

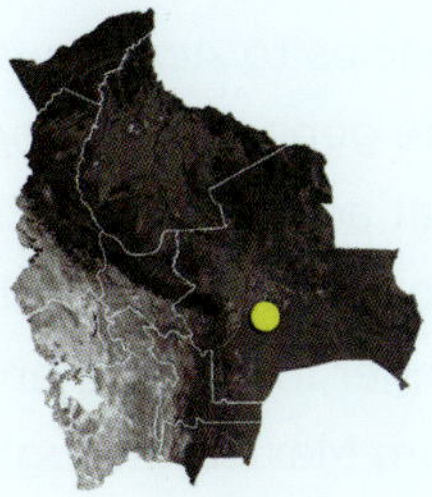

Pyrianoreina piranga Martins & Galileo, 2008

Lamiinae: Acanthoderini

This small (6–8 mm), reddish-brown acanthoderine is unusual in that the legs and antennae (at least the basal segments) are coated with long ochraceous setae. The pronotum and elytra have shorter, more appressed setae of similar color. The pronotum in the species is very short, more than twice as wide as long. The elytral suture is very dark and mostly lacking setae. This species is known only from Ecuador and Bolivia, where it has been collected in Santa Cruz Department at the Flora and Fauna Hotel and Buena Vista.

Steirastoma breve (Sulzer, 1776)

Lamiinae: Acanthoderini

This is a large (15–23 mm), rugged-looking, common species with a shining piceous integument and a distinct white pubescent pattern interrupted by small, dark, shining tubercles at the elytral base and circular, glabrous black spots on the dorsum. Like other species in the genus, each elytron has a longitudinal, glabrous carina. Both sexes have wide front tarsal segments fringed in long, silky, black hairs. It is the most widespread species in the genus and one of the most widespread of all neotropical lamiines, having been recorded from throughout Central and South America and the West Indies. It is widespread in Bolivia, having been collected in the Riberalta area (Beni and Pando Departments), Cristal Mayu (Cochabamba Department), and Potrerillo del Guenda, Refugio Los Volcanes, the Flora and Fauna Hotel near Buena Vista, and Camiri (Santa Cruz Department). It has also been recorded from La Paz Department.

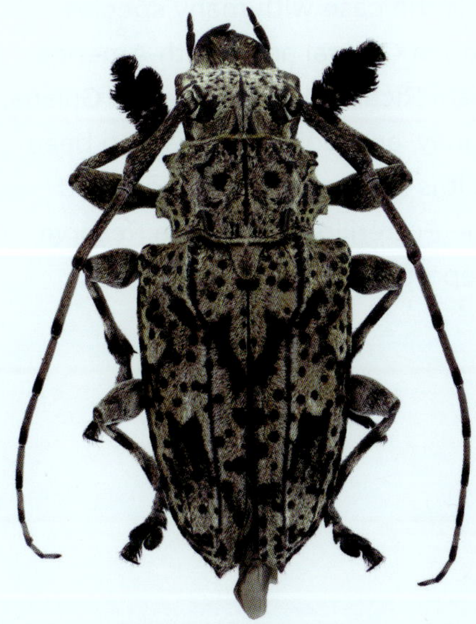

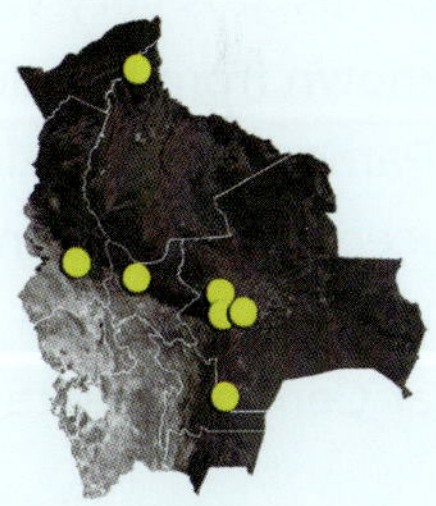

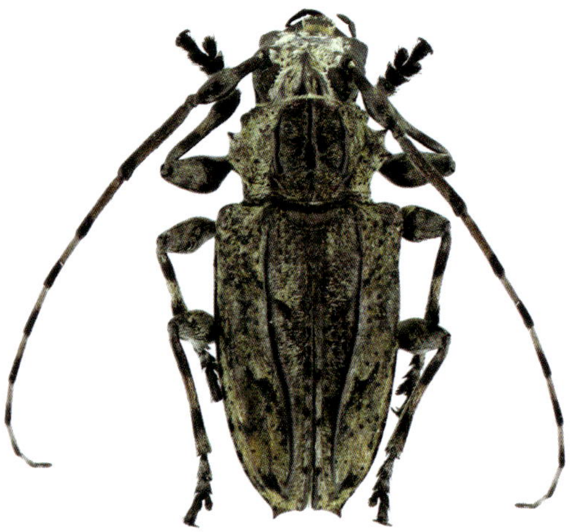

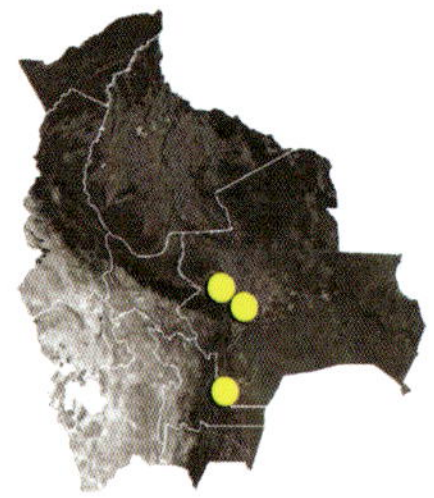

Steirastoma stellio Pascoe, 1866

Lamiinae: Acanthoderini

This is one of the smallest (13–16 mm), but still robust, *Steirastoma* species. It has three glabrous carinae on the pronotum and one on each elytron. The pronotum has two acute tubercles on each side. The head is quite broad, as is the case with many species of the genus. It is very widespread in Central and South America, being known from Honduras, Costa Rica, Colombia, French Guiana, Brazil, Paraguay, Argentina, Uruguay, and Bolivia. It has also been introduced into Africa. In Bolivia it is known from the Flora and Fauna Hotel near Buena Vista, the Potrerillo del Guenda Reserve, and near Camiri in Santa Cruz Department.

Acrocinus longimanus (Linnaeus, 1758)

Lamiinae: Acrocinini

This is one of the most recognized species of Cerambycidae in the world. It received its common name, the harlequin beetle, due to the unique pattern of orange, yellow, and black pubescence on the elytra and pronotum. Specimens can be quite large, exceeding 70 mm. The males have disproportionately elongate front legs that can be twice the length of the body. There are no closely related species to *Acrocinus,* and it is the sole member of the Acrocinini. The genus *Macropophora* (in the tribe Acanthoderini) is similar in that adults of some species are large, have striking patterns (but more subdued), and also have very long front legs that are relatively longer in males than females. This widespread species is known from Mexico through Central America and south to Argentina and southern Brazil. In Bolivia it is known from the grounds of the Flora and Fauna Hotel near Buena Vista, Refugio Los Volcanes, and Potrerillo del Guenda in Santa Cruz Department. It has also been collected in Rurrenabaque and Riberalta in Beni Department.

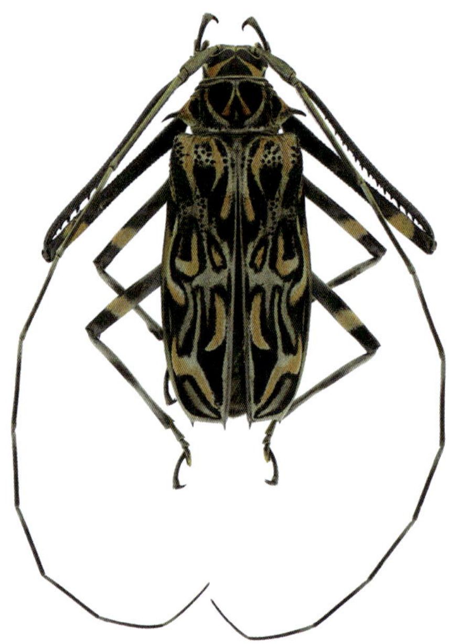

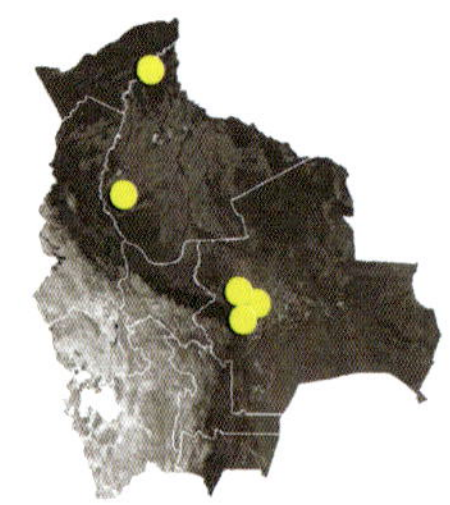

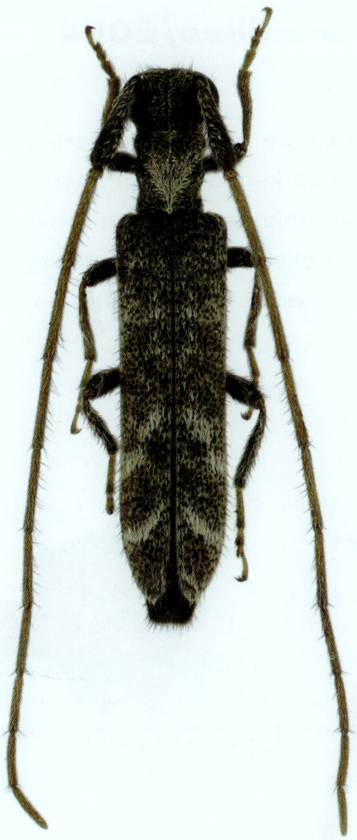

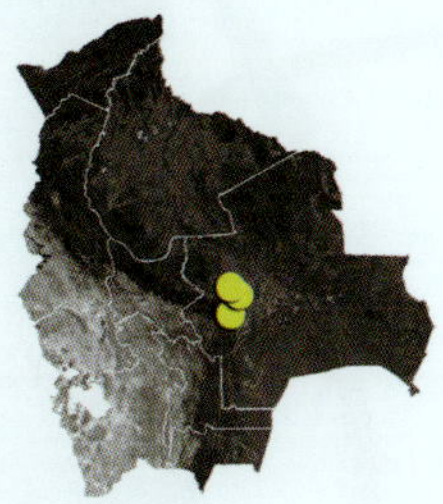

Aerenica canescens (Klug, 1825)

Lamiinae: Aerenicini

Like many species of the tribe Aerenicini, this moderate-sized species (10–14 mm) is very elongate and narrow bodied. The antennae, beyond the dark scape, are pale yellow. The elytra have two distinctive, obliquely angled white maculae postmedially. Occasional at lights, this species is known from Brazil, Paraguay, Argentina, and Bolivia. In Bolivia it is known from the Flora and Fauna Hotel near Buena Vista, Potrerillo del Guenda, and Refugio Los Volcanes in Santa Cruz Department.

Aerenomera boliviensis Gilmour, 1962

Lamiinae: Aerenicini

This somewhat robust member of the tribe Aerenicini (10–14 mm long) is distinguished by its pattern of white and yellow-green pubescence on the elytra. The pronotum has three vague longitudinal bands of yellow-green pubescence but otherwise is mostly covered in white setae. This species is rare at lights and is known from Brazil, Paraguay, Argentina, and Bolivia. In Bolivia it is known from near Roboré (the type locality), the Flora and Fauna Hotel near Buena Vista, Potrerillo del Guenda, Refugio Los Volcanes, and the Camiri environs, all in Santa Cruz Department.

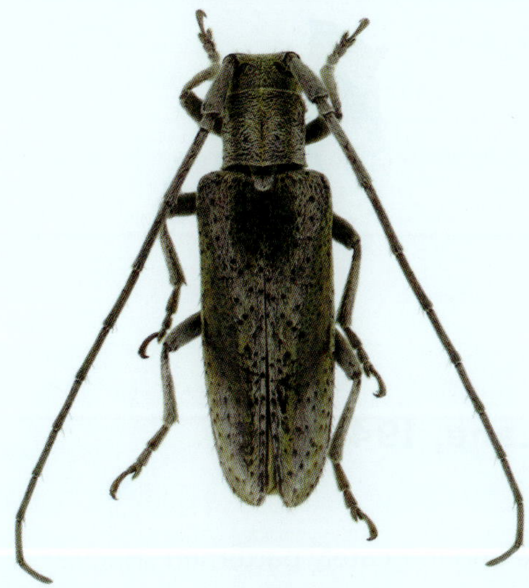

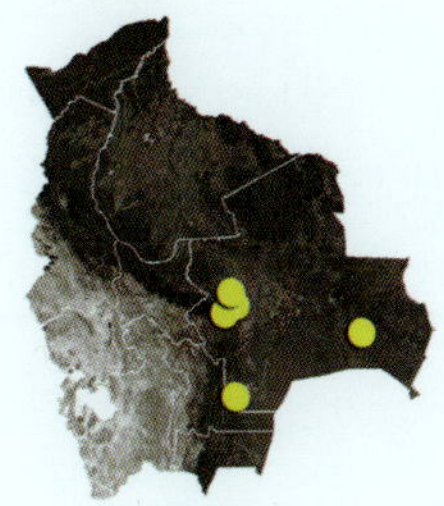

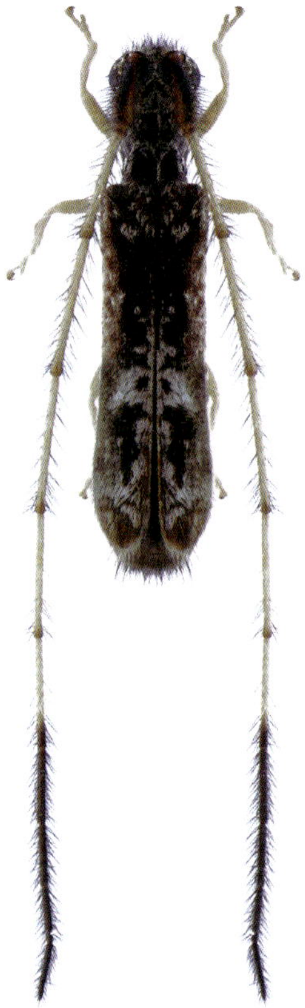

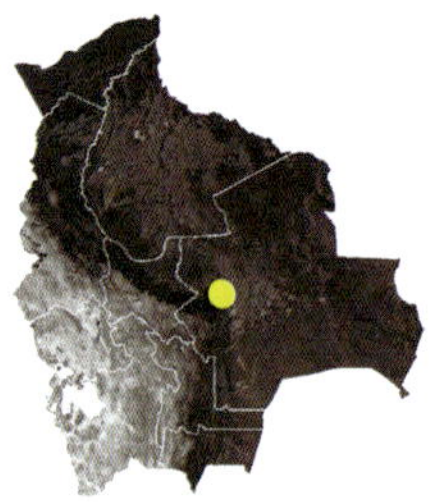

Antodice abstrusa Lane, 1940

Lamiinae: Aerenicini

This is an exceedingly delicate and intricately patterned species (7–8 mm long and less than 2 mm broad) with very long antennae that extend, in males, nearly twice the length of the body. The basal seven antennomeres are pale, and the apical four are nearly black. The eyes are very large and occupy most of the head. This species is rarely encountered, almost exclusively at lights, and is known from Brazil, French Guiana, and Bolivia. In Bolivia it is known from the grounds of the Flora and Fauna Hotel near Buena Vista in Santa Cruz Department.

Antodice opena Martins & Galileo, 2004

Lamiinae: Aerenicini

This species (6–9 mm) is recognized by the cryptic, mottled pubescence of the elytra. The apically darkened antennomeres, except for the last four, which are uniformly dark reddish brown to black, are distinctive. It is endemic to Bolivia. The type locality is Buena Vista, and it has also been collected at lights in Refugio Los Volcanes in Santa Cruz Department.

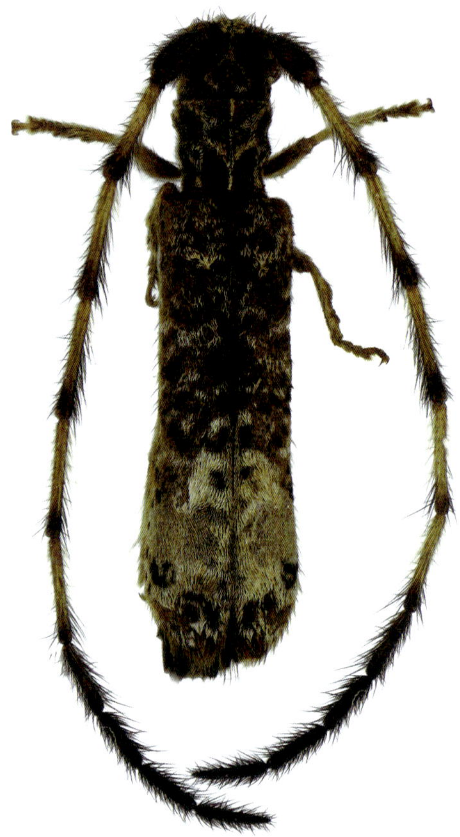

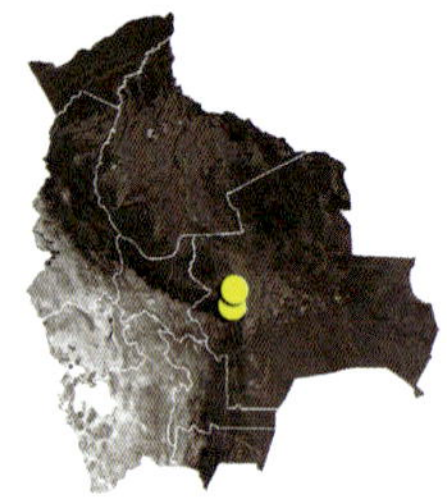

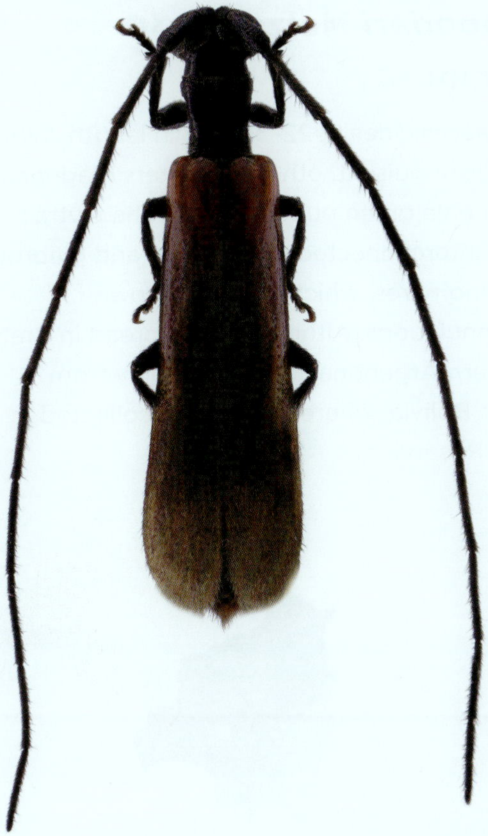

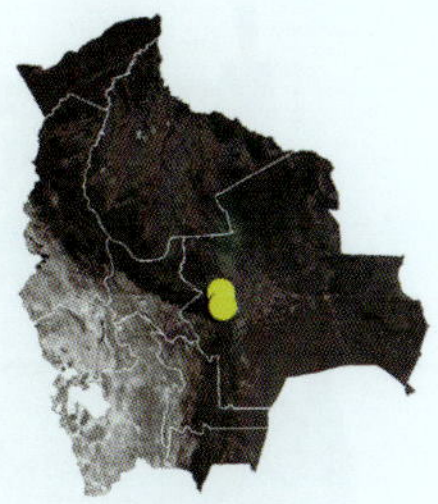

Antodice tricolor Martins & Galileo, 1985

Lamiinae: Aerenicini

This is an unusual species of *Antodice* (7–9 mm) in that it is mostly glabrous on the head, pronotum, and basal half of the elytra. The head and pronotum are black, and the elytra are red to orange. The antennae are at least one and one-half times the length of the body. This species is also a Bolivian endemic and has been collected at lights during the rainy season from the grounds of the Flora and Fauna Hotel near Buena Vista and Refugio Los Volcanes in Santa Cruz Department.

Apoaerenica martinsi Monné, 1979

Lamiinae: Aerenicini

This is one of the most strikingly colored and distinctly patterned aerenicines. Specimens average about 12 mm long. The base and lateral margins of the elytra are vividly red to orange pubescent, while the remainder is a dark, ashy gray. The anterior two-thirds of the pronotum and most of the head are colored in vibrant red to orange pubescence as well. Another endemic to Bolivia, it has been collected at lights from the grounds of the Flora and Fauna Hotel near Buena Vista, Potrerillo del Guenda, Refugio Los Volcanes, and Cuevas in Santa Cruz Department.

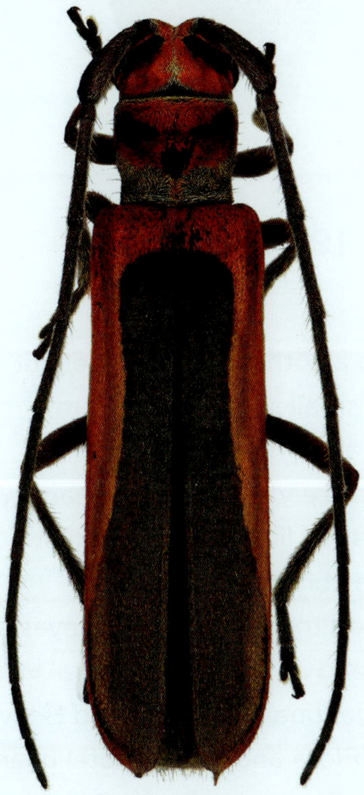

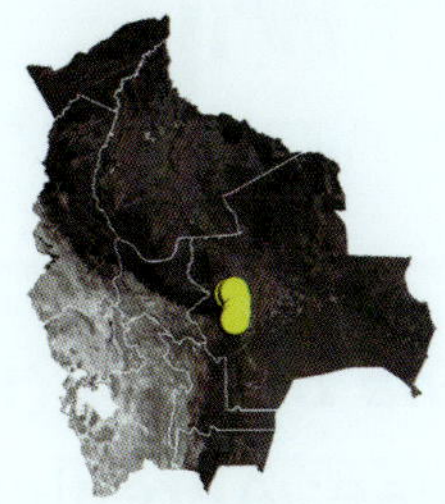

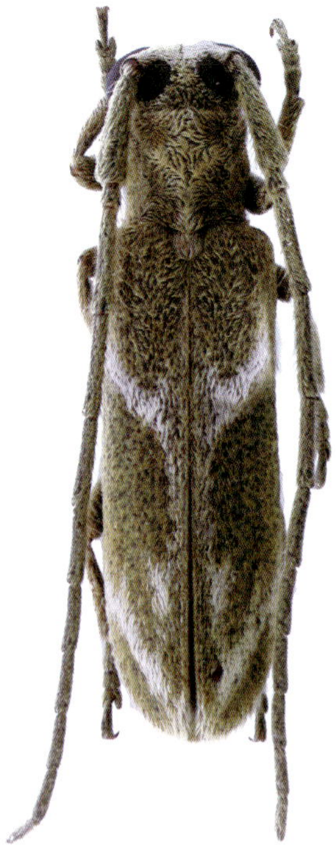

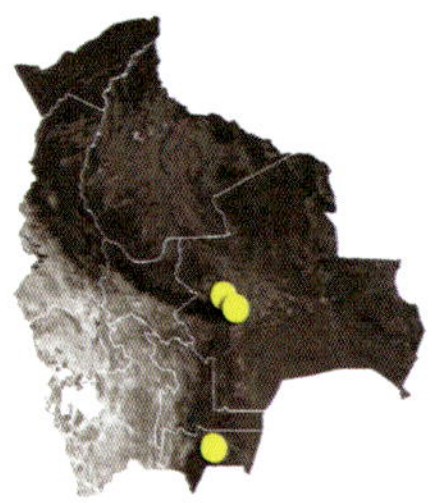

Eponina breyeri (Prosen, 1954)

Lamiinae: Aerenicini

This aerenicine may not be as striking as others in the tribe, but with its subtle colors, it is an elegant, small species (8–9 mm). Most of the body is covered in moderately dense, light tan pubescence, which is interrupted on the elytra by several narrow fasciae of white setae. The antennae are a little lighter in color than most of the body, although not as bright as the elytral maculae. The elytral apices are rounded to the suture. The eyes are quite large and occupy most of the head as seen from lateral view. This species is known from Brazil, Argentina, Paraguay, and Bolivia. In Bolivia it has been collected at the Flora and Fauna Hotel near Buena Vista and Potrerillo del Guenda Reserve in Santa Cruz Department and near Villamontes in Tarija Department.

Hoplistonychus bondari Melzer, 1930

Lamiinae: Aerenicini

This is one of the largest aerenicines at 22–30 mm in length. While its substantial size helps distinguish it, other characters diagnostic to the species include the pale green pubescence of the elytra interrupted by a series of interconnected white spots and glabrous black spots and the antennomeres, which are gray-brown pubescent and without annulations. Although widespread in Brazil, French Guiana, and northern Argentina, it is known only from Santa Cruz Department in Bolivia, where it has been collected in the Potrerillo del Guenda Reserve.

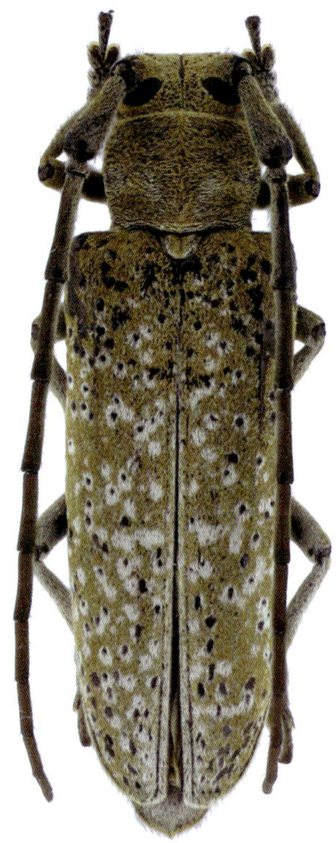

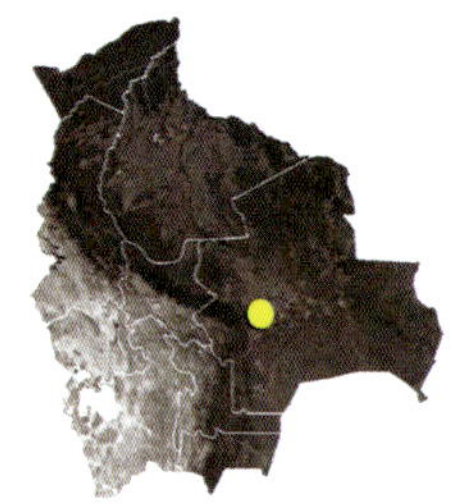

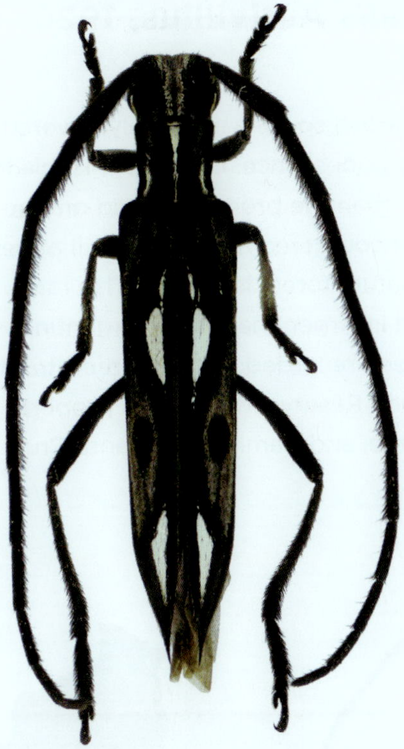

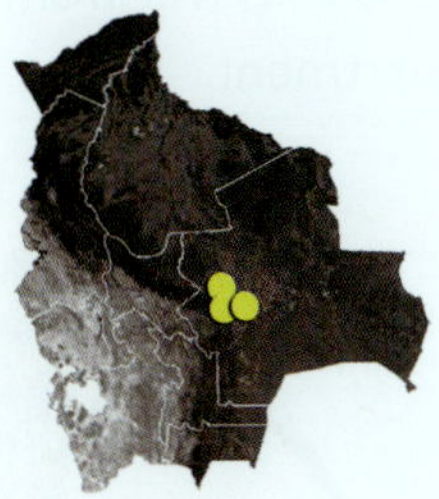

Hydraschema leptostyla Lane, 1938

Lamiinae: Aerenicini

This is a medium-sized (15–24 mm), narrow, velvety black and gray species with three linear white maculae on the pronotum and two pairs on the elytra. It has black, heavily fringed antennae, and the elytral apices are extended into long, sharp, singular spines. This is among the most strikingly beautiful Cerambycidae that come to lights in South America. It is recorded from Argentina, southeastern Brazil, and Bolivia, where it is active during the rainy season in Santa Cruz Department from the vicinity of the Flora and Fauna Hotel near Buena Vista, the Potrerillo del Guenda Reserve, and Refugio Los Volcanes, adjacent to Amboró National Park.

Melzerella huedepohli Monné, 1979

Lamiinae: Aerenicini

This is another of the beautifully colored and strikingly patterned aerenicine longhorned beetles (12–14 mm). There are large triangular patches of dense yellow pubescence that converge at the middle of the elytra. These are narrowly bordered by a fiery red margin. Elsewhere, the integument is dark brown, particularly at the base and apex of the elytra. The pronotum is mostly yellow pubescent, but this is interrupted by a narrow, inverted V-shaped macula across the disk. Another of the Bolivian endemics, it is rarely encountered at lights in lowland tropical forests. It is known from the vicinity of the Flora and Fauna Hotel near Buena Vista and the Potrerillo del Guenda Reserve in Santa Cruz Department.

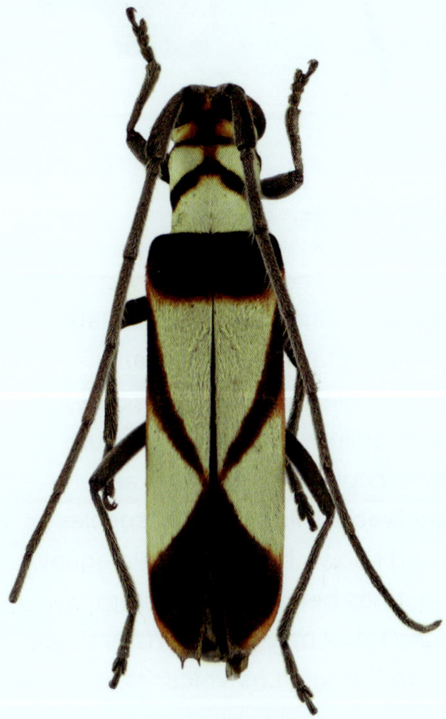

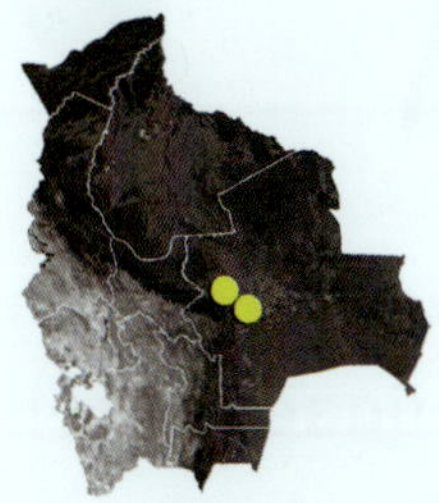

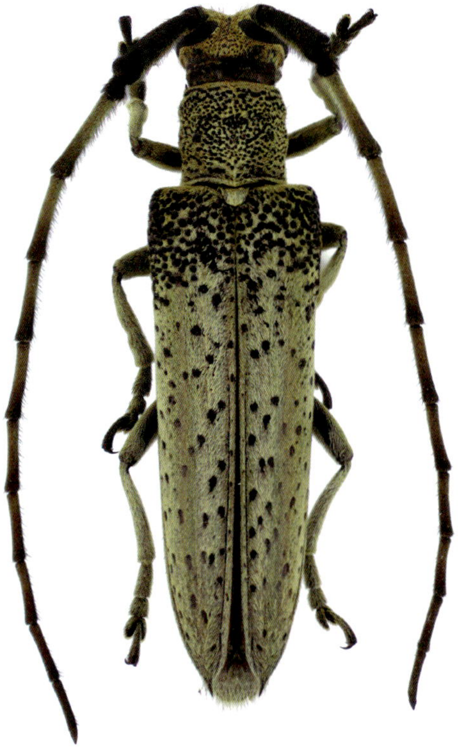

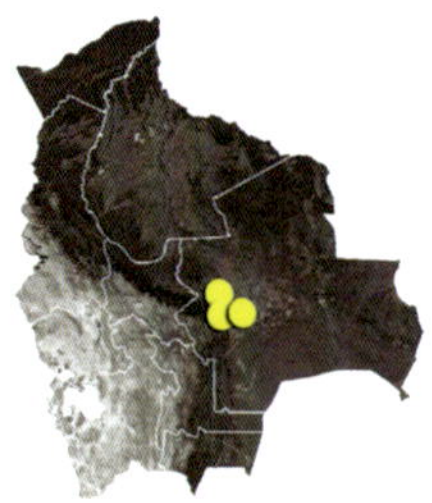

Phaula thomsoni Lacordaire, 1872

Lamiinae: Aerenicini

This large (20–28 mm) aerenicine is distinct with its granulate and punctate elytral base, heavily punctate pronotum, and strongly spined elytral apex. Most of the integument (except the punctures) is covered in a pale yellow to greenish mixture of pubescent vestiture. The only exception is the scape and second antennomeres, which are black. This nocturnal species is occasionally common at lights and known from Brazil, Paraguay, Argentina, and Bolivia. In Bolivia it has been collected from the vicinity of the Flora and Fauna Hotel near Buena Vista, the Potrerillo del Guenda Reserve, and Refugio Los Volcanes in Santa Cruz Department.

Pseudomecas femoralis Aurivillius, 1920

Lamiinae: Aerenicini

This small (11–14 mm), parallel-sided species is densely covered in short, recumbent, whitish-gray pubescence. It is recognizable by its unusually wide head (wider than the pronotum) and orange femora. This species is rarely encountered although it will appear late at night at light traps in mature forest locations. This rarely collected species is distributed in Argentina, Brazil, Argentina, Paraguay, and Bolivia. It has been recorded at only three sites in Bolivia: the Potrerillo del Guenda Reserve, the Huaico property across the river from the Reserve, and Camiri, all in Santa Cruz Department.

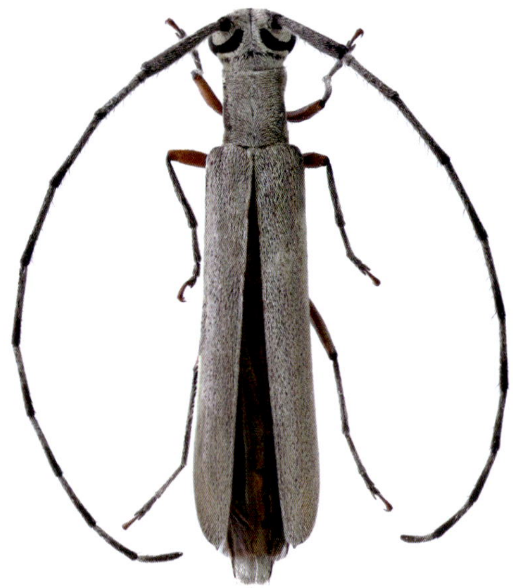

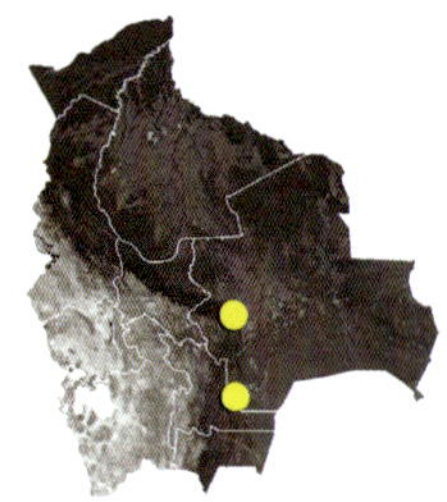

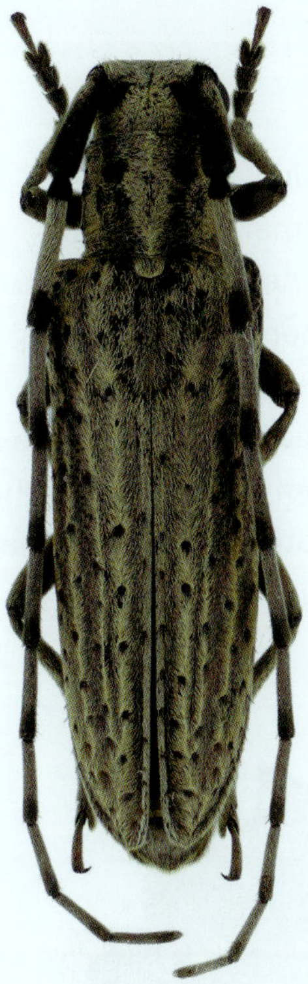

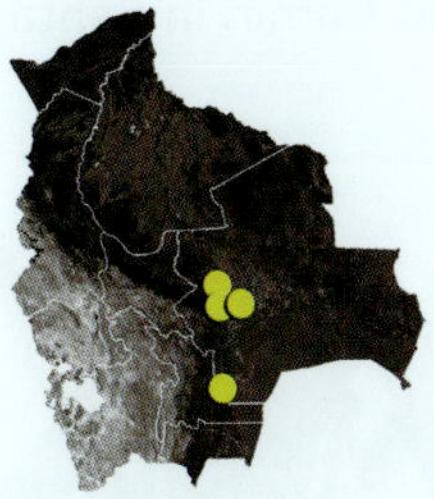

Pseudophaula strigulata Lane, 1973

Lamiinae: Aerenicini

This moderate-sized species (14–19 mm) is densely covered in yellow, pale ochraceous, and gray pubescence. The pubescence forms three to four longitudinal lines down each elytron. These lines are partially interrupted by sparse, large, glabrous punctures. The antennal segments, starting with the third, are pale pubescent over the basal two-thirds but black at the apices. Occasionally encountered at lights, this species is recorded only from Argentina and Bolivia. In Bolivia it is known from the vicinity of the Flora and Fauna Hotel near Buena Vista, the Potrerillo del Guenda Reserve, Bermejo, Refugio Los Volcanes, and Camiri in Santa Cruz Department.

Recchia albicans (Guérin-Méneville, 1831)

Lamiinae: Aerenicini

Even though most *Recchia* have vivid markings, many are difficult to identify to species. This moderate-sized (10–13 mm) species has contrasting zigzag fasciae of white and dark brown pubescence at the apical half of the elytra. The base of the elytra has more subtle patterns of white and pale green pubescence. The pronotum has a distinct, small, black macula at the center. This species is known from Argentina, Paraguay, Brazil, and Bolivia. In Bolivia it is found occasionally by beating along forest edges or by mercury vapor or blacklights from the vicinity of the Flora and Fauna Hotel near Buena Vista, the Potrerillo del Guenda Reserve, Bermejo, Refugio Los Volcanes, and Camiri in Santa Cruz Department.

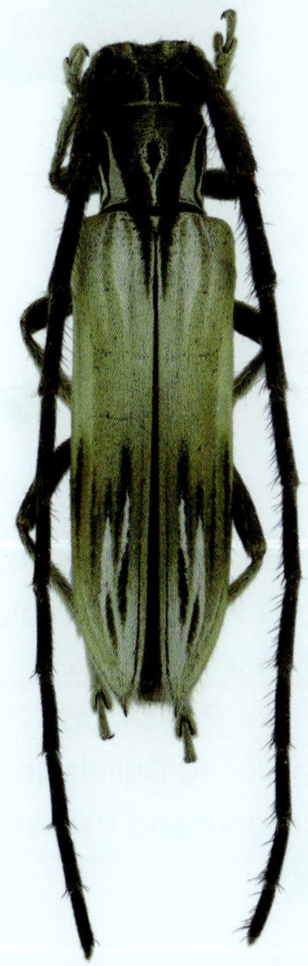

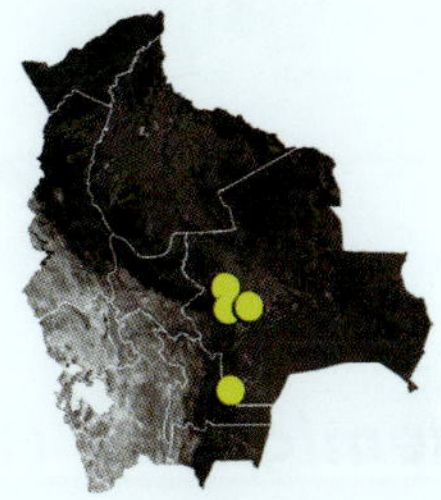

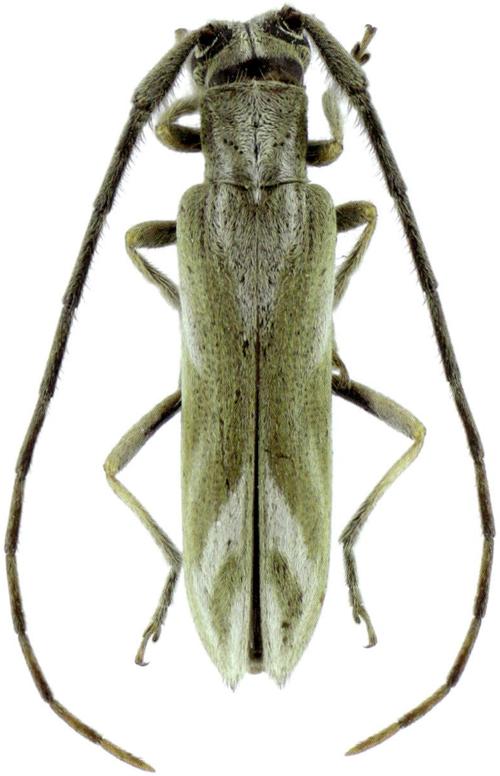

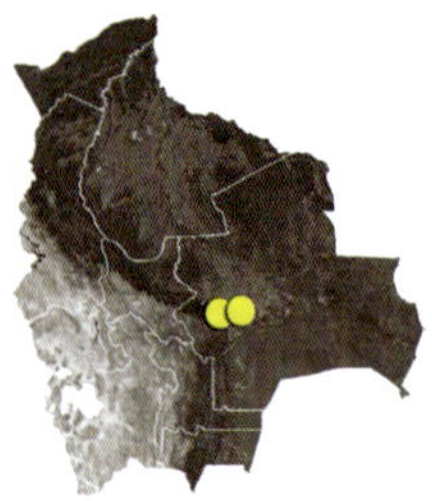

Recchia gemignanii (Lane, 1939)

Lamiinae: Aerenicini

This species is similar in size to the previous one, *Recchia albicans*, at 10–13 mm in length. It lacks the bold zigzag markings of the elytra, instead having a more subtle combination of white and pale green pubescence on the pronotum and elytral base and apex. The pronotum lacks any distinct dark maculae. Like *R. albicans*, this one is known from Argentina, Paraguay, Brazil, and Bolivia. In Bolivia it is known from Potrerillo del Guenda Reserve and Refugio Los Volcanes in Santa Cruz Department.

Recchia hirticornis (Klug, 1825)

Lamiinae: Aerenicini

This 8–12 mm long species has markings very similar to *Recchia albicans*, and their identification is very problematic. This is yet another group of longhorned beetles that is in need of revision to better clarify the species limits. Like its congeners, it has similar bold and contrasting zigzag pubescent patterning on the elytral apex and pronotum. It is known from Bolivia and Brazil. In Bolivia it has been collected in Beni Department and from the vicinity of the Flora and Fauna Hotel near Buena Vista, the Potrerillo del Guenda Reserve, and Camiri in Santa Cruz Department.

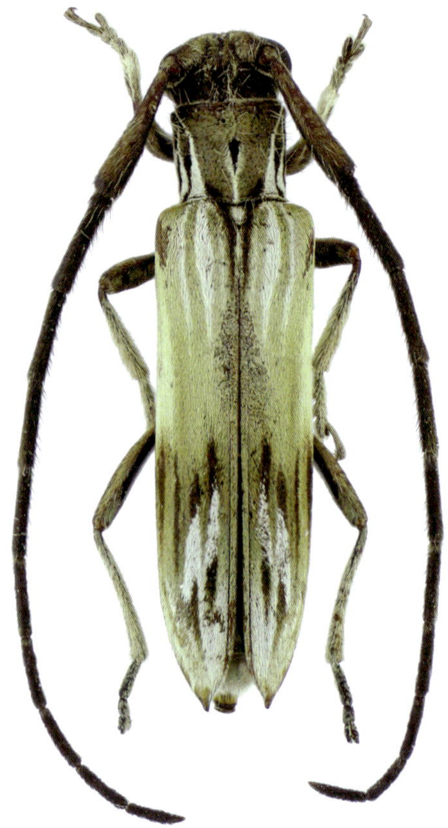

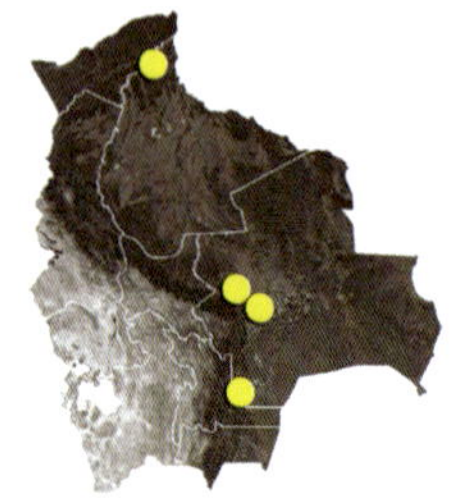

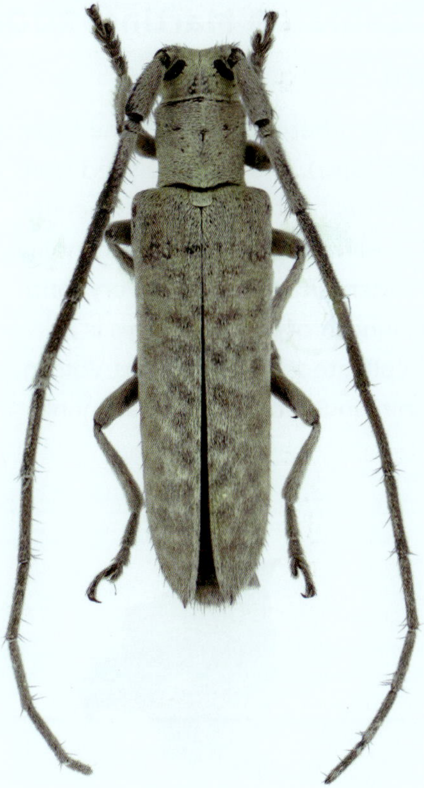

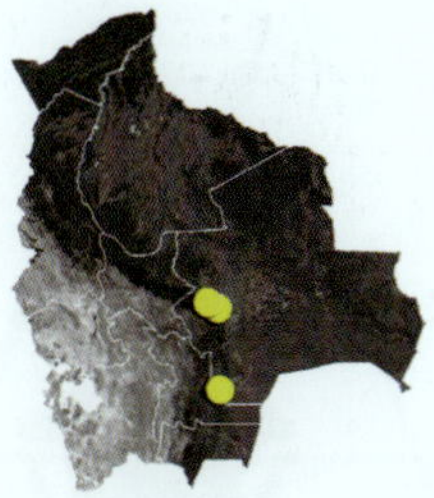

Rumacon canescens (Bruch, 1926)

Lamiinae: Aerenicini

With its subtly mottled white pubescence over the entire body and appendages, this 11–14 mm long species is easily recognized and always a pleasure to encounter. The pubescence of the elytra is slightly more dense in patches, giving the appearance of vaguely defined, darker maculae in areas where it is less dense. This rarely collected nocturnal species is known from Argentina, Paraguay, and Bolivia. In Bolivia it is recorded from Refugio Los Volcanes, Camiri, the road to Bella Vista near Cuevas, and the Achira resort, near Samaipata, all in Santa Cruz Department.

Helvina strandi (Breuning, 1942)

Lamiinae: Agapanthiini

At 16–19 mm long, this is one of the largest and broadest of all agapanthiines. It is densely covered with short, recumbent, golden-brown pubescence with very small, scattered flecks of white setae over the elytra. The antennae, from the third antennomere, are densely fringed with long, erect setae. Occasionally collected by beating dead branches or by running light traps, this species is known from Ecuador, Peru, and Bolivia, where it has been collected in Cochabamba Department and Santa Cruz Department at Refugio Los Volcanes.

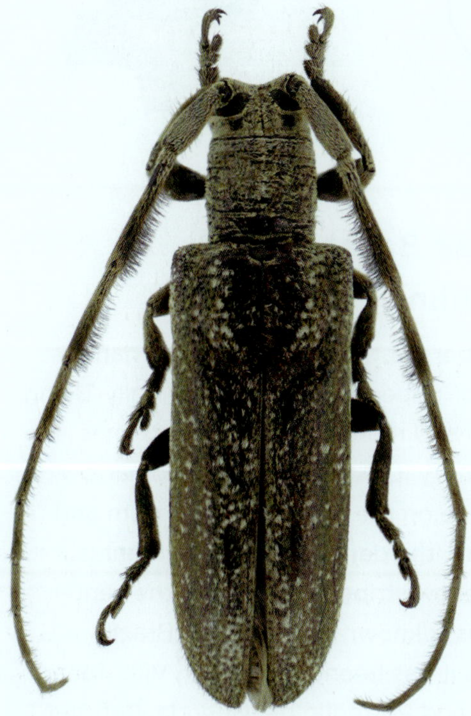

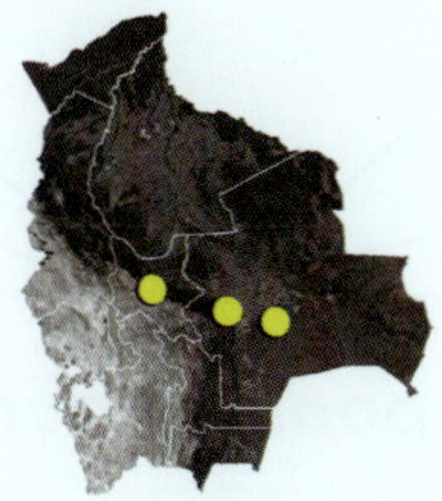

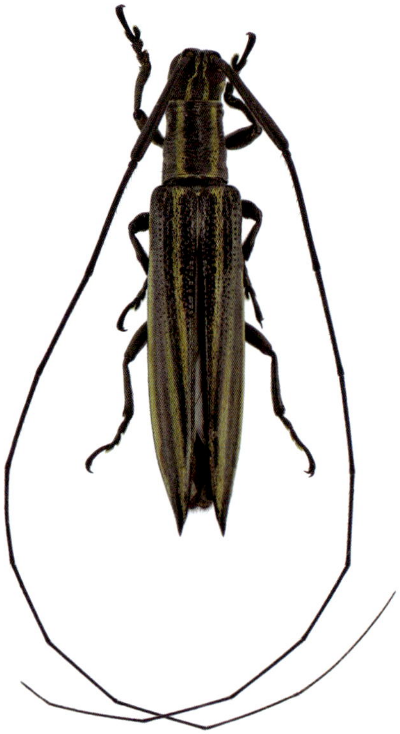

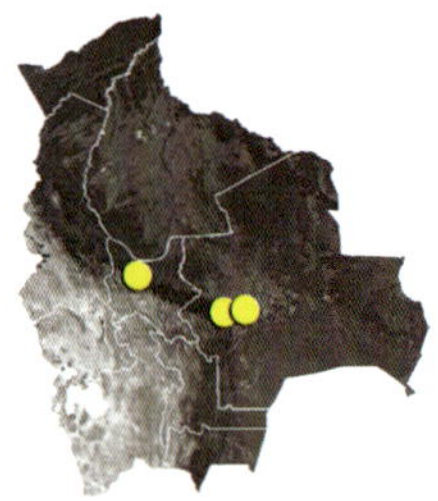

Hippopsis fractilinea Bates, 1866

Lamiinae: Agapanthiini

This species, like most members of the Agapanthiini, is narrow with long antennae and linear pubescent patterns dorsally. Being one of the larger (12–17 mm) and more robust agapanthiines found in Bolivia helps to identify it. Dorsally, it has two narrow, longitudinal pubescent fasciae on the head and pronotum and three on the elytra. Elsewhere, the dark brown integument is exposed. Ventrally, a bold yellow stripe is present at the center of the abdomen. This species is known from Bolivia, Brazil, Ecuador, and Peru. In Bolivia it has been collected in Villa Tunari (Cochabamba Department) and Potrerillo del Guenda, Refugio Los Volcanes, and the road to Bella Vista, near Cuevas (Santa Cruz Department).

Megacera acuminata Galileo & Martins, 2006

Lamiinae: Agapanthiini

This species has similar markings to other members of the tribe. It is also relatively large (13–19 mm long). It is distinguished from similar-looking members of *Hippopsis* by its orange pubescent metepisternum and, below that, a narrow, orange, pubescent line along the lateral part of the metasternum. Males have very long antennae, nearly three times the length of the body. This is a Bolivian endemic that has been collected at Refugio Los Volcanes and the Amboró road above Achira in Santa Cruz Department.

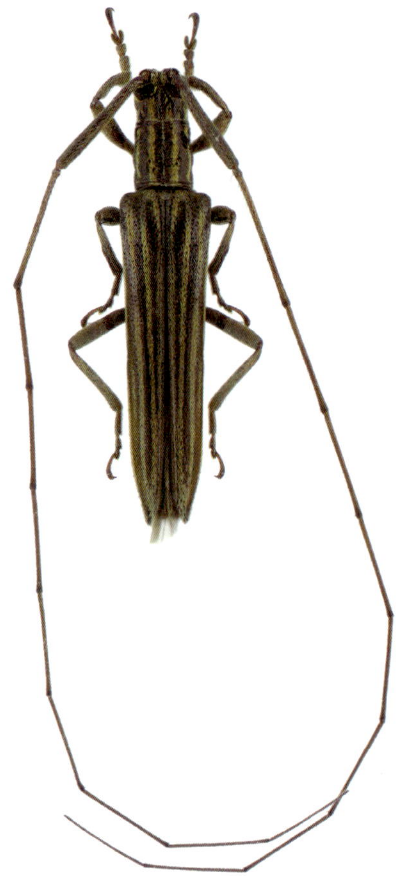

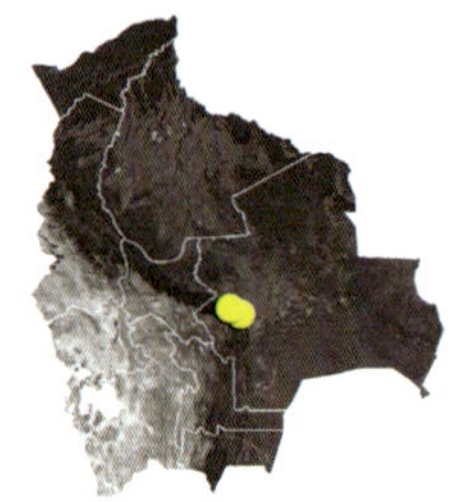

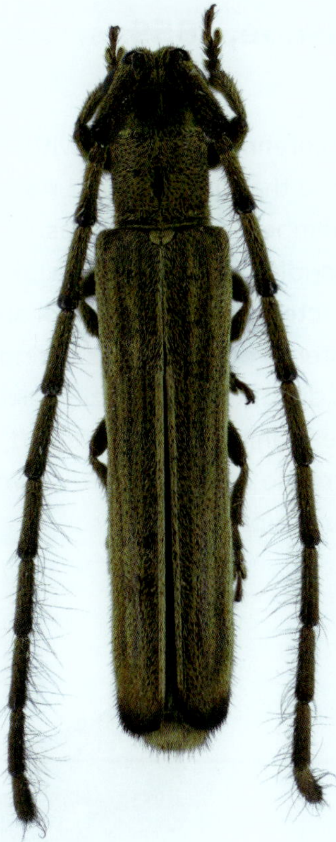

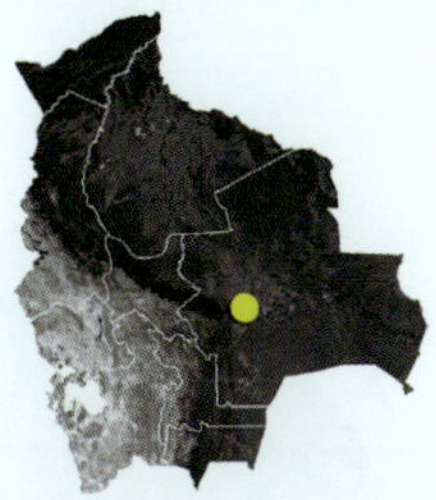

Trichohippopsis magna Martins & Carvalho, 1983

Lamiinae: Agapanthiini

This is an atypical member of Agapanthiini since the body and antennae are more densely coated in pubescence and the antennae are stouter and shorter than in other members of the tribe. The elytra are broadly rounded and darkened at the apices. Most specimens are 12–15 mm long. Known only from Brazil and Bolivia, this rarely collected species has been found only in the area in and around the Potrerillo del Guenda Reserve, Santa Cruz Department.

Anisocerus stellatus Guérin-Méneville, 1855

Lamiinae: Anisocerini

This boldly patterned cerambycid (12–16 mm) is quite distinct. It is recognized by the numerous black spots on the elytra, the large tuft of pubescence at the apex of the third antennomere, and the strongly clavate scape. This species is known from Ecuador, Peru, Brazil, and Bolivia. It has been rarely collected in Bolivia and is known only from Pando Department and from Rurrenabaque in Beni Department.

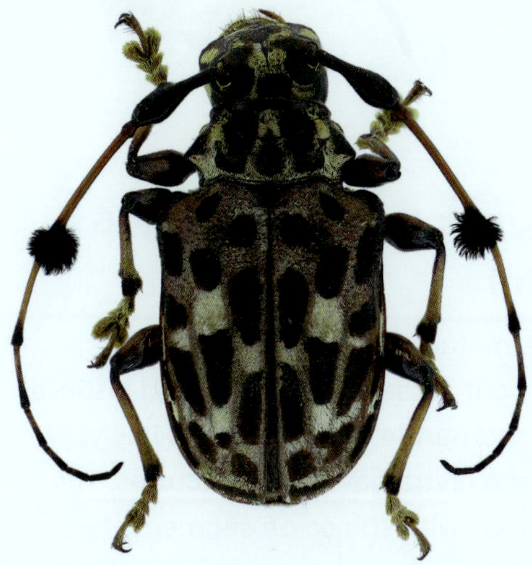

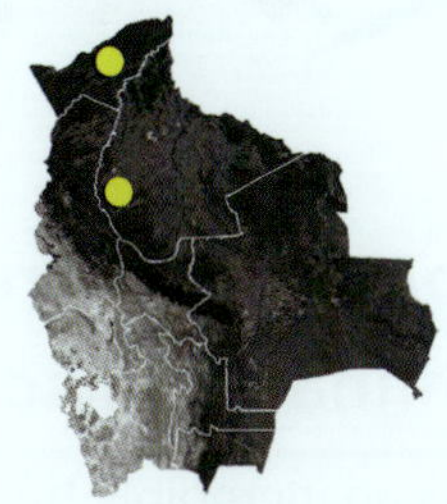

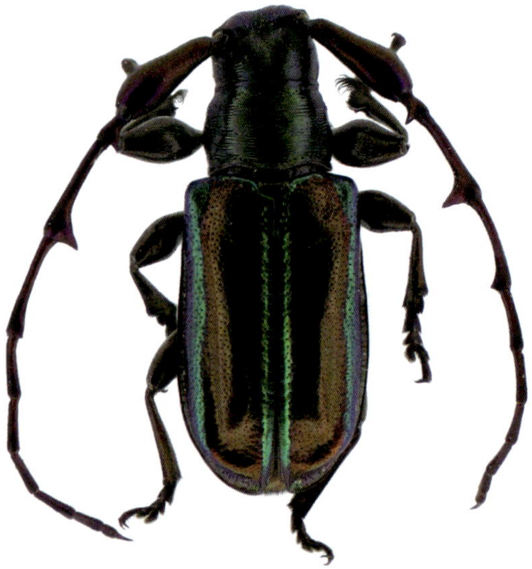

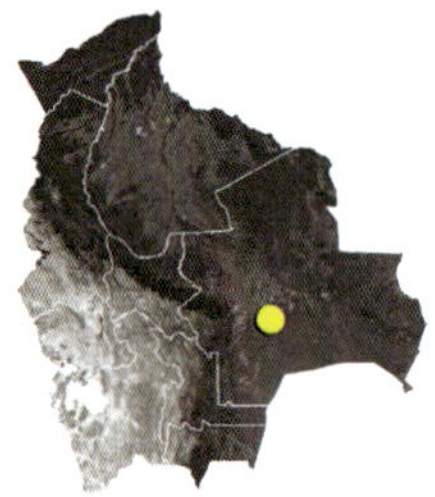

Hoplistocerus refulgens Blanchard, 1846

Lamiinae: Anisocerini

This is an incredibly beautiful, metallic-colored longhorned beetle. The integument is without setae, exposing a vivid luster of metallic green, blue, copper, and purple. The pronotum has very fine transverse wrinkles, and the elytra have fine punctures. The antennae are internally produced with spines on segments two to four, and a modified stinger-like ultimate segment. Specimens are 7–10 mm long. Widely distributed but rarely collected, this species is known from Ecuador, French Guiana, Brazil, and Bolivia. In Bolivia it has been collected in only the Santa Cruz Botanical Garden in the city of Santa Cruz de la Sierra.

Onychocerus albitarsis Pascoe, 1859

Lamiinae: Anisocerini

This is the only known beetle that has antennae modified with stingers, morphologically quite similar to that of scorpions. As attested to by the first two authors, members of this species can inflict a painful stab if carelessly handled. This species is moderately sized (11–18 mm) and characterized by bold black and white pubescence. It is usually encountered only at lights during the beginning of the rainy season. It is distributed in Brazil, Peru, and Bolivia, where it is known from the vicinity of the Flora and Fauna Hotel near Buena Vista, the Potrerillo del Guenda Reserve, and Refugio Los Volcanes in Santa Cruz Department.

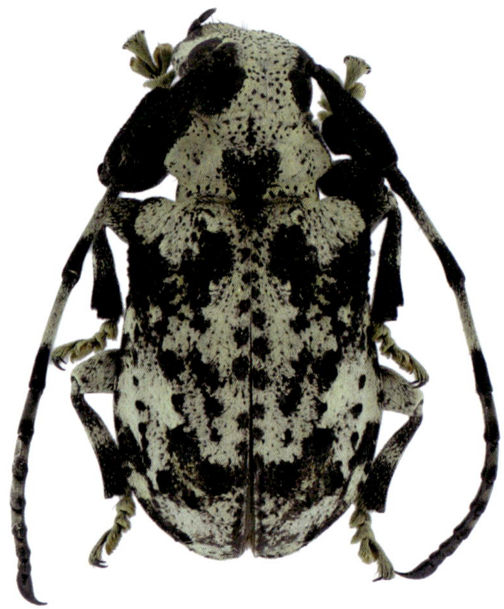

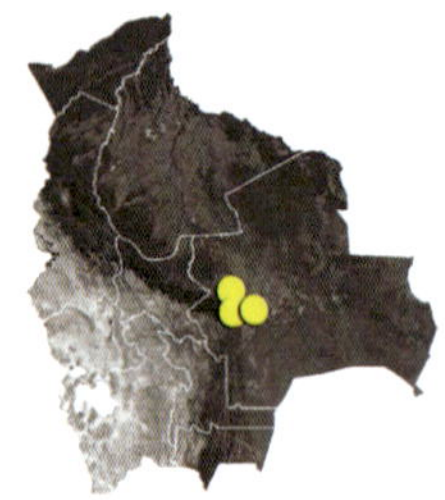

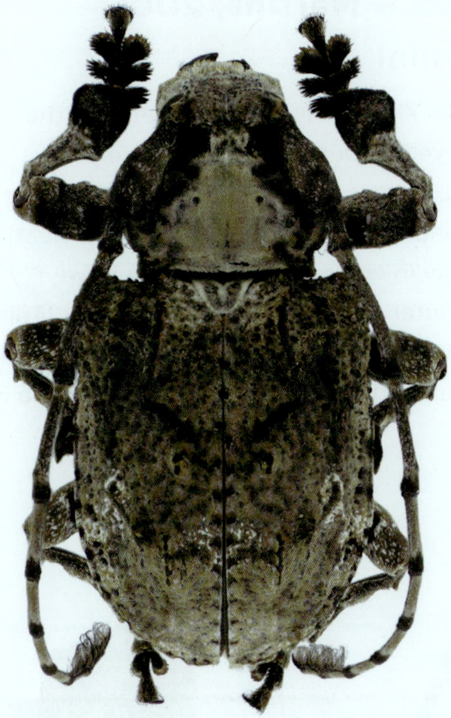

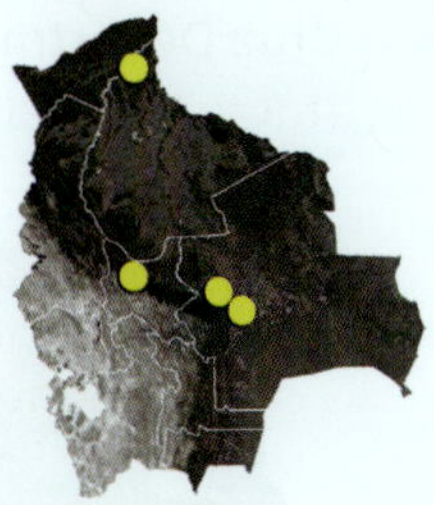

Onychocerus scorpio (Fabricius, 1781)

Lamiinae: Anisocerini

At 15–22 mm, this is the largest species of this peculiar genus found in Bolivia. Until recently, it was known by the name *Onychocerus crassus* (Voet). Like most members of the genus, it is very short and broad bodied, almost toad-like. Unlike the previous species, *Onychocerus albitarsus*, the antennae of this species are heavily pubescent at the apices, concealing the acute terminal segment. The protibiae are strongly expanded. This species is typically seen only at lights. In Bolivia it has been collected at Cristal Mayu (Cochabamba Department), the vicinity of the Flora and Fauna Hotel near Buena Vista and the Potrerillo del Guenda Reserve (Santa Cruz Department), and Beni Department.

Adetus cacapira Martins & Galileo, 2005

Lamiinae: Apomecynini

Like many species of *Adetus*, this species is very bullet shaped, with elytral apices rounded, pronotum without lateral margins, and short antennae that extend just a little beyond the midpoint of the elytra. Specimens are 10–13 mm in length. The pronotum and elytra have large, sparse punctures. This species is known only from Bolivia, where it has been collected at the Flora and Fauna Hotel near Buena Vista (the type locality) and in Refugio Los Volcanes in Santa Cruz Department.

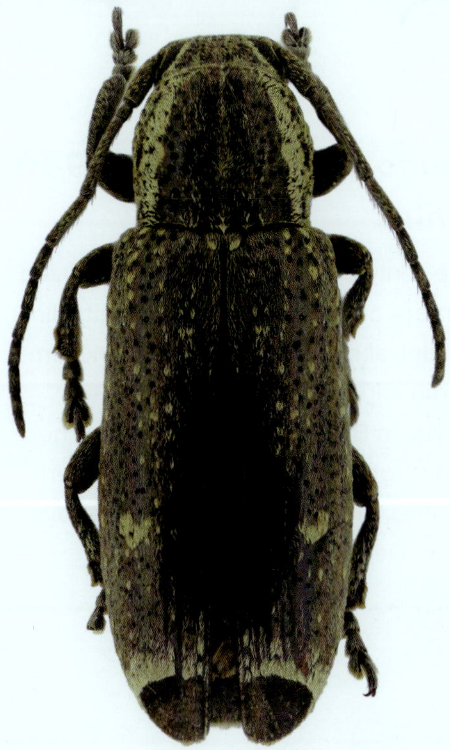

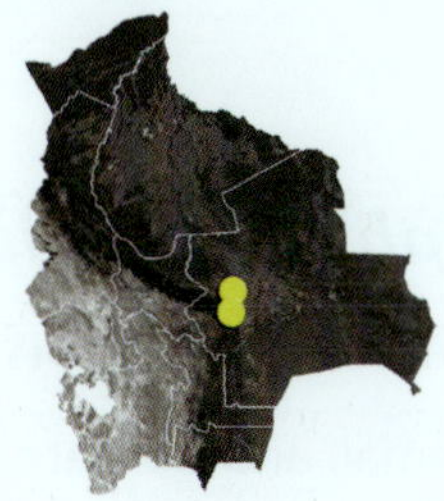

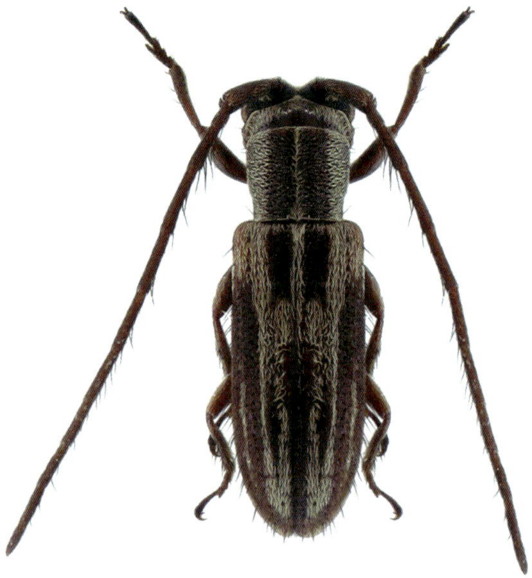

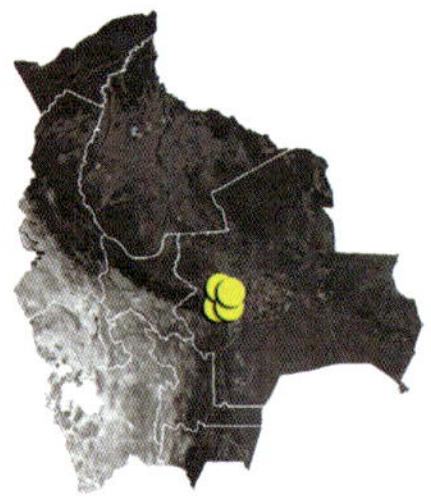

Amphicnaeia flavofemorata Breuning, 1940

Lamiinae: Apomecynini

This small and delicate lamiine (only 3–4 mm long) is recognized by the pale reddish-brown femora and the pattern of longitudinal bands and maculae of yellow elytral pubescence. The pronotum has a narrow, longitudinal median line. This species has been recorded from Brazil and Bolivia. In Bolivia it is known from the grounds of the Flora and Fauna Hotel near Buena Vista, Potrerillo del Guenda, Refugio Los Volcanes, and the Vicoquin area above Achira at 2,000 m in Santa Cruz Department.

Bebelis tagua Galileo & Martins, 2006

Lamiinae: Apomecynini

This small but stout lamiine (4–7 mm long) is recognized by the pattern of brown, white, and yellow pubescence on the elytra. There is a vague, longitudinal, yellowish band projecting from the base to beyond the middle. The apical fifth of the elytra are nearly uniformly pale brown to yellow in color. The mesotibiae have pronounced grooves on the outer margins toward the apices. This species is known only from Brazil and Bolivia. In Bolivia it is known from the grounds of the Flora and Fauna Hotel near Buena Vista in Santa Cruz Department.

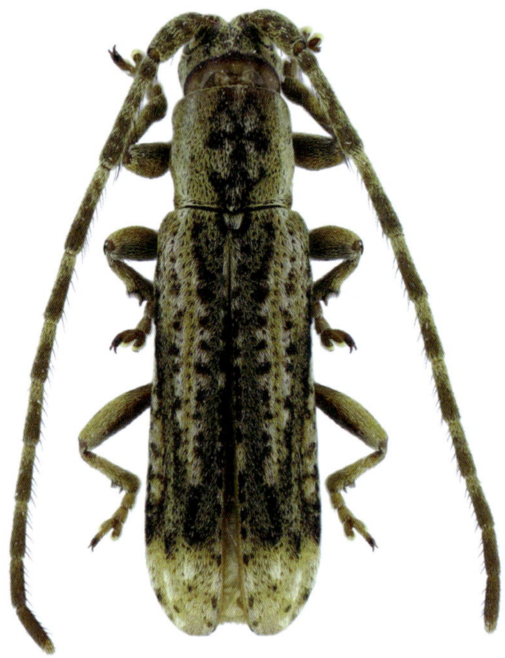

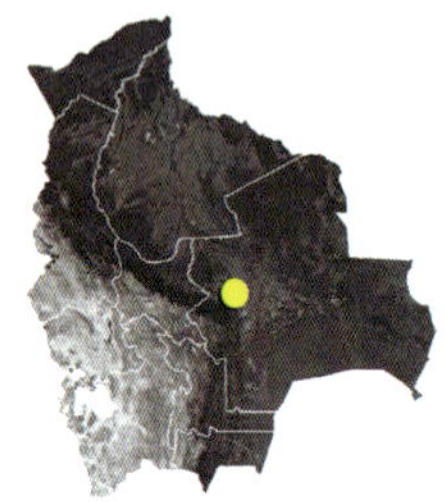

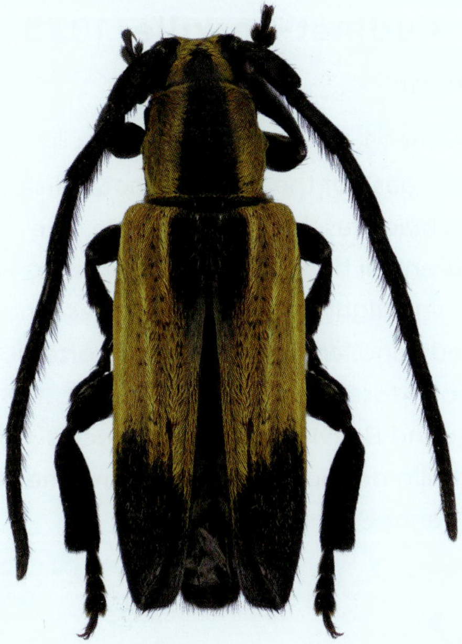

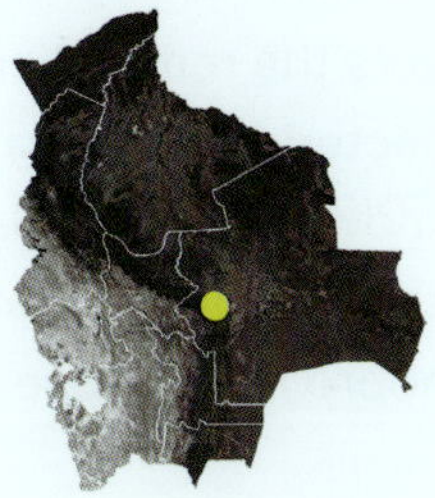

Bisaltes adustus (Burmeister, 1865)

Lamiinae: Apomecynini

This is, perhaps, the most striking species of *Bisaltes* (6–9 mm), with vivid and contrasting orange and black pubescence on the elytra, pronotum, and head. The apical third and periscutellar region of the elytra are black, along with the middle of the pronotum and vertex of the head. The legs and antennae are all black. This species is known only from Argentina and Bolivia. In Bolivia it has been collected from the Vicoquin area above Achira at 2,000 m in Santa Cruz Department.

Dorcasta implicata Melzer, 1934

Lamiinae: Apomecynini

This very small (4–6 mm), narrow-bodied longhorned beetle is a member of the morphologically diverse tribe, Apomecynini. It is recognized by the slanted head with posteroventrally directed mouthparts, pointed elytral apices, and longitudinal but subtle rows of orange pubescence on the elytra. Numerous stout, erect, black setae extend over the surface of the body and appendages. In Bolivia it is known from the Camiri environs in Santa Cruz Department and near Villamontes in Tarija Department.

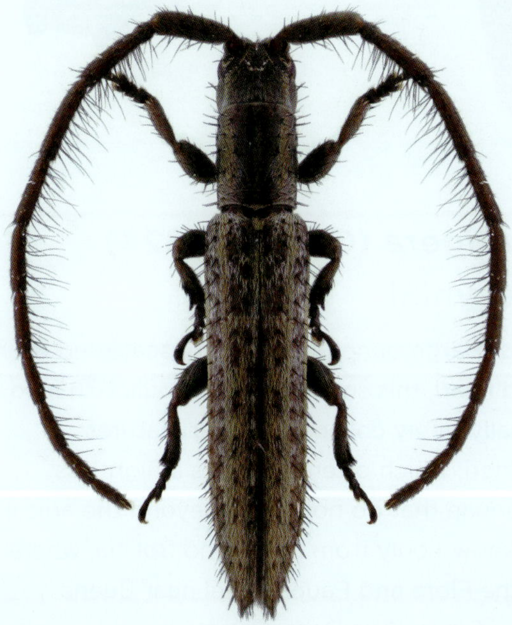

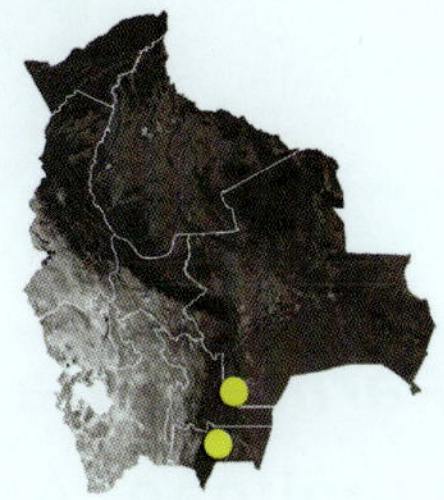

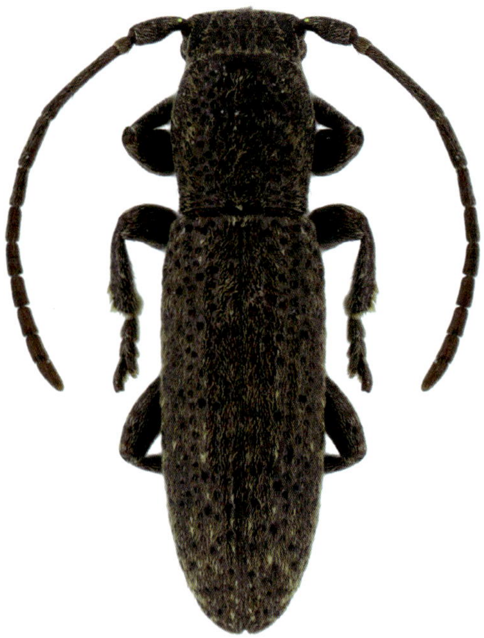

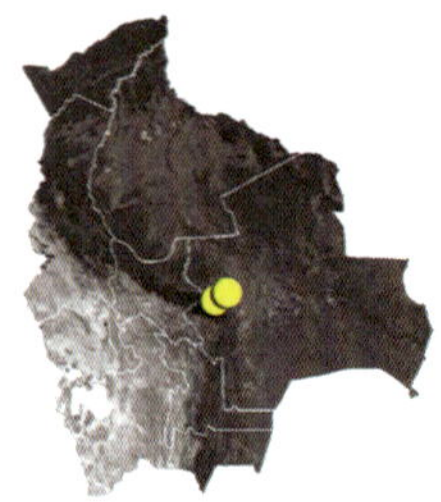

Parmenonta punctigera (Germar, 1824)

Lamiinae: Apomecynini

Aptly named for the dense, large punctures that are scattered over the elytra, pronotum, and head, this small (5–7 mm) bullet-shaped longhorned beetle is usually rarely collected. Other features are the very rounded humeri, which often suggests a flightless condition, and short antennae that do not reach beyond the apical fourth of the elytra. It is known only from Brazil and Bolivia, where it has been collected at the Flora and Fauna Hotel near Buena Vista and Pampagrande in Santa Cruz Department.

Ptericoptus dorsalis Audinet-Serville, 1835

Lamiinae: Apomecynini

Although not brightly colored, this 10–13 mm long species is beautiful with its patterning. The base of the head, pronotal disk, and elytral suture have a broad black fascia of pubescence. The elytral apex is broadly black, as are all but the terminal three to four antennal segments. Elsewhere, light tan or yellow pubescence covers the remainder of the body and legs. This species is often collected by beating foliage along forest openings or trails. It is known from Brazil, Paraguay, and Bolivia, where this species has been collected at the Potrerillo del Guenda Reserve and the Amboró road above Achira in Santa Cruz Department.

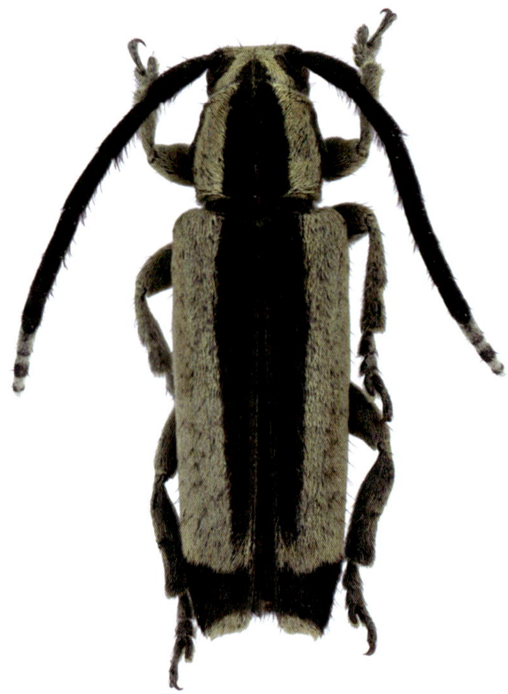

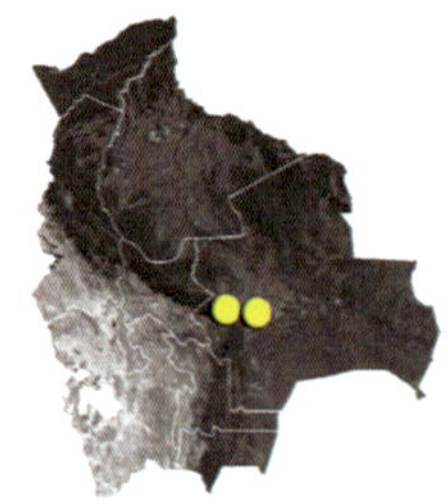

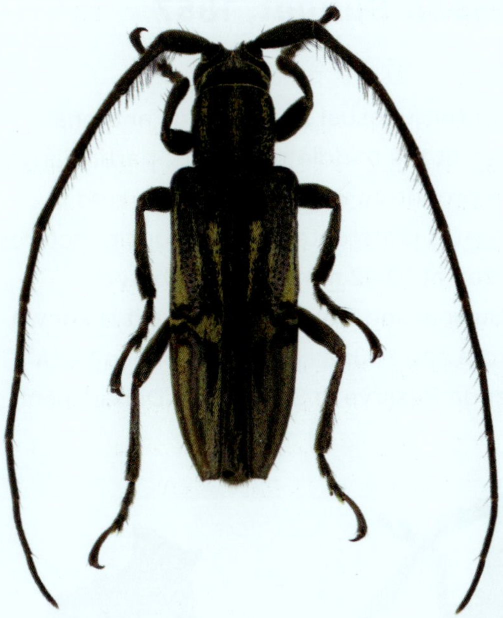

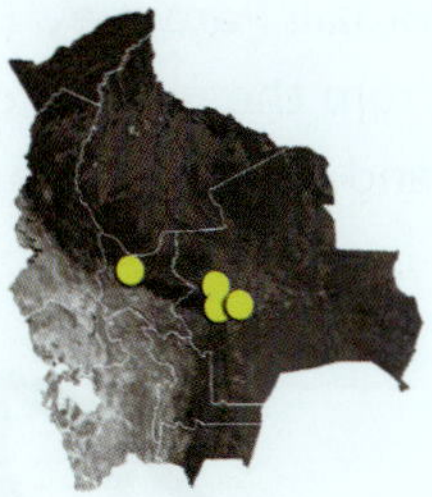

Rosalba inscripta (Bates, 1866)

Lamiinae: Apomecynini

This species can be commonly collected by beating foliage and branches in forest clearings. It is rather small at 8–11 mm long. The pronotum is adorned by three diffuse, longitudinal, yellow pubescent fasciae. The elytra have a transverse yellow fascia at middle, expanded toward the suture, and radiating from it are several longitudinal yellow fasciae extending toward the base and apex of the elytra. This species is known from Brazil, Venezuela, French Guiana, and Bolivia. In Bolivia it has been collected at Cristal Mayu (Cochabamba Department) and the Flora and Fauna Hotel near Buena Vista, the Potrerillo del Guenda Reserve, and Refugio Los Volcanes (Santa Cruz Department).

Callia boliviana Belon, 1903

Lamiinae: Calliini

This boldly colored calliine (4–6 mm) is recognized by the bright orange pubescence on the head, pronotum, and elytra. The apical third and basal-sutural portion of the elytra are black. This species is known only from Peru and Bolivia. In Bolivia it is known from Cristal Mayu in Cochabamba Department and Potrerillo del Guenda and Refugio Los Volcanes in Santa Cruz Department.

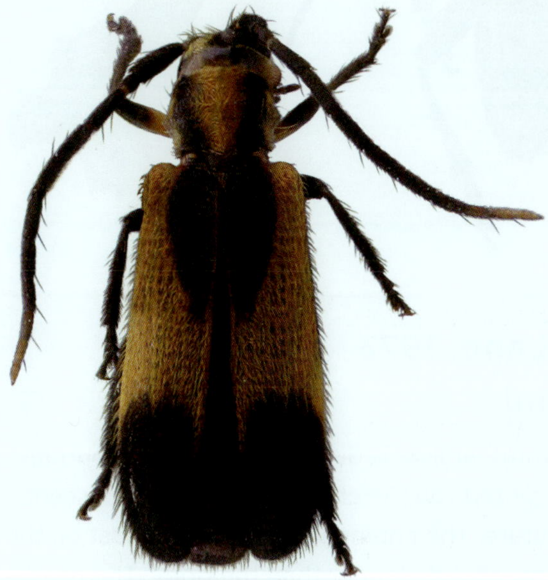

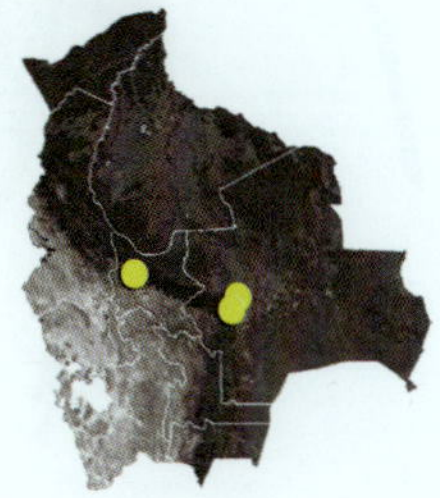

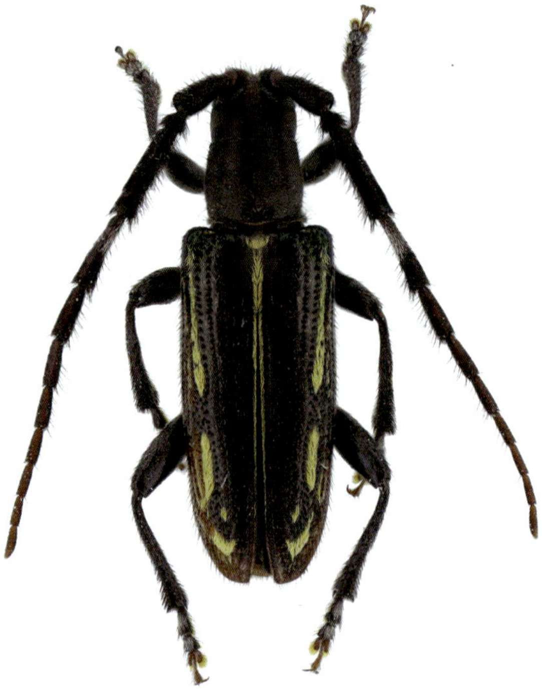

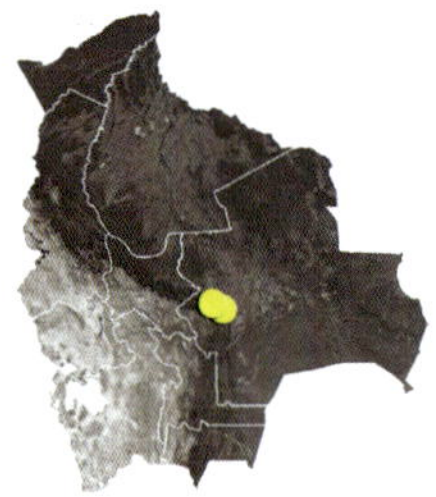

Gryllica prava Lane, 1973

Lamiinae: Calliini

This small-sized (9–11 mm) lamiine is distinct with its dark purple-brown color that is accented (on the elytra) by yellow pubescent fasciae and a yellow suture. The antennae are stout, at least on the basal segments, but only slightly longer than the body. The head and pronotum are very smooth, without punctures, but the elytra have several rows of large, conspicuous punctures. This species is distributed in Brazil and Bolivia, where it has been collected by beating foliage and branches in Refugio Los Volcanes and the Amboró road above Achira, both in Santa Cruz Department.

Hemicladus thomsonii Buquet, 1857

Lamiinae: Calliini

The most striking aspect of this unusual species is its antennae, which are greatly expanded at the middle segments, particularly segments five to nine. The pronotum and elytra are covered in dirty yellow pubescence that is punctuated by glabrous punctures. It is small- to moderate-sized at 10–12 mm. It is distributed in Brazil, Paraguay, French Guiana, and Bolivia. In Bolivia it is known from the vicinity of the Flora and Fauna Hotel near Buena Vista and the Potrerillo del Guenda Reserve in Santa Cruz Department.

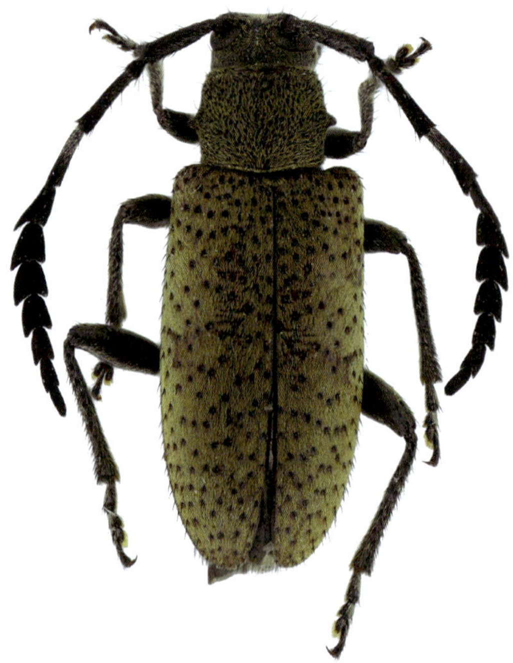

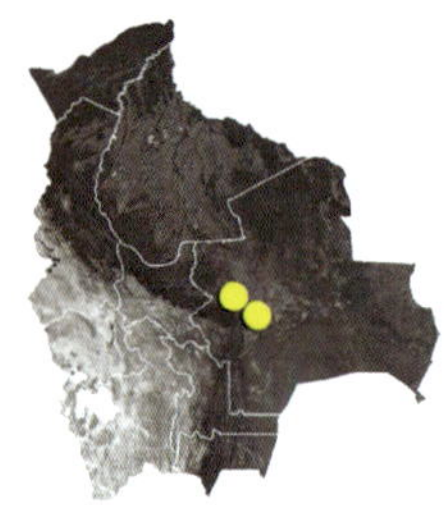

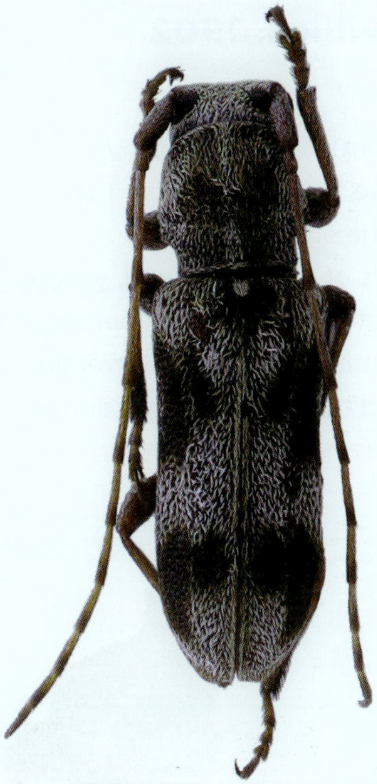

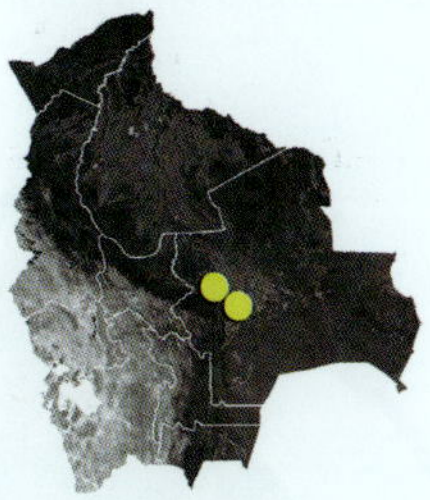

Xenocallia punctatissima Galileo & Martins, 1990

Lamiinae: Calliini

This is a very small (4–5 mm) species that is, as the name implies, covered in punctures, particularly evident (under illumination and microscopic examination) on the elytra. The color is reddish brown with darker brown maculae separated by white pubescent maculae on the elytra. This species emerges with the beginning of the rainy season in September or October and can then be locally abundant at lights but is rarely encountered after that. This is a Bolivian endemic species known only from the vicinity of the Flora and Fauna Hotel near Buena Vista and the Potrerillo del Guenda Reserve in Santa Cruz Department.

Colobothea bicuspidata (Latreille, 1811)

Lamiinae: Colobotheini

This large colobotheine (19–24 mm) is mostly black with a few scattered spots of white pubescence on the elytra. The pronotum has four narrow white bands of pubescence dorsally and two laterally (like the following species, *Colobothea brullei*). The antennae are black except for the sixth antennomere, which is mostly covered in white pubescence. This widespread South American species is known from Colombia, Ecuador, Peru, French Guiana, Brazil, and Bolivia. It is known from the grounds of the Flora and Fauna Hotel near Buena Vista and Potrerillo del Guenda Reserve in Santa Cruz Department in Bolivia.

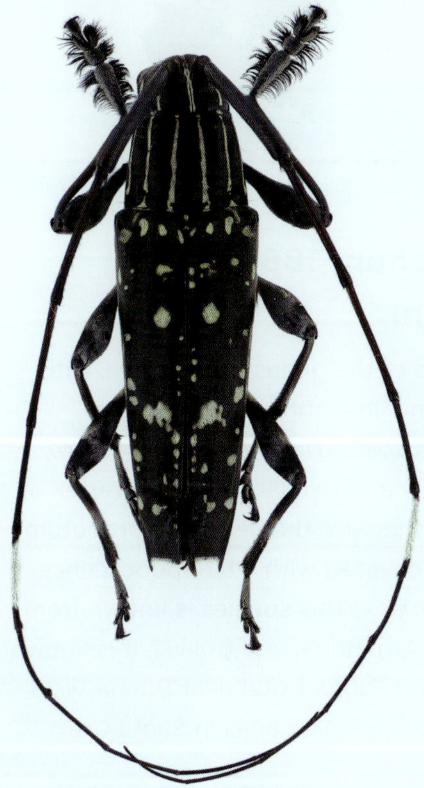

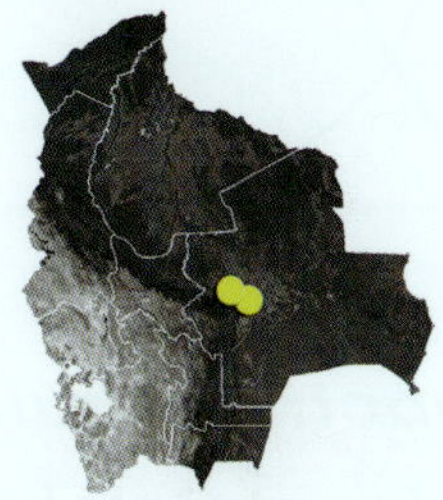

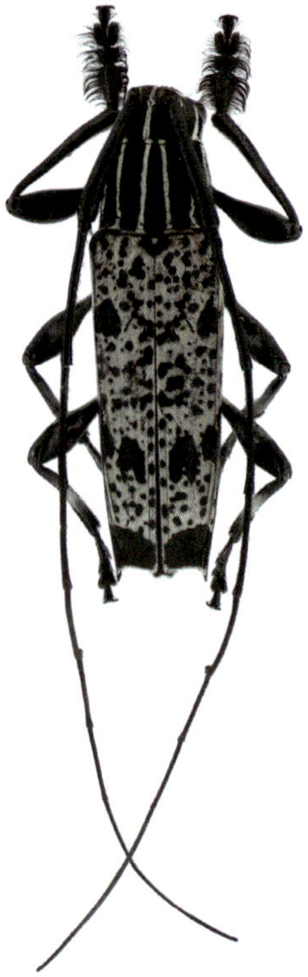

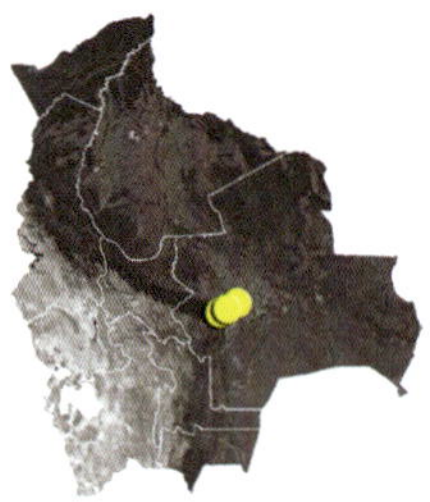

Colobothea brullei Gahan, 1889

Lamiinae: Colobotheini

This is one of the largest species of the genus *Colobothea*, with most specimens exceeding 20 mm in length. The pronotum (like the previous species, *C. bicuspidata*) has four very narrow, longitudinal, white bands dorsally and two laterally. The vertex of the head has one white band. Otherwise the head and pronotum are black. The elytra are mostly covered with white pubescence except at the apices, which are black. This species is known from Ecuador, Brazil, Peru, Paraguay, Argentina, and Bolivia. It is known from the grounds of the Flora and Fauna Hotel near Buena Vista, Potrerillo del Guenda, and Refugio Los Volcanes in Santa Cruz Department in Bolivia.

Colobothea declivis Aurivillius, 1902

Lamiinae: Colobotheini

This endemic Bolivian species (11–13 mm) is recognized by the partial band of golden-yellow setae that extend from the pronotum across the basal fourth of the elytra. There is a pronounced humeral carina. Three sets of poorly defined, yellow-bordered, black maculae are present at the middle, just before the middle, and at the apex of the elytra. It is known from Potrerillo del Guenda, Refugio Los Volcanes, and the Vicoquin area above Achira at 2,000 m in Santa Cruz Department.

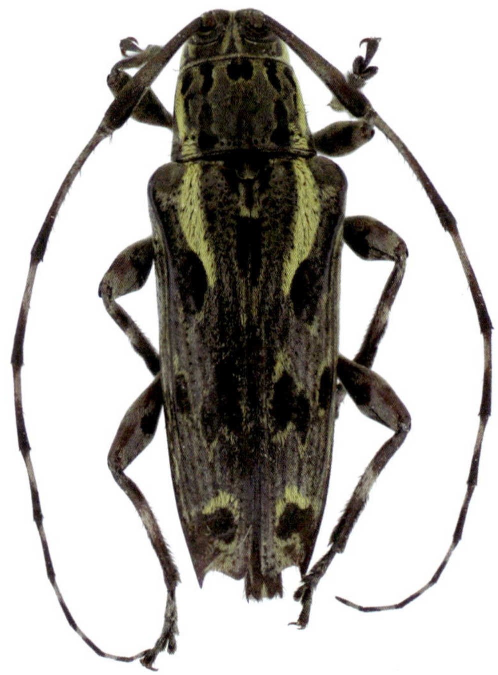

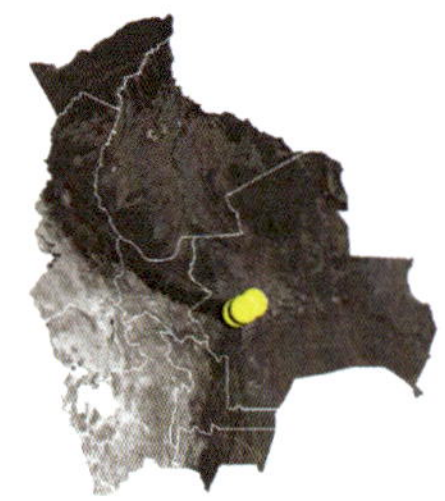

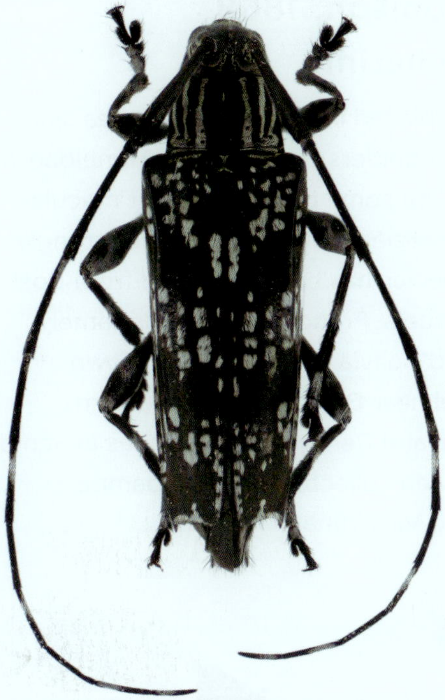

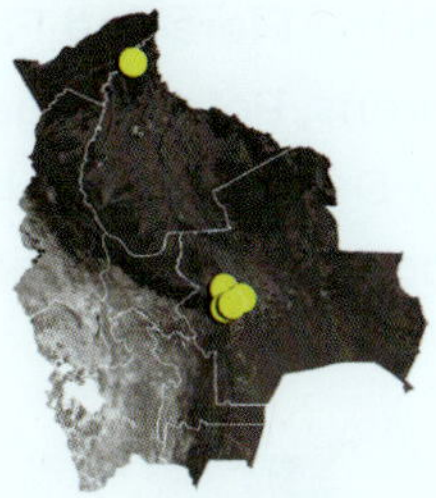

Colobothea plebeja Aurivillius, 1902

Lamiinae: Colobotheini

This species, averaging 14 mm long, is distinct with its mostly black integument that is not obscured by pubescence. There are three vaguely defined bands on the elytra that are composed of small, separate spots of white pubescence. There are six narrow, longitudinal bands of white pubescence on the pronotum. Known only from Peru and Bolivia, it is the most commonly encountered member of the genus. In Bolivia it is widespread in many localities, including the grounds of the Flora and Fauna Hotel near Buena Vista, Potrerillo del Guenda, Refugio Los Volcanes, and the Vicoquin area above Achira at 2,000 m (all in Santa Cruz Department); it has also been recorded from near Riberalta in Beni Department.

Colobothea rubroornata Zajciw, 1962

Lamiinae: Colobotheini

This species of *Colobothea* (9–12 mm) has a unique pattern of white-bordered orange spots of pubescence on the elytra. Three black maculae are evenly spaced for the length of each elytron. The pronotum has a broad, longitudinal, black band at middle. This species is known from Brazil, Paraguay, and Bolivia. It is known from the grounds of the Flora and Fauna Hotel near Buena Vista, Potrerillo del Guenda, Refugio Los Volcanes, and the Camiri environs in Santa Cruz Department in Bolivia.

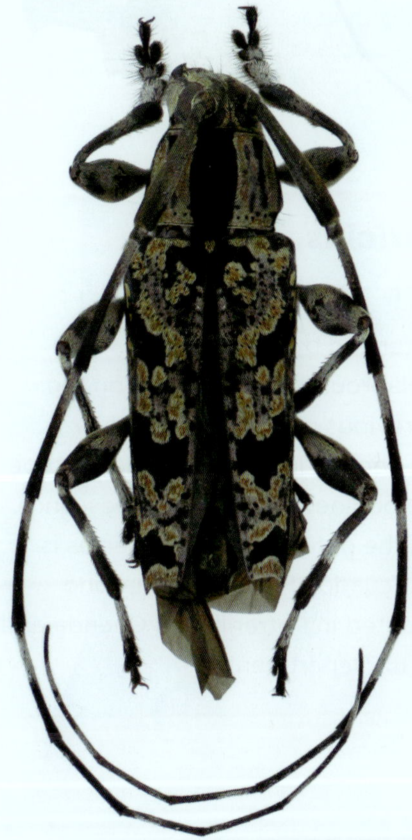

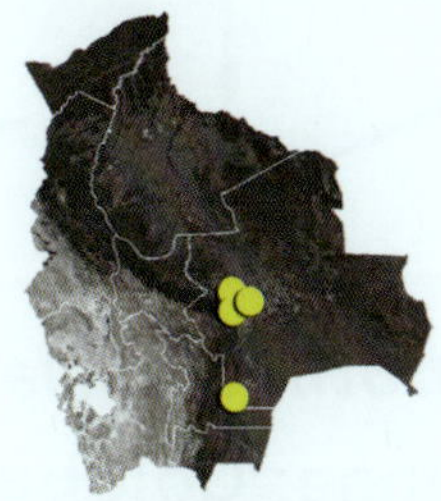

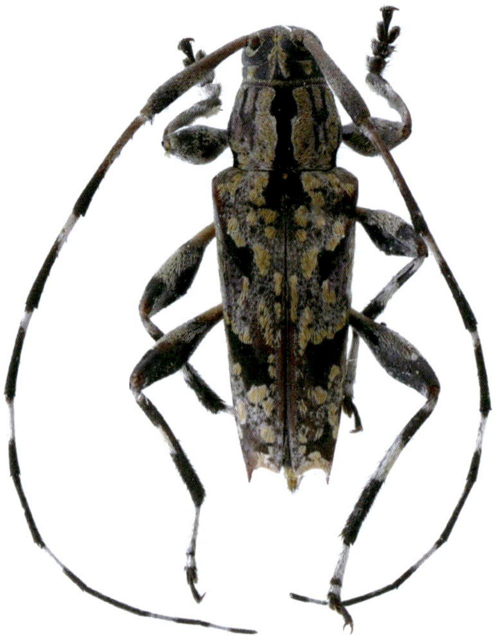

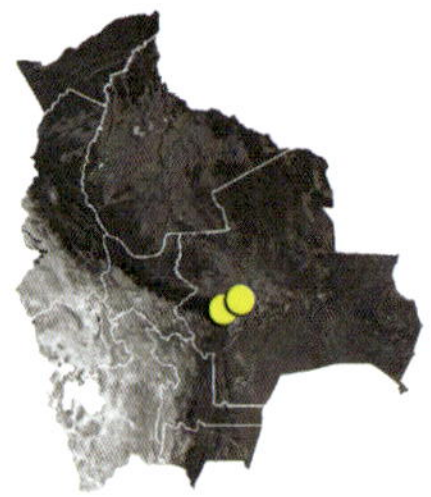

Colobothea varia (Fabricius, 1787)

Lamiinae: Colobotheini

This species (8–10 mm) is characterized by its intricate pattern of orange pubescent spots and glabrous, black maculae on the elytra, somewhat similar to the previous species, *Colobothea rubroornata*. The pronotum has a central, longitudinal, black stripe (narrower than in *C. rubroornata*) outlined by orange pubescence. The elytra are strongly spined on the outer apex. This species is known from Panama, Venezuela, Ecuador, French Guiana, and Bolivia. In Bolivia it has been collected in Potrerillo del Guenda and Refugio Los Volcanes in Santa Cruz Department.

Aerenea brunnea Thomson, 1868

Lamiinae: Compsosomatini

This species (6–11 mm) is aptly placed in the genus *Aerenea* since it resembles various orb-weaver spiders of the family Araneidae. It is mostly dark brown in color with some pale pubescent maculae, particularly on the pronotal tubercles that are acute and strongly projecting. This widespread species has been recorded from Costa Rica, Panama, Colombia, Venezuela, French Guiana, Suriname, Brazil, Paraguay, Argentina, and Bolivia. In Bolivia it is known from the Flora and Fauna Hotel near Buena Vista, the Camiri environs, Potrerillo del Guenda, and Refugio Los Volcanes in Santa Cruz Department. It has also been collected in Cochabamba Department in Cristal Mayu and Villa Tunari.

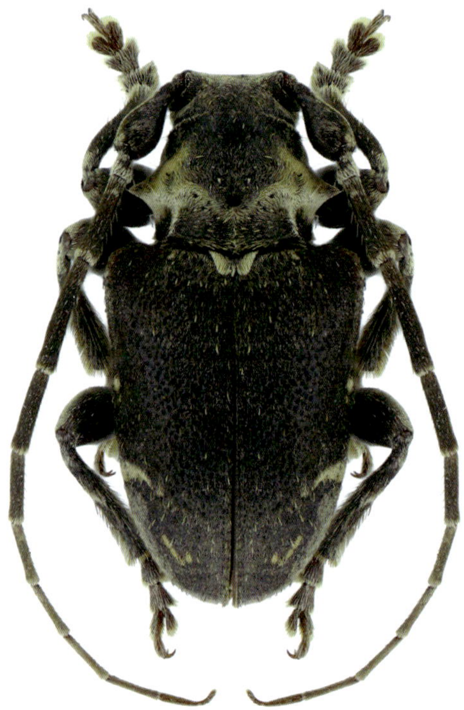

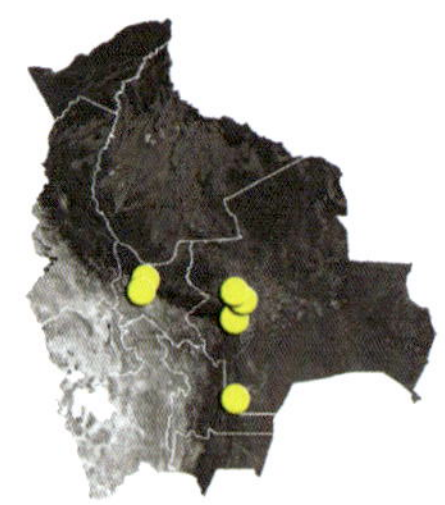

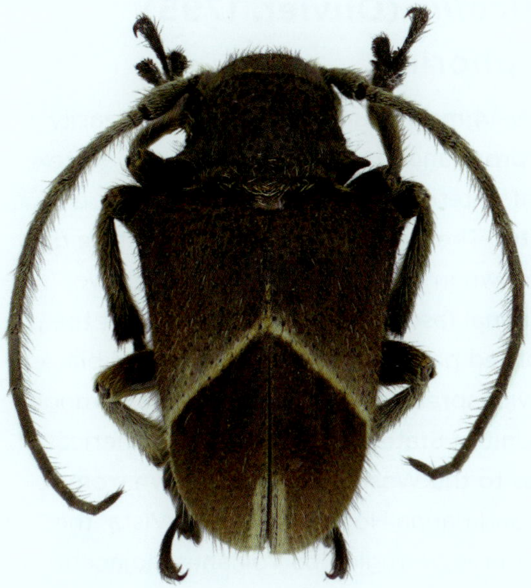

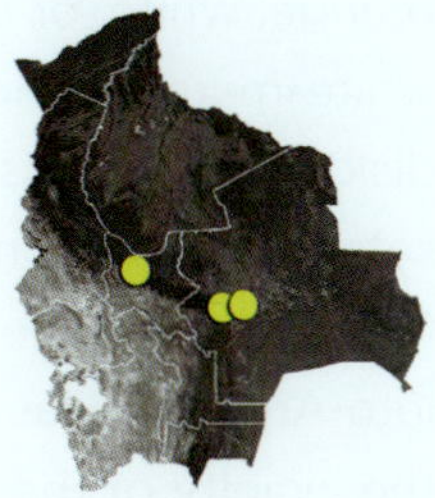

Compsosoma monnei Martins & Galileo, 1996

Lamiinae: Compsosomatini

Many members of the tribe Compsosomatini, like this one, are broad and robust in proportions. This species (13–18 mm long) has a characteristic pattern of dense reddish-brown pubescence over most of the elytra that is interrupted by a narrow, angled, postmedian band of off-white pubescence. The apical fifth of the elytral suture is also marked by off-white setae. It is very similar to *Compsosoma oculata* Lingafelter & Wappes (see frontispiece). This species is known only from Bolivia, where it has been collected in Villa Tunari, the type locality in Cochabamba Department, and Potrerillo del Guenda and Refugio Los Volcanes in Santa Cruz Department.

Blabicentrus alberti Galileo & Moysés, 2013

Lamiinae: Desmiphorini

This Bolivian endemic is small in size (8–11 mm) but grand in appearance with a unique pattern of rust-red and whitish-gray pubescence. The entire dorsum and appendages also have long, erect setae rather evenly intermixed with its short and dense pubescent covering. Although collected several times, to date it has been found at only the type locality, Refugio Los Volcanes in Santa Cruz Department, where it has been attracted to mercury vapor and ultraviolet lights.

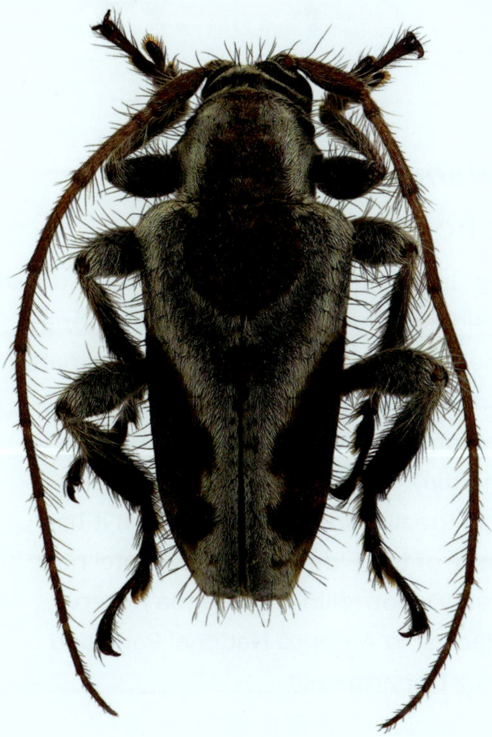

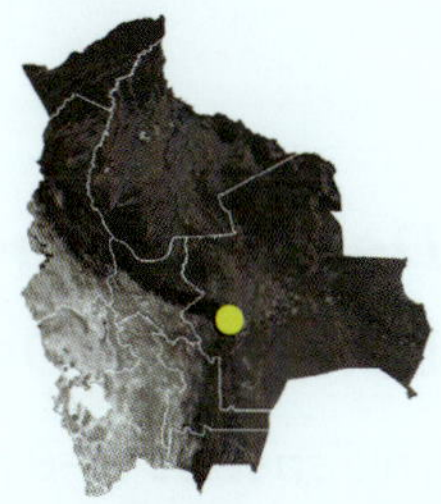

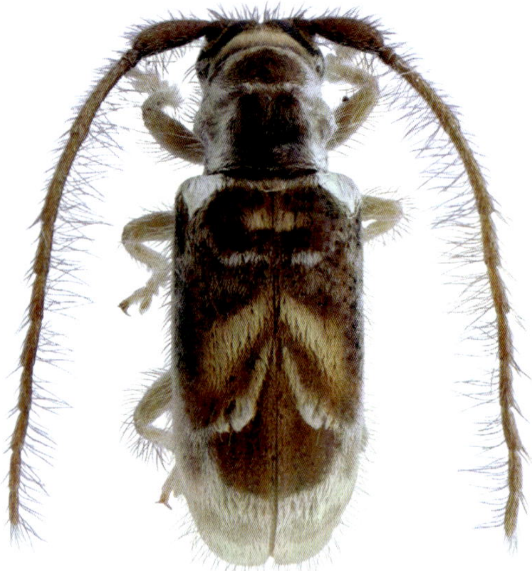

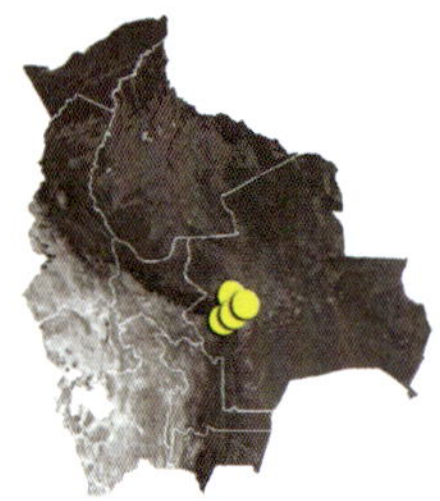

Desmiphora boliviana Breuning, 1948

Lamiinae: Desmiphorini

This species (5–8 mm long) is recognized by the patches of dense, white pubescence on the sides of the pronotum and base, apex, and sides of the elytra. The central elytral disk comprises a pattern of reddish-brown and orange pubescence forming alternating bands and shapes. The antennae are very densely fringed (especially on the inner margin) with long setae. This species is known from French Guiana, Brazil, and Bolivia. In Bolivia it has been collected in the vicinity of the Flora and Fauna Hotel near Buena Vista (the type locality), Potrerillo del Guenda Reserve, Refugio Los Volcanes, adjacent to Amboró National Park, and Samaipata (all in Santa Cruz Department).

Desmiphora hirticollis (Olivier, 1795)

Lamiinae: Desmiphorini

This well-known, large (9–14 mm) *Desmiphora* species is easily recognized by the very pronounced tufts of setae on the vertex of the head, the adjacent anterior margin of the pronotum, and postmedially on the elytra. These tufts of hairs are a mixture of orange, white, or light brown in color. The pronotum has two white-margined, longitudinal fasciae (one on either side of the disk). The elytra is a mottled pattern of white, brown, and black setae. This is the most widespread species in the genus. It ranges from the south-central United States through Central America into Argentina, as well as to the West Indies. It is known from the vicinity of the Flora and Fauna Hotel near Buena Vista, the Potrerillo del Guenda Reserve, Refugio Los Volcanes, adjacent to Amboró National Park, and the Camiri environs in Santa Cruz Department in Bolivia.

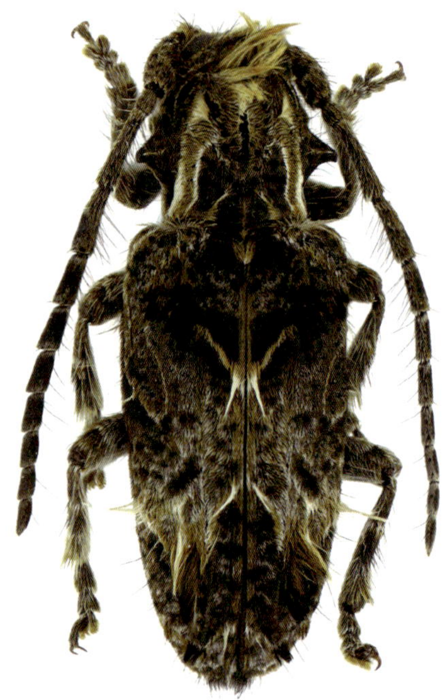

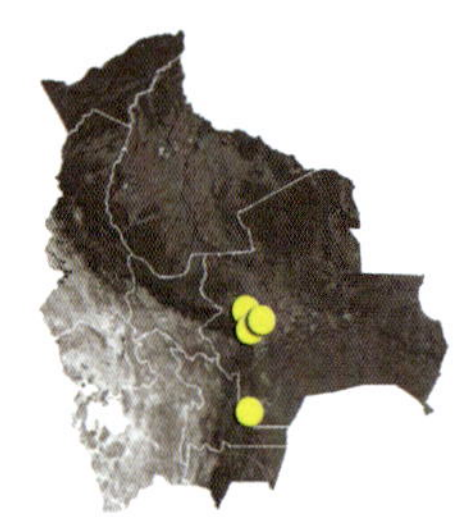

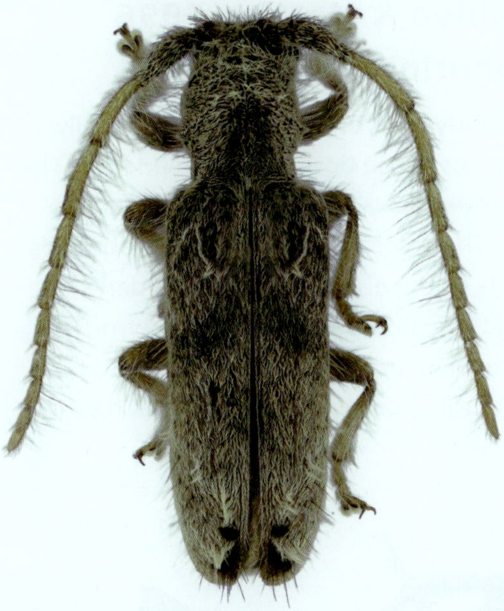

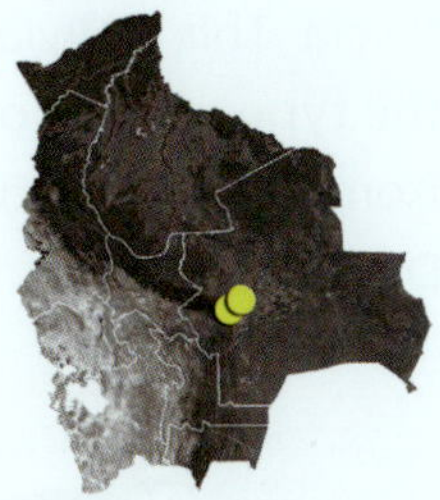

Desmiphora intonsa (Germar, 1824)

Lamiinae: Desmiphorini

This is one of the smaller, narrower species of *Desmiphora* at 6–8 mm long. It is densely covered in a mix of white and brown pubescence with the elytral apices tipped in black. The antennae, beyond the scape, are very pale yellow in color and covered with long flying hairs. This widespread but relatively uncommon species is known from Brazil, Paraguay, Argentina, Uruguay, and Bolivia. It has been collected from the Potrerillo del Guenda Reserve and Refugio Los Volcanes, adjacent to Amboró National Park, in Santa Cruz Department in Bolivia.

Desmiphora lateralis Thomson, 1868

Lamiinae: Desmiphorini

This species (6–9 mm long) is distinct with its very stout, dark brown to black antennae and broad, lateral, black fasciae on the outer margin of the elytra at the middle. Just behind these fasciae are four tufts of long setae (two on either side) that extend vertically from the elytral surface. This species has been collected from foliage of young *Ficus* trees and occasionally at lights. This species ranges from Brazil, Paraguay, Uruguay, into Bolivia. It is known from Refugio Los Volcanes and the Vicoquin area above Achira at 2,000 m, adjacent to Amboró National Park, in Santa Cruz Department in Bolivia.

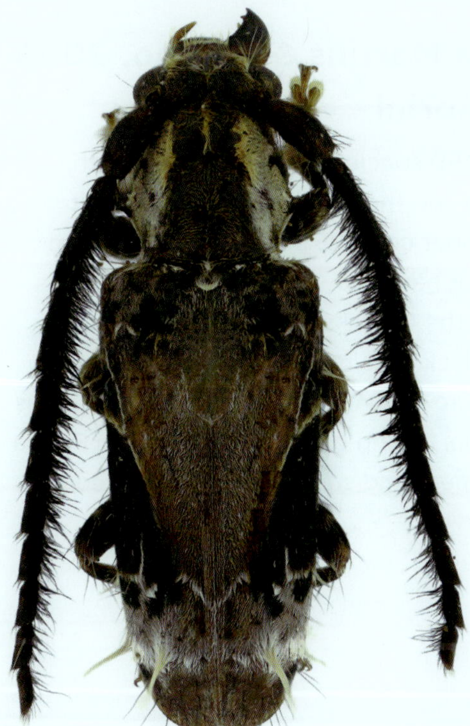

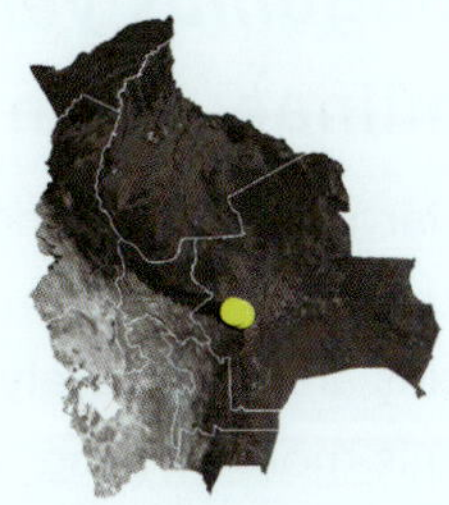

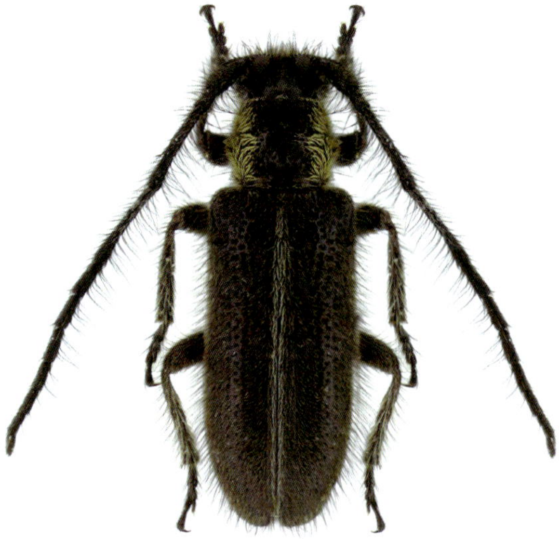

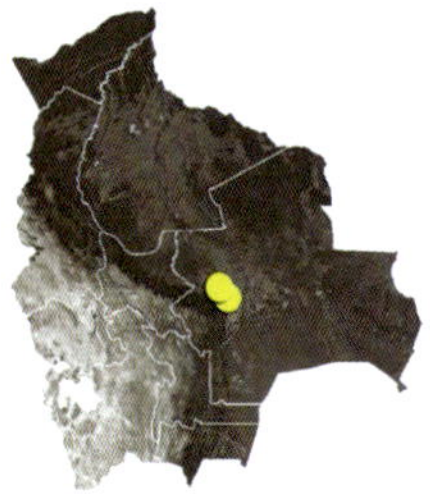

Eupogonius yeiuba Martins & Galileo, 2005

Lamiinae: Desmiphorini

This large genus has nearly 50 species scattered over North, Central, and South America and the Caribbean Islands. Most species are completely somber colored, but this one (3–4 mm long) is more striking since it has yellowish fasciae on the sides of the pronotum and along the elytral suture. Otherwise the coloration is dark gray to black. Like most species of the genus, the body and appendages are covered with very long setae. This species is endemic to Bolivia, having been described from Buena Vista in Santa Cruz Department and subsequently collected in Potrerillo del Guenda Reserve.

Ischnolea piim Galileo & Martins, 1996

Lamiinae: Desmiphorini

This is a small (7–9 mm), subtly colored species. It is generally covered in gray pubescence over the body and appendages. The pronotum has a narrow, yellow fascia at the middle, extending from the apex to the base. The sides have small, acute tubercles. Just posterior to the middle of the elytra is an oblique dark fascia. This Bolivia endemic species has been collected from the type locality of Chulumani in La Paz Department and from Pampagrande and Refugio Los Volcanes in Santa Cruz Department.

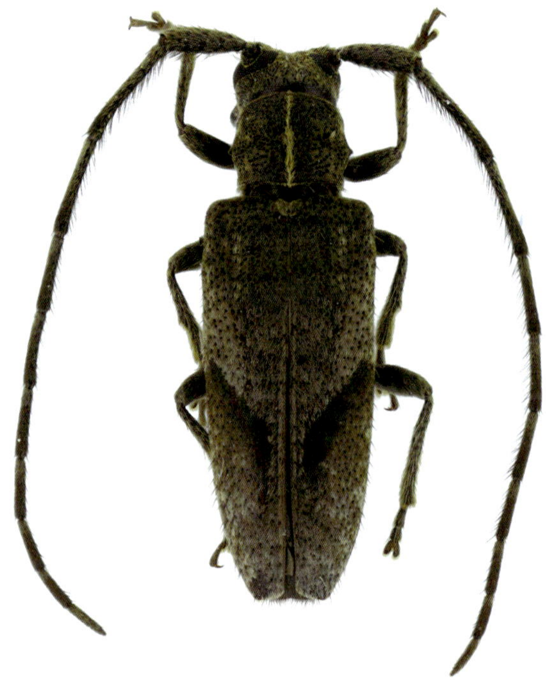

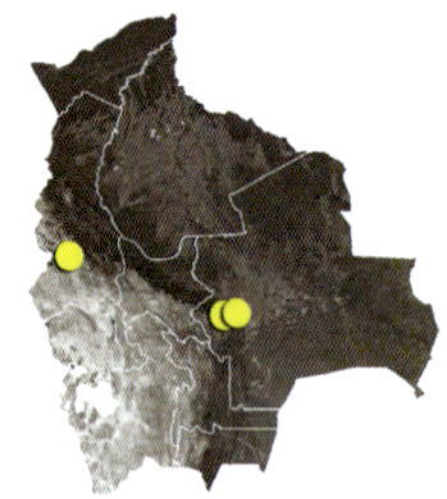

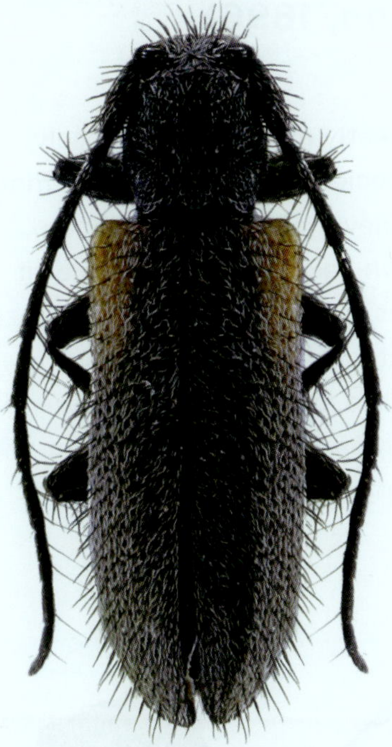

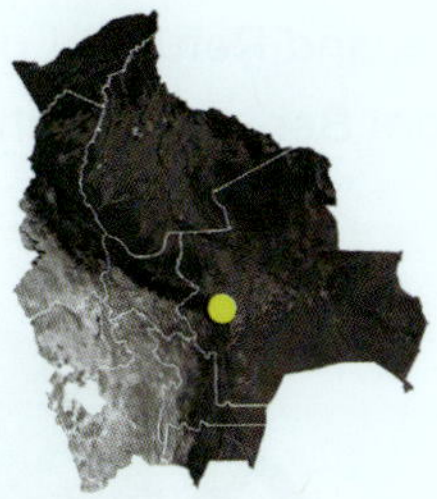

Mimasyngenes ytu Galileo & Martins, 1996

Lamiinae: Desmiphorini

This small species (3–4 mm) is distinct from others in the genus by the dense pubescence on the pronotum, especially at the sides where it hides the lateral tubercle and its very short tooth. The humeri are orange, but otherwise the color of this densely pubescent species is dark blue-gray. In South America, this species has an unusual, disparate distribution, being known from eastern Brazil (Minas Gerais) and Bolivia. This species has been collected from Huaico (across the river from the Potrerillo del Guenda Reserve) and Refugio Los Volcanes in Santa Cruz Department.

Unelcus lineatus Bates, 1885

Lamiinae: Desmiphorini

This small (7–9 mm) species is very attractive with the three longitudinal yellow pubescent fasciae on the pronotum and the single one on each elytron. In addition, several smaller, less complete white fasciae are present. The pronotum has a very small, acute, lateral tubercle at each side. This species occurs from Mexico through central South America. In Bolivia, it is known from the vicinity of the Flora and Fauna Hotel near Buena Vista, the Potrerillo del Guenda Reserve, the vicinity of Cuevas, and Refugio Los Volcanes in Santa Cruz Department.

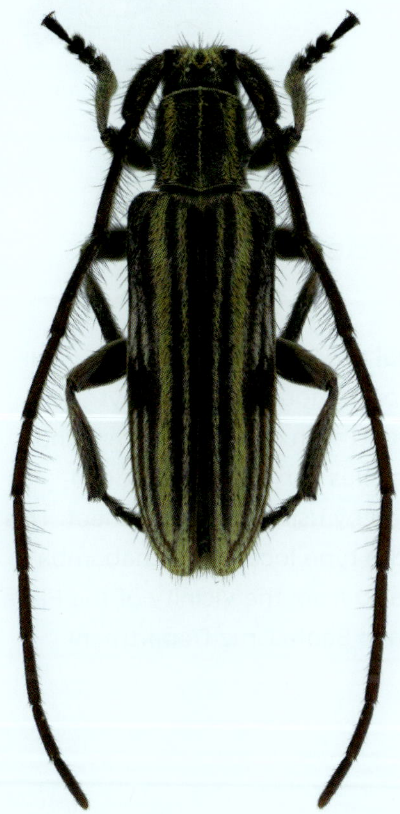

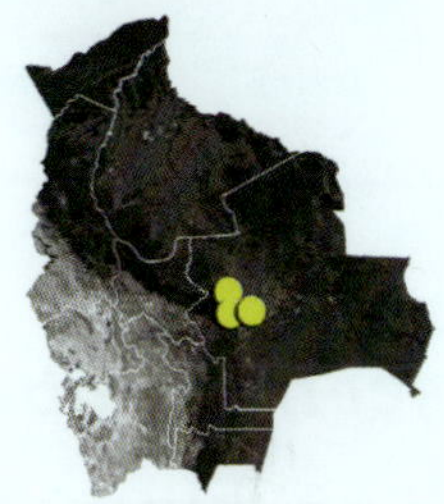

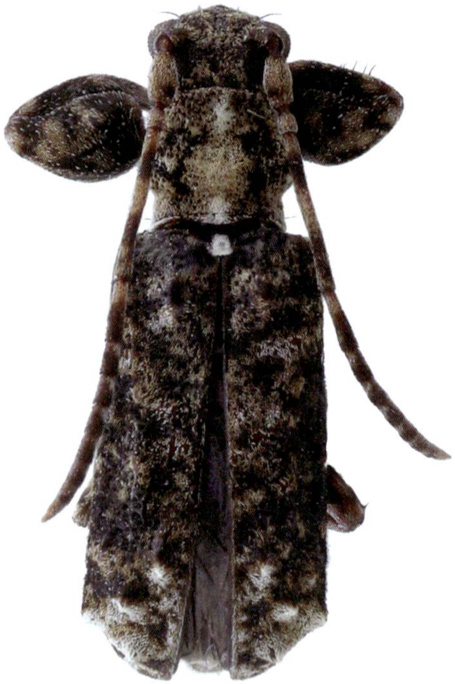

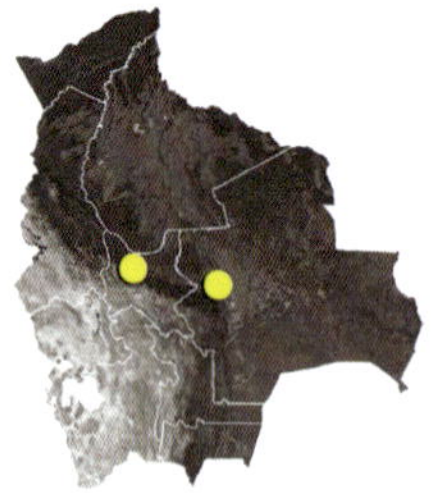

Iquiracetima tuberosa (Belon, 1896)

Lamiinae: Eupromerini

Like most members of this unusual tribe, Eupromerini, this small (3–5 mm) species has very large profemora. The legs, elytra, pronotum, and antennae are mottled with dark brown, tan, and white pubescence, making this species very cryptic on dead branches, where it can be collected by using a beating sheet. This Bolivian endemic is known from the type locality, Cochabamba (Cochabamba Department) and also from the vicinity of the Flora and Fauna Hotel near Buena Vista in Santa Cruz Department.

Pibanga ochropyga (Belon, 1896)

Lamiinae: Eupromerini

The profemora of this species are greatly enlarged and closely articulated with the protibae. The species is quite small (4–6 mm), like most members of the tribe. It is never commonly collected, probably because of its cryptic brown coloration that enables it to stay hidden among dead branches. It is known only from Brazil and Bolivia, where it has been collected from the vicinity of the Flora and Fauna Hotel near Buena Vista, the Potrerillo del Guenda Reserve, and Refugio Los Volcanes in Santa Cruz Department and also from Beni and Cochabamba Departments.

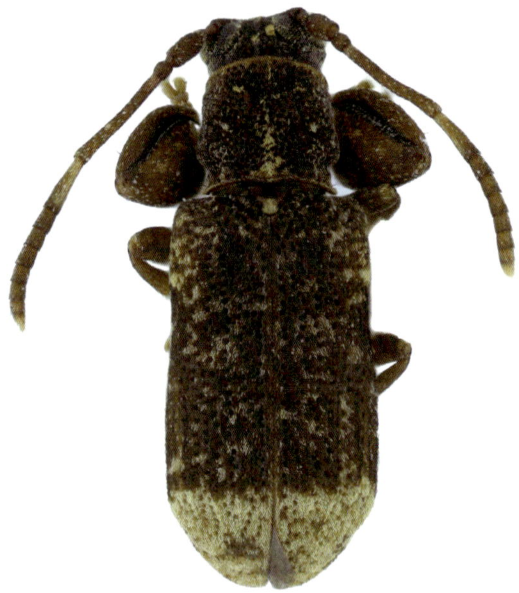

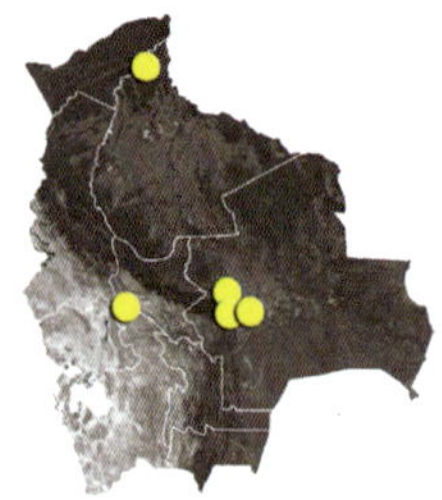

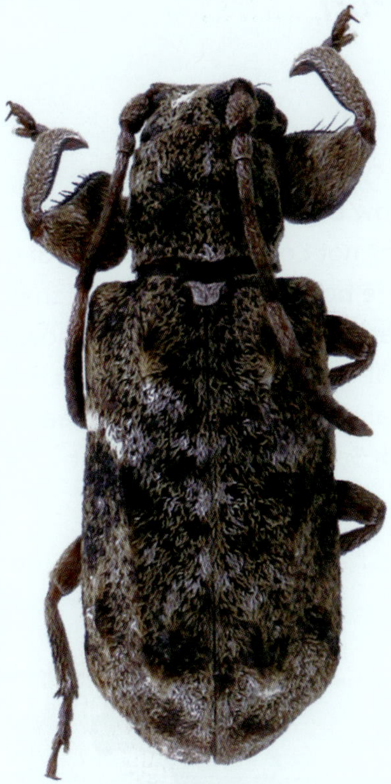

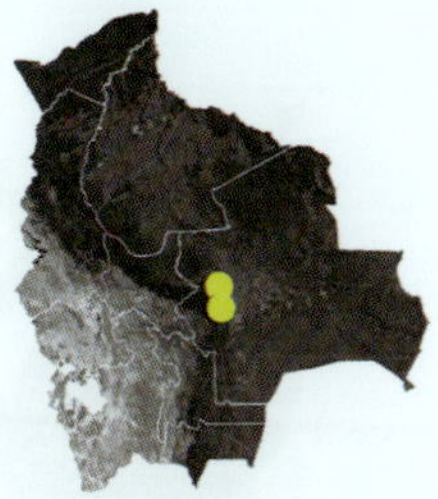

Puanama sara Galileo & Martins, 1998

Lamiinae: Eupromerini

This small species (3–4 mm) is very similar to others in the tribe in its mottled coloration and enlarged profemora. The predominant coloration is slightly more reddish than in other species, however. It is endemic to Bolivia, where it has been collected by beating vegetation on the road to Bella Vista and in the vicinity of the Flora and Fauna Hotel near Buena Vista in Santa Cruz Department.

Falsamblesthis unguicularis (Tippmann, 1960)

Lamiinae: Forsteriini

This small (9–10 mm) and narrow-bodied species is generally only rarely encountered by beating vegetation. The coloration is dark brown on the head, pronotum, and elytra but covered in lighter-colored pubescence. The antennae and legs are paler brown in color. Most antennomeres are darker apically. This endemic Bolivian species was originally described from Coroico in La Paz Department. Since then, additional specimens have been collected at Refugio Los Volcanes, Samaipata, and the Amboró road above Achira in Santa Cruz Department.

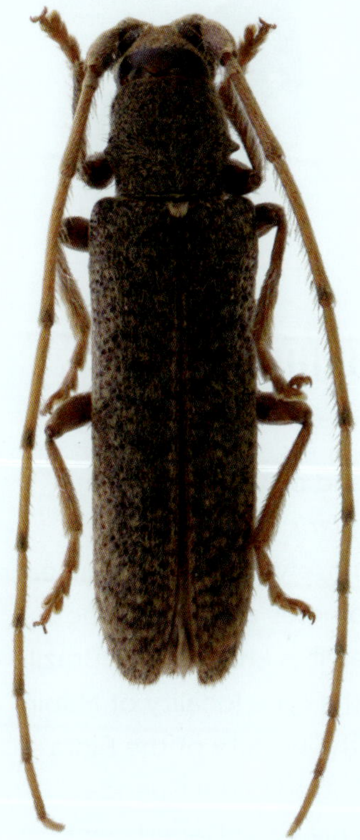

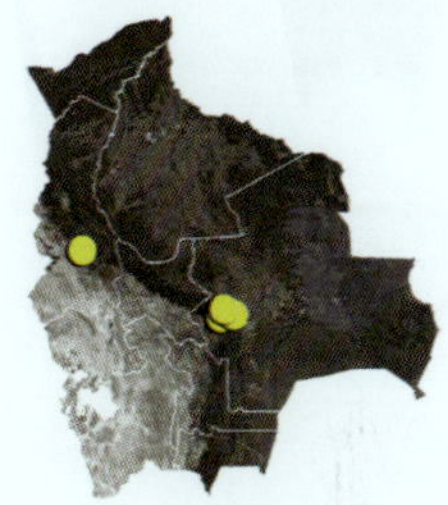

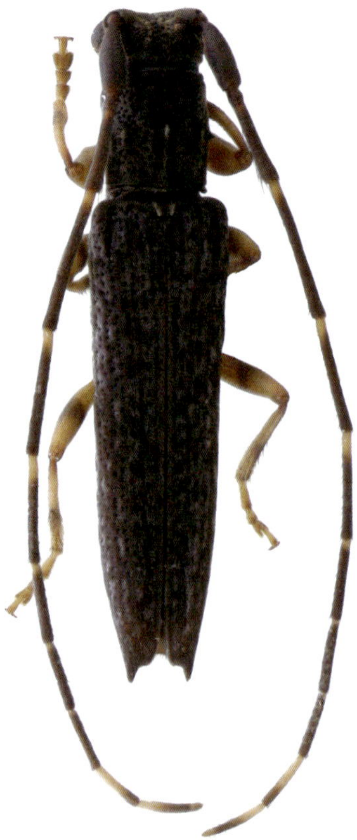

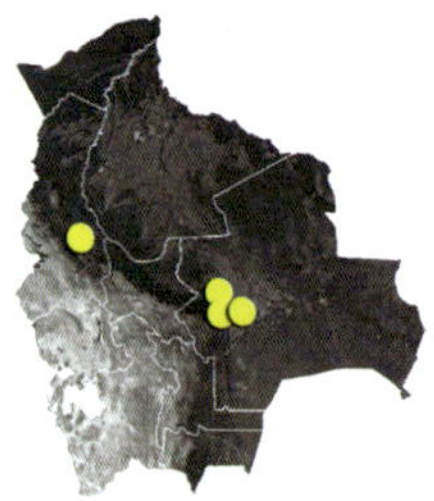

Nyctonympha flavipes Aurivillius, 1920

Lamiinae: Forsteriini

This species is similar in size (7–9 mm) and proportions to the previous species, *Falsamblesthis unguicularis*. The integument is darker, however, on the head, pronotum, elytra, and antennae. Unlike the previous species, the antennomeres are pale at the extreme base and dark for the remainder. It is known from Brazil, Peru, and Bolivia. In Bolivia, aside from the type locality of Mapiri in La Paz Department, it is known from the vicinity of the Flora and Fauna Hotel near Buena Vista, the Potrerillo del Guenda Reserve, and Refugio Los Volcanes in Santa Cruz Department.

Pseudogisostola reichardti Fontes & Martins, 1977

Lamiinae: Forsteriini

This species, 9–12 mm long, is very typical of the small tribe Forsteriini in its small size and narrow body. The pubescence is mostly uniform pale brown but does not completely hide the dark brown integument. The antennae are boldly pale annulate at the bases of most segments and dark at the apical halves. It is known only from Brazil and Bolivia, where it has been collected from the vicinity of the Flora and Fauna Hotel near Buena Vista and near the Potrerillo del Guenda Reserve in Santa Cruz Department.

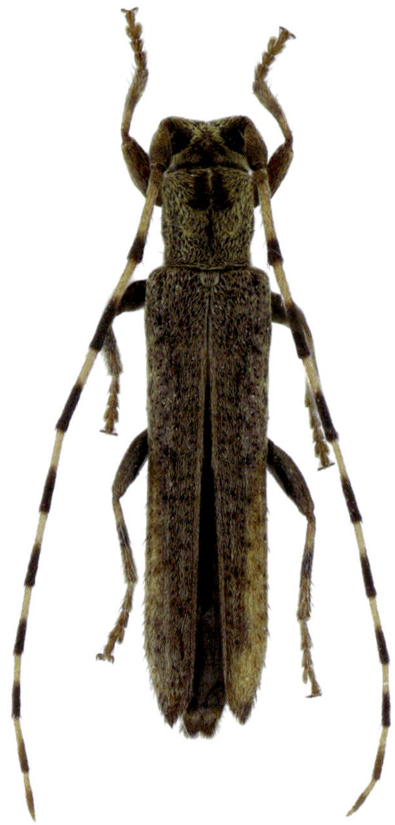

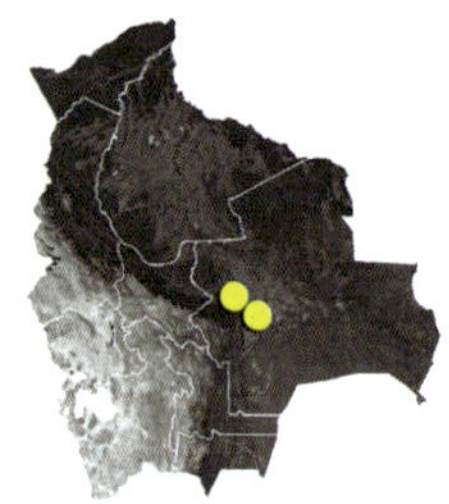

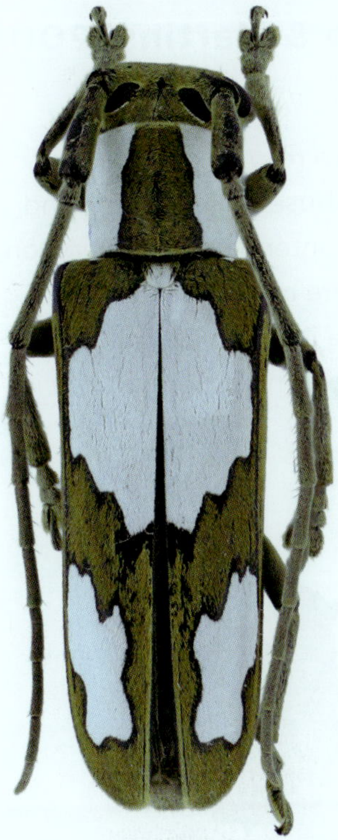

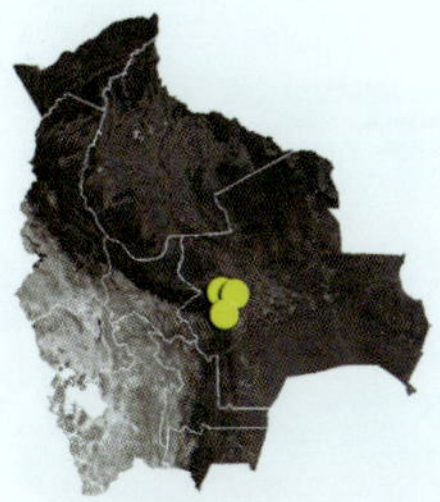

Adesmus divus (Chabrillac, 1857)

Lamiinae: Hemilophini

The genus *Adesmus*, as well as the entire tribe Hemilophini, includes some of the most beautiful Cerambycidae. This species (19–22 mm long) is very distinct with its extremely dense patches of white pubescence at the sides of the pronotum and in the middle half and apex of the elytra. The antennae are covered in pale green or gray pubescence, unlike the similar species *Adesmus nevisi*, which has darker antennae without pale pubescence. This species is known from Brazil, French Guiana, Paraguay, Argentina, and Bolivia. In Bolivia it is known from the Flora and Fauna Hotel near Buena Vista, Potrerillo del Guenda, and Refugio Los Volcanes in Santa Cruz Department.

Adesmus nevisi (Gounelle, 1909)

Lamiinae: Hemilophini

Like the preceding species, *Adesmus divus*, this species (15–17 mm long) is also recognized by its patches of dense, white pubescence on the elytra and pronotum. The patches are smaller than in *A. divus*, however. The antennae are dark reddish brown and not covered in gray pubescence as in *A. divus*. This species is known only from Brazil and Bolivia. It is known from the grounds of the Flora and Fauna Hotel near Buena Vista, Refugio Los Volcanes, and the Concepción area in Santa Cruz Department in Bolivia.

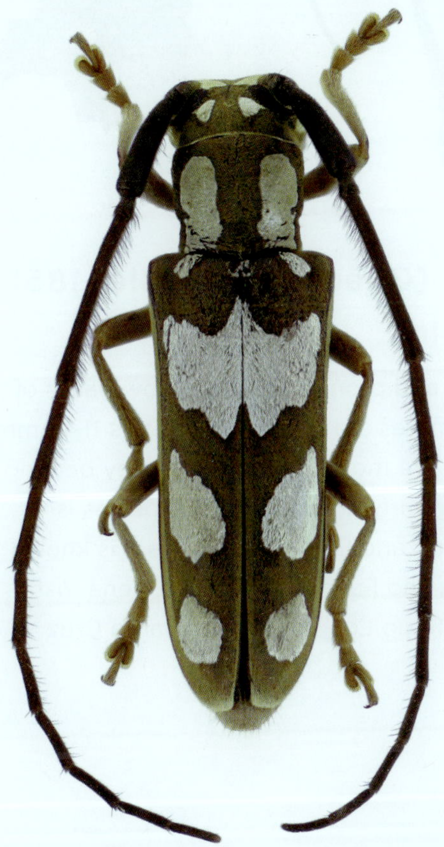

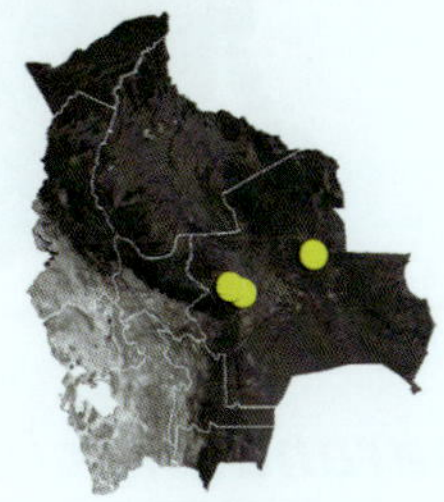

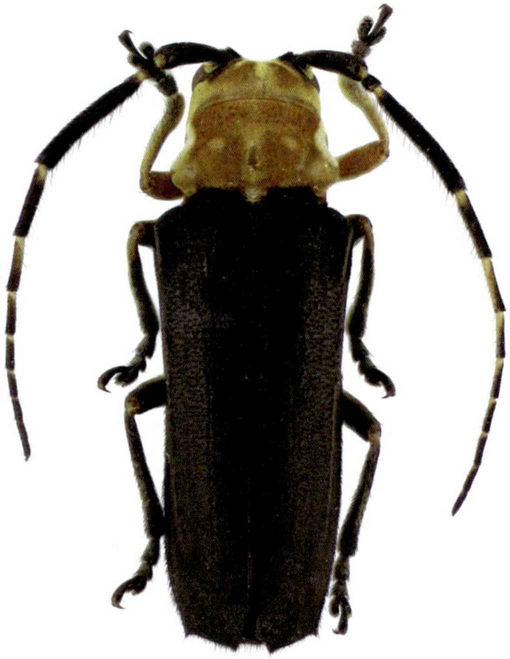

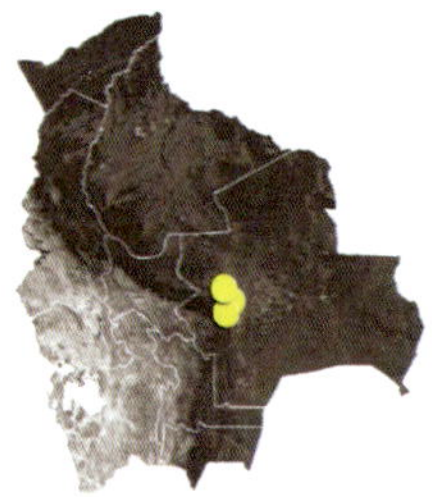

Canarana brachialis (Guérin-Méneville, 1855)

Lamiinae: Hemilophini

The black elytra, yellow head and pronotum, and yellow bases of the otherwise black antennomeres are distinctive for this 11–14 mm long hemilophine. Specimens are most often collected by beating foliage during late morning or early afternoon. This species is known only from Ecuador, Peru, and Bolivia. In Bolivia it is known from the grounds of the Flora and Fauna Hotel near Buena Vista, Potrerillo del Guenda, and Refugio Los Volcanes in Santa Cruz Department.

Canarana roseicollis Galileo & Martins, 2004

Lamiinae: Hemilophini

This hemilophine (9–11 mm) has distinctive rosy or pink bands of pubescence on the pronotum and a diffuse white sutural and epipleural margin of the elytra. The third antennomere is very long, and the subsequent antennal segments are much shorter. This species is known only from Bolivia, where it has been collected at the type locality at Flora and Fauna Hotel near Buena Vista in Santa Cruz Department.

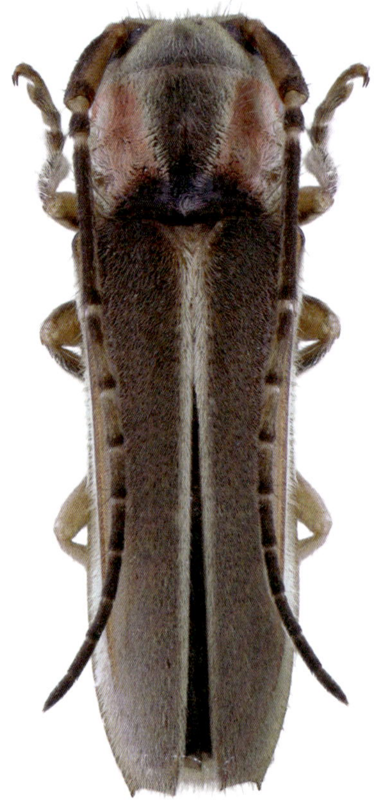

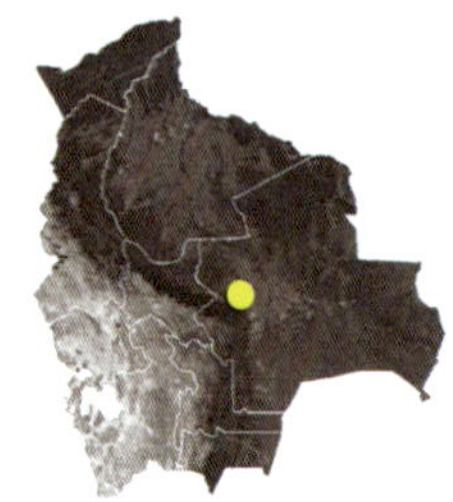

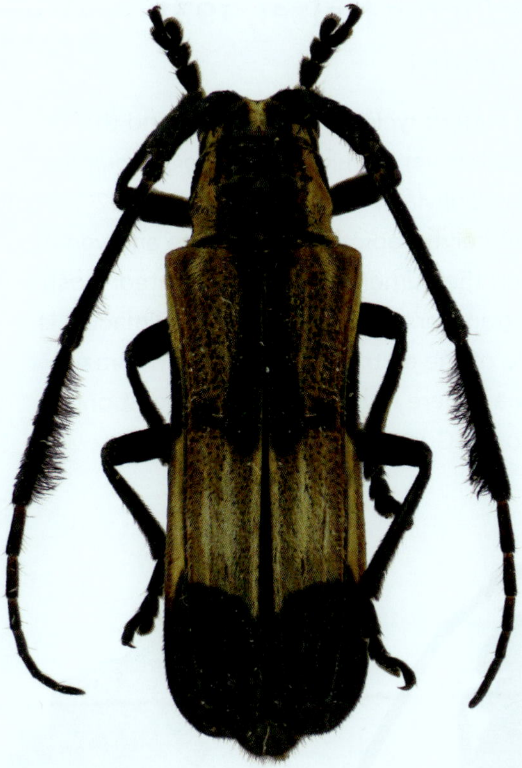

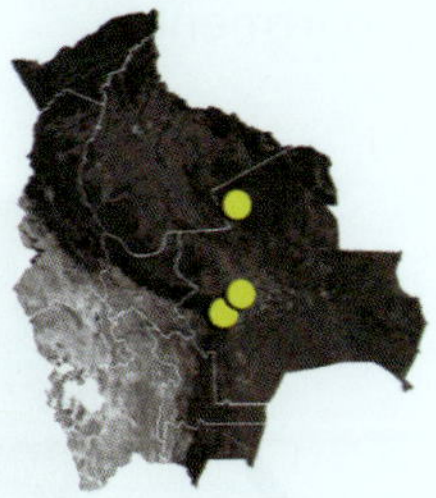

Cuicirama spectabilis (Blanchard, 1846)

Lamiinae: Hemilophini

This beautiful hemilophine (12–16 mm) is recognized by the bold pattern of orange and black on the elytra and pronotum. The antennae have a very long third segment, and the fourth, though not as long, has a dense fringe of pubescence along the inner margin. The remaining segments are much shorter. This species is known only from Peru and Bolivia. In Bolivia it is known from Guarayos (the type locality), Potrerillo del Guenda, and Refugio Los Volcanes in Santa Cruz Department.

Hemiloapis ybyra Galileo & Martins, 2004

Lamiinae: Hemilophini

Small to moderate sized (9–11 mm), this species has a very dark brown to black head and pronotum with distinct patterning of golden pubescence. The basal half of the elytra is light brown, the apex is dark brown, and these are separated by a transverse fascia of golden pubescence. This rarely collected, endemic Bolivian species is known only from the vicinity of the Flora and Fauna Hotel near Buena Vista in Santa Cruz Department.

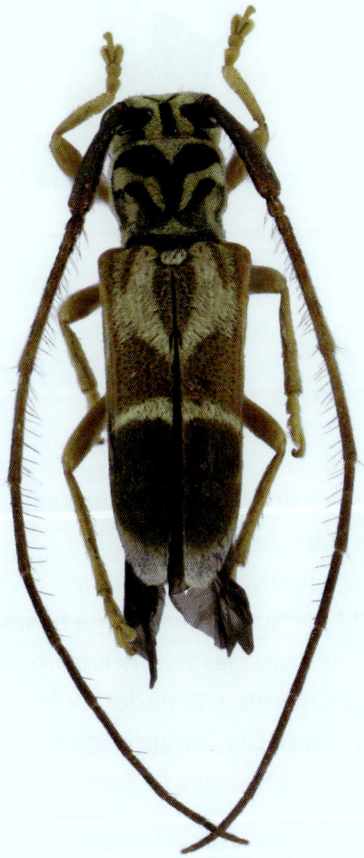

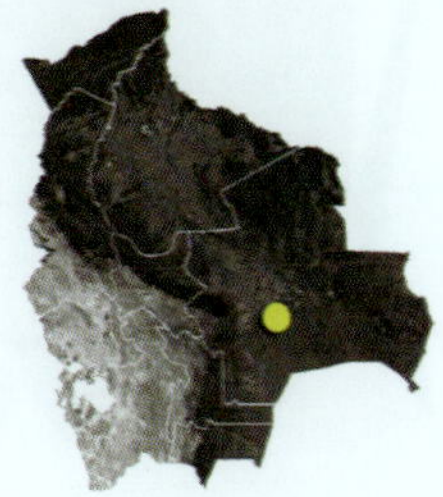

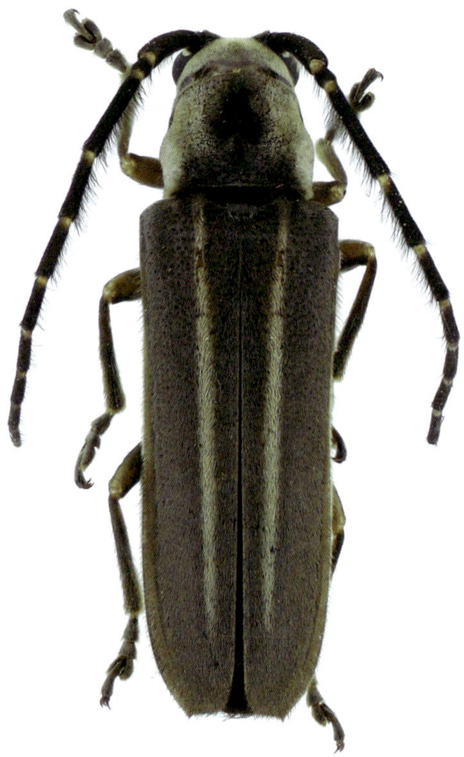

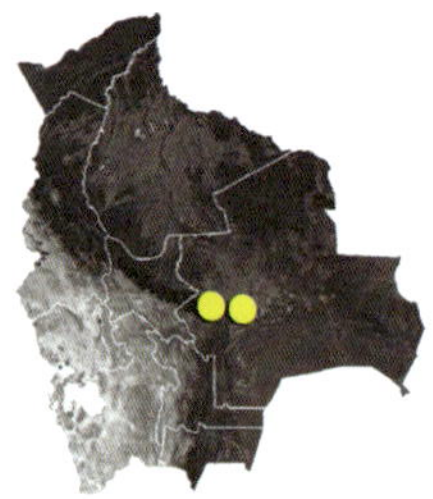

Isomerida lineata Bates, 1874
Lamiinae: Hemilophini

This 13–15 mm long hemilophine is a very good mimic of lightning bugs of the family Lampyridae. It has dark integument, but it is covered by pale pubescence that is concentrated into a stripe on each elytron. The antennomeres are black but have extreme bases of pale yellow. This is a widely distributed species that is known from throughout Central America into Colombia, Ecuador, Brazil, Peru, and Bolivia. In Bolivia it is known from the Amboró road above Achira and the Potrerillo del Guenda Reserve in Santa Cruz Department.

Lycomimus albocinctus Melzer, 1931
Lamiinae: Hemilophini

Several hemilophines are superb lycid beetle mimics, and this species is among the best. It is very similar to *Lycidola simulatrix* Bates, which also occurs in Bolivia. It is moderately large (12–15 mm) with very greatly expanded elytra apically. The costae and venation are quite exaggerated, and the humeri are highly elevated. Mostly purple to black, a broad, transverse yellow to off-white macula is present across the middle of the elytra. It is known from Brazil, French Guiana, and Bolivia, where it has been collected at only one locality: the Potrerillo del Guenda Reserve in Santa Cruz Department.

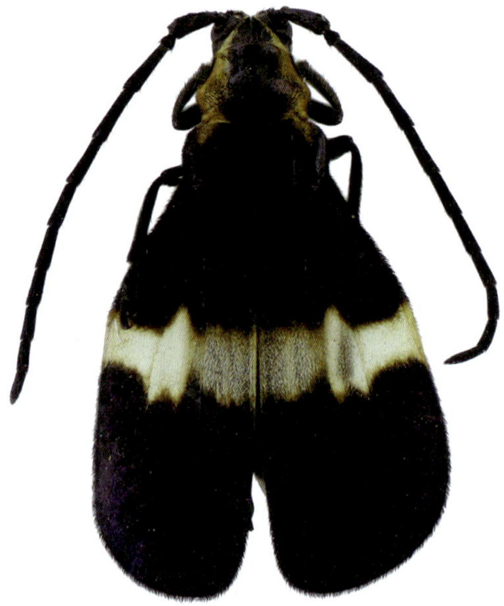

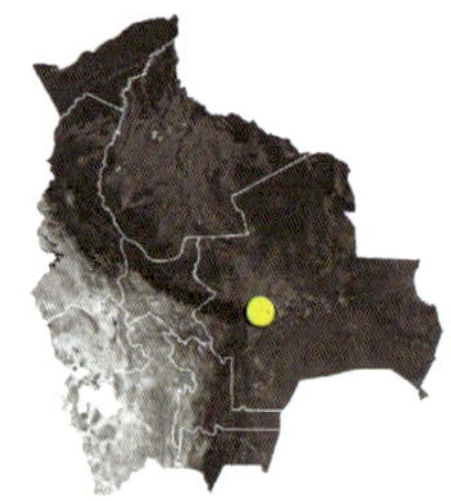

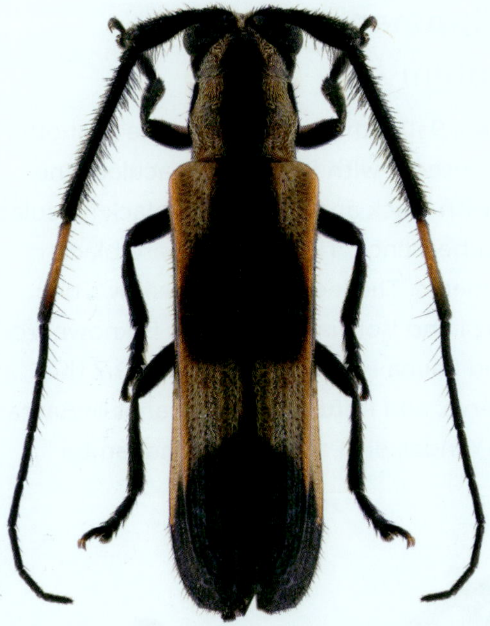

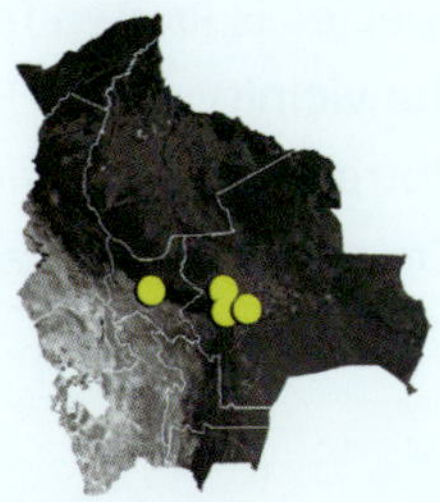

Malacoscylus cinctulus Bates, 1881

Lamiinae: Hemilophini

This is a small- to medium-sized (5–9 mm), black and orange longhorned beetle. The head and pronotum have a black macula at the center; the elytra have a large black macula at the basal half, extending up to the scutellum, along with the apical third black. Elsewhere, the elytra are orange. The antennae have a very long third segment and are all black except for the fourth segment, which is orange. Specimens are diurnal, often found on the underside of large leaves along forest trails. This species is known from Ecuador, Peru, and Bolivia, where it has been collected from the vicinity of the Flora and Fauna Hotel near Buena Vista, the Potrerillo del Guenda Reserve, Cuevas, and Refugio Los Volcanes in Santa Cruz Department. It is also known from the Yungas in Cochabamba Department.

Mariliana cicadellida Galileo & Martins, 2004

Lamiinae: Hemilophini

This small hemilophine (5–6 mm) is beautifully decorated with black-bordered white maculae at the base of the pronotum and on each elytron. Bright orange pubescence covers the head and elsewhere on the pronotum and elytra. The color combination of this delicate species renders this as one of the more boldly patterned longhorned beetles known from Bolivia. It is rarely collected and known from Buena Vista, Santa Cruz Department, Bolivia, but is also recorded from Peru.

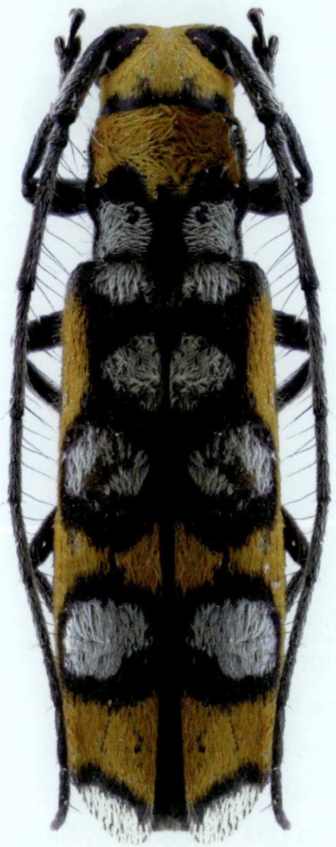

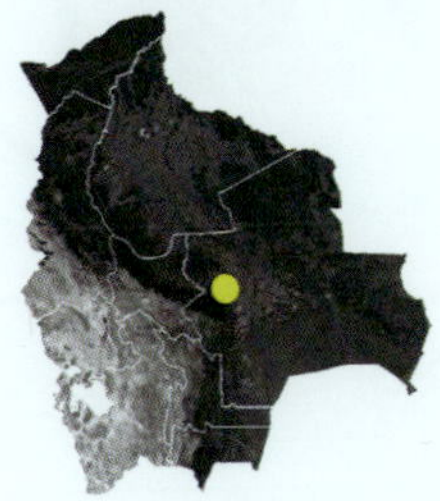

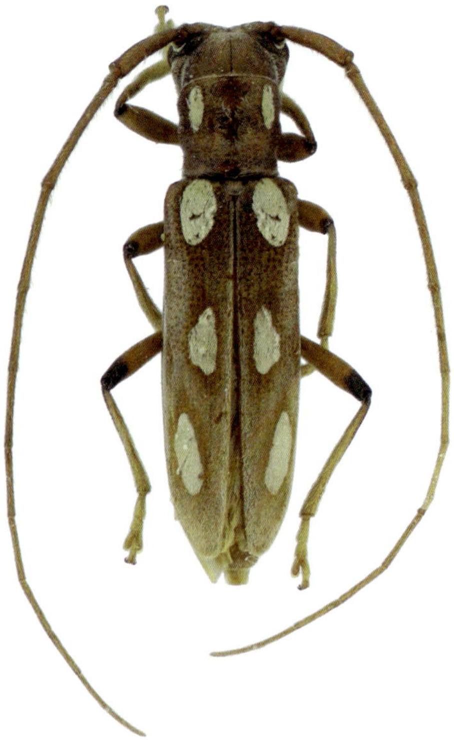

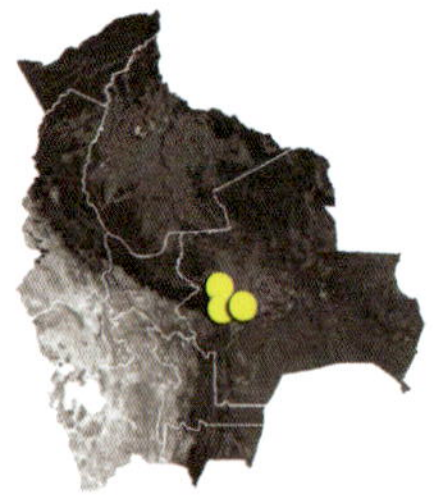

Murupeaca pinimatinga Martins & Galileo, 1992

Lamiinae: Hemilophini

This moderate-sized species (8–12 mm long) is distinct because of the two very dense, white, pubescent maculae on the anterior margin of the pronotum and the three widely spaced, oval, white maculae on each elytron. Elsewhere the integument is orange except for apices of the femora, which are black. This species is known from Brazil and Bolivia, where it has been collected from the vicinity of the Flora and Fauna Hotel near Buena Vista, the Potrerillo del Guenda Reserve, and Refugio Los Volcanes in Santa Cruz Department.

Phoebe ornator Tippmann, 1960

Lamiinae: Hemilophini

This is a strikingly patterned, 9–11 mm long species. The pronotum is densely, bright white pubescent with five black maculae. The elytra are each margined with black and have three black maculae in a field of dense, white pubescence. The antennae are black except for paler basal segments. The legs are pale yellow. This species is known from Brazil and Bolivia. In Bolivia it is known from the vicinity of the Flora and Fauna Hotel near Buena Vista, the Potrerillo del Guenda Reserve, and Refugio Los Volcanes in Santa Cruz Department and the Yungas del Palmar in Cochabamba Department.

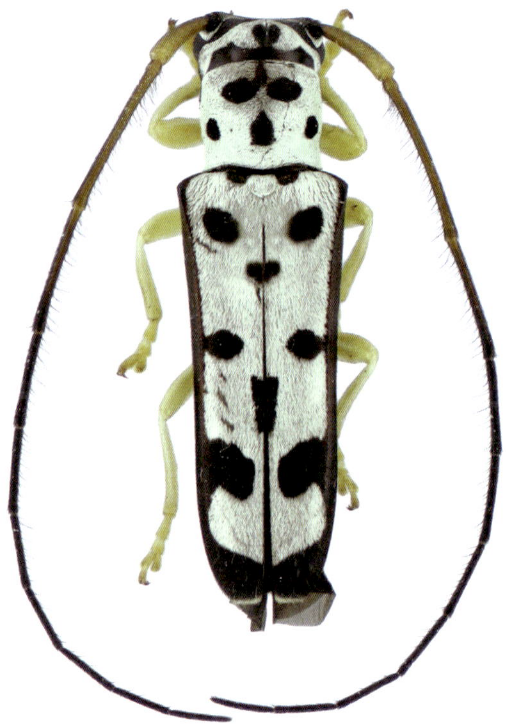

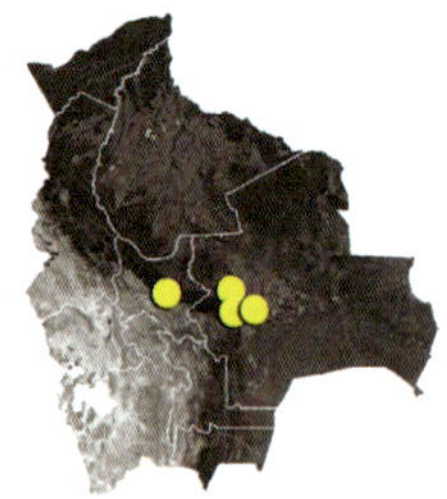

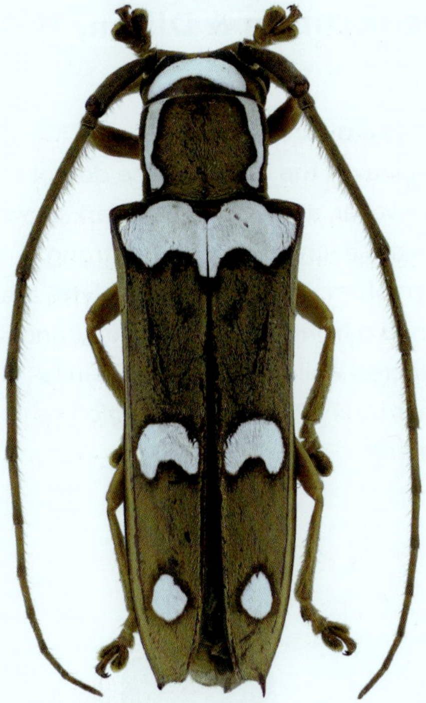

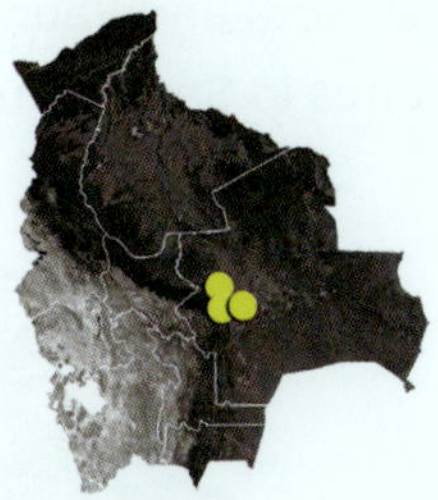

Porangonycha princeps (Bates, 1872)

Lamiinae: Hemilophini

This is one of the larger species of Hemilophini at 18–24 mm in length. The vertex of the head, sides of pronotum, base, middle, and apex of the elytra have patches of extremely dense, bright white pubescence. Elsewhere, the integument is covered with tan pubescence. These are not commonly collected but are occasionally attracted to lights at the beginning of the rainy season. This species ranges from Nicaragua, Costa Rica, Colombia, Peru, French Guiana, and Bolivia. In Bolivia it has been collected at two sites in Santa Cruz Department: the vicinity of the Flora and Fauna Hotel near Buena Vista and the Potrerillo del Guenda Reserve.

Purusia acreana Lane, 1956

Lamiinae: Hemilophini

This species (14–21 mm long) is very similar in coloration to the previous species, *Porangonycha princeps*. It differs in having the white macula of the head divided and the patches on the elytra more expansive. The elytral apices are strongly spinose. This species is also usually encountered at lights. Ranging from Costa Rica to Ecuador, Brazil, and Bolivia, this species has been collected in Bolivia at the Potrerillo del Guenda Reserve in Santa Cruz Department.

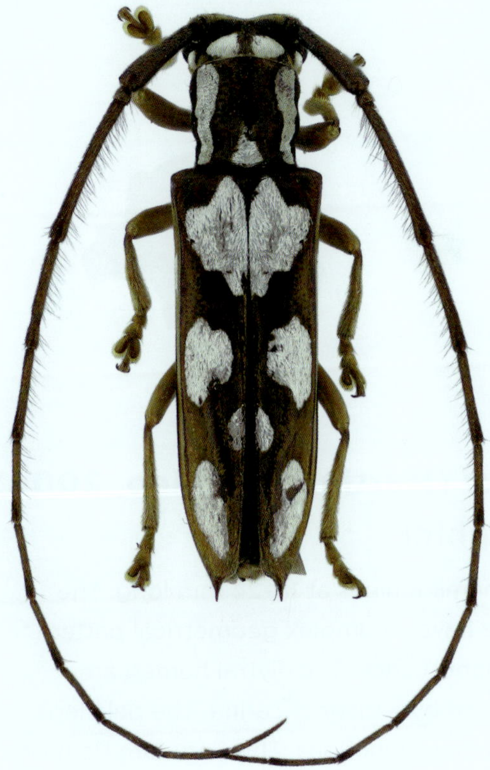

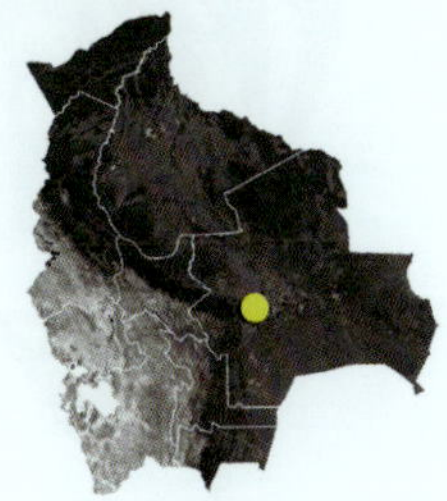

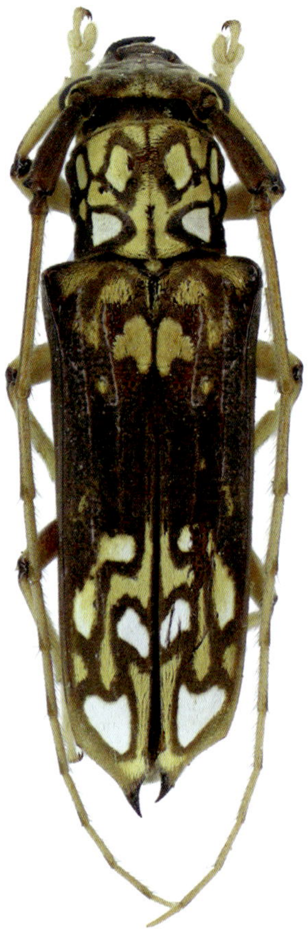

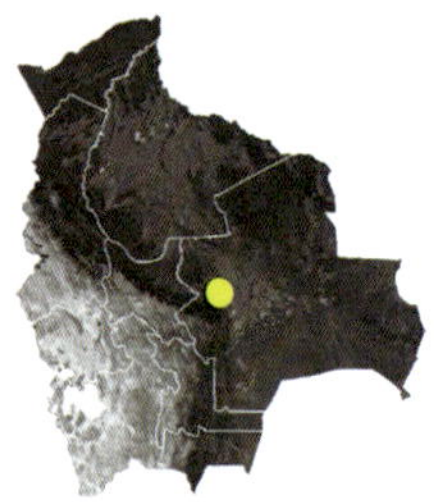

Purusiella wappesi (Martins & Galileo, 2004)

Lamiinae: Hemilophini

This is one of the largest hemilophines at 16–24 mm long. The pronotum and elytral apex have a complex geometrical pattern of white, yellow, and tan pubescence. The elytral humeri are clearly projected and defined by a distinct carina. The pale legs have black-tipped femoral–tibial junctions. This very rare Bolivian endemic species was described from Buena Vista in Santa Cruz Department.

Plagiohammus confusor Dillon & Dillon, 1941

Lamiinae: Lamiini

This large lamiine (25–30 mm) has a distinctive network of irregular, yellow pubescent maculae on the elytra. These do not completely cover the surface, however, and expose the dark brown integument beneath. The apices are each armed with a strong tooth, as are the sides of the pronotum. The base of the elytra are granulate. This rare species is known from Bolivia, Paraguay, and Brazil. In Bolivia it has been collected at lights from the vicinity of the Flora and Fauna Hotel near Buena Vista and Refugio Los Volcanes in Santa Cruz Department.

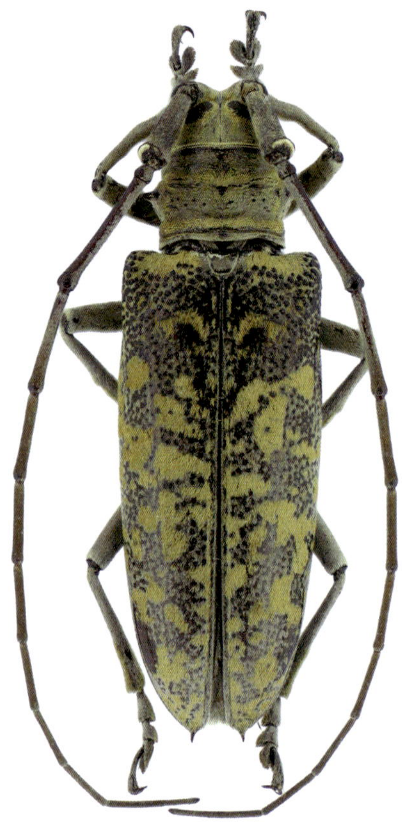

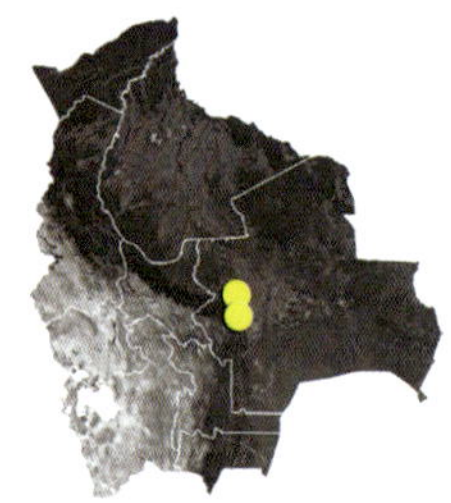

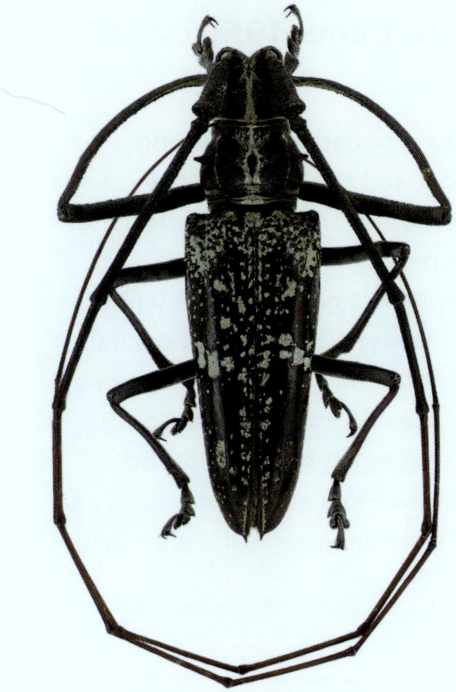

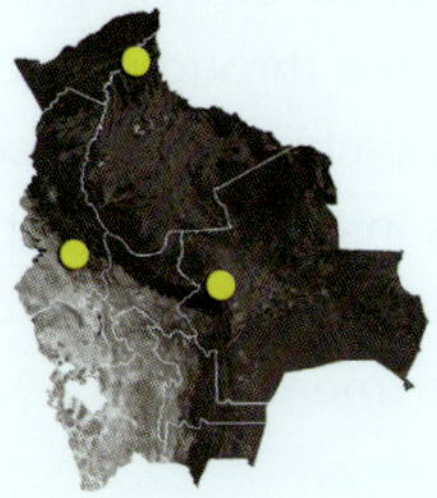

Taeniotes boliviensis Dillon & Dillon, 1941

Lamiinae: Lamiini

This large (30–40 mm) cerambycid, particularly in males like the one shown here, is remarkable in having exceedingly long antennae, nearly three times the length of the body. The third antennomere alone is almost half the body length. The front legs of the males are also quite long, especially compared with the other legs. The coloration is black but with small patches of white or off-white pubescence on the elytra and forming vittae on the pronotum. Although the type locality is in Bolivia, this species has been recorded also from Ecuador and Argentina. In Bolivia it is known from the vicinity of the Flora and Fauna Hotel near Buena Vista (Santa Cruz Department), Larecaja Province (La Paz Department), and near Rurrenabaque (Beni Department).

Taeniotes orbignyi Guérin-Méneville, 1844

Lamiinae: Lamiini

This is another large member of the tribe Lamiini, with specimens 26–29 mm long. This species is black but has a bold pattern of very bright yellow patches forming a nearly complete line at the middle of the pronotum, connecting to the scutellum, and forming an approximate row of spots down the middle of each elytron. This species has been recorded from Panama through northern South America, into Bolivia. In Bolivia it has been collected at Potrerillo del Guenda and surrounding property, Refugio Los Volcanes, and the road to Bella Vista near Cuevas (all in Santa Cruz Department) and also in Beni, La Paz, and Pando Departments.

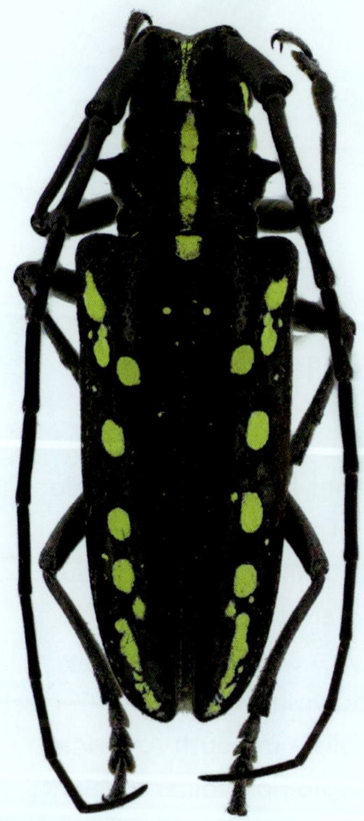

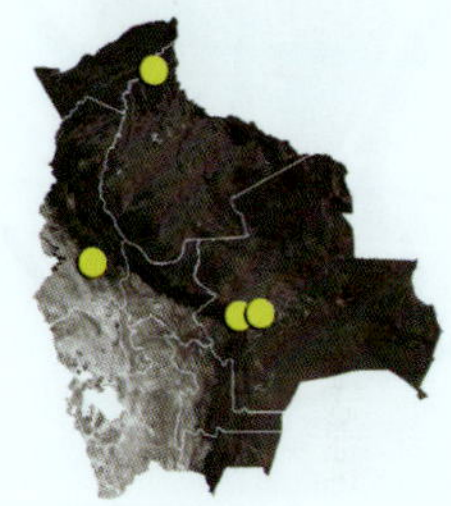

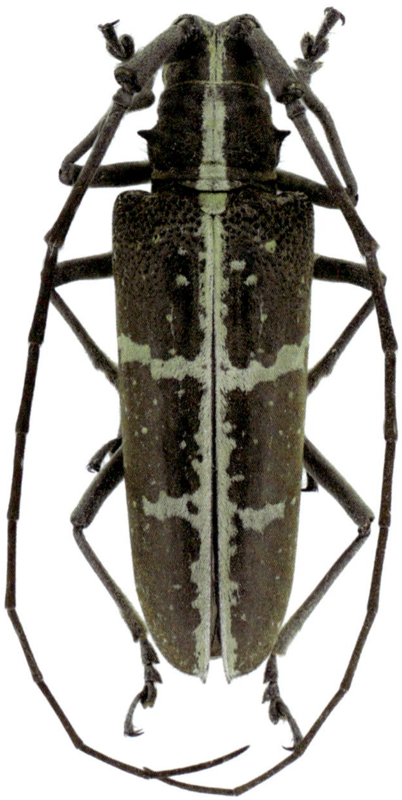

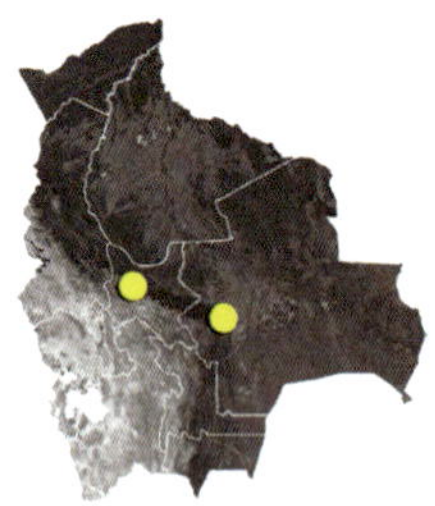

Taeniotes praeclarus Bates, 1872

Lamiinae: Lamiini

This large (24–27 mm), dark brown species has distinctive pale pubescence extending from the vertex of the head, across the middle of the pronotum, and along the suture of the elytra. Just before and after the middle of this band are two lateral bands that extend across each elytron. The elytra are rather strongly granulate at the base. This species is known from southern Central America and Colombia, Ecuador, and Bolivia in South America. In Bolivia this has been collected at Refugio Los Volcanes (Santa Cruz Department) and Alto Chaparé (Cochabamba Department).

Laticranium mandibulare Lane, 1959

Lamiinae: Laticraniini

This very large (32–36 mm) species is remarkable in having a swollen head and mandibles larger and broader than the pronotum, from which the name is based. The elytral patterning is variable, but typically they each have a large antemedial macula of white pubescence. Elsewhere, small specks of white and ochraceous setae occur over the elytra but do not conceal the dark brown or black integument below. This species has been recorded throughout Central America and into much of South America but is never commonly collected. It has been found at lights in Bolivia from the vicinity of the Flora and Fauna Hotel near Buena Vista and the Potrerillo del Guenda Reserve in Santa Cruz Department.

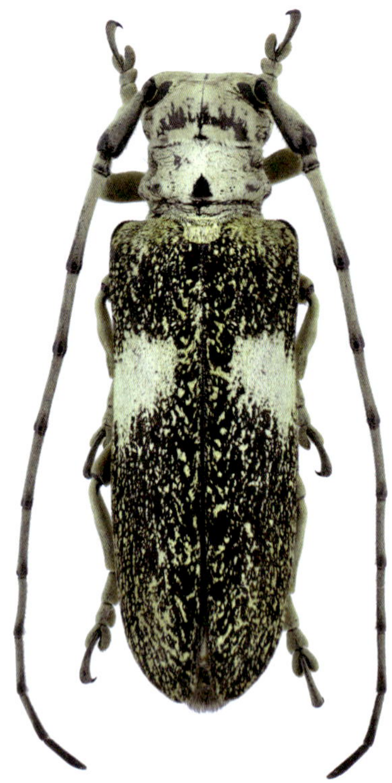

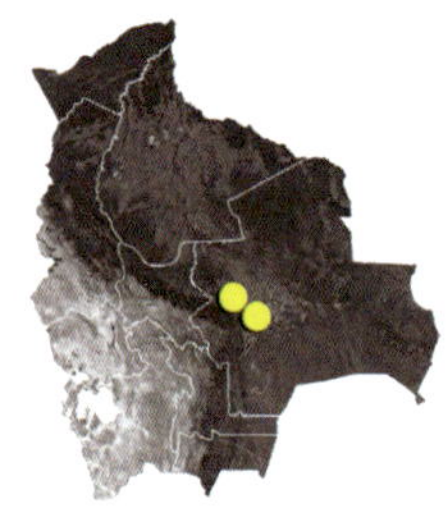

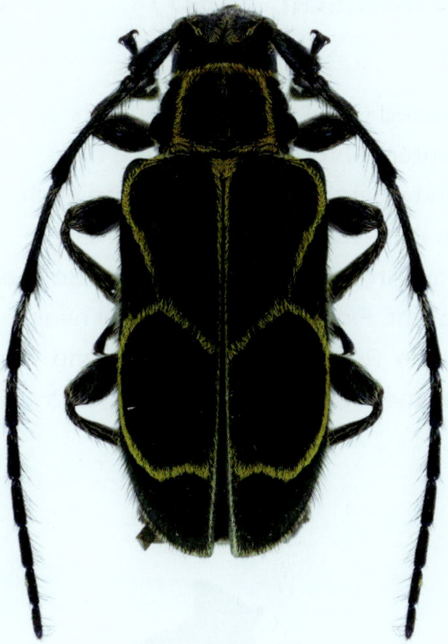

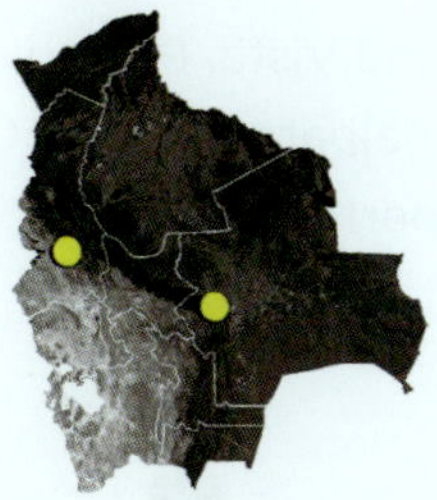

Coroicoia ligata (Schwarzer, 1930)

Lamiinae: Mauesiini

This species is a member of the very small tribe Mauesiini. Averaging about 8–9 mm long, it has a unique pattern of golden-yellow pubescence that forms narrow, interconnected shapes on the elytra and pronotum. The surface integument is otherwise a subtle, iridescent bluish green. Specimens are very rarely encountered and typically only by beating vegetation; when collected, they curl their antennae in a most unusual and distinctive manner. This species is known from Bolivia and Peru. In Bolivia it has been collected at Coroico, the type locality, in La Paz Department and in the Vicoquin area above Achira at 2,000 m in Santa Cruz Department.

Taurolema cicatricosa Lane, 1966

Lamiinae: Mauesiini

The beautiful colors of this small (5–7 mm) species are best revealed under microscopic illumination. The metallic purple and green integument on the elytra, head, pronotum, and legs is magnificent. The antennae of this species are covered with very long, erect setae and a dense brush on the fifth and sixth segments. A pronounced scape cicatrix is present as well. This species is known from French Guiana, Guyana, Brazil, and Bolivia. In Bolivia it is known from the vicinity of the Flora and Fauna Hotel near Buena Vista and the Potrerillo del Guenda Reserve in Santa Cruz Department.

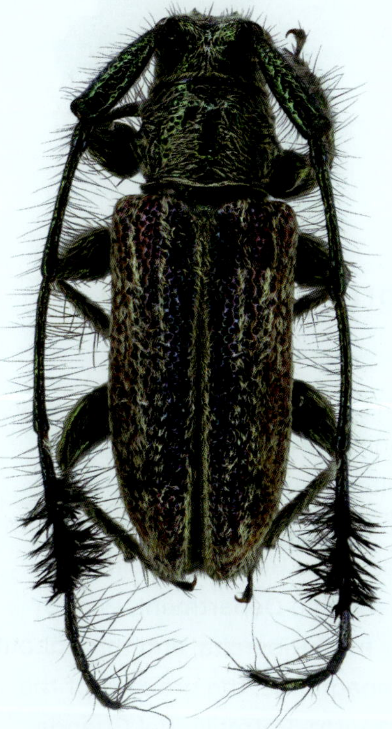

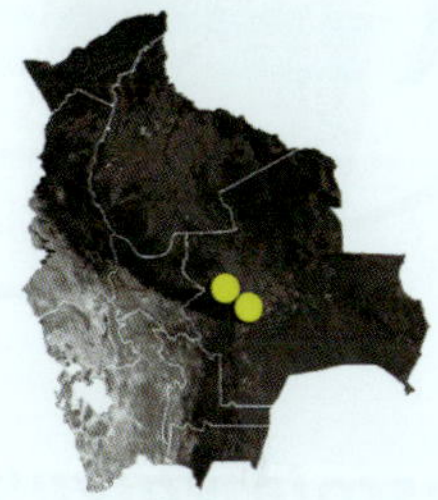

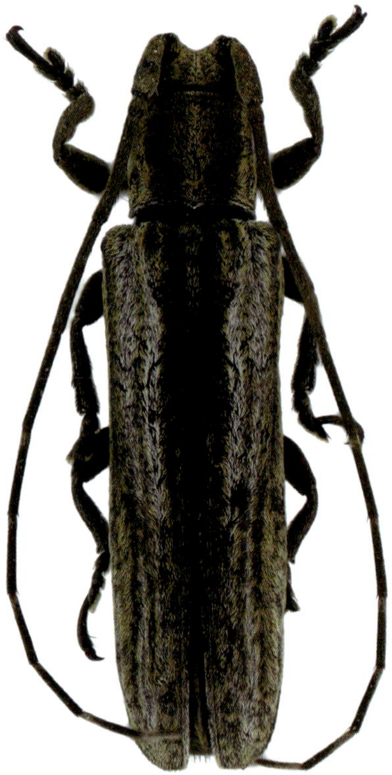

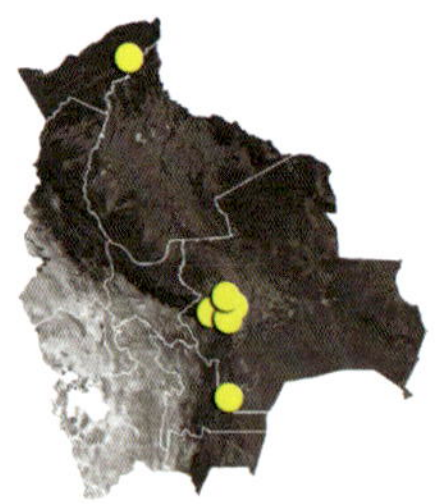

Cacostola volvula (Fabricius, 1781)

Lamiinae: Onciderini

This is a member of a genus of very narrow and elongate onciderines (9–11 mm). It is recognized by the three indistinct vittae on the pronotum and the alternating white and yellow-brown, longitudinal, pubescent stripes on the elytra. This species is known from French Guiana, Brazil, and Bolivia. In Bolivia it is known from Cristal Mayu in Cochabamba Department, Pando Department across the Beni River from Riberalta, and throughout Santa Cruz Department from Pampagrande, the grounds of the Flora and Fauna Hotel near Buena Vista, Potrerillo del Guenda, Refugio Los Volcanes, and the Camiri environs.

Cipriscola fasciata (Thomson, 1860)

Lamiinae: Onciderini

This distinctive and moderate-sized (11–18 mm) onciderine is recognized by the prominent antennal tubercles, arcuate third antennal segment, and bold white patch of pubescence on each elytron. This widespread South American species is known from Colombia, Brazil, Peru, Paraguay, Argentina, and Bolivia. In Bolivia it is known from the grounds of the Flora and Fauna Hotel near Buena Vista, Potrerillo del Guenda, Refugio Los Volcanes, and Samaipata in Santa Cruz Department and Villamontes in Tarija Department.

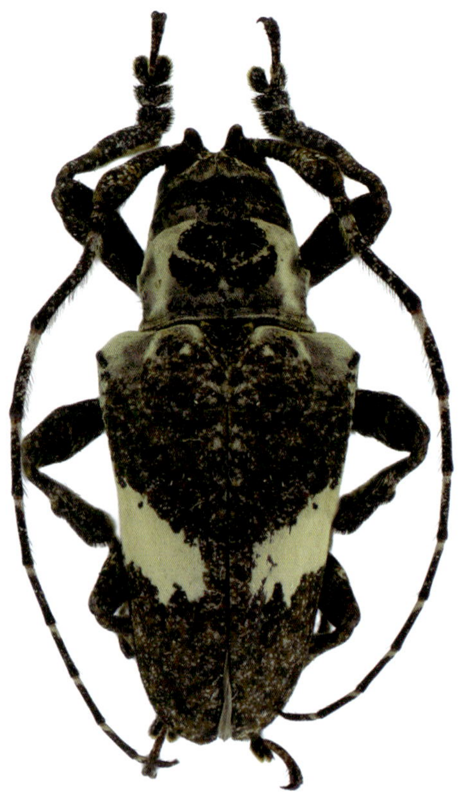

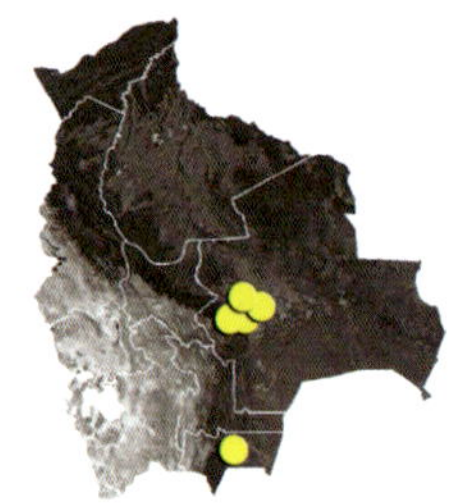

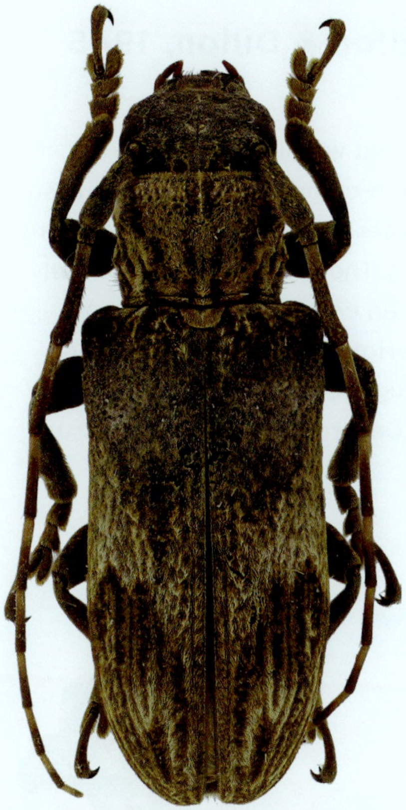

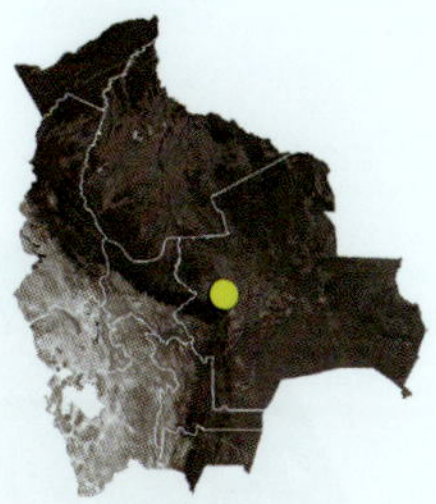

Eudesmus ferrugineus (Thomson, 1860)

Lamiinae: Onciderini

This moderate-sized onciderine (12–19 mm) is recognized by its subtle, diffuse pattern of reddish-brown (some specimens more rust-red than others), white, and tan pubescence over most of the integument. The middle of the elytra have an indistinct white fascia that angles downward toward the suture. The scape is short and the eyes are small in this species. This species ranges from Mexico, through Central America, and into the Guianas, Brazil, and Bolivia. In Bolivia it is known only from Santa Cruz Department near the Flora and Fauna Hotel near Buena Vista.

Eudesmus grisescens Audinet-Serville, 1835

Lamiinae: Onciderini

Similar in size (13–15 mm) and proportions to the previous species, *Eudesmus ferrugineus*, this onciderine has a bolder pattern of pubescence on the elytra. It has a similarly placed white fascia but much more contrast to the surrounding brown pubescence than in *E. ferrugineus*. The antennae are distinctly white annulate at the base, and the third antennomere is broadly swollen in this species. It ranges in the following countries in Central and South America: Nicaragua, Costa Rica, Panama, French Guiana, Trinidad and Tobago, Venezuela, Ecuador, Brazil, Peru, and Bolivia. In Bolivia it is known from the vicinity of the Flora and Fauna Hotel near Buena Vista and the Potrerillo del Guenda Reserve in Santa Cruz Department.

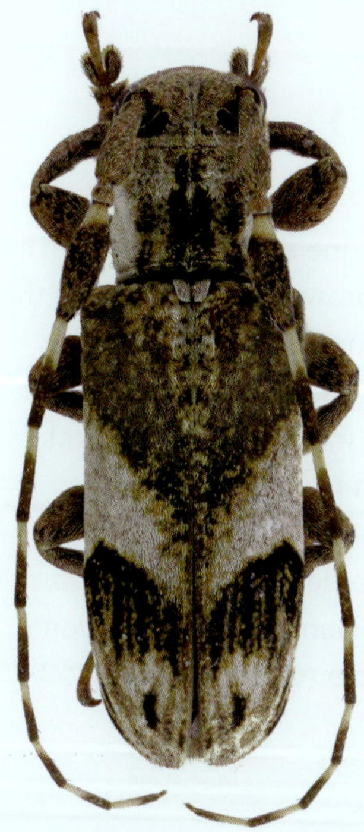

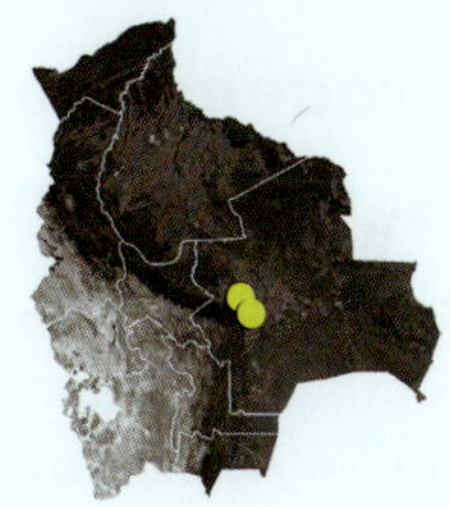

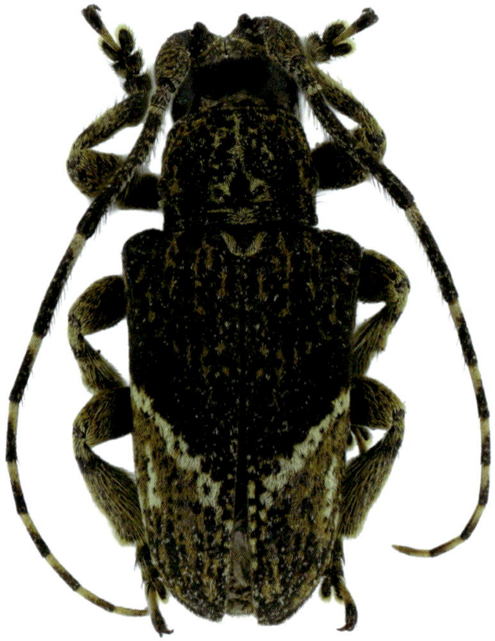

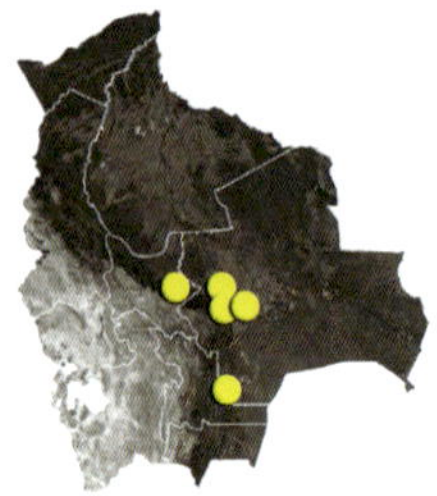

Euthima rodens (Bates, 1865)

Lamiinae: Onciderini

This is a rather small, stout-bodied onciderine at 8–11 mm long. The apical third of the elytra are mostly light brown pubescent, and the remainder are very dark brown with the angled border between the two colors marked by an oblique, narrow, white pubescent fascia. The basal half of the elytra contain many large punctures. This is a widespread species that is known from Ecuador, Peru, French Guiana, Brazil, and Bolivia. It has been collected from the following localities in Bolivia: Potrerillo del Guenda Reserve, Refugio Los Volcanes, and the Flora and Fauna Hotel near Buena Vista (Santa Cruz Department) and El Sacta (Cochabamba Department).

Hesychotypa aeropa Dillon & Dillon, 1945

Lamiinae: Onciderini

This species, 10–17 mm long, is distinct with its incomplete longitudinal fasciae of pink- or peach-colored pubescence. These are bordered narrowly by white pubescence. Elsewhere the elytra and remaining integument is brown. The pronotum is moderately tuberculate at the sides. This Bolivian endemic is known from the vicinity of the Flora and Fauna Hotel near Buena Vista, the Potrerillo del Guenda Reserve, Refugio Los Volcanes, adjacent to Amboró National Park, and the Camiri environs in Santa Cruz Department.

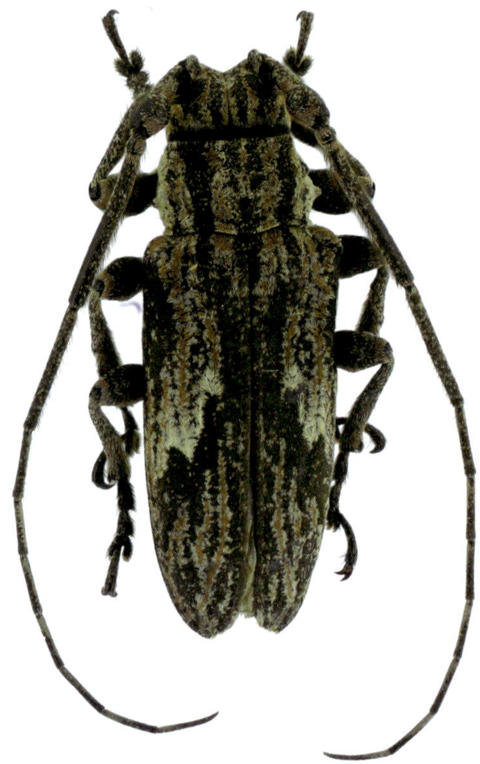

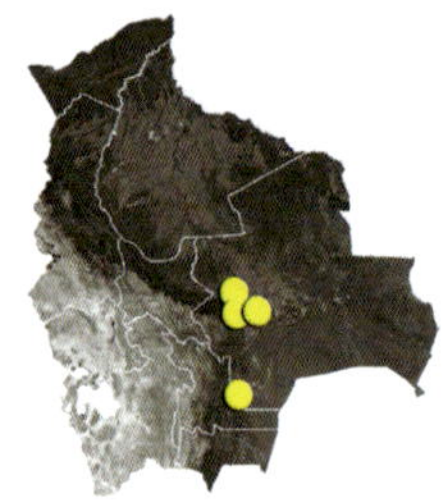

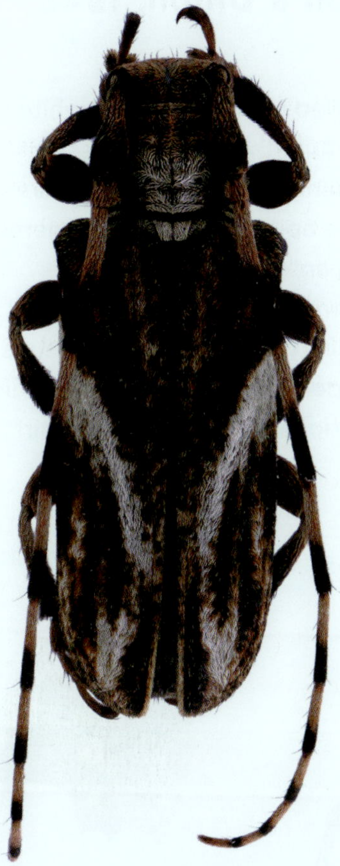

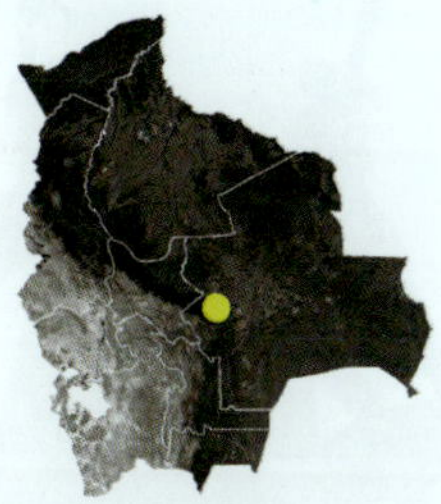

Hesychotypa aotinga Martins & Galileo, 2008

Lamiinae: Onciderini

This 12–15 mm long species is recognized by the clavate scape, rusty-red basal halves of the antennomeres, oblique white fascia angling across the middle third of each elytron, and diffuse, white, pubescent patch on the posterior half of the pronotal disk. This species is endemic to Bolivia and has been collected only on the Amboró road above Achira (Santa Cruz Department).

Hesychotypa balia Martins & Galileo, 2009

Lamiinae: Onciderini

This is a beautiful and rather large member of the genus at 16–20 mm in length. It has very dark brown integument, but on the elytra it is covered by a reticulating network of yellow pubescence. This pubescence continues in a more organized fashion over the laterally tuberculate pronotum. The antennal tubercles of the males are equipped with a very long and acute, anteriorly directed spine. This rare Bolivian endemic is known only from the Flora and Fauna Hotel near Buena Vista in Santa Cruz Department.

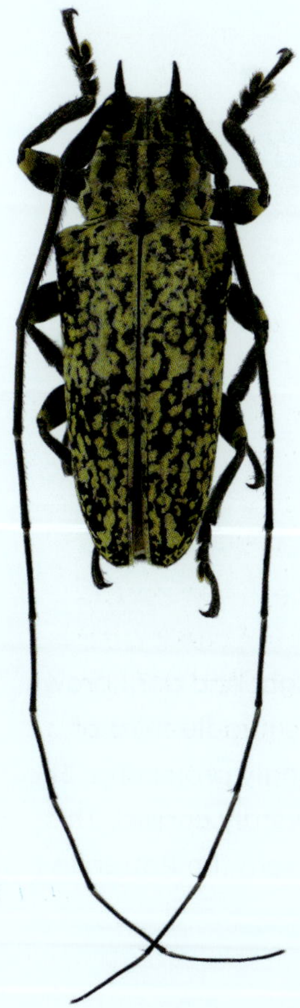

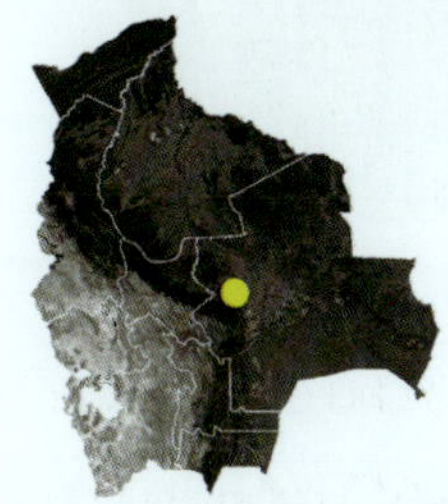

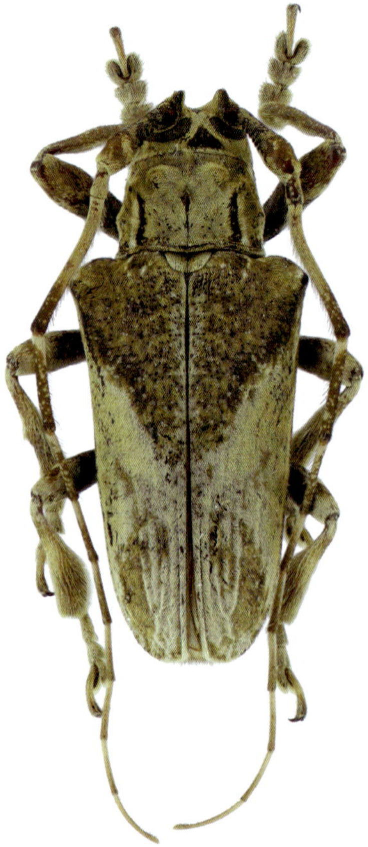

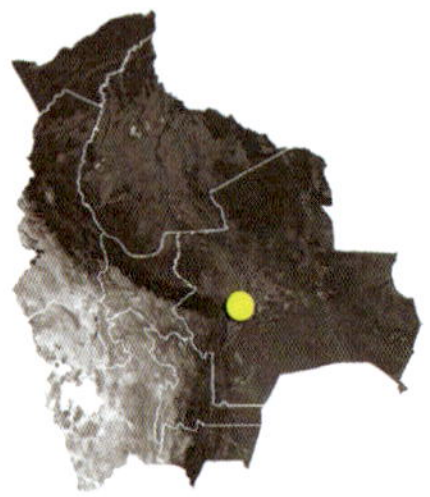

Hesychotypa magnifica Martins & Galileo, 2007

Lamiinae: Onciderini

This 14–19 mm species has subtle white and light and dark brown pubescence over most of the integument. The middle third of elytra is very pale. The elytral humeri are strongly projecting. The third antennomere is quite elongate and outwardly curved. This rare Bolivian endemic species is known only from the Potrerillo del Guenda Reserve in Santa Cruz Department.

Hypsioma steinbachi Dillon & Dillon, 1945

Lamiinae: Onciderini

This species (15–19 mm long) has mottled ochraceous and white pubescence over the elytra and pronotum, though this does not completely conceal the black integument. The elytral humeri are strongly projecting and glabrous at the apices. The antennomeres are basally white pubescent. The third antennomere is sinuate. This common, nocturnal species is distributed in Brazil, Peru, and Bolivia, where it is known from the vicinity of the Flora and Fauna Hotel near Buena Vista, the Potrerillo del Guenda Reserve, Refugio Los Volcanes, adjacent to Amboró National Park, and the Camiri environs in Santa Cruz Department.

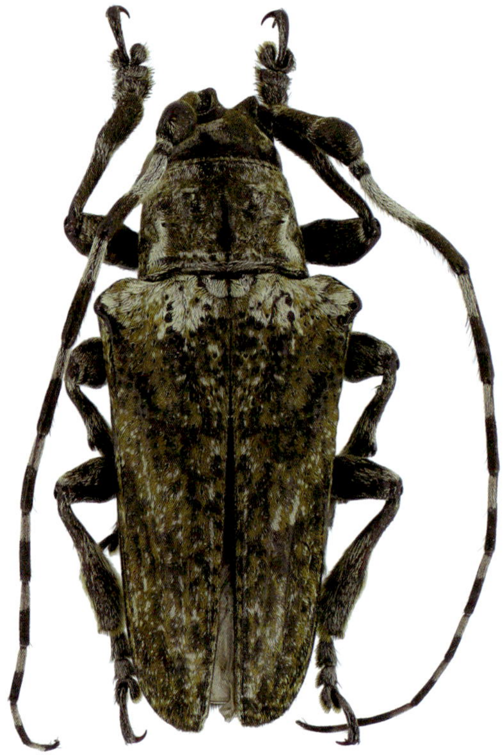

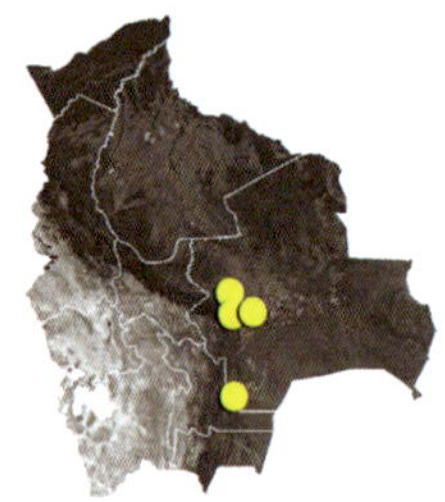

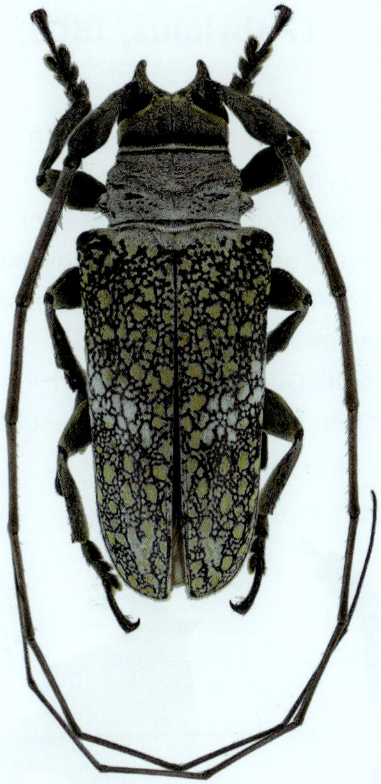

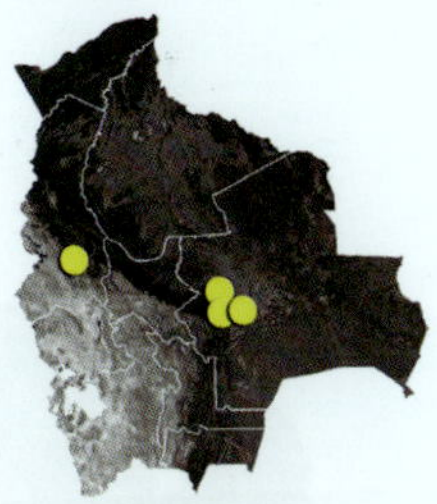

Lochmaeocles callidryas (Bates, 1865)

Lamiinae: Onciderini

This large (17–24 mm) onciderine is attractively decorated with small, closely spaced ochraceous and white pubescent maculae on the elytra. The pronotum is mostly covered in white pubescence. The antennal scape is clavate, and the third antennomere is straight. The antennal tubercles are armed with variably developed, anterior processes. This common species ranges throughout northern South America into Bolivia. In Bolivia it is known from the vicinity of the Flora and Fauna Hotel near Buena Vista, the Potrerillo del Guenda Reserve, and Refugio Los Volcanes in Santa Cruz Department and also from La Paz Department.

Lochmaeocles pulcher Dillon & Dillon, 1946

Lamiinae: Onciderini

This 19–23 mm long species is very similar to the previous species, *Lochmaeocles callidryas*, in size, shape, and coloration. It differs in having the maculae of the elytra smaller, more widely separated, and virtually all yellow or ochraceous in color. Some maculae toward the apex are outlined in white setae. The extreme base of the elytra, including the scutellum, is covered in white pubescence, as is the pronotum. This species is known from Peru, Ecuador, French Guiana, Brazil, and Bolivia, where it is known only from Beni Department.

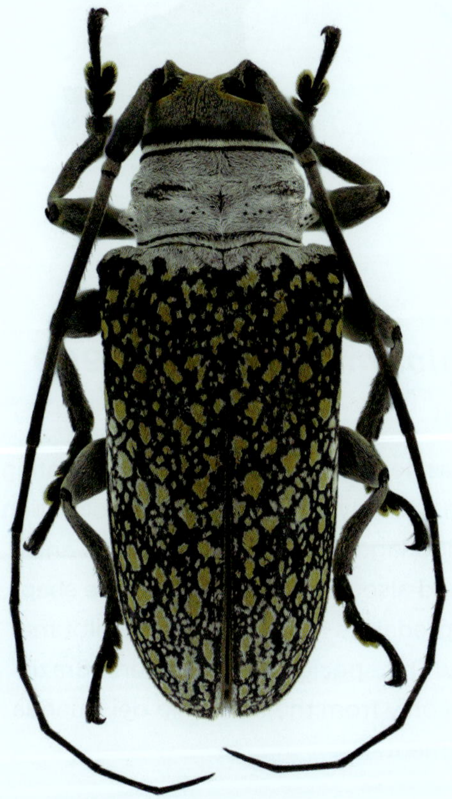

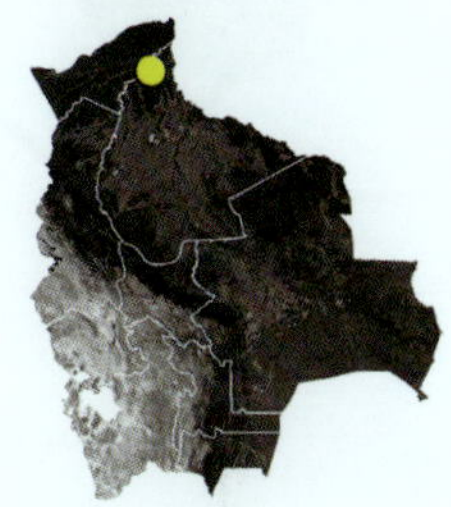

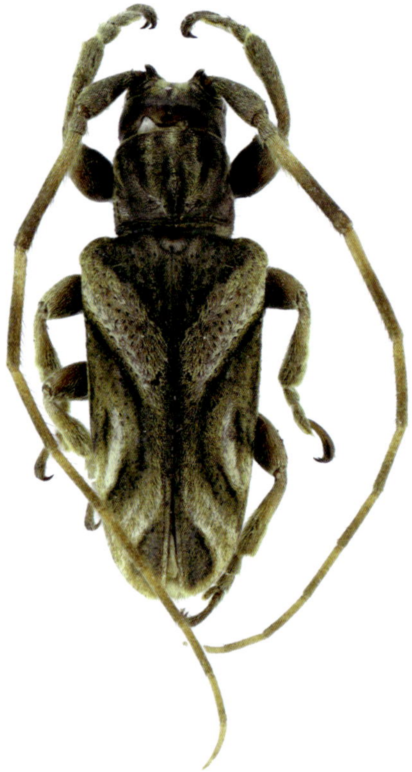

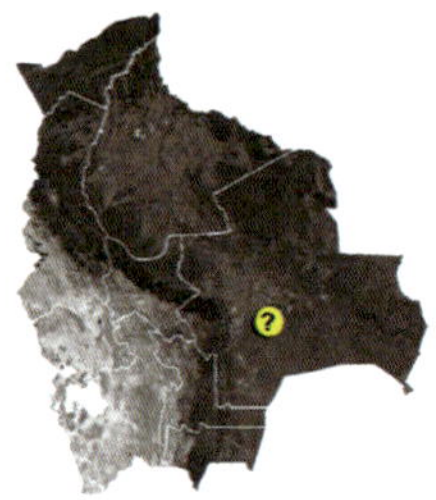

Neocherentes dilloniorum Tippmann, 1960

Lamiinae: Onciderini

This 12–15 mm long species has hemispherical bands of alternating light and dark pubescence on each elytron. This type of coloration provides a very effective camouflage in the shadowy leaves and branches where adults hide and also serves to break up the shape of the beetle visually so that predators have a more difficult time perceiving it as potential prey. This species occurs in Peru, Brazil, and Bolivia, where it is known only from the Potrerillo del Guenda Reserve in Santa Cruz Department.

Neolampedusa obliquator (Fabricius, 1801)

Lamiinae: Onciderini

The most notable feature of this small (7–9 mm) onciderine is the narrow, poorly defined, white fascia extending from the humerus toward the suture and curving around the apex of each elytron. Otherwise, the coloration is very dark brown. The legs and antennae are mottled with light and dark pubescence. This species has been collected in French Guiana, Brazil, Peru, and Bolivia, where it is known from the Potrerillo del Guenda Reserve and Refugio Los Volcanes in Santa Cruz Department and northwest of Riberalta in Pando Department.

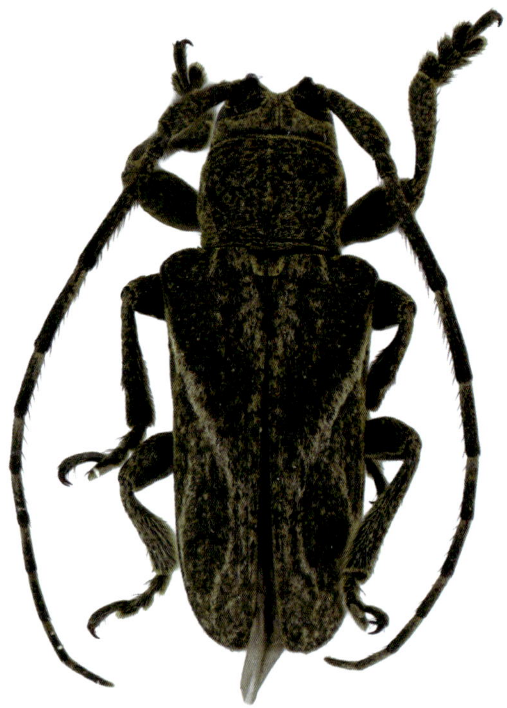

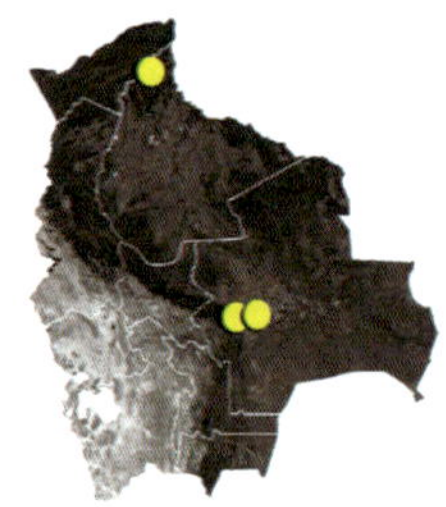

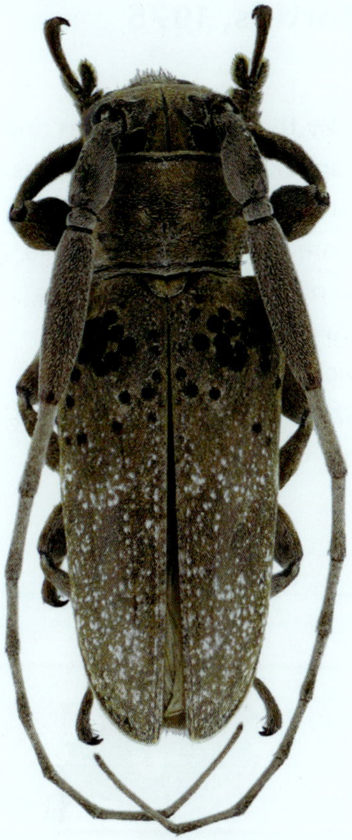

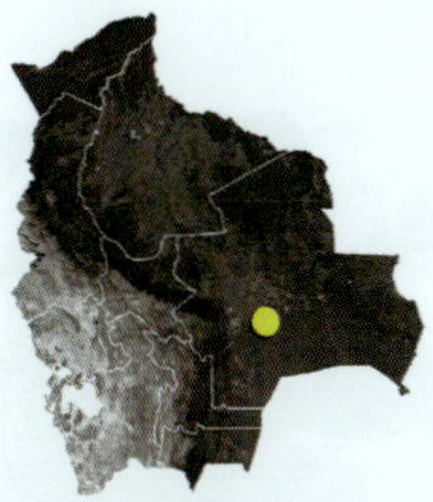

Oncideres ulcerosa (Germar, 1824)

Lamiinae: Onciderini

This is distinctive member of the very large genus *Oncideres*. Specimens are 13–17 mm long. The third antennomere is considerably swollen and as broad (and much longer) than the scape. The base of each elytron has 10–15 large, shiny black granules. Otherwise, the elytra lack granules or notable punctures. This species occurs in Brazil, Bolivia, and Paraguay. In Bolivia it has been recorded only from Santa Cruz de la Sierra in Santa Cruz Department.

Oncideres wappesi Martins & Galileo, 2005

Lamiinae: Onciderini

This small (8–10 mm) onciderine has light sandy-brown pubescence over most of the body. The pronotum has a transverse row of five shiny, glabrous calli. The elytra each have 10–12 large, shiny black granules. The black, shiny humeri are completely denuded. This Bolivian endemic species is known from the vicinity of the Flora and Fauna Hotel near Buena Vista and the Potrerillo del Guenda Reserve in Santa Cruz Department.

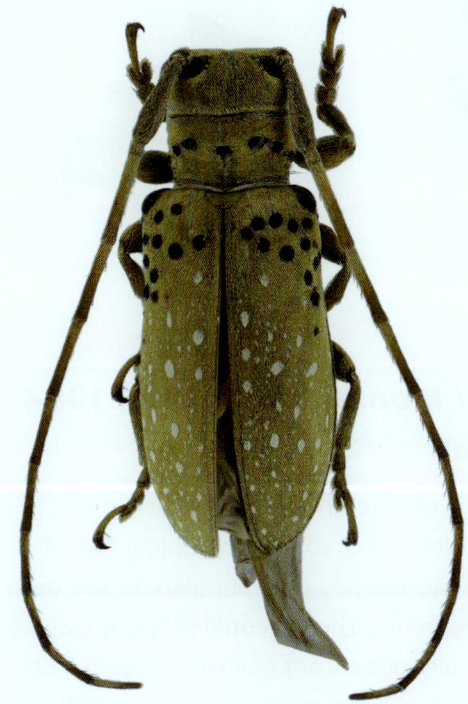

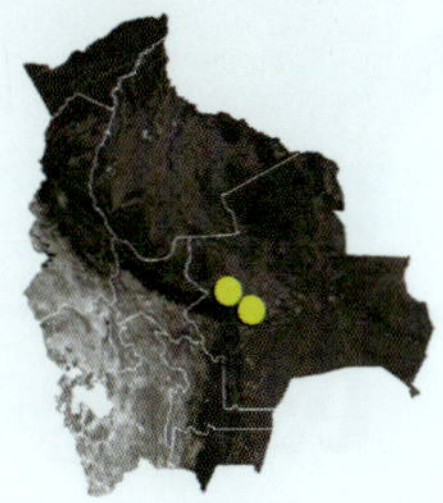

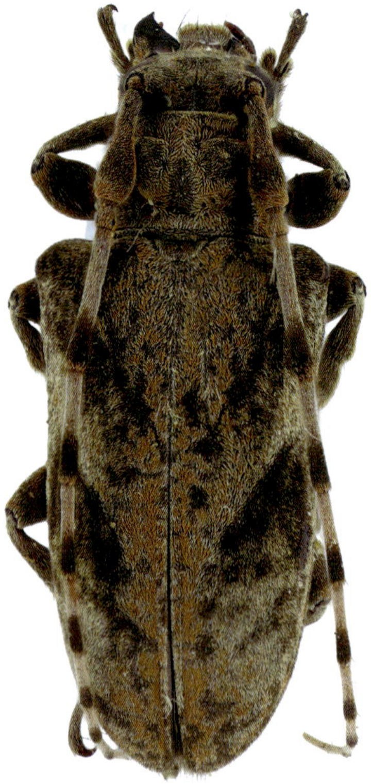

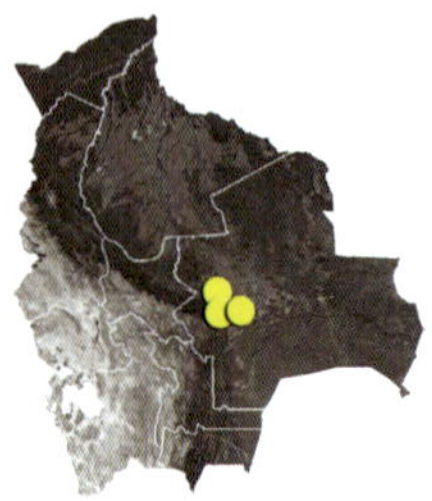

Pseudobeta seabrai Monné & Fragoso, 1984

Lamiinae: Onciderini

This species (9–12 mm long) is characterized by the rusty-red pubescence along the suture (most notable at the apical half of the elytra). The antennae are moderately pale annulate at the base and dark apically on most segments. The antennal scape is clavate. The pronotum has three dorsally pubescent calli. It is known from the vicinity of the Flora and Fauna Hotel near Buena Vista, the Potrerillo del Guenda Reserve, and Refugio Los Volcanes in Santa Cruz Department.

Trachysomus cavigibba Martins, 1975

Lamiinae: Onciderini

The genus *Trachysomus* is recognized by the warty tubercles and projections over the elytral base. This species (10–15 mm long) is no exception with its highly pronounced multituberculate basal elevations. The basal half of the elytra are heavily punctate. The coloration is very cryptic and made up of light and dark brown pubescence. It is known only from Peru and Bolivia, where this species has been collected from the vicinity of the Flora and Fauna Hotel near Buena Vista, the Potrerillo del Guenda Reserve, and Refugio Los Volcanes in Santa Cruz Department.

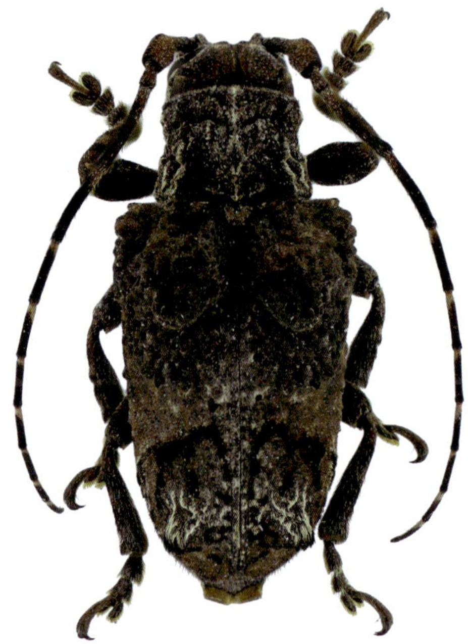

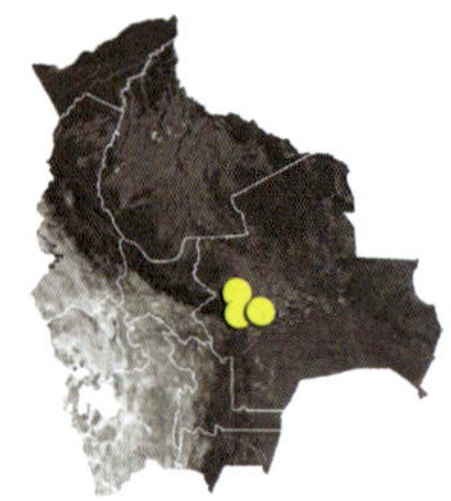

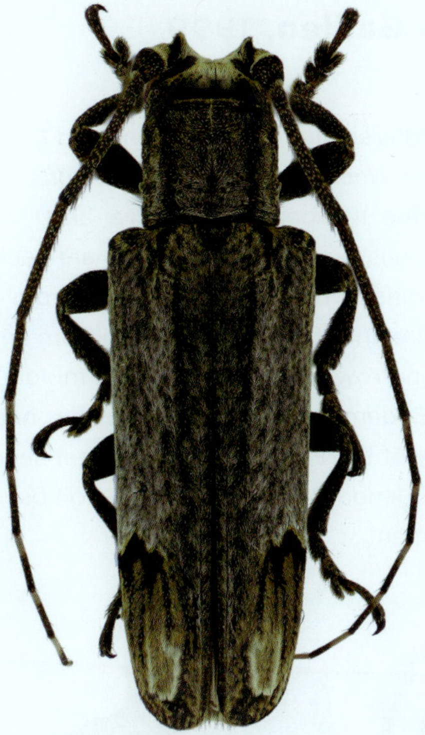

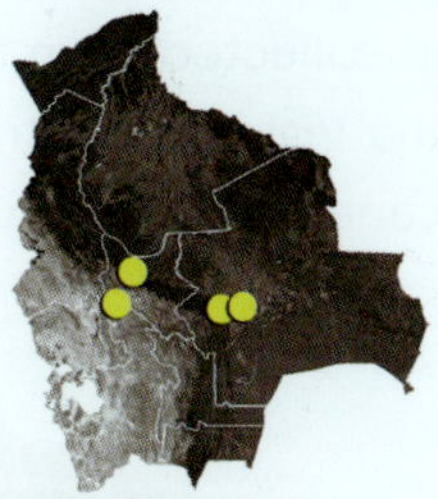

Trestonia frontalis (Erichson, 1847)

Lamiinae: Onciderini

This elongate onciderine (15–18 mm) has distinctive white-specked setal patches on the antennae and legs, which are otherwise brown. A mixture of white setae and brown integument are visible on the basal two-thirds of the elytra while the apical third are coated with ochraceous tan pubescence. The junction of these two elytral colors is a wavy, transverse black macula. It is known from Colombia, French Guiana, Peru, Ecuador, Brazil, and Bolivia. In Bolivia this species has been collected in Cristal Mayu (Cochabamba Department) and Potrerillo del Guenda and Refugio Los Volcanes (Santa Cruz Department).

Tulcus pigrus (Martins & Galileo, 1990)

Lamiinae: Onciderini

This onciderine (9–12 mm long) has pronounced humeri and an angled band of white pubescence originating on the outside middle of each elytron, extending posteriorly toward the suture and then along the suture to the apex. Anterior to this band, the elytra are rather heavily punctate. Another distinctive feature is the tarsal color: the fifth tarsomere is apically dark but otherwise pale. This is another of the endemic Bolivian cerambycids. It is known from the vicinity of the Flora and Fauna Hotel near Buena Vista, the Potrerillo del Guenda Reserve, and west of Samaipata, all in Santa Cruz Department, and from Beni Department.

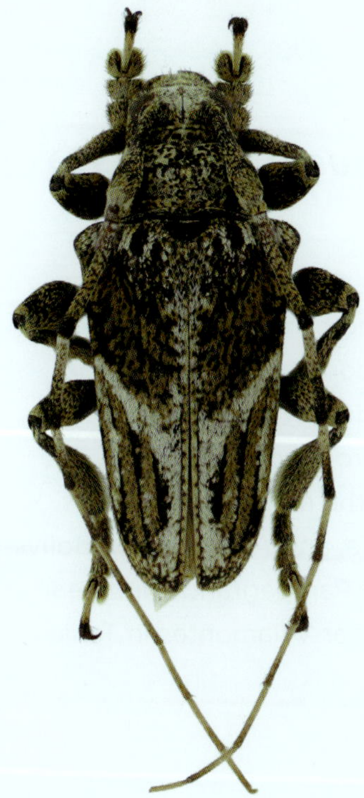

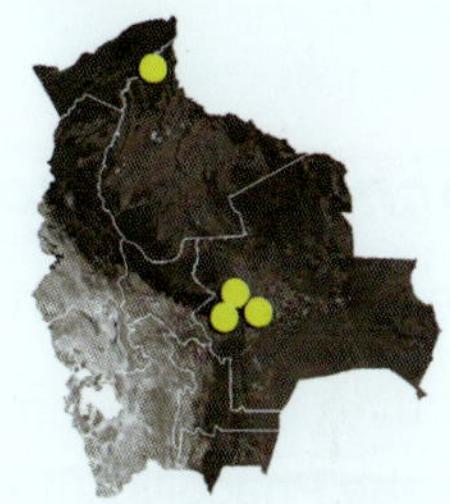

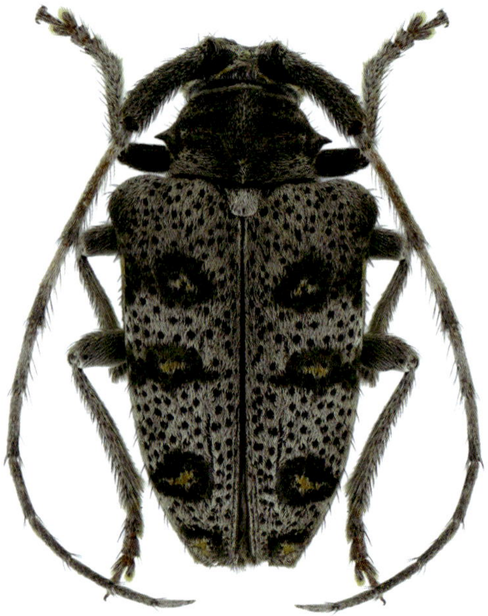

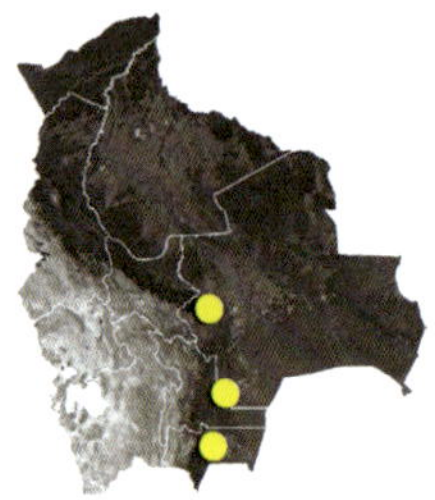

Neobrachychilus consobrinus (Lane, 1939)

Lamiinae: Phacellini

This distinctive, small (5–9 mm) species is a member of a very small tribe, the Phacellini. It is recognized by its gray pubescence over the body and appendages and decoration of four orange maculae on each elytron, each surrounded by a black ring. Large, widely spaced punctures are present throughout the elytra. The pronotum is acutely spined laterally, at the middle of the sides. This species is known from Brazil, Bolivia, and Argentina. In Bolivia this species has been collected north of Pampagrande and near Camiri in Santa Cruz Department and near Villamontes in Tarija Department.

Piola rubra Martins & Galileo, 1999

Lamiinae: Phacellini

This small (6–8 mm) cerambycid is similar in proportions and structure to the previous species, *Neobrachychilus consobrinus*, and is a member of the same tribe. It is perhaps narrower and more elongate. Overall coloration is brown, but the elytra are each covered with an arcing, longitudinal, white fascia extending from behind the humerus to the suture and the apex. The pronotum and scutellum at the middle have a narrow band (expanded at middle) of golden-yellow pubescence. Endemic to Bolivia, this species has been collected from the vicinity of the Flora and Fauna Hotel near Buena Vista, the Potrerillo del Guenda Reserve, and Refugio Los Volcanes in Santa Cruz Department.

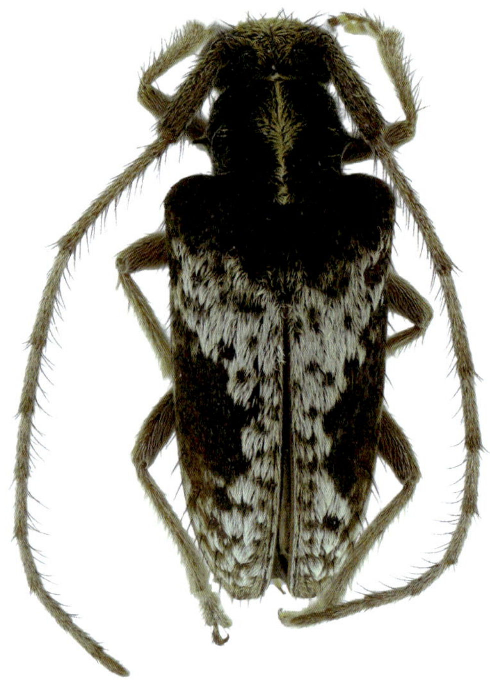

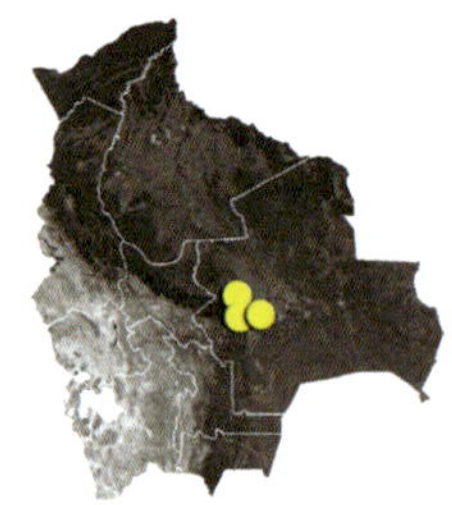

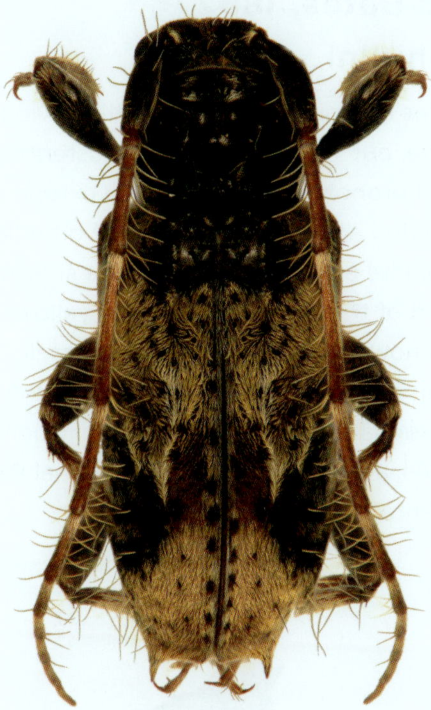

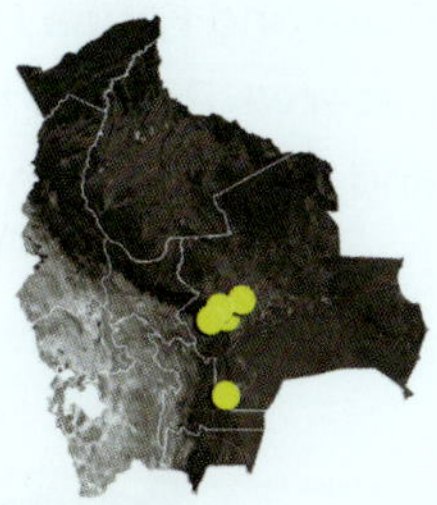

Cosmotomidius egregius (Martins & Galileo, 2007)

Lamiinae: Pogonocherini

This lamiine has patches of dense pubescence somewhat more typical of desmiphorines rather than pogonocherines, where it is currently classified. It is recognized by the rusty-orange fasciae on the apical two-thirds of the elytra, interrupted by a brown macula postmedially. Specimens are 5–7 mm in length. This species is known only from Bolivia. It is known from Potrerillo del Guenda (the type locality), Refugio Los Volcanes, Samaipata, the Vicoquin area above Achira at 2,000 m, and the Camiri environs in Santa Cruz Department.

Lypsimena fuscata Haldeman, 1847

Lamiinae: Pogonocherini

This small (5–6 mm) species is distinct by its brown coloration and series of three to four short, longitudinal, white maculae on each elytron. The pronotum is slightly vittate with alternating white and light and darker brown fasciae. This is a very widespread species that occurs from the eastern United States, through the Caribbean Islands, Mexico, and into much of South America. In Bolivia it has been collected along the Amboró road above Achira and near Mairano, both in Santa Cruz Department.

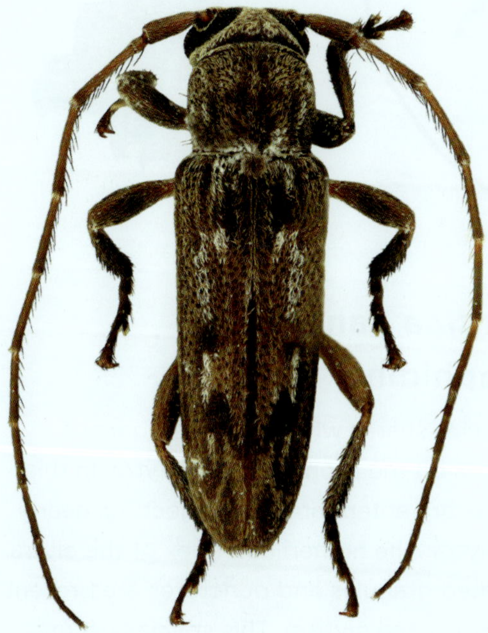

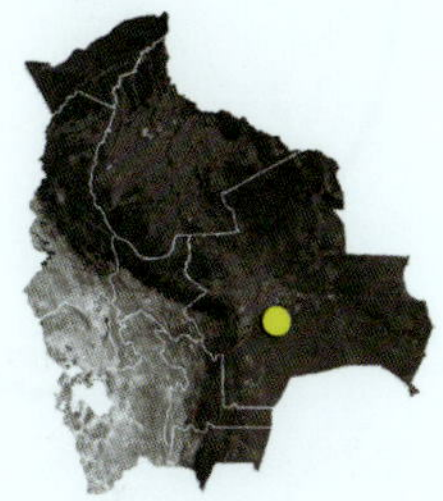

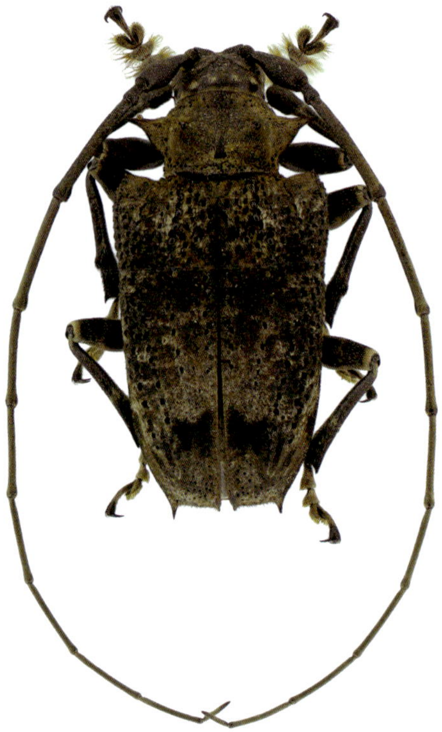

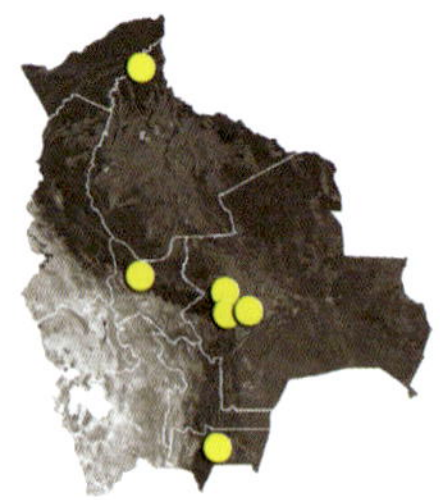

Polyrhaphis argentina Lane, 1978

Lamiinae: Polyrhaphidini

This large (19–29 mm) species shares with other members of this tribe pronounced spines on the pronotum and elytra. In this species, the pronotal spines are anterolaterally projecting, nearly beyond the plane of the elytra. The humeri and apex of the elytra are spinose. Rows of elevated granules and punctures are present for nearly the entire length of each elytron. This species occurs in Peru, Bolivia, Argentina, Brazil, and Paraguay. In Bolivia it is known from Villa Tunari (Cochabamba Department), Riberalta (Beni Department), near Villamontes (Tarija Department), and throughout Santa Cruz Department.

Polyrhaphis gracilis Bates, 1862

Lamiinae: Polyrhaphidini

This large (22–30 mm) species is similar in shape to the previous species, *Polyrhaphis argentina*, but with less tapering posteriorly. It also has pronounced lateral pronotal spines, but they extend more laterally and are not angled anteriorly. The rows of granulate punctures on the elytra terminate at the apical third. The sides of the pronotum and apical third of the elytra have diffuse yellow pubescence. This species is known from French Guiana, Brazil, and Bolivia, where it has been collected from the vicinity of the Flora and Fauna Hotel near Buena Vista, the Potrerillo del Guenda Reserve, Refugio Los Volcanes, adjacent to Amboró National Park, and the Camiri environs in Santa Cruz Department.

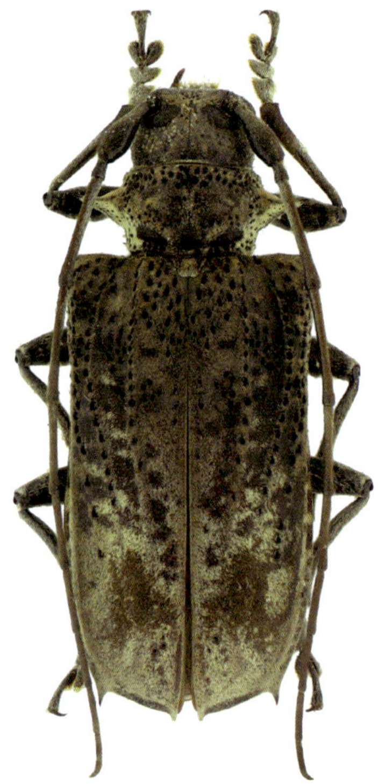

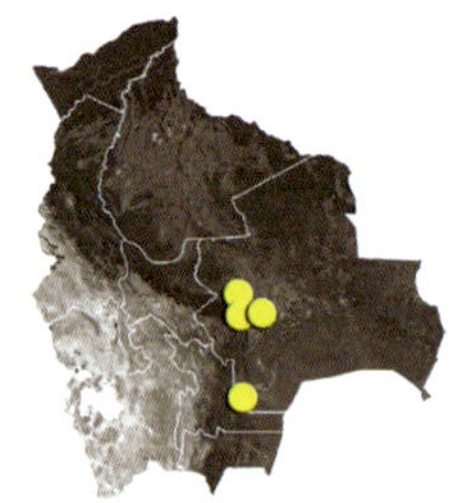

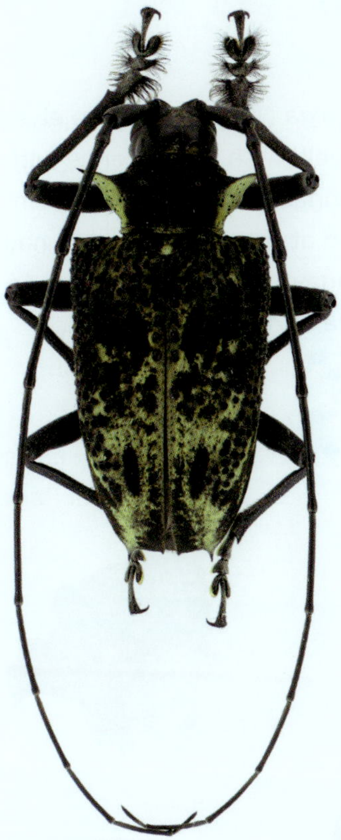

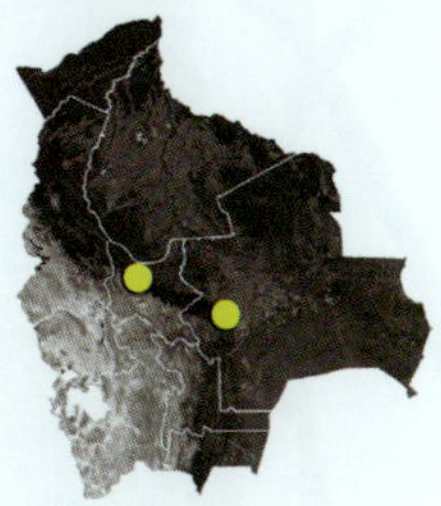

Polyrhaphis pilosa Lane, 1965

Lamiinae: Polyrhaphidini

This large species (24–28 mm long) is most similar in shape to *Polyrhaphis argentina*. The pronotal spines, long and anteriorly projecting, are as in that species as well. It differs in having these pronotal spines covered posteriorly with bright yellow pubescence. This pubescence also occurs on the apical half of the elytra but not densely so. The large, granulate punctures occur throughout the elytra, except for the extreme apex. In Bolivia this species has been collected at Villa Tunari (Cochabamba Department) and Refugio Los Volcanes (Santa Cruz Department).

Anobrium fasciatum Galileo & Martins, 2002

Lamiinae: Pteropliini

This small lamiine (6–8 mm) is somewhat similar to *Amphicnaeia* species but has thicker legs and antennae and is more robust. The body is covered with erect to suberect, long setae. A vague and narrow longitudinal middle stripe of pale pubescence is present on the pronotum. The antennae are shorter than the body. This species is known only from Brazil, French Guiana, and Bolivia, where it has been collected only from Potrerillo del Guenda in Santa Cruz Department.

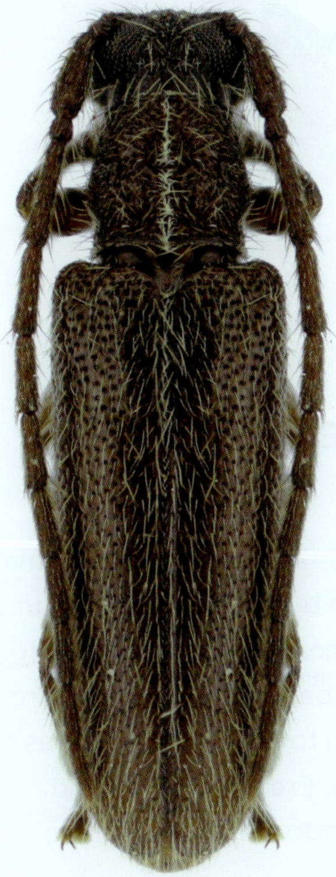

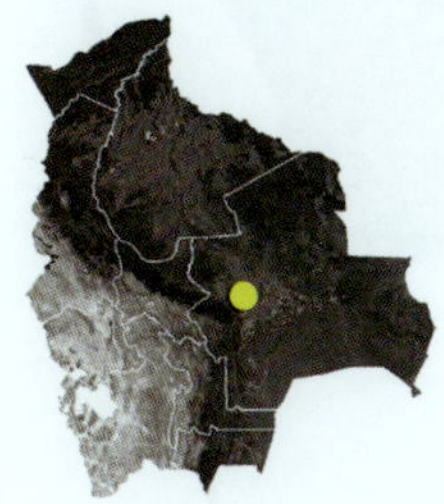

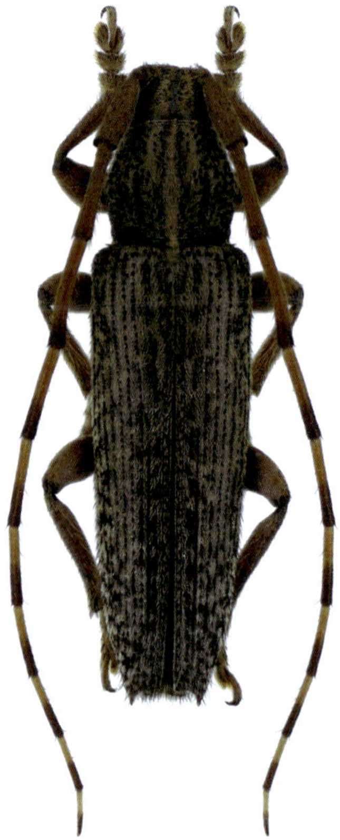

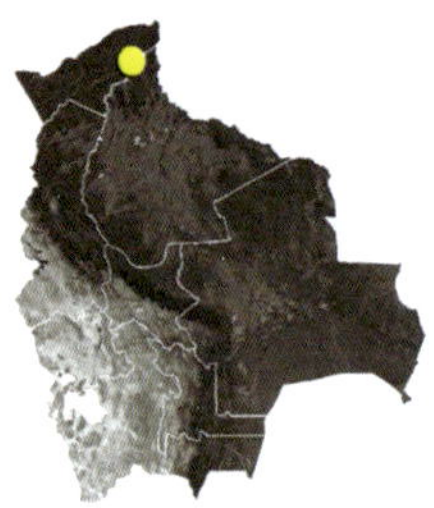

Ataxia obscura (Fabricius, 1801)

Lamiinae: Pteropliini

This species of *Ataxia* (12–14 mm) is recognized by the longitudinal rows of punctures on the elytra, separated by rows of mottled gray and brown pubescence. The pronotum is coarsely punctured with an impunctate, central, golden pubescent band. The antennae are reddish brown at the base and dark brown to black at the apices. This species is known from Brazil, Ecuador, French Guiana, and Bolivia. In Bolivia it is known only from Pando Department across the Beni River from Riberalta in Bolivia.

Ataxia obtusa (Bates, 1866)

Lamiinae: Pteropliini

This is a very dark-colored species of *Ataxia* (12–14 mm). The elytra and pronotum have sparse, appressed, yellowish pubescence with scattered suberect dark setae. The antennae and legs are reddish brown. Most antennomeres are darkened at the apical third to half. This species is known from Brazil, Paraguay, and Bolivia. In Bolivia it is known from the grounds of the Flora and Fauna Hotel near Buena Vista in Santa Cruz Department.

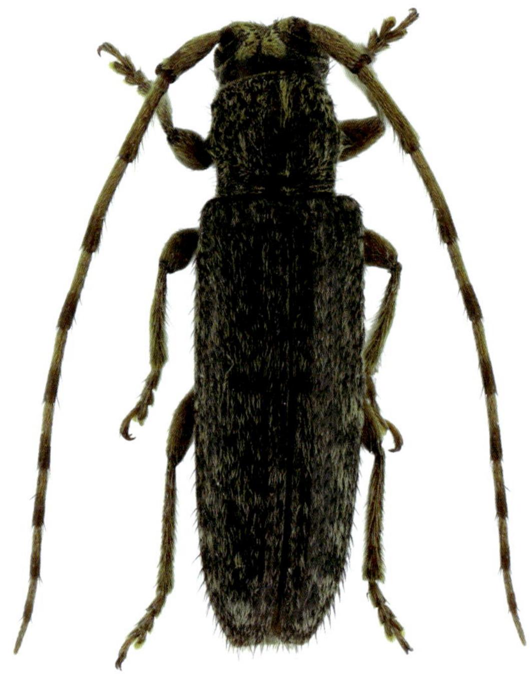

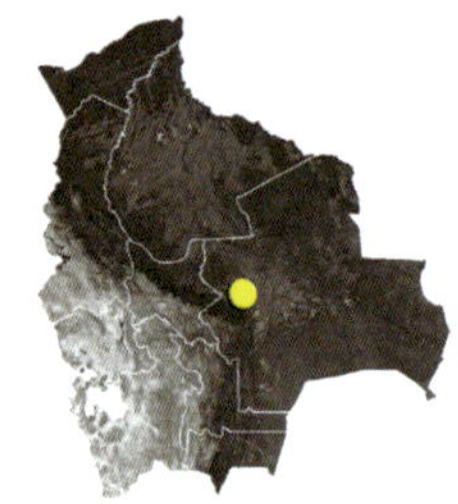

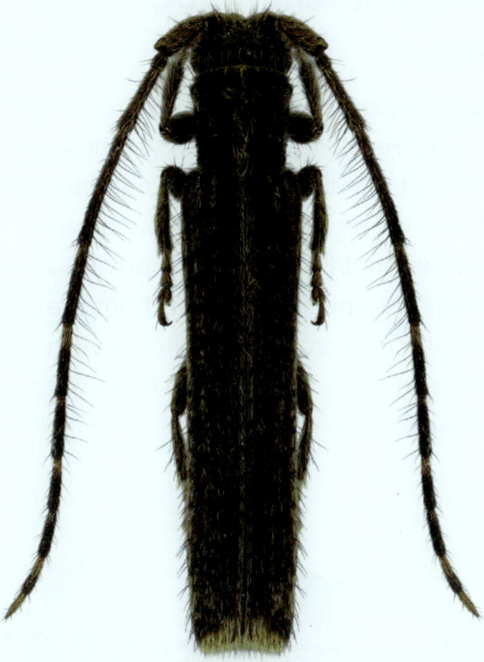

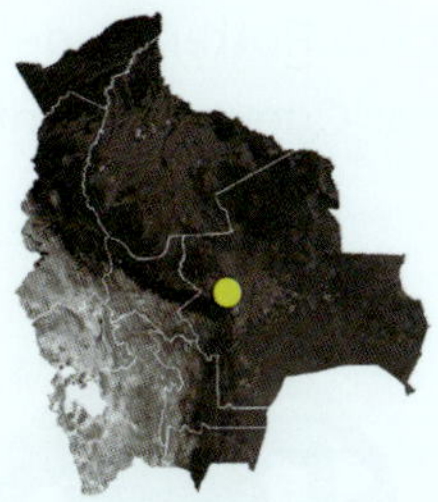

Epectasis juncea (Newman, 1840)

Lamiinae: Pteropliini

This elegant species is recognizable by its small size (6–8 mm long), narrow but parallel-sided body, faint, pale sutural stripe and pale apices of the elytra, and scattered long setae over the body and antennae. The pronotum is slightly wider anteriorly and lacks lateral spines or tubercles. This species is widespread from Mexico through Central America and into South America, extending as far south as Brazil, Peru, and Bolivia. In Bolivia it is known from the vicinity of the Flora and Fauna Hotel near Buena Vista in Santa Cruz Department.

Esthlogena foveolata Aurivillius, 1920

Lamiinae: Pteropliini

This pteropliine (11–17 mm long) is distinct with its pale green pubescence on the pronotum and elytra that is interrupted by separate, black spots. The antennae and legs are black. The elytral apices are angled but not spinose. Known only from Bolivia, it has been collected from the vicinity of the Flora and Fauna Hotel near Buena Vista, Refugio Los Volcanes, and the Vicoquin area above Achira at 2,000 m adjacent to Amboró National Park in Santa Cruz Department.

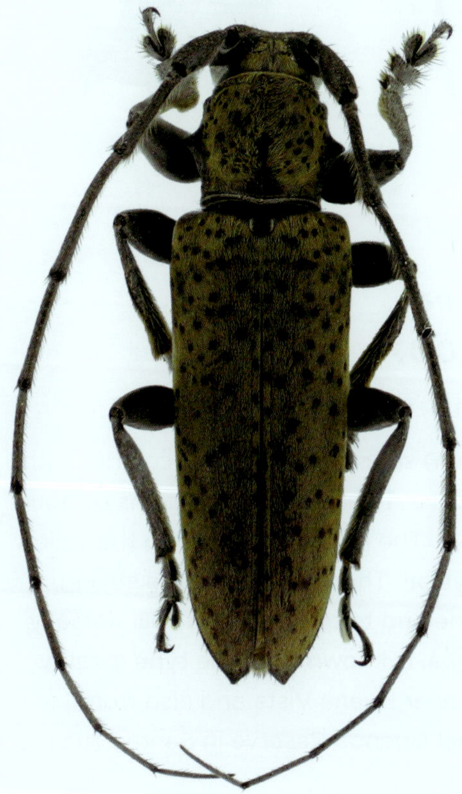

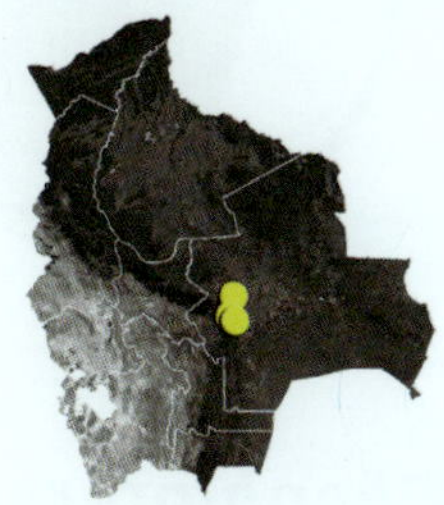

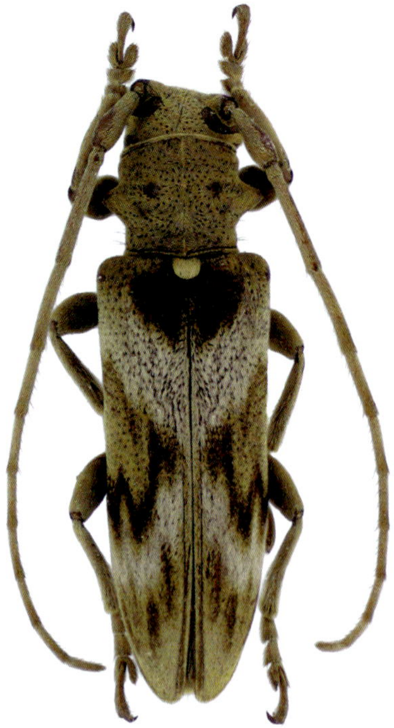

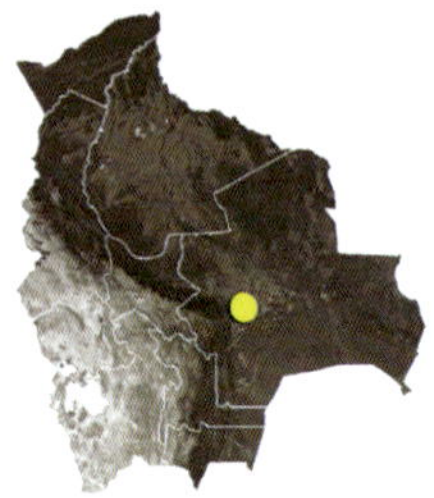

Rhaphiptera boliviana Galileo & Martins, 2007

Lamiinae: Pteropliini

This moderate-sized (12–14 mm) species is elegantly colored with white, tan, and black pubescence. This pubescence forms a vague X-marked pattern on the elytra. The region posterior to the yellow scutellum is black on each elytron. The pronotum has a strong, subacute tubercle on each side and two small black calli dorsally. This species, endemic to Bolivia, is known from the type locality of the Flora and Fauna Hotel near Buena Vista and also from the vicinity of the Potrerillo del Guenda Reserve in Santa Cruz Department.

Rhaphiptera oculata Gounelle, 1908

Lamiinae: Pteropliini

This relatively narrow-bodied, medium-sized (12–14 mm) cerambycid is distinct with its eye-spot-like markings at the apex of each elytron. The remainder of the elytron (except for the base) is covered in pale ochraceous or white pubescence. The post-scutellar region is ochraceous and not black as in the previous species, *Rhaphiptera boliviana*. This species is known from Brazil, Argentina, and Bolivia, where it has been collected in the vicinity of the Potrerillo del Guenda Reserve and the road to Bella Vista near Cuevas in Santa Cruz Department.

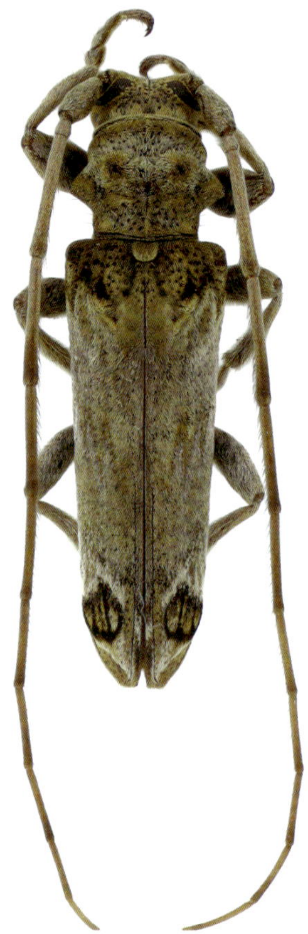

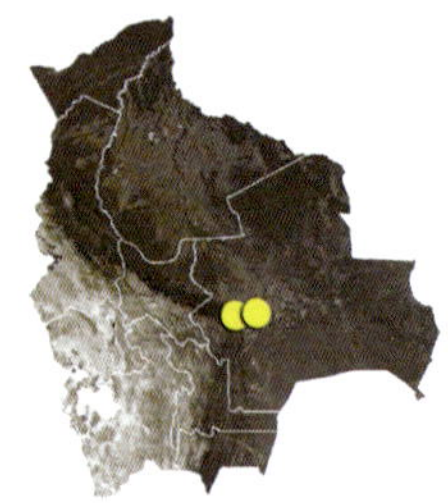

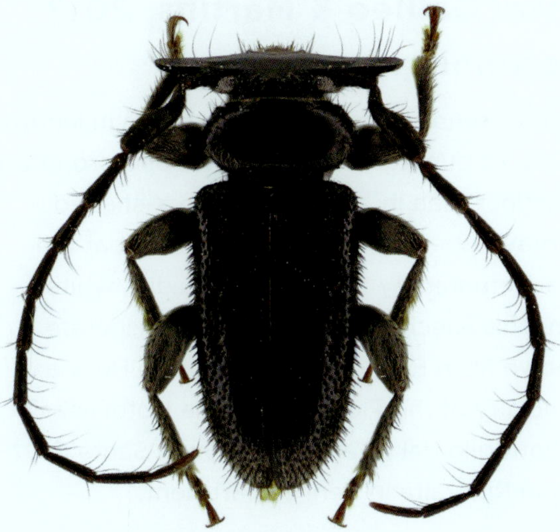

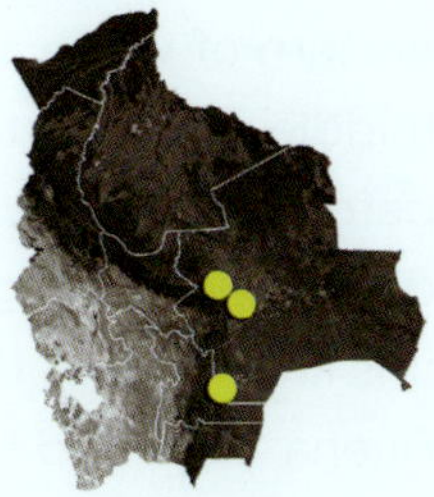

Tapeina paulista Marinoni, 1972

Lamiinae: Tapeinini

This is a member of the distinctive tribe, Tapeinini. Members of this tribe typically have the frons of the head, including the antennal tubercles, greatly expanded. This small species (5–7 mm) is no exception. The head is at least one and one-half times the width of the pronotum. The antennae arise from the curved scape at the base of the expansion. Coloration is uniformly black, but the legs have a somewhat gray sheen of pubescence. This species is always rare and usually collected only at lights. It occurs in Brazil and Bolivia, where this species is known from the vicinity of the Flora and Fauna Hotel near Buena Vista, the Potrerillo del Guenda Reserve, and the Camiri environs in Santa Cruz Department.

Phaea coccinea Bates, 1866

Lamiinae: Tetraopini

This is a small (9 mm) reddish-orange species with distally black appendages (scape and legs basally reddish orange). The elytra are densely punctate, with the punctures arranged in rows and less distinct apically. The entire insect is covered in long hairs (one-half to one time the length of the scape), longer on the appendages, and short, dense recumbent pubescence. The pronotum has a central umbone and laterally a rounded protuberance on each side. This rare species is known from northern Brazil and Santa Cruz Department, Bolivia, being collected only once there in Potrerillo del Guenda Reserve in November.

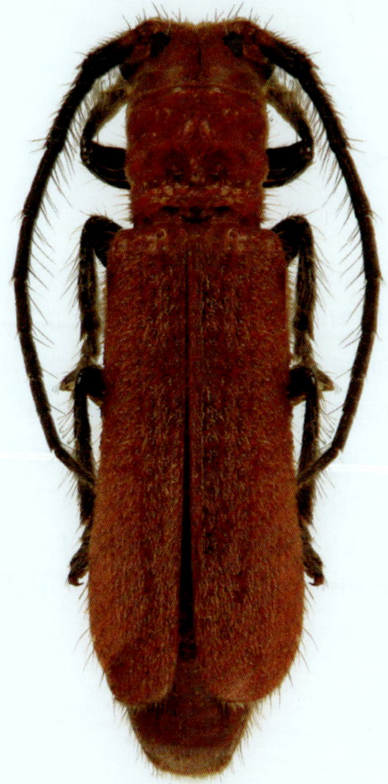

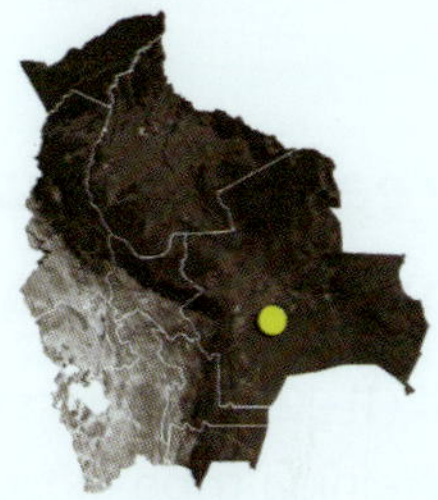

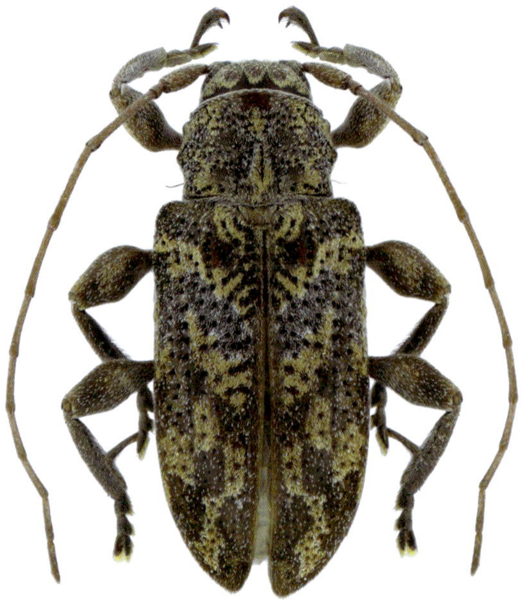

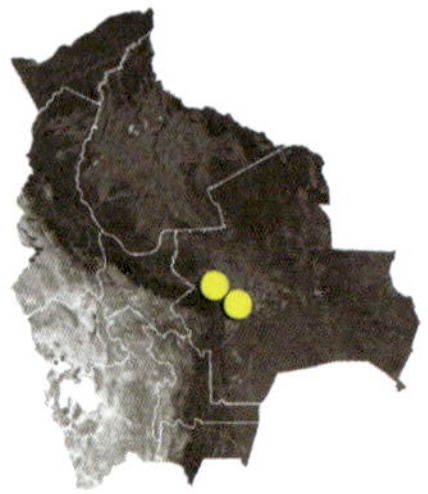

Xenofrea punctata Galileo & Martins, 2005

Lamiinae: Xenofreini

This small (11–12 mm) but stout-bodied species is a member of the Xenofreini, a tribe closely related to the Acanthoderini. It is recognized by the large but widely spaced punctures over the elytra and pronotum. Patches of orange pubescent maculae are present on each elytron. At the basal third is an inverted V-shaped fascia of sparse gray pubescence. This species is known from Brazil and Bolivia where it has been taken from the vicinity of the Flora and Fauna Hotel near Buena Vista and the Potrerillo del Guenda Reserve in Santa Cruz Department.

Xenofrea zischkai Galileo & Martins, 2012

Lamiinae: Xenofreini

This species, smaller than *Xenofrea punctata* at 7–9 mm in length, is similarly stout bodied. Its punctures on the elytra and pronotum are smaller and less distinct than in *X. punctata*. An X-shaped pattern of white and gray pubescent fasciae are present at the middle of the elytra, punctuated by a small, postmedial, white patch on each elytron. This species has a somewhat disparate distribution, being known from Ecuador and Bolivia. In Bolivia it is known from the vicinity of the Flora and Fauna Hotel near Buena Vista and the Potrerillo del Guenda Reserve in Santa Cruz Department and around Riberalta in Beni Department.

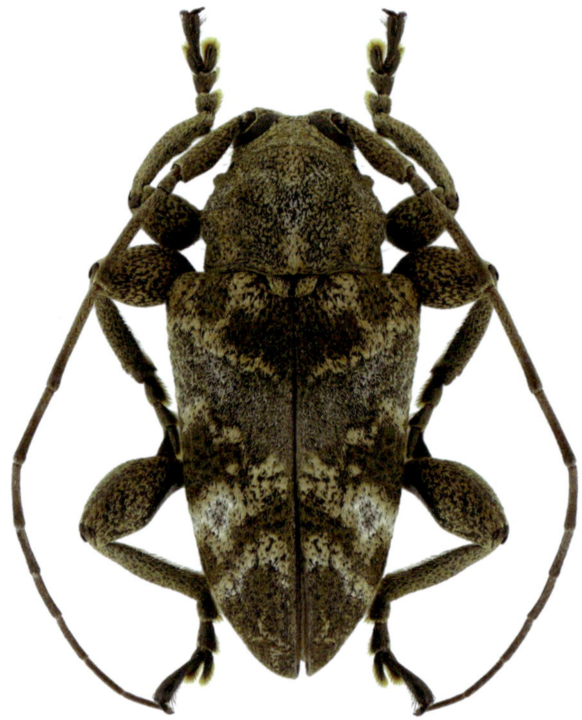

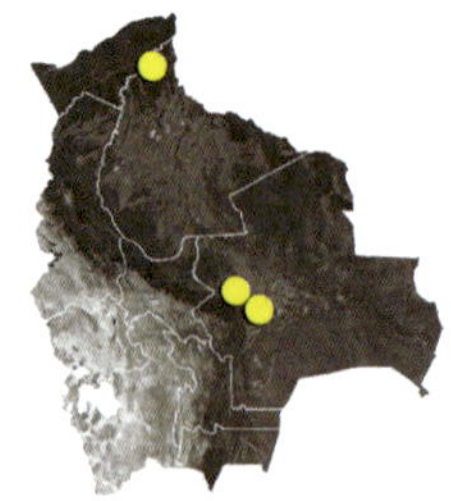

LITERATURE CITED

Aramayo B., J. L., and J. Ledezma A. 2004. "Orden Coleoptera (Scarabaeidae, Cerambycidae, Cicindelidae)." In *Historia Natural del Municipio de Pampagrande: Una localidad típica de los valles interandinos de Bolivia*, ed. H. Azurduy F., J. L. Aramayo B., Ma. J. Ledezma A., and F. A. Ma. Langer, pp. 133–141. Santa Cruz, Bolivia: Museo de Historia Natural de Noel Kempff Mercado.

Clarke, R. O. S. 2007. New Species of Rhinotragini and Heteropsini and a Note on *Trachelissa maculicollis* (Trachyderini) (Coleoptera, Cerambycidae, Cerambycinae). *Papé*is Avulsos de Zoologia, 47(11):153–158.

———. 2012. Bolivian Rhinotragini V: New Species of *Erythroplatys* White, 1855, *Rhinotragus* Germar, 1824, *Ornistomus* Thomson, 1864, and *Aechmutes* Bates, 1867 (Coleoptera, Cerambycidae). *Papé*is Avulsos de Zoologia, 52(5):55–79.

Evans, A. V. 2014. Beetles of Eastern North America. Princeton, N.J.: Princeton University Press.

Herzog, S. K., A. Caroli Hamel-Leigue, T. H. Larsen, D. J. Mann, R. W. Soria-Auza, B. D. Gill, W. D. Edmonds, and S. Spector. 2013. Elevational Distribution and Conservation Biogeography of Phanaeine Dung Beetles (Coleoptera: Scarabaeinae) in Bolivia. *PLOS ONE*, 8(5):1–11.

Ibisch, P. L., and G. Mérida, eds. 2004. *Biodiversity: The Richness of Bolivia*. Santa Cruz de la Sierra, Bolivia: Editorial FAN.

Killeen, T., E. Garcia, and S. Beck, eds. 1993. *Guía de Árboles de Bolivia*. La Paz, Bolivia: Herbario Nacional de Bolivia; St. Louis, Mo.: Missouri Botanical Garden.

MacCreagh, G. 2001. *White Waters and Black*. Chicago, Ill.: University of Chicago Press.

Monné, M. A. 2016a. *Catalogue of the Cerambycidae (Coleoptera) of the Neotropical Region. Part I. Subfamily Cerambycinae*. http://www.cerambyxcat.com/Parte1_Cerambycinae.pdf (accessed 14 December 2016).

Monné, M. A. 2016b. *Catalogue of the Cerambycidae (Coleoptera) of the Neotropical Region. Part II. Subfamily Lamiinae*. http://www.cerambyxcat.com/Parte2_Lamiinae.pdf (accessed 14 December 2016).

Monné, M. A. 2016c. *Catalogue of the Cerambycidae (Coleoptera) of the Neotropical Region. Part III. Subfamilies Lepturinae, Necydalinae, Parandrinae, Prioninae, Spondylidinae and Families Oxypeltidae, Vesperidae and Disteniidae*. http://www.cerambyxcat.com/Parte3_Lepturinae_e_outros.pdf (accessed 14 December 2016).

Perger, R., and F. Guerra. 2013. Longhorn Beetles (Coleoptera: Cerambycidae) of the Tucuman-Bolivian Forest in the Tariquıa Flora and Fauna National Reserve, Southern Bolivian Andes, with Notes on Ecoregion Endemism and Conservation. *The Pan-Pacific Entomologist*, 89(4):209–221.

Svacha, P., and J. F. Lawrence. 2014. "2.4: Cerambycidae Latreille, 1802." In *Coleoptera, Beetles. Volume 3: Morphology and Systematics (Phytophaga)*, ed. R. A. B. Leschen and R. G. Beutel, pp. 77–177. *Handbook of Zoology, Arthropoda: Insecta*. Berlin: De Gruyter.

Tavakilian, G., and H. Chevillotte. 2016. Titan: base de données internationales sur les Cerambycidae ou Longicornes. Version 3.0. http://titan.gbif.fr/aff_type.php (accessed 1 October 2016).

Wappes, J. E., S. W. Lingafelter, M. A. Monné, and J. Ledezma-Arias. 2013. Additions to the Known Vesperidae and Cerambycidae (Coleoptera) of Bolivia. *Insecta Mundi*, 0319:1–25.

Wappes, J. E., S. W. Lingafelter, and R. Perger. 2011. Additions and Deletions to the Known Cerambycidae (Coleoptera) of Bolivia. *Insecta Mundi*, 150:1–8.

Wappes, J. E., R. F. Morris II, E. H. Nearns, and M. C. Thomas. 2006. Preliminary Checklist of Bolivian Cerambycidae (Coleoptera). *Insecta Mundi*, 20(1–2):1–45.

World Wildlife Fund. 2016. "Threats | Deforestation." http://www.worldwildlife.org/threats/deforestation (accessed 14 December 2016).

ABOUT THE AUTHORS

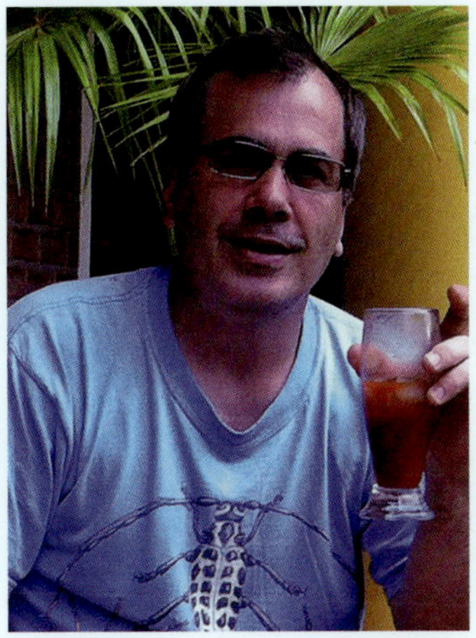

Steve Lingafelter was a research entomologist for the U.S. Department of Agriculture based at the Smithsonian Institution's National Museum of Natural History in Washington, D.C., for nearly 20 years. Last year, a relocation opportunity arose in Arizona, and he became an insect identifier on the Mexican border. Steve received his PhD in entomology from the University of Kansas in 1996. Steve has made seven expeditions to Bolivia starting in 2002. He has published over 60 papers and four books on beetle systematics and taxonomy, with an emphasis on neotropical longhorned beetles.

Jim Wappes is the founder and curator of the American Coleoptera Museum in San Antonio, Texas. Jim received his BS in agriculture (entomology major) from Purdue University in 1968. He has collected Cerambycidae and studied cerambycid taxonomy for more than 50 years. Jim worked extensively in Mexico and Central America prior to the Bolivia Cerambycidae Project. Since 2000 he has made 30 expeditions to Bolivia and published more than 25 papers on neotropical longhorned beetles.

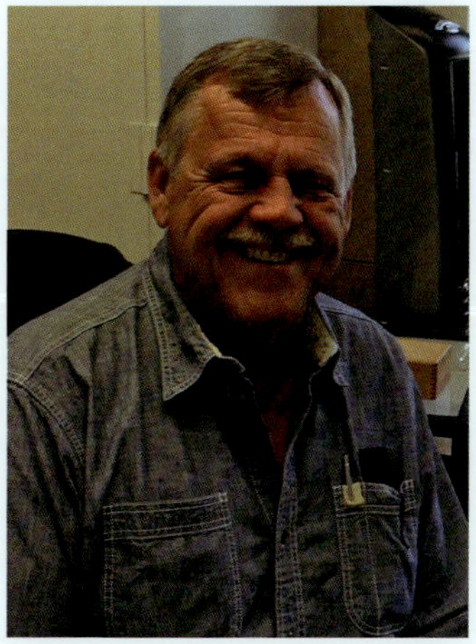

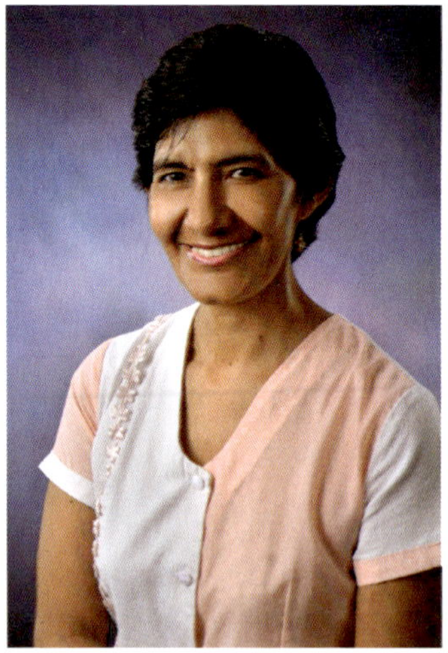

Julieta Ledezma Arias is a research entomologist and head of the Department of Invertebrates at the Noel Kempff Mercado Natural History Museum in Santa Cruz de la Sierra, Bolivia. She is also Professor of Ecology in the Faculty of Pharmacy and Biochemistry at the Universidad Autónoma Gabriel Rene Moreno. Julieta is a specialist on butterflies and has published over 20 papers on the Bolivian fauna, including several books.

Julieta Ledezma Arias es Bióloga y jefe del Área de Invertebrados del Museo de Historia Natural Noel Kempff Mercado, profesor de la asignatura de Ecología de la Facultad de Ciencias Farmacéuticas y Bioquímica de la Universidad Autónoma Gabriel Rene Moreno, Santa Cruz de la Sierra, Bolivia. Julieta es especialista en mariposas y ha publicado más de 20 artículos sobre la Entomofauna Boliviana, incluyendo varios libros.

INDEX